Vauxhall Nova Owners Workshop Manual

John S Mead

Models covered
All Vauxhall Nova models, Saloon and Hatchback, including GTE and most special/limited editions
993 cc, 1196 cc, 1297 cc & 1598 cc

Does not cover all aspects of the limited edition Sport

(909-7U11)　　ABCDE FGH

Haynes Publishing Group
Sparkford Nr Yeovil
Somerset BA22 7JJ England

Haynes Publications, Inc
861 Lawrence Drive
Newbury Park
California 91320 USA

Acknowledgements

Thanks are due to Champion Spark Plug, who supplied the illustrations showing spark plug conditions, to Holt Lloyd Limited who supplied the illustrations showing bodywork repair, and to Duckhams Oils, who provided lubrication data. Certain other illustrations are the copyright of Vauxhall Motors Limited, and are used with their permission. Thanks are also due to Sykes-Pickavant, who provided some of the workshop tools, and all those people at Sparkford who helped in the production of this manual.

© **Haynes Publishing Group 1992**

A book in the **Haynes Owners Workshop Manual Series**

Printed by J. H. Haynes & Co. Ltd., Sparkford, Nr Yeovil, Somerset BA22 7JJ, England

All rights reserved. No part of this book may be reproduced or transmitted in any form or by any means, electronic or mechanical, including photocopying, recording or by any information storage or retrieval system, without permission in writing from the copyright holder.

ISBN 1 85010 809 9

British Library Cataloguing in Publication Data
A catalogue record for this book is available from the British Library

We take great pride in the accuracy of information given in this manual, but vehicle manufacturers make alterations and design changes during the production run of a particular vehicle of which they do not inform us. No liability can be accepted by the authors or publishers for loss, damage or injury caused by any errors in, or omissions from, the information given.

Restoring and Preserving our Motoring Heritage

Few people can have had the luck to realise their dreams to quite the same extent and in such a remarkable fashion as John Haynes, Founder and Chairman of the Haynes Publishing Group.

Since 1965 his unique approach to workshop manual publishing has proved so successful that millions of Haynes Manuals are now sold every year throughout the world, covering literally thousands of different makes and models of cars, vans and motorcycles.

A continuing passion for cars and motoring led to the founding in 1985 of a Charitable Trust dedicated to the restoration and preservation of our motoring heritage. To inaugurate the new Museum, John Haynes donated virtually his entire private collection of 52 cars.

Now with an unrivalled international collection of over 210 veteran, vintage and classic cars and motorcycles, the Haynes Motor Museum in Somerset is well on the way to becoming one of the most interesting Motor Museums in the world.

A 70 seat video cinema, a cafe and an extensive motoring bookshop, together with a specially constructed one kilometre motor circuit, make a visit to the Haynes Motor Museum a truly unforgettable experience.

Every vehicle in the museum is preserved in as near as possible mint condition and each car is run every six months on the motor circuit.

Enjoy the picnic area set amongst the rolling Somerset hills. Peer through the William Morris workshop windows at cars being restored, and browse through the extensive displays of fascinating motoring memorabilia.

From the 1903 Oldsmobile through such classics as an MG Midget to the mighty 'E' Type Jaguar, Lamborghini, Ferrari Berlinetta Boxer, and Graham Hill's Lola Cosworth, there is something for everyone, young and old alike, at this Somerset Museum.

Haynes Motor Museum

Situated mid-way between London and Penzance, the Haynes Motor Museum is located just off the A303 at Sparkford, Somerset (home of the Haynes Manual) and is open to the public 7 days a week all year round, except Christmas Day and Boxing Day.

Contents

	Page
Acknowledgements	
About this manual	2
Introduction to the Vauxhall Nova	4
General dimensions, weights and capacities *(also see Chapter 13, page 229)*	
Buying spare parts and vehicle identification numbers	
General repair procedures	
Tools and working facilities	
Jacking, towing and wheel changing	
Recommended lubricants and fluids	12
Safety first!	13
Routine maintenance *(also see Chapter 13, page 229)*	14
Fault diagnosis	20
Chapter 1 Engine *(also see Chapter 13, page 229)*	23
Chapter 2 Cooling system *(also see Chapter 13, page 229)*	57
Chapter 3 Fuel and exhaust systems *(also see Chapter 13, page 229)*	64
Chapter 4 Ignition system *(also see Chapter 13, page 229)*	85
Chapter 5 Clutch *(also see Chapter 13, page 229)*	96
Chapter 6 Transmission *(also see Chapter 13, page 229)*	104
Chapter 7 Driveshafts *(also see Chapter 13, page 229)*	127
Chapter 8 Steering gear	131
Chapter 9 Braking system *(also see Chapter 13, page 229)*	142
Chapter 10 Electrical system *(also see Chapter 13, page 229)*	160
Chapter 11 Suspension *(also see Chapter 13, page 229)*	193
Chapter 12 Bodywork *(also see Chapter 13, page 229)*	209
Chapter 13 Supplement: Revisions and information on later models	229
Conversion factors	328
Index	329

Spark plug condition and bodywork repair colour pages between pages 32 and 33

About this manual

Its aim

The aim of this manual is to help you get the best value from your vehicle. It can do so in several ways. It can help you decide what work must be done (even should you choose to get it done by a garage), provide information on routine maintenance and servicing, and give a logical course of action and diagnosis when random faults occur. However, it is hoped that you will use the manual by tackling the work yourself. On simpler jobs it may even be quicker than booking the car into a garage and going there twice, to leave and collect it. Perhaps most important, a lot of money can be saved by avoiding the costs a garage must charge to cover its labour and overheads.

The manual has drawings and descriptions to show the function of the various components so that their layout can be understood. Then the tasks are described and photographed in a step-by-step sequence so that even a novice can do the work.

Its arrangement

The manual is divided into thirteen Chapters, each covering a logical sub-division of the vehicle. The Chapters are each divided into Sections, numbered with single figures, eg 5; and the Sections into paragraphs (or sub-sections), with decimal numbers following on from the Section they are in, eg 5.1, 5.2, 5.3 etc.

It is freely illustrated, especially in those parts where there is a detailed sequence of operations to be carried out. There are two forms of illustration: figures and photographs. The figures are numbered in sequence with decimal numbers, according to their position in the Chapter – eg Fig. 6.4 is the fourth drawing/illustration in Chapter 6. Photographs carry the same number (either individually or in related groups) as the Section or sub-section to which they relate.

There is an alphabetical index at the back of the manual as well as a contents list at the front. Each Chapter is also preceded by its own individual contents list.

References to the 'left' or 'right' of the vehicle are in the sense of a person in the driver's seat facing forwards.

Unless otherwise stated, nuts and bolts are removed by turning anti-clockwise, and tightened by turning clockwise.

Vehicle manufacturers continually make changes to specifications and recommendations, and these, when notified, are incorporated into our manuals at the earliest opportunity.

We take great pride in the accuracy of information given in this manual, but vehicle manufacturers make alterations and design changes during the production run of a particular vehicle of which they do not inform us. No liability can be accepted by the authors or publishers for loss, damage or injury caused by any errors in, or omissions from, the information given.

Introduction to the Vauxhall Nova

After a controversial launch early in 1983, owing to its Spanish origins, the Nova quickly established itself as a leading contender in the small car market.

The Vauxhall Nova is General Motors UK version of the established Opel Corsa and shares many mechanical components with the Astra/Cavalier range.

The model line-up includes two- and four-door Saloon and three- and five-door Hatchback versions, available with a choice of 1.0 litre ohv and 1.2, 1.3 or 1.6 litre ohc engines. Various levels of trim and optional equipment are available depending upon the model selected from the range.

In keeping with current design trends, the Nova is of straightforward design and construction with easy access to major components and service points.

Vauxhall Nova Saloon

General dimensions, weights and capacities

For information applicable to later models, see Supplement at end of manual

Dimensions
Overall length:
- Saloon .... 3955 mm (155.7 in)
- Hatchback .... 3622 mm (142.6 in)

Overall width:
- Saloon .... 1540 mm (60.6 in)
- Hatchback .... 1532 mm (60.3 in)

Overall height:
- Saloon .... 1360 mm (53.5 in)
- Hatchback .... 1365 mm (53.7 in)

Wheelbase .... 2343 mm (92.2 in)

Track:
- Front .... 1320 mm (52.0 in)
- Rear .... 1307 mm (51.6 in)

Ground clearance .... 144 mm (5.7 in)

Weights
Kerb weight:
- Saloon models .... 740 kg (1631 lbs)
- Hatchback models .... 735 kg (1620 lbs)
- L Saloon models .... 750 kg (1653 lbs)
- L Hatchback models .... 750 kg (1653 lbs)
- SR models .... 770 kg (1698 lbs)
- GTE models .... 834 kg (1839 lbs)

Maximum roof rack load .... 80 kg (176 lbs)

Maximum caravan/trailer towing weight: **Braked trailer**   **Unbraked trailer**
- 1.0 models .... 400 kg (882 lbs)   400 kg (882 lbs)
- 1.2 models .... 650 kg (1433 lbs)   400 kg (882 lbs)
- 1.3 and 1.6 models .... 800 kg (1764 lbs)   400 kg (882 lbs)

Maximum tow ball socket vertical load .... 50 kg (110 lbs)

Capacities
Fuel tank .... 42 litres (9.2 gal)

Cooling system:
- 1,0 models .... 5.5 litres (9.7 pt)
- 1.2, 1.3 and 1.6 models .... 6.1 litres (10.7 pt)

Engine oil (with filter change):
- 1.0 models .... 2.5 litres (4.4 pt)
- 1.2 and 1.3 models .... 3.0 litres (5.3 pt)
- 1.6 models .... 3.5 litres (6.2 pt)

Difference between MAX and MIN marks on dipstick:
- 1.0, 1.2 and 1.3 models .... 0.75 litres (1.3 pt)
- 1.6 models .... 1.0 litre (1.8 pt)

Transmission:
- Four-speed models .... 1.75 litres (3.1 pt)
- Five-speed models .... 1.85 litres (3.3 pt)

Buying spare parts and vehicle identification numbers

Buying spare parts

Spare parts are available from many sources, for example: Vauxhall garages, other garages and accessory shops, and motor factors. Our advice regarding spare part sources is as follows:

Officially appointed Vauxhall garages – This is the best source of parts which are peculiar to your vehicle and are otherwise not generally available (eg complete cylinder heads, internal gearbox components, badges, interior trim etc). It is also the only place at which you should buy parts if your vehicle is still under warranty; non-Vauxhall components may invalidate the warranty. To be sure of obtaining the correct parts it will always be necessary to give the storeman your vehicle's engine and chassis number, and if possible, to take the 'old' parts along for positive identification. Remember that some parts are available on a factory exchange scheme – any parts returned should always be clean. It obviously makes good sense to go straight to the specialists on your vehicle for this type of part for they are best equipped to supply you.

Other garages and accessory shops – These are often very good places to buy materials and components needed for the maintenance of your vehicle (eg spark plugs, bulbs, drivebelts, oils and greases, touch-up paint, filler paste, etc). They also sell general accessories, usually have convenient opening hours, charge lower prices and can often be found not far from home.

Motor factors – Good factors will stock all of the more important components which wear out relatively quickly (eg clutch components, pistons, valves, exhaust systems, brake cylinders/ /pipes/hoses/seals/shoes and pads etc). Motor factors will often provide new or reconditioned components on a part exchange basis – this can save a considerable amount of money.

Vehicle identification numbers

The *Vehicle Identification Number* is located inside the engine compartment on top of the front end panel. The plate is marked with the vehicle chassis and designation number and the colour code. Also shown is the maximum gross weight for the car.

The *engine number* is stamped on a flat machined on the engine cylinder block.

The *chassis number* is stamped on the body floor panel between the driver's seat and the door sill.

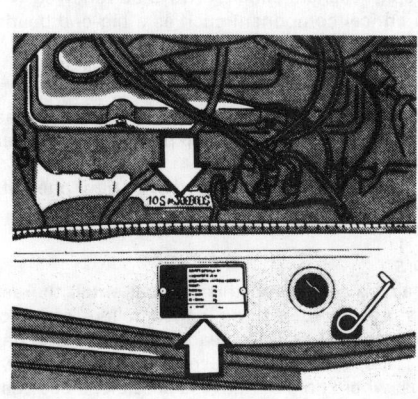

Vehicle identification number plate and engine number locations

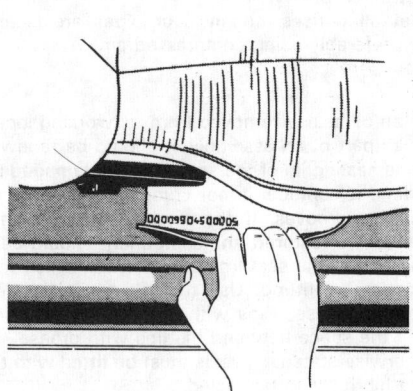

Chassis number location

General repair procedures

Whenever servicing, repair or overhaul work is carried out on the car or its components, it is necessary to observe the following procedures and instructions. This will assist in carrying out the operation efficiently and to a professional standard of workmanship.

Joint mating faces and gaskets

Where a gasket is used between the mating faces of two components, ensure that it is renewed on reassembly, and fit it dry unless otherwise stated in the repair procedure. Make sure that the mating faces are clean and dry with all traces of old gasket removed. When cleaning a joint face, use a tool which is not likely to score or damage the face, and remove any burrs or nicks with an oilstone or fine file.

Make sure that tapped holes are cleaned with a pipe cleaner, and keep them free of jointing compound if this is being used unless specifically instructed otherwise.

Ensure that all orifices, channels or pipes are clear and blow through them, preferably using compressed air.

Oil seals

Whenever an oil seal is removed from its working location, either individually or as part of an assembly, it should be renewed.

The very fine sealing lip of the seal is easily damaged and will not seal if the surface it contacts is not completely clean and free from scratches, nicks or grooves. If the original sealing surface of the component cannot be restored, the component should be renewed.

Protect the lips of the seal from any surface which may damage them in the course of fitting. Use tape or a conical sleeve where possible. Lubricate the seal lips with oil before fitting and, on dual lipped seals, fill the space between the lips with grease.

Unless otherwise stated, oil seals must be fitted with their sealing lips toward the lubricant to be sealed.

Use a tubular drift or block of wood of the appropriate size to install the seal and, if the seal housing is shouldered, drive the seal down to the shoulder. If the seal housing is unshouldered, the seal should be fitted with its face flush with the housing top face.

Screw threads and fastenings

Always ensure that a blind tapped hole is completely free from oil, grease, water or other fluid before installing the bolt or stud. Failure to do this could cause the housing to crack due to the hydraulic action of the bolt or stud as it is screwed in.

When tightening a castellated nut to accept a split pin, tighten the nut to the specified torque, where applicable, and then tighten further to the next split pin hole. Never slacken the nut to align a split pin hole unless stated in the repair procedure.

When checking or retightening a nut or bolt to a specified torque setting, slacken the nut or bolt by a quarter of a turn, and then retighten to the specified setting.

Locknuts, locktabs and washers

Any fastening which will rotate against a component or housing in the course of tightening should always have a washer between it and the relevant component or housing.

Spring or split washers should always be renewed when they are used to lock a critical component such as a big-end bearing retaining nut or bolt.

Locktabs which are folded over to retain a nut or bolt should always be renewed.

Self-locking nuts can be reused in non-critical areas, providing resistance can be felt when the locking portion passes over the bolt or stud thread.

Split pins must always be replaced with new ones of the correct size for the hole.

Special tools

Some repair procedures in this manual entail the use of special tools such as a press, two or three-legged pullers, spring compressors etc. Wherever possible, suitable readily available alternatives to the manufacturer's special tools are described, and are shown in use. In some instances, where no alternative is possible, it has been necessary to resort to the use of a manufacturer's tool and this has been done for reasons of safety as well as the efficient completion of the repair operation. Unless you are highly skilled and have a thorough understanding of the procedure described, never attempt to bypass the use of any special tool when the procedure described specifies its use. Not only is there a very great risk of personal injury, but expensive damage could be caused to the components involved.

Tools and working facilities

Introduction

A selection of good tools is a fundamental requirement for anyone contemplating the maintenance and repair of a motor vehicle. For the owner who does not possess any, their purchase will prove a considerable expense, offsetting some of the savings made by doing-it-yourself. However, provided that the tools purchased meet the relevant national safety standards and are of good quality, they will last for many years and prove an extremely worthwhile investment.

To help the average owner to decide which tools are needed to carry out the various tasks detailed in this manual, we have compiled three lists of tools under the following headings: *Maintenance and minor repair, Repair and overhaul,* and *Special*. The newcomer to practical mechanics should start off with the *Maintenance and minor repair* tool kit and confine himself to the simpler jobs around the vehicle. Then, as his confidence and experience grow, he can undertake more difficult tasks, buying extra tools as, and when, they are needed. In this way, a *Maintenance and minor repair* tool kit can be built-up into a *Repair and overhaul* tool kit over a considerable period of time without any major cash outlays. The experienced do-it-yourselfer will have a tool kit good enough for most repair and overhaul procedures and will add tools from the *Special* category when he feels the expense is justified by the amount of use to which these tools will be put.

It is obviously not possible to cover the subject of tools fully here. For those who wish to learn more about tools and their use there is a book entitled *How to Choose and Use Car Tools* available from the publishers of this manual.

Maintenance and minor repair tool kit

The tools given in this list should be considered as a minimum requirement if routine maintenance, servicing and minor repair operations are to be undertaken. We recommend the purchase of combination spanners (ring one end, open-ended the other); although more expensive than open-ended ones, they do give the advantages of both types of spanner.

Combination spanners - 10, 11, 12, 13, 14 & 17 mm
Adjustable spanner - 9 inch
Spark plug spanner (with rubber insert)
Spark plug gap adjustment tool
Set of feeler gauges
Brake bleed nipple spanner
Screwdriver - 4 in long x $\frac{1}{4}$ in dia (flat blade)
Screwdriver - 4 in long x $\frac{1}{4}$ in dia (cross blade)
Combination pliers - 6 inch
Hacksaw (junior)
Tyre pump
Tyre pressure gauge
Oil can
Fine emery cloth (1 sheet)
Wire brush (small)
Funnel (medium size)

Repair and overhaul tool kit

These tools are virtually essential for anyone undertaking any major repairs to a motor vehicle, and are additional to those given in the *Maintenance and minor repair* list. Included in this list is a comprehensive set of sockets. Although these are expensive they will be found invaluable as they are so versatile - particularly if various drives are included in the set. We recommend the $\frac{1}{2}$ in square-drive type, as this can be used with most proprietary torque wrenches. If you cannot afford a socket set, even bought piecemeal, then inexpensive tubular box spanners are a useful alternative.

The tools in this list will occasionally need to be supplemented by tools from the *Special* list.

Sockets (or box spanners) to cover range in previous list
Reversible ratchet drive (for use with sockets)
Extension piece, 10 inch (for use with sockets)
Universal joint (for use with sockets)
Torque wrench (for use with sockets)
'Mole' wrench - 8 inch
Ball pein hammer
Soft-faced hammer, plastic or rubber
Screwdriver - 6 in long x $\frac{5}{16}$ in dia (flat blade)
Screwdriver - 2 in long x $\frac{5}{16}$ in square (flat blade)
Screwdriver - 1$\frac{1}{2}$ in long x $\frac{1}{4}$ in dia (cross blade)
Screwdriver - 3 in long x $\frac{1}{8}$ in dia (electricians)
Pliers - electricians side cutters
Pliers - needle nosed
Pliers - circlip (internal and external)
Cold chisel - $\frac{1}{2}$ inch
Scriber
Scraper
Centre punch
Pin punch
Hacksaw
Valve grinding tool
Steel rule/straight-edge
Allen keys
Selection of files
Wire brush (large)
Axle-stands
Jack (strong scissor or hydraulic type)

Special tools

The tools in this list are those which are not used regularly, are expensive to buy, or which need to be used in accordance with their manufacturers' instructions. Unless relatively difficult mechanical jobs are undertaken frequently, it will not be economic to buy many of these tools. Where this is the case, you could consider clubbing together with friends (or joining a motorists' club) to make a joint purchase, or borrowing the tools against a deposit from a local garage or tool hire specialist.

The following list contains only those tools and instruments freely available to the public, and not those special tools produced by the vehicle manufacturer specifically for its dealer network. You will find occasional references to these manufacturers' special tools in the text of this manual. Generally, an alternative method of doing the job without the vehicle manufacturers' special tool is given. However, sometimes, there is no alternative to using them. Where this is the case and the relevant tool cannot be bought or borrowed, you will have to entrust the work to a franchised garage.

Valve spring compressor
Piston ring compressor
Balljoint separator
Universal hub/bearing puller
Impact screwdriver
Micrometer and/or vernier gauge
Dial gauge
Stroboscopic timing light

Tools and working facilities

Dwell angle meter/tachometer
Universal electrical multi-meter
Cylinder compression gauge
Lifting tackle
Trolley jack
Light with extension lead

Buying tools

For practically all tools, a tool factor is the best source since he will have a very comprehensive range compared with the average garage or accessory shop. Having said that, accessory shops often offer excellent quality tools at discount prices, so it pays to shop around.

There are plenty of good tools around at reasonable prices, but always aim to purchase items which meet the relevant national safety standards. If in doubt, ask the proprietor or manager of the shop for advice before making a purchase.

Care and maintenance of tools

Having purchased a reasonable tool kit, it is necessary to keep the tools in a clean serviceable condition. After use, always wipe off any dirt, grease and metal particles using a clean, dry cloth, before putting the tools away. Never leave them lying around after they have been used. A simple tool rack on the garage or workshop wall, for items such as screwdrivers and pliers is a good idea. Store all normal wrenches and sockets in a metal box. Any measuring instruments, gauges, meters, etc, must be carefully stored where they cannot be damaged or become rusty.

Take a little care when tools are used. Hammer heads inevitably become marked and screwdrivers lose the keen edge on their blades from time to time. A little timely attention with emery cloth or a file will soon restore items like this to a good serviceable finish.

Working facilities

Not to be forgotten when discussing tools, is the workshop itself. If anything more than routine maintenance is to be carried out, some form of suitable working area becomes essential.

It is appreciated that many an owner mechanic is forced by circumstances to remove an engine or similar item, without the benefit of a garage or workshop. Having done this, any repairs should always be done under the cover of a roof.

Wherever possible, any dismantling should be done on a clean, flat workbench or table at a suitable working height.

Any workbench needs a vice: one with a jaw opening of 4 in (100 mm) is suitable for most jobs. As mentioned previously, some clean dry storage space is also required for tools, as well as for lubricants, cleaning fluids, touch-up paints and so on, which become necessary.

Another item which may be required, and which has a much more general usage, is an electric drill with a chuck capacity of at least $\frac{5}{16}$ in (8 mm). This, together with a good range of twist drills, is virtually essential for fitting accessories such as mirrors and reversing lights.

Last, but not least, always keep a supply of old newspapers and clean, lint-free rags available, and try to keep any working area as clean as possible.

Spanner jaw gap comparison table

Jaw gap (in)	Spanner size
0.250	$\frac{1}{4}$ in AF
0.276	7 mm
0.313	$\frac{5}{16}$ in AF
0.315	8 mm
0.344	$\frac{11}{32}$ in AF; $\frac{1}{8}$ in Whitworth
0.354	9 mm
0.375	$\frac{3}{8}$ in AF
0.394	10 mm
0.433	11 mm
0.438	$\frac{7}{16}$ in AF
0.445	$\frac{3}{16}$ in Whitworth; $\frac{1}{4}$ in BSF
0.472	12 mm
0.500	$\frac{1}{2}$ in AF
0.512	13 mm
0.525	$\frac{1}{4}$ in Whitworth; $\frac{5}{16}$ in BSF
0.551	14 mm
0.563	$\frac{9}{16}$ in AF
0.591	15 mm
0.600	$\frac{5}{16}$ in Whitworth; $\frac{3}{8}$ in BSF
0.625	$\frac{5}{8}$ in AF
0.630	16 mm
0.669	17 mm
0.686	$\frac{11}{16}$ in AF
0.709	18 mm
0.710	$\frac{3}{8}$ in Whitworth; $\frac{7}{16}$ in BSF
0.748	19 mm
0.750	$\frac{3}{4}$ in AF
0.813	$\frac{13}{16}$ in AF
0.820	$\frac{7}{16}$ in Whitworth; $\frac{1}{2}$ in BSF
0.866	22 mm
0.875	$\frac{7}{8}$ in AF
0.920	$\frac{1}{2}$ in Whitworth; $\frac{9}{16}$ in BSF
0.938	$\frac{15}{16}$ in AF
0.945	24 mm
1.000	1 in AF
1.010	$\frac{9}{16}$ in Whitworth; $\frac{5}{8}$ in BSF
1.024	26 mm
1.063	$1\frac{1}{16}$ in AF; 27 mm
1.100	$\frac{5}{8}$ in Whitworth; $\frac{11}{16}$ in BSF
1.125	$1\frac{1}{8}$ in AF
1.181	30 mm
1.200	$\frac{11}{16}$ in Whitworth; $\frac{3}{4}$ in BSF
1.250	$1\frac{1}{4}$ in AF
1.260	32 mm
1.300	$\frac{3}{4}$ in Whitworth; $\frac{7}{8}$ in BSF
1.313	$1\frac{5}{16}$ in AF
1.390	$\frac{13}{16}$ in Whitworth; $\frac{15}{16}$ in BSF
1.417	36 mm
1.438	$1\frac{7}{16}$ in AF
1.480	$\frac{7}{8}$ in Whitworth; 1 in BSF
1.500	$1\frac{1}{2}$ in AF
1.575	40 mm; $\frac{15}{16}$ in Whitworth
1.614	41 mm
1.625	$1\frac{5}{8}$ in AF
1.670	1 in Whitworth; $1\frac{1}{8}$ in BSF
1.688	$1\frac{11}{16}$ in AF
1.811	46 mm
1.813	$1\frac{13}{16}$ in AF
1.860	$1\frac{1}{8}$ in Whitworth; $1\frac{1}{4}$ in BSF
1.875	$1\frac{7}{8}$ in AF
1.969	50 mm
2.000	2 in AF
2.050	$1\frac{1}{4}$ in Whitworth; $1\frac{3}{8}$ in BSF
2.165	55 mm
2.362	60 mm

Jacking, towing and wheel changing

Jacking

Use the jack supplied with the vehicle only for wheel changing during roadside emergencies (photo). Chock the wheel diagonally opposite the one being removed.

When raising the vehicle for repair or maintenance, preferably use a trolley or hydraulic jack with a wooden block as an insulator to prevent damage to the underbody. Place the jack under a structural member at the points indicated, never raise the vehicle by jacking up under the engine sump, transmission casing or rear axle. If both front or both rear wheels are to be raised, jack up one side first and securely support it on an axle stand before raising the other side.

To avoid repetition, the procedure for raising the vehicle in order to carry out work under it is not included before each relevant operation described in this manual.

It is to be preferred and is certainly recommended that the vehicle is positioned over an inspection pit or raised on a lift. Where such equipment is not available, use ramps or jack up the vehicle as previously described, but always supplement the lifting device with axle stands.

Towing

Towing hooks are welded to the front and rear of the vehicle and should only be used in an emergency, as their designed function is as lash-down hooks, for use during transportation.

When being towed, remember to insert the ignition key and turn it to Position I. Expect to apply greater pressure to the footbrake, as servo assistance will not be available after the first few brake applications.

Wheel changing

To change a roadwheel, first prise off the wheel trim or remove the roadwheel bolt plastic caps.

If the car is fairly new, the roadwheels and tyres will have been balanced on the vehicle during production. In order to maintain this balance then the position of the roadwheel in relation to the mounting hub must be marked before removing the wheel.

Release but do not remove each roadwheel bolt and then raise the vehicle with the jack. Remove the bolts and take off the wheel.

Tool kit jack in use

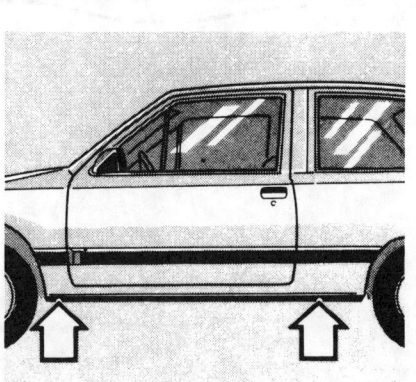

Jacking point locations

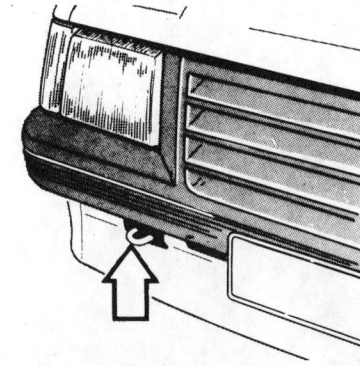

Front towing hook

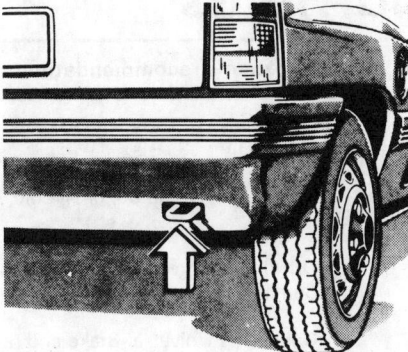

Rear towing hook

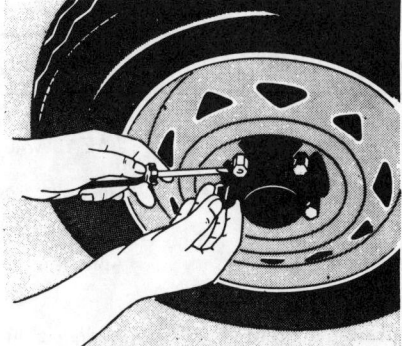

Prise off the wheel trim or the wheel bolt plastic caps to gain access to the wheel retaining bolts

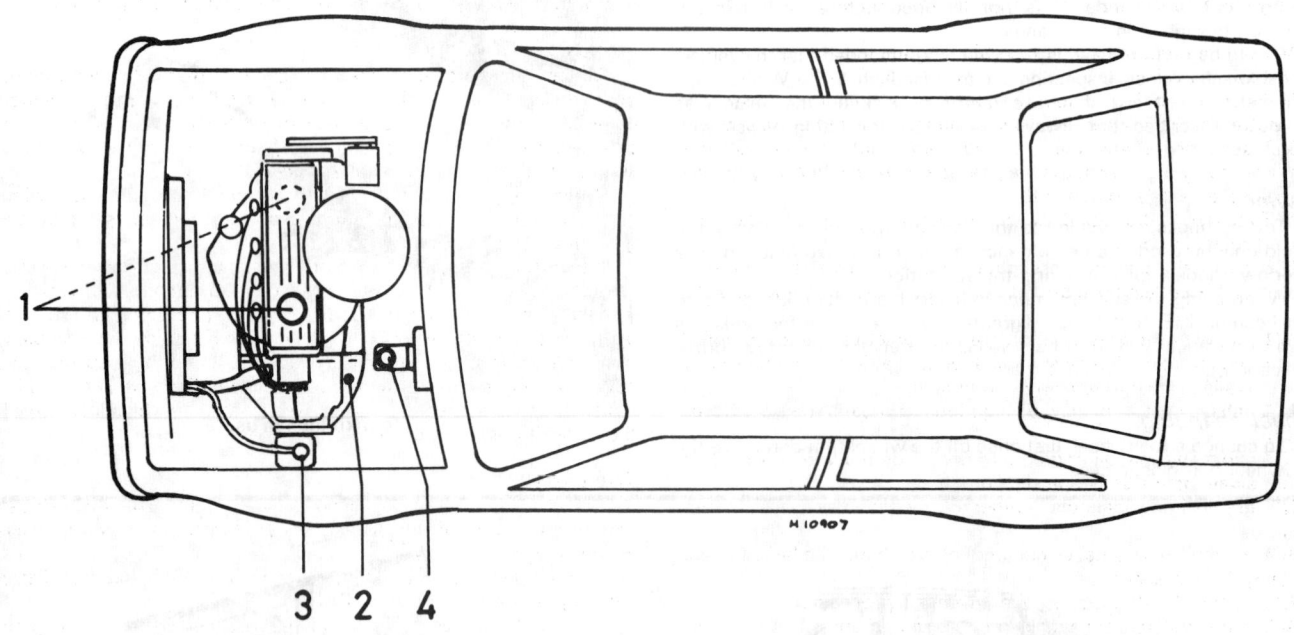

Recommended lubricants and fluids

Component or system	Lubricant type/specification	Duckhams recommendation
Engine (1)	Multigrade engine oil, viscosity range SAE 10W/40 to 20W/50, to API-SG	Duckhams QXR, Hypergrade, or 10W/40 Motor Oil
Transmission (2)	Gear oil, viscosity SAE 80, to API-GL3 or GL4, or GM special oil 90 188 629	Duckhams Hypoid 80, or Hypoid 75W/90S
Cooling system (3)	Antifreeze to GME L 6368	Duckhams Universal Antifreeze and Summer Coolant
Brake hydraulic system (4)	Hydraulic fluid to SAE J1703 or DOT 4	Duckhams Universal Brake and Clutch Fluid

Safety first!

Professional motor mechanics are trained in safe working procedures. However enthusiastic you may be about getting on with the job in hand, do take the time to ensure that your safety is not put at risk. A moment's lack of attention can result in an accident, as can failure to observe certain elementary precautions.

There will always be new ways of having accidents, and the following points do not pretend to be a comprehensive list of all dangers; they are intended rather to make you aware of the risks and to encourage a safety-conscious approach to all work you carry out on your vehicle.

Essential DOs and DON'Ts

DON'T rely on a single jack when working underneath the vehicle. Always use reliable additional means of support, such as axle stands, securely placed under a part of the vehicle that you know will not give way.

DON'T attempt to loosen or tighten high-torque nuts (e.g. wheel hub nuts) while the vehicle is on a jack; it may be pulled off.

DON'T start the engine without first ascertaining that the transmission is in neutral (or 'Park' where applicable) and the parking brake applied.

DON'T suddenly remove the filler cap from a hot cooling system – cover it with a cloth and release the pressure gradually first, or you may get scalded by escaping coolant.

DON'T attempt to drain oil until you are sure it has cooled sufficiently to avoid scalding you.

DON'T grasp any part of the engine, exhaust or catalytic converter without first ascertaining that it is sufficiently cool to avoid burning you.

DON'T allow brake fluid or antifreeze to contact vehicle paintwork.

DON'T syphon toxic liquids such as fuel, brake fluid or antifreeze by mouth, or allow them to remain on your skin.

DON'T inhale dust – it may be injurious to health (see *Asbestos* below).

DON'T allow any spilt oil or grease to remain on the floor – wipe it up straight away, before someone slips on it.

DON'T use ill-fitting spanners or other tools which may slip and cause injury.

DON'T attempt to lift a heavy component which may be beyond your capability – get assistance.

DON'T rush to finish a job, or take unverified short cuts.

DON'T allow children or animals in or around an unattended vehicle.

DO wear eye protection when using power tools such as drill, sander, bench grinder etc, and when working under the vehicle.

DO use a barrier cream on your hands prior to undertaking dirty jobs – it will protect your skin from infection as well as making the dirt easier to remove afterwards; but make sure your hands aren't left slippery. Note that long-term contact with used engine oil can be a health hazard.

DO keep loose clothing (cuffs, tie etc) and long hair well out of the way of moving mechanical parts.

DO remove rings, wristwatch etc, before working on the vehicle – especially the electrical system.

DO ensure that any lifting tackle used has a safe working load rating adequate for the job.

DO keep your work area tidy – it is only too easy to fall over articles left lying around.

DO get someone to check periodically that all is well, when working alone on the vehicle.

DO carry out work in a logical sequence and check that everything is correctly assembled and tightened afterwards.

DO remember that your vehicle's safety affects that of yourself and others. If in doubt on any point, get specialist advice.

IF, in spite of following these precautions, you are unfortunate enough to injure yourself, seek medical attention as soon as possible.

Asbestos

Certain friction, insulating, sealing, and other products – such as brake linings, brake bands, clutch linings, torque converters, gaskets, etc – contain asbestos. *Extreme care must be taken to avoid inhalation of dust from such products since it is hazardous to health.* If in doubt, assume that they *do* contain asbestos.

Fire

Remember at all times that petrol (gasoline) is highly flammable. Never smoke, or have any kind of naked flame around, when working on the vehicle. But the risk does not end there – a spark caused by an electrical short-circuit, by two metal surfaces contacting each other, by careless use of tools, or even by static electricity built up in your body under certain conditions, can ignite petrol vapour, which in a confined space is highly explosive.

Always disconnect the battery earth (ground) terminal before working on any part of the fuel or electrical system, and never risk spilling fuel on to a hot engine or exhaust.

It is recommended that a fire extinguisher of a type suitable for fuel and electrical fires is kept handy in the garage or workplace at all times. Never try to extinguish a fuel or electrical fire with water.

Note: *Any reference to a 'torch' appearing in this manual should always be taken to mean a hand-held battery-operated electric lamp or flashlight. It does NOT mean a welding/gas torch or blowlamp.*

Fumes

Certain fumes are highly toxic and can quickly cause unconsciousness and even death if inhaled to any extent. Petrol (gasoline) vapour comes into this category, as do the vapours from certain solvents such as trichloroethylene. Any draining or pouring of such volatile fluids should be done in a well ventilated area.

When using cleaning fluids and solvents, read the instructions carefully. Never use materials from unmarked containers – they may give off poisonous vapours.

Never run the engine of a motor vehicle in an enclosed space such as a garage. Exhaust fumes contain carbon monoxide which is extremely poisonous; if you need to run the engine, always do so in the open air or at least have the rear of the vehicle outside the workplace.

If you are fortunate enough to have the use of an inspection pit, never drain or pour petrol, and never run the engine, while the vehicle is standing over it; the fumes, being heavier than air, will concentrate in the pit with possibly lethal results.

The battery

Never cause a spark, or allow a naked light, near the vehicle's battery. It will normally be giving off a certain amount of hydrogen gas, which is highly explosive.

Always disconnect the battery earth (ground) terminal before working on the fuel or electrical systems.

If possible, loosen the filler plugs or cover when charging the battery from an external source. Do not charge at an excessive rate or the battery may burst.

Take care when topping up and when carrying the battery. The acid electrolyte, even when diluted, is very corrosive and should not be allowed to contact the eyes or skin.

If you ever need to prepare electrolyte yourself, always add the acid slowly to the water, and never the other way round. Protect against splashes by wearing rubber gloves and goggles.

When jump starting a car using a booster battery, for negative earth (ground) vehicles, connect the jump leads in the following sequence: First connect one jump lead between the positive (+) terminals of the two batteries. Then connect the other jump lead first to the negative (–) terminal of the booster battery, and then to a good earthing (ground) point on the vehicle to be started, at least 18 in (45 cm) from the battery if possible. Ensure that hands and jump leads are clear of any moving parts, and that the two vehicles do not touch. Disconnect the leads in the reverse order.

Mains electricity and electrical equipment

When using an electric power tool, inspection light etc, always ensure that the appliance is correctly connected to its plug and that, where necessary, it is properly earthed (grounded). Do not use such appliances in damp conditions and, again, beware of creating a spark or applying excessive heat in the vicinity of fuel or fuel vapour. Also ensure that the appliances meet the relevant national safety standards.

Ignition HT voltage

A severe electric shock can result from touching certain parts of the ignition system, such as the HT leads, when the engine is running or being cranked, particularly if components are damp or the insulation is defective. Where an electronic ignition system is fitted, the HT voltage is much higher and could prove fatal.

Routine maintenance

For modifications, and information applicable to later models, see Supplement at end of manual

Maintenance is essential for ensuring safety and desirable for the purpose of getting the best in terms of performance and economy from your car. Over the years the need for periodic lubrication has been greatly reduced if not totally eliminated. This has unfortunately tended to lead some owners to think that, because no such action is required, the items either no longer exist, or will last forever. This is certainly not the case, it is essential to carry out regular visual examination as comprehensively as possible in order to spot any possible defects at an early stage before they develop into major expensive repairs.

The following service schedules are a list of the maintenance requirements and the intervals at which they should be carried out, as recommended by the manufacturers. Where applicable these procedures are covered in greater detail throughout this manual, near the beginning of each Chapter.

Weekly or before a long journey

Check engine oil level (photos).
Check operation of all lights, flashers and wipers
Check coolant level (photo)
Check washer fluid level(s), adding a screen wash such as Turtle Wax High Tec Screen Wash
Check tyre pressures (cold), not forgetting the spare (photo)

Every 9000 miles (15 000 km) or six months, whichever comes first

Renew engine oil and filter (photos)
Renew air cleaner element (photo)
Check and adjust valve clearances (1.0 models only)
Clean, adjust or renew distributor contact breaker points and lubricate distributor cam (1.0 models only)
Renew the spark plugs
Lubricate controls, hinges and locks (photo)
Adjust rear brakes and check lining wear
Renew carburettor fuel filter (10S engine)
Inspect tyres for damage and wear (photo)
Check front disc pads for wear
Check brake hydraulic fluid level (photo)
Check ignition timing
Check carburettor adjustment
Check steering and suspension for wear, and gaiters and bellows for damage
Check transmission oil level
Check drivebelt tension and condition
Check brake hydraulic hoses and pipes for damage or corrosion
Check the operation of all electrical equipment, also check the wiring and connectors
Check the condition of the screen wash/wipe systems
Inspect all joint faces and seals for damage, deterioration or leakage

Every 18 000 miles (30 000 km) or 12 months, whichever comes first

In addition to, or instead of, the work specified in the previous schedule

Check exhaust system condition and security of mountings
Check rear wheel bearing adjustment
Check the front wheel alignment
Check clutch pedal adjustment
Check handbrake adjustment and condition of linkage
Check headlamp beam alignment
Renew brake hydraulic fluid (annually, regardless of mileage)
Check coolant antifreeze concentration

Every 36 000 miles (60 000 km) or 2 years, whichever comes first

In addition to the work specified in the previous schedules
On ohc engines, check the condition of the timing belt and, if necessary, adjust the belt tension. It is recommended that the belt is renewed if its condition is in any way suspect.

Every 2 years, regardless of mileage

Renew coolant

Check the engine oil level on the dipstick

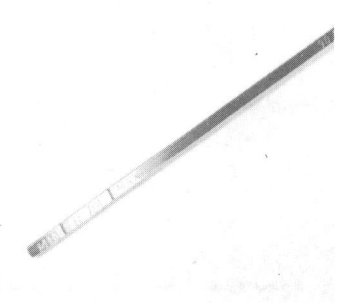
0.75 litre (1.3 pt) separates the MAX and MIN marks on the dipstick

Top up or refill the oil through the filler on the rocker cover

Maintain the level in the cooling system up to the KALT level, when the engine is cold

Check the tyre pressures with an accurate gauge

Engine oil drain plug location

Engine oil filter location

Renewing the air filter element

Lubricating the door locks and hinges

Checking tyre tread depth

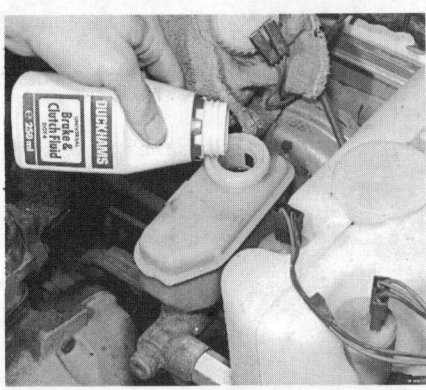
Top up the brake hydraulic fluid through the filler neck in the master cylinder reservoir

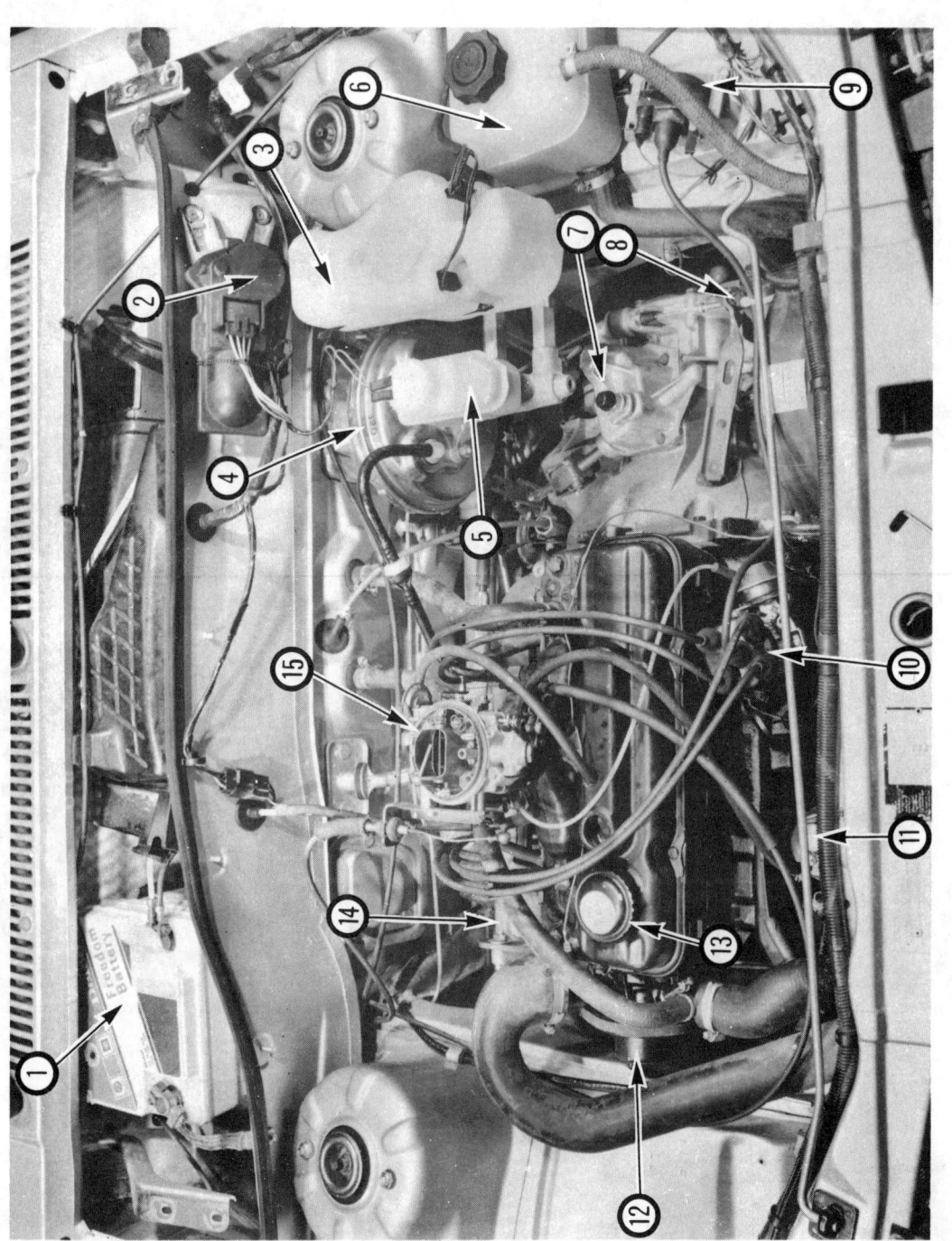

Engine and underbonnet component locations on 1.0 models (air cleaner removed for photographic access)

1 Battery
2 Windscreen wiper motor
3 Screen washer reservoir
4 Brake servo
5 Brake master cylinder reservoir
6 Cooling system expansion tank
7 Transmission breather/filler plug
8 Clutch cable adjustment point
9 Ignition coil
10 Distributor
11 Fuel pump
12 Water pump
13 Oil filler cap
14 Alternator
15 Carburettor

Engine and underbonnet component locations on 1.2 and 1.3 models (air cleaner removed for photographic access)

1 Battery
2 Windscreen wiper motor
3 Brake servo unit
4 Windscreen washer reservoir
5 Cooling system expansion tank
6 Ignition coil
7 Distributor
8 Brake master cylinder reservoir
9 Oil filler cap
10 Camshaft toothed belt cover
11 Fuel pump
12 Alternator
13 Carburettor

Front underbody view of a 1.2 model

1 Exhaust front pipe
2 Engine/transmission rear mounting
3 Suspension control arm
4 Anti-roll bar
5 Suspension tie-bar front mounting
6 Oil drain plug
7 Oil filter
8 Driveshaft inner constant velocity joint
9 Clutch access cover
10 Transmission oil level plug
11 Differential cover plate

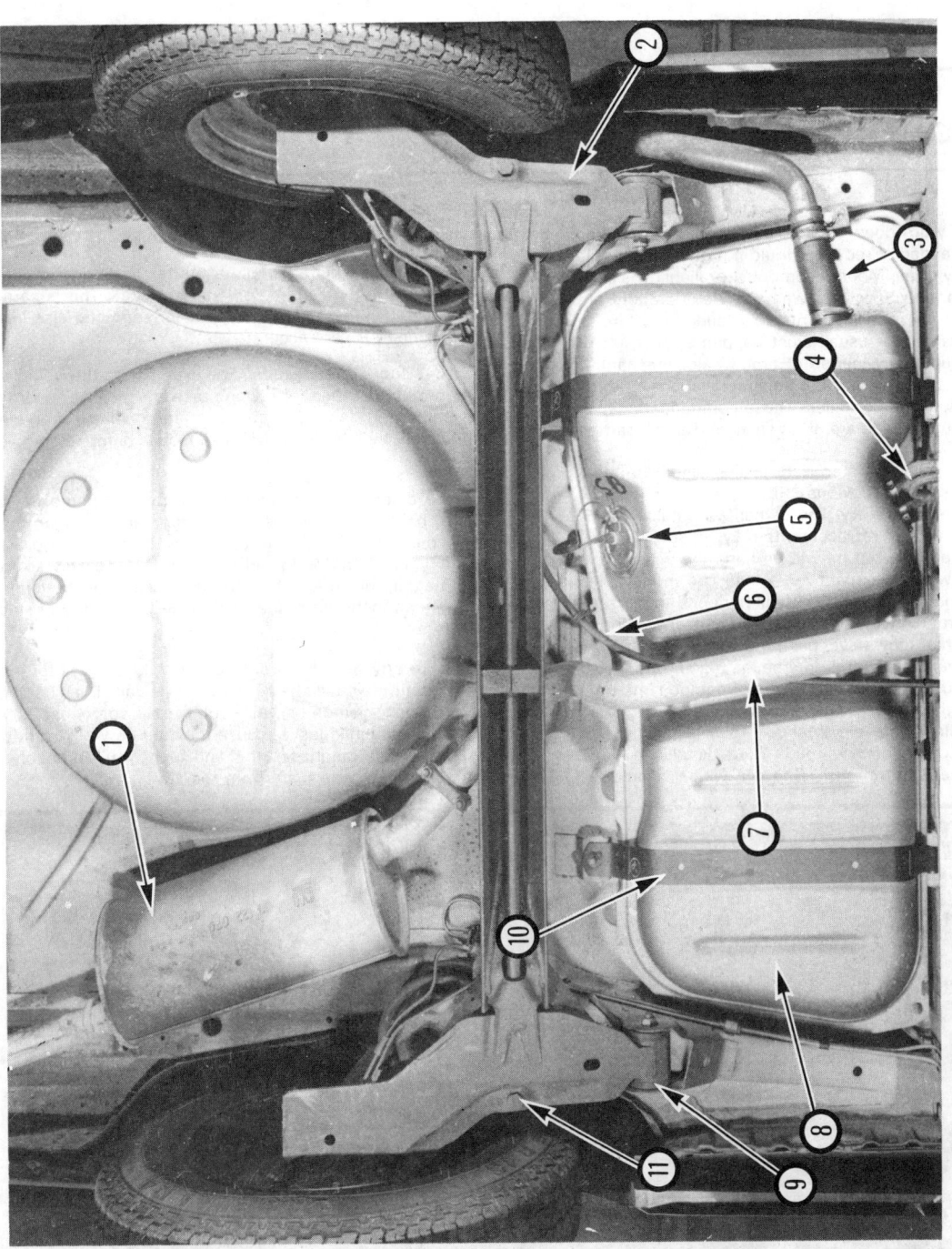

Rear underbody view of a 1.2 model

1 Exhaust silencer
2 Trailing arm
3 Fuel tank filler hose
4 Fuel feed and return hoses
5 Fuel gauge sender unit
6 Handbrake cable
7 Exhaust intermediate pipe
8 Fuel tank
9 Rear suspension pivot/mounting bolt
10 Fuel tank retaining strap
11 Anti-roll bar attachment bolt

Fault diagnosis

Introduction

The vehicle owner who does his or her own maintenance according to the recommended schedules should not have to use this section of the manual very often. Modern component reliability is such that, provided those items subject to wear or deterioration are inspected or renewed at the specified intervals, sudden failure is comparatively rare. Faults do not usually just happen as a result of sudden failure, but develop over a period of time. Major mechanical failures in particular are usually preceded by characteristic symptoms over hundreds or even thousands of miles. Those components which do occasionally fail without warning are often small and easily carried in the vehicle.

With any fault finding, the first step is to decide where to begin investigations. Sometimes this is obvious, but on other occasions a little detective work will be necessary. The owner who makes half a dozen haphazard adjustments or replacements may be successful in curing a fault (or its symptoms), but he will be none the wiser if the fault recurs and he may well have spent more time and money than was necessary. A calm and logical approach will be found to be more satisfactory in the long run. Always take into account any warning signs or abnormalities that may have been noticed in the period preceding the fault – power loss, high or low gauge readings, unusual noises or smells, etc – and remember that failure of components such as fuses or spark plugs may only be pointers to some underlying fault.

The pages which follow here are intended to help in cases of failure to start or breakdown on the road. There is also a Fault Diagnosis Section at the end of each Chapter which should be consulted if the preliminary checks prove unfruitful. Whatever the fault, certain basic principles apply. These are as follows:

Verify the fault. This is simply a matter of being sure that you know what the symptoms are before starting work. This is particularly important if you are investigating a fault for someone else who may not have described it very accurately.

Don't overlook the obvious. For example, if the vehicle won't start, is there petrol in the tank? (Don't take anyone else's word on this particular point, and don't trust the fuel gauge either!) If an electrical fault is indicated, look for loose or broken wires before digging out the test gear.

Cure the disease, not the symptom. Substituting a flat battery with a fully charged one will get you off the hard shoulder, but if the underlying cause is not attended to, the new battery will go the same way. Similarly, changing oil-fouled spark plugs for a new set will get you moving again, but remember that the reason for the fouling (if it wasn't simply an incorrect grade of plug) will have to be established and corrected.

Don't take anything for granted. Particularly, don't forget that a 'new' component may itself be defective (especially if it's been rattling round in the boot for months), and don't leave components out of a fault diagnosis sequence just because they are new or recently fitted. When you do finally diagnose a difficult fault, you'll probably realise that all the evidence was there from the start.

Carrying a few spares can save you a long walk!

Fault diagnosis

Electrical faults

Electrical faults can be more puzzling than straightforward mechanical failures, but they are no less susceptible to logical analysis if the basic principles of operation are understood. Vehicle electrical wiring exists in extremely unfavourable conditions – heat, vibration and chemical attack – and the first things to look for are loose or corroded connections and broken or chafed wires, especially where the wires pass through holes in the bodywork or are subject to vibration.

All metal-bodied vehicles in current production have one pole of the battery 'earthed', ie connected to the vehicle bodywork, and in nearly all modern vehicles it is the negative (–) terminal. The various electrical components – motors, bulb holders etc – are also connected to earth, either by means of a lead or directly by their mountings. Electric current flows through the component and then back to the battery via the bodywork. If the component mounting is loose or corroded, or if a good path back to the battery is not available, the circuit will be incomplete and malfunction will result. The engine and/or gearbox are also earthed by means of flexible metal straps to the body or subframe; if these straps are loose or missing, starter motor, generator and ignition trouble may result.

Assuming the earth return to be satisfactory, electrical faults will be due either to component malfunction or to defects in the current supply. Individual components are dealt with in Chapter 10. If supply wires are broken or cracked internally this results in an open-circuit, and the easiest way to check for this is to bypass the suspect wire temporarily with a length of wire having a crocodile clip or suitable connector at each end. Alternatively, a 12V test lamp can be used to verify the presence of supply voltage at various points along the wire and the break can be thus isolated.

If a bare portion of a live wire touches the bodywork or other earthed metal part, the electricity will take the low-resistance path thus formed back to the battery: this is known as a short-circuit. Hopefully a short-circuit will blow a fuse, but otherwise it may cause burning of the insulation (and possibly further short-circuits) or even a fire. This is why it is inadvisable to bypass persistently blowing fuses with silver foil or wire.

Spares and tool kit

Most vehicles are supplied only with sufficient tools for wheel changing; the *Maintenance and minor repair* tool kit detailed in *Tools and working facilities*, with the addition of a hammer, is probably sufficient for those repairs that most motorists would consider attempting at the roadside. In addition a few items which can be fitted without too much trouble in the event of a breakdown should be carried. Experience and available space will modify the list below, but the following may save having to call on professional assistance:

Spark plugs, clean and correctly gapped
HT lead and plug cap – long enough to reach the plug furthest from the distributor
Distributor rotor, condenser and contact breaker points (where applicable)
Drivebelt(s) – emergency type may suffice
Spare fuses
Set of principal light bulbs
Tin of radiator sealer and hose bandage
Exhaust bandage
Roll of insulating tape
Length of soft iron wire
Length of electrical flex
Torch or inspection lamp (can double as test lamp)
Battery jump leads
Tow-rope
Ignition water dispersant aerosol
Litre of engine oil
Sealed can of hydraulic fluid
Worm drive clips

If spare fuel is carried, a can designed for the purpose should be used to minimise risks of leakage and collision damage. A first aid kit and a warning triangle, whilst not at present compulsory in the UK, are obviously sensible items to carry in addition to the above.

When touring abroad it may be advisable to carry additional spares which, even if you cannot fit them yourself, could save having to wait while parts are obtained. The items below may be worth considering:

Clutch and throttle cables
Cylinder head gasket
Alternator brushes
Tyre valve core

One of the motoring organisations will be able to advise on availability of fuel etc in foreign countries.

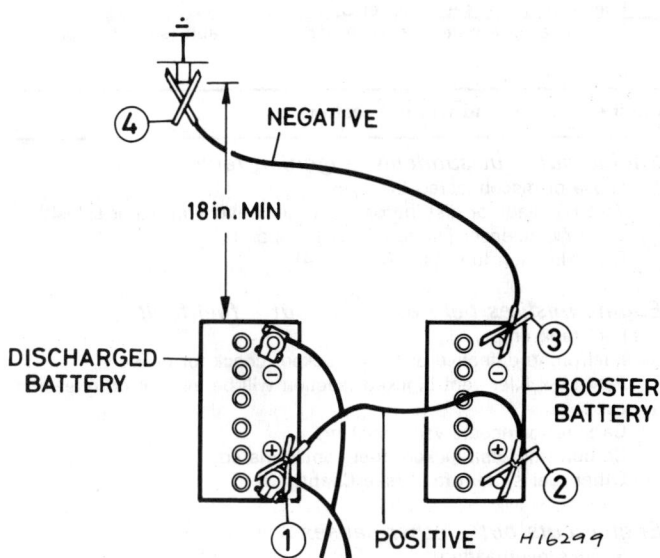

Jump start lead connections for negative earth vehicles – connect leads in order shown

---

Engine will not start

Engine fails to turn when starter operated
Flat battery (recharge, use jump leads, or push start)
Battery terminals loose or corroded
Battery earth to body defective
Engine earth strap loose or broken
Starter motor (or solenoid) wiring loose or broken
Automatic transmission selector in wrong position, or inhibitor switch faulty
Ignition/starter switch faulty
Major mechanical failure (seizure)
Starter or solenoid internal fault (see Chapter 10)

Starter motor turns engine slowly
Partially discharged battery (recharge, use jump leads, or push start) – see Chapter 10
Battery terminals loose or corroded
Battery earth to body defective
Engine earth strap loose
Starter motor (or solenoid) wiring loose
Starter motor internal fault (see Chapter 10)

Starter motor spins without turning engine
Flat battery
Starter motor pinion sticking on sleeve
Flywheel gear teeth damaged or worn
Starter motor mounting bolts loose

Engine turns normally but fails to start
Damp or dirty HT leads and distributor cap (crank engine and check for spark) – try moisture dispersant such as Holts Wet Start
Dirty or incorrectly gapped distributor points (if applicable)
No fuel in tank (check for delivery at carburettor)
Excessive choke (hot engine) or insufficient choke (cold engine)

Fouled or incorrectly gapped spark plugs (remove and regap, or renew)
Other ignition system fault (see Chapter 4)
Other fuel system fault (see Chapter 3)
Poor compression (see Chapter 1)
Major mechanical failure (eg camshaft drive)

Engine fires but will not run
Insufficient choke (cold engine)
Air leaks at carburettor or inlet manifold
Fuel starvation (see Chapter 3)
Ballast resistor defective, or other ignition fault (see Chapter 4)

Engine cuts out and will not restart

Engine cuts out suddenly – ignition fault
Loose or disconnected LT wires
Wet HT leads or distributor cap (after traversing water splash)
Coil or condenser failure (check for spark)
Other ignition fault (see Chapter 4)

Engine misfires before cutting out – fuel fault
Fuel tank empty
Fuel pump defective or filter blocked (check for delivery)
Fuel tank filler vent blocked (suction will be evident on releasing cap)
Carburettor needle valve sticking
Carburettor jets blocked (fuel contaminated)
Other fuel system fault (see Chapter 3)

Engine cuts out – other causes
Serious overheating
Major mechanical failure (eg camshaft drive)

Engine overheats

Ignition (no-charge) warning light illuminated
Slack or broken drivebelt – retension or renew (Chapter 2)

Ignition warning light not illuminated
Coolant loss due to internal or external leakage (see Chapter 2)
Thermostat defective
Low oil level
Brakes binding
Radiator clogged externally or internally
Electric cooling fan not operating correctly
Engine waterways clogged
Ignition timing incorrect or automatic advance malfunctioning
Mixture too weak

Note: *Do not add cold water to an overheated engine or damage may result*

Low engine oil pressure

Gauge reads low or warning light illuminated with engine running
Oil level low or incorrect grade
Defective gauge or sender unit
Wire to sender unit earthed
Engine overheating
Oil filter clogged or bypass valve defective
Oil pressure relief valve defective
Oil pick-up strainer clogged
Oil pump worn or mountings loose
Worn main or big-end bearings

Note: *Low oil pressure in a high-mileage engine at tickover is not necessarily a cause for concern. Sudden pressure loss at speed is far more significant. In any event, check the gauge or warning light sender before condemning the engine.*

Engine noises

Pre-ignition (pinking) on acceleration
Incorrect grade of fuel
Ignition timing incorrect
Distributor faulty or worn
Worn or maladjusted carburettor
Excessive carbon build-up in engine

Whistling or wheezing noises
Leaking vacuum hose
Leaking carburettor or manifold gasket
Blowing head gasket

Tapping or rattling
Incorrect valve clearances (where applicable)
Worn valve gear
Worn timing chain or belt
Broken piston ring (ticking noise)

Knocking or thumping
Worn fanbelt
Peripheral component fault (generator, water pump etc)
Worn big-end bearings (regular heavy knocking, perhaps less under load)
Worn main bearings (rumbling and knocking, perhaps worsening under load)
Piston slap (most noticeable when cold)

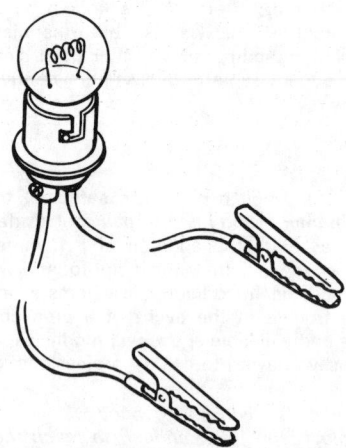

Simple test lamp is useful for tracing electrical faults

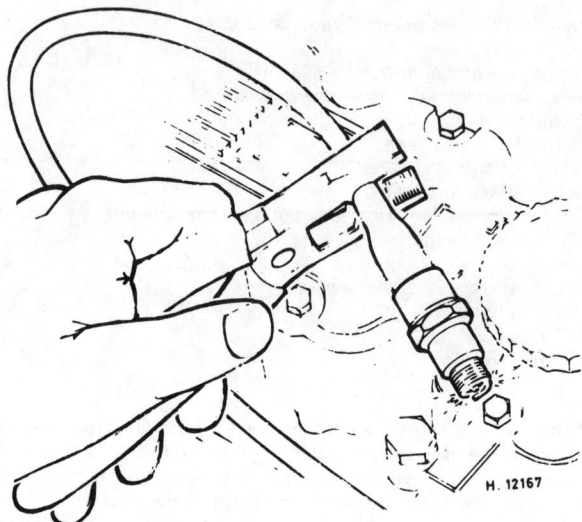

Crank engine and check for a spark. Note use of insulated tool

Chapter 1 Engine

For modifications, and information applicable to later models, see Supplement at end of manual

Contents

Part A: 1.0 litre engine

Ancillary components – removal and refitting	6
Camshaft and tappets – removal and refitting	18
Crankshaft and main bearings – removal and refitting	20
Crankshaft rear oil seal – removal and refitting	19
Cylinder head – overhaul	9
Cylinder head – removal and refitting	8
Cylinder head and pistons – decarbonising	10
Engine – removal and refitting	22
Engine components – examination and renovation	21
Engine dismantling and reassembly – general	5
Engine lubrication system – general description	23
Engine/transmission mountings – removal and refitting	17
Flywheel – removal and refitting	16
General description	1
Maintenance and inspection	2
Oil pump – overhaul	13
Oil pump – removal and refitting	12
Operations possible with the engine in the car	3
Operations requiring engine removal	4
Pistons and connecting rods – removal and refitting	15
Sump – removal and refitting	11
Timing gear components – removal and refitting	14
Valve clearances – adjustment	7

Part B: 1.2 and 1.3 litre engines

Ancillary components – removal and refitting	29
Camshaft housing and camshaft – dismantling and reassembly	33
Camshaft toothed belt – removal, refitting and adjustment	31
Crankshaft and main bearings – removal and refitting	44
Crankshaft front oil seal – removal and refitting	41
Crankshaft rear oil seal – removal and refitting	43
Cylinder head – overhaul	34
Cylinder head – removal and refitting	32
Cylinder head and pistons – decarbonising	35
Engine – removal and refitting	46
Engine components – examination and renovation	45
Engine dismantling and reassembly – general	28
Engine lubrication and crankcase ventilation systems – general description	47
Engine/transmission mountings – removal and refitting	42
Flywheel – removal and refitting	40
General description	24
Maintenance and inspection	25
Oil pressure regulator valve – removal and refitting	30
Oil pump – overhaul	38
Oil pump – removal and refitting	37
Operations possible with the engine in the car	26
Operations requiring engine removal	27
Pistons and connecting rods – removal and refitting	39
Sump – removal and refitting	36

Part C: All engines

Fault diagnosis – engine	48

Specifications

Note: *The engines fitted to Nova models are manufactured to metric dimensions and tolerances and therefore only these are quoted for critical sizes, to avoid the risk of conversion error.*

Part A: 1.0 litre ohv engine
General

Type	Four cylinder, in-line, overhead valve
Designation	10S (S denotes high compression)
Bore	72.0 mm
Stroke	61.0 mm
Capacity	993 cc
Compression ratio	9.2:1
Firing order	1-3-4-2
Location of No 1 cylinder	At timing cover end
Output (DIN)	33 kW (45 bhp) at 5400 rpm

Chapter 1 Engine

Cylinder block (crankcase)
Type	Cast iron, cylinders cast integrally with upper half of crankcase
Maximum cylinder ovality	0.013 mm
Maximum cylinder taper	0.013 mm
Standard production bore available in 16 grades	71.95 to 72.10 mm
Oversize bore size (nominal 0.5 mm)	72.47 to 72.50 mm

Crankshaft
Endfloat	0.09 to 0.20 mm
No of main bearings	3
Main bearing diameter – standard:	
No 1 journal	53.997 to 54.010 mm
Nos 2 and 3 journals	54.007 to 54.020 mm
Main bearing play:	
Bearing 1	0.020 to 0.046 mm
Bearing 2	0.010 to 0.036 mm
Bearing 3	0.00 to 0.0315 mm
Main bearing journals – undersize diameters available	0.25 and 0.50 mm
Big-end journal diameter (standard)	44.971 to 44.987 mm
Big-end journal – undersize diameters available	0.25 and 0.50 mm
Big-end clearance	0.015 to 0.059 mm
Maximum ovality	0.006 mm
Maximum taper	0.01 mm

Camshaft
Endfloat	0.17 to 0.43 mm
Camshaft drive	Single row chain
Number of bearings	3
Bearing journal diameters – standard:	
No 1 journal	40.960 to 40.975 mm
No 2 journal	40.460 to 40.475 mm
No 3 journal	39.960 to 39.975 mm
Maximum undersize diameters:	
No 1 journal	40.460 to 40.475 mm
No 2 journal	39.960 to 39.975 mm
No 3 journal	39.460 to 39.475 mm

Pistons and rings
Piston type	Recessed head
Number of rings	2 compression and 1 oil control
Piston diameter	71.94 to 72.09 mm
Piston clearance in bore	0.01 to 0.03 mm
Oversize pistons	+ 0.5 mm
Gudgeon pin location	Interference fit in connecting rod
Piston ring/groove clearance:	
Top compression	0.060 to 0.087 mm
Second compression	0.033 to 0.063 mm
Ring end gap:	
Compression rings	0.30 to 0.45 mm
Oil control ring	0.40 to 1.40 mm

Cylinder head
Material	Cast iron
Maximum permissible distortion of sealing face	0.015 mm over 150 mm or 0.05 mm over total surface area
Valve seat width:	
Inlet	1.25 to 1.50 mm
Exhaust	1.60 to 1.85 mm

Valve
Valve clearances – hot:	
Inlet	0.15 mm
Exhaust	0.25 mm
Valve timing:	
Inlet opens	46° BTDC
Inlet closes	90° ABDC
Exhaust opens	70° BBDC
Exhaust closes	30° ATDC
Valve seat angle	44°
Inlet valve head diameter	32 mm
Exhaust valve head diameter	27 mm
Valve stem diameter	7.00 to 7.010 mm
Oversizes available	0.075 mm, 0.150 mm, 0.250 mm
Valve overall length:	
Inlet	99.3 mm
Exhaust	101.1 mm

Chapter 1 Engine

Maximum permissible valve stem clearance:	
Inlet	0.015 to 0.045 mm
Exhaust	0.035 to 0.065 mm
Valve guide internal diameter:	
Standard	7.025 to 7.045 mm
Oversizes	0.075 mm, 0.150 mm, 0.250 mm
Valve springs:	
Length at load of 15 kg	32.5 mm
Length at load of 45 kg	23.0 mm

Lubrication system

Oil pump tolerances:	
Permissible tooth surface play	0.1 to 0.2 mm
Gear endfloat and gear-to-housing clearance	0.04 to 0.1 mm
Oil capacity:	
With filter change	2.5 litres (4.4 pts)
Without filter change	2.25 litres (4.0 pts)
Oil type/specification	Multigrade engine oil, viscosity range SAE 10W/40 to 20W/50, to API-SG (Duckhams QXR, Hypergrade, or 10W/40 Motor Oil)
Oil pressure at idle	Not less than 0.3 bar (4.4 lbf/in^2)
Oil filter element	Champion C103

Torque wrench settings

	Nm	lbf ft
Cylinder head bolts (cold):		
Stage 1	25	18
Stage 2	Turn further through 60°	Turn further through 60°
Stage 3	Turn further through 60°	Turn further through 60°
Stage 4	Turn further through 60°	Turn further through 60°
Main bearing cap bolts	62	45
Big-end cap bolts	27	20
Flywheel bolts	35	26
Oil pump mounting bolts	20	15
Sump bolts (use locking compound)	5	4
Camshaft sprocket bolt	40	30
Crankshaft pulley bolt	40	30
Oil drain plug	45	33
Engine/transmission mountings and mounting bracket bolts*	40	30

Use sealant on right-hand bracket-to-block bolts

Part B: 1.2 and 1.3 litre ohc engines

General

Type	Four cylinder, in-line overhead camshaft
Designation:	
1.2 models	12ST S denotes high compression, T or B
1.3 models	13SB denotes power output variation
Bore:	
1.2 models	77.8 mm
1.3 models	75.0 mm
Stroke:	
1.2 models	62.9 mm
1.3 models	73.4 mm
Capacity:	
1.2 models	1196 cc
1.3 models	1297 cc
Compression ratio	9.2:1
Firing order	1-3-4-2
Location of No 1 cylinder	At camshaft belt end
Output (DIN):	
1.2 models	40 kW (55 bhp) at 5600 rpm
1.3 models	51 kW (70 bhp) at 5800 rpm

Cylinder block (crankcase)

Material	Cast iron
Maximum cylinder bore out-of-round	0.013 mm
Maximum permissible taper	0.013 mm
Maximum rebore oversize	0.5 mm

Crankshaft

Number of main bearings	5
Main bearing journal diameter	54.972 to 54.985 mm
Crankpin diameter	42.971 to 42.987 mm
Undersizes	0.25 and 0.50 mm
Crankshaft endfloat	0.1 to 0.2 mm
Main bearing running clearance	0.025 to 0.05 mm
Big-end running clearance	0.019 to 0.071 mm
Big-end side-play	0.11 to 0.24 mm

Camshaft
Identification code:
- 1.2 models ... C
- 1.3 models ... B

Endfloat ... 0.09 to 0.21 mm

Camlift:
- 1.2 models ... 5.1 mm (inlet), 5.5 mm (exhaust)
- 1.3 models ... 6.0 mm

Camshaft journal diameters:
- No 1 ... 39.435 to 39.450 mm
- No 2 ... 39.685 to 39.700 mm
- No 3 ... 39.935 to 39.950 mm
- No 4 ... 40.125 to 40.200 mm
- No 5 ... 40.435 to 40.450 mm

Camshaft bearing (direct in housing) diameters:
- No 1 ... 39.500 to 39.525 mm
- No 2 ... 39.750 to 39.775 mm
- No 3 ... 40.000 to 40.025 mm
- No 4 ... 40.250 to 40.275 mm
- No 5 ... 40.550 to 40.525 mm

Pistons and rings
- Type ... Alloy, recessed head
- Piston-to-bore clearance ... 0.02 mm
- Number of piston rings ... 2 compression, 1 oil control

Ring end gap:
- Compression ... 0.3 to 0.5 mm
- Oil control (rail) ... 0.40 to 1.40 mm

Ring gap offset ... 120°

Gudgeon pin:
- Length ... 65.0 mm
- Diameter ... 20.0 mm
- Fit ... Interference in connecting rod
- Clearance in piston ... 0.007 to 0.0115 mm

Cylinder head
- Material ... Light alloy
- Maximum permissible distortion of sealing face ... 0.025 mm
- Overall height of cylinder head ... 95.9 to 96.1 mm

Valve seat width:
- Inlet ... 1.3 to 1.4 mm
- Exhaust ... 1.7 to 1.8 mm

Valves
Valve clearance ... Automatic by hydraulic valve lifters (tappets)

Valve stem-to-guide clearance:
- Inlet ... 0.02 to 0.05 mm
- Exhaust ... 0.04 to 0.07 mm

- Valve seat angle ... 44°
- Valve guide installed height ... 80.85 to 81.25 mm

Valve stem diameter:
- Inlet ... 7.00 to 7.010 mm
- Exhaust ... 6.980 to 6.990 mm

- Oversizes ... 0.075, 0.150, 0.250 mm
- Valve guide bore ... 7.030 to 7.050 mm

Flywheel
Maximum thickness reduction at driven plate and pressure plate cover contact surfaces ... 0.3 mm

Oil pump
- Tooth play (gear to gear) ... 0.1 to 0.2 mm
- Clearance (outer gear to housing) and gear endfloat ... 0.08 to 0.15 mm
- Oil pressure at idle with engine at operating temperature ... 1.5 bar (21.75 lbf/in²)
- Oil capacity with filter change ... 3.0 litres (5.2 pts)
- Dipstick MIN to MAX quantity ... 0.75 litres (1.3 pts)

Torque wrench settings

	Nm	lbf ft
Flywheel to crankshaft	60	44
Main bearing cap bolts:	80	59
Stage 1	50	37
Stage 2	Turn through further 45 to 60°	
Oil pump mounting bolts	6	4
Oil pump relief valve cap	30	22
Alternator bracket bolts	40	29

Chapter 1 Engine

	Nm	lbf ft
Oil pump relief valve cap	30	22
Alternator bracket bolts	40	29
Big-end cap bolts	25 plus 30° to 45°	18 plus 30° to 45°
Sump bolts	5	4
Cylinder head bolts:		
Stage 1	25	18
Stage 2	Turn through further 60°	Turn through further 60°
Stage 3	Turn through further 60°	Turn through further 60°
Stage 4	Turn through further 30°	Turn through further 30°
Stage 5 (after the engine has reached normal operating temperature)	Turn through further 30°	Turn through further 30°
Camshaft sprocket bolt	45	33
Crankshaft pulley bolt (use locking compound)	55	40
Oil pump housing bolts	6	4
Oil drain plug	45	33
Engine/transmission mountings and mounting bracket bolts	40	30

PART A: 1.0 LITRE ENGINE

1 General description

The engine is of four cylinder, in-line overhead valve type, mounted transversely at the front of the car.

The crankshaft is supported in three shell type main bearings. Thrust flanges are incorporated in the centre main bearing to control crankshaft endfloat.

The connecting rods are attached to the crankshaft by horizontally split shell type big-end bearings, and to the pistons by gudgeon pins which are an interference fit in the connecting rod small-end bore. The aluminium alloy postons are of the slipper type and are fitted with three piston rings: two compression rings and an oil control ring.

The camshaft is chain driven from the crankshaft and operates the rocker arms via tappets and short pushrods. The inlet and exhaust valves are each closed by a single valve spring and operate in guides, integral with the cylinder head. The valves are actuated directly by the rocker arms.

Engine lubrication is by a gear type oil pump. The pump is mounted beneath the crankcase and is driven by a camshaft, as are the distributor and fuel pump.

Many of the engine component retaining bolts are of the socket-headed type and require the use of special Torx type or multi-tooth keys or socket bits for removal. These are readily available from retail outlets and should be obtained if major dismantling or repair work is to be carried out on the engine.

2 Maintenance and inspection

1 Every 9000 miles (15 000 km) or 6 months, whichever occurs first, carry out the following maintenance operations on the engine.
2 Visually inspect the engine joint faces, gaskets and seals for any sign of water or oil leaks. Pay particular attention to the areas around the rocker cover, cylinder head, timing cover and sump joint faces. Rectify any leaks by referring to the appropriate Sections of this Chapter.
3 Carefully inspect the condition of the engine breather hoses and renew them if there are any signs of cracking or deterioration of the rubber.
4 Place a suitable container beneath the oil drain plug at the rear of the sump. Unscrew the plug using a ring spanner or socket and allow the oil to drain. Inspect the condition of the drain plug sealing washer and renew it if necessary. refit and tighten the plug after draining.
5 Refill the engine using the correct grade of oil through the filler orifice on the rocker cover. Fill until the level reaches the MAX mark on the dipstick.
6 Move the bowl to the front of the engine under the oil filter.
7 Using a strap wrench or filter removal tool, slacken the filter and then unscrew it from the housing and discard.
8 Wipe the mating face on the housing with a rag and then lubricate the rubber seal on the filter using clean engine oil.
9 Screw the filter into position and tighten it by hand only, do not use any tools.
10 With the engine running, check for leaks around the filter seal. Switch off the engine and top up the oil level.

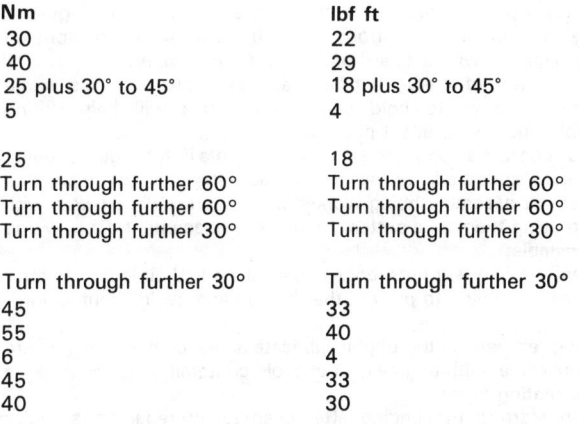

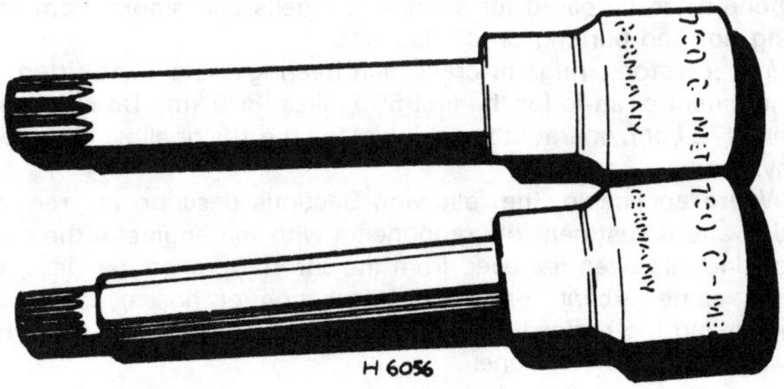

Fig. 1.1 Typical multi-tooth socket bits (Sec 1)

11 Adjust the valve clearances using the procedure described in Section 7.

3 Operations possible with the engine in the car

The following operations may be carried out without having to remove the engine from the car:

(a) Adjustment of the valve clearances
(b) Removal and refitting of cylinder head
(c) Removal and refitting of sump
(d) Removal and refitting of oil pump
(e) Removal and refitting of the timing gear components
(f) Removal and refitting of pistons and connecting rods
(g) Removal and refitting of the flywheel
(h) Removal and refitting of the engine/transmission mountings

4 Operations requiring engine removal

The following operations can only be carried out after removal of the engine from the car:

(a) Removal and refitting of the camshaft and tappets
(b) Removal and refitting of the crankshaft and main bearings
(c) Removal and refitting of the crankshaft rear oil seal

5 Engine dismantling and reassembly – general

1 If the engine has been removed from the car for major overhaul or if individual components have been removed for repair or renewal, observe the following general hints on dismantling and reassembly.
2 Drain the oil into a suitable container and then thoroughly clean the exterior of the engine using a degreasing solvent or paraffin. Clean away as much of the external dirt and grease as possible before dismantling.
3 As parts are removed, clean them in a paraffin bath. However, do not immerse parts with internal oilways in paraffin as it is difficult to remove, usually requiring a high pressure hose. Clean oilways with nylon pipe cleaners.
4 Avoid working with the engine or any of the components directly on a concrete floor, as grit presents a real source of trouble.

5 Wherever possible work should be carried out with the engine or individual components on a strong bench. If the work must be done on the floor, cover it with a board or sheets of newspaper.
6 Have plenty of clean, lint-free rags available and also some containers or trays to hold small items. This will help during reassembly and also prevent possible losses.
7 Always obtain a complete set of new gaskets if the engine is being completely dismantled, or all those necessary for the individual component or assembly being worked on. Keep the old gaskets with a view to using them as a pattern to make a replacement if a new one is not available.
8 When possible refit nuts, bolts and washers in their locations after removal as this helps to protect the threads and avoids confusion or loss.
9 During reassembly thoroughly lubricate all the components, where this is applicable, with engine oil, but avoid contaminating the gaskets and joint mating faces.
10 When starting the engine after overhaul or repair to a major component, be prepared for some odd smells and smoke from parts getting hot and burning off oil deposits.
11 If new pistons, rings or crankshaft bearings have been fitted, the engine must be run-in for the first 500 miles (800 km). Do not exceed 45 mph (72 kph), operate the engine at full throttle or allow it to labour in any gear.
12 Where applicable, the following Sections describe the removal, refitting and adjustment of components with the engine in the car. If the engine has been removed from the car the procedures described are the same except for the disconnection of hoses, cables and linkages, and the removal of components necessary for access, which will already have been done.

7.4 Remove the engine breather hose from the rocker cover

6 Ancillary components – removal and refitting

If the engine has been removed from the car for complete dismantling, the following externally mounted ancillary components should be removed. When the engine has been reassembled these components can be refitted before the engine is installed in the car, as setting up and adjustment is often easier with the engine removed. The removal and refitting sequence need not necessarily follow the order given:

Alternator (Chapter 10)
Distributor and spark plugs (Chapter 4)
Inlet and exhaust manifolds and carburettor (Chapter 3)
Fuel pump (Chapter 3)
Water pump and thermostat (Chapter 2)
Clutch assembly (Chapter 5)
Oil filter (Section 2 of this Chapter)
Dipstick

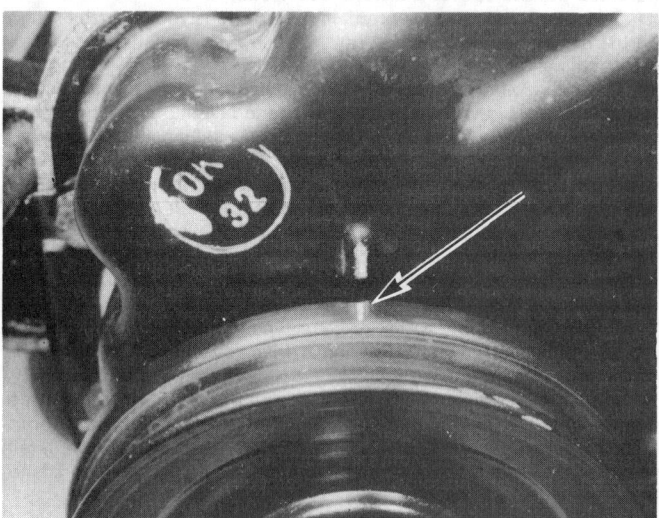
7.7 Ignition timing marks (arrowed) in alignment

7 Valve clearances – adjustment

1 This adjustment should be carried out with the engine at its normal operating temperature. If it is being done after overhaul when the engine is cold, repeat the adjustment after the car has been driven a few miles when the engine will then be hot.
2 Begin by removing the air cleaner, as described in Chapter 3.
3 Mark the spark plug leads to ensure correct refitting and then pull them off the spark plugs.
4 Disconnect the engine breather hoses at the rocker cover (photo).
5 Undo the four bolts securing the rocker cover to the cylinder head and lift off the shaped spreader washers.
6 Withdraw the rocker cover from the cylinder head. If it is stuck give it a tap with the palm of your hand to free it.
7 Turn the engine by means of the crankshaft pulley bolt, or by engaging top gear and pulling the car forward, until No 1 piston is rising on the firing stroke. This can be checked by removing No 1 spark plug and feeling for compression with your finger as the engine is turned, or by removing the distributor cap and checking the position of the rotor arm which should be pointing to the No 1 spark plug lead segment in the cap. The ignition timing marks on the pulley and timing cover must be aligned (photo).

7.10 Adjusting valve clearances

8   With the engine in this position the following valves can be adjusted – counting from the timing cover end of the engine

   1 exhaust
   2 inlet
   3 inlet
   5 exhaust

9   Now turn the engine crankshaft through one complete revolution and adjust the following remaining valves:

   4 exhaust
   6 inlet
   7 inlet
   8 exhaust

10   As each clearance is being checked, slide a feeler blade of the appropriate size, as given in the Specifications, between the end of the valve stem and the rocker arm (photo). Adjust the clearance by turning the rocker arm retaining nut using a socket or ring spanner until the blade is a stiff sliding fit.

11   It is also possible to check and adjust the clearances with the engine running. This is done in the same way, but each valve is checked in turn. It will of course be necessary to refit the plug leads and No 1 spark plug if this method is adopted. To reduce oil splash place a piece of cardboard, suitably cut to shape, between the pushrod side of the rocker arms and the edge of the cylinder head.

12   After adjustment remove all traces of old gasket from the cylinder head mating face and renew the rocker cover gasket if it is cracked or perished.

13   Refit the rocker cover and secure with the retaining bolts and shaped spreader washers.

14   Refit the spark plug and plug leads, reconnect the engine breather hoses and refit the air cleaner, as described in Chapter 3.

8   Cylinder head – removal and refitting

1   Make sure that the engine is cold before commencing operations to avoid any chance of the head distorting.
2   Disconnect the battery negative terminal.
3   Drain the cooling system, as described in Chapter 2, and remove the air cleaner, as described in Chapter 3.
4   From behind the engine undo the two bolts securing the exhaust front pipe to the manifold (photo). Remove the bolts and tension springs, then separate the pipe joint from the manifold.
5   Slacken the retaining clip and disconnect the radiator top hose from the thermostat housing in the water pump (photo).
6   Slacken the alternator mounting and adjustment arm bolts, move the alternator towards the engine and slip the drivebelt off the pulleys.
7   Slacken the retaining clips and disconnect the heater hose (photo) and radiator bottom hose from the water pump (photo).
8   Disconnect the other heater hose at the cylinder head outlet after slackening the retaining clip (photo).
9   Undo the union nut and disconnect the brake servo vacuum hose from the inlet manifold (photo).
10   Note the location of the plug leads to aid refitting and pull them off the spark plugs. Disconnect the HT lead at the coil, undo the distributor cap retaining screws and remove the cap and leads.
11   Refer to Chapter 3 and disconnect the choke and accelerator cables from the carburettor. Detach the distributor vacuum advance pipe.
12   Disconnect the fuel hose from the carburettor and plug its end after removal.
13   Disconnect the engine breather hoses from the rocker cover.
14   Undo the three socket-headed screws securing the inlet manifold to the cylinder head (photo). Note the spark plug lead support brackets fitted to the two end retaining bolts.
15   Lift the inlet manifold complete with carburettor from the cylinder head and recover the gasket (photo).
16   Undo the four bolts and shaped spreader washers and lift off the rocker cover (photo).
17   Slacken the rocker arm retaining nuts, move the rocker arms to one side and lift out the pushrods (photo). Keep the pushrods in order after removal.
18   Undo the cylinder head retaining bolts, half a turn at a time in the reverse sequence to that shown in photo 8.26. Unscrew the bolts fully and remove them. Discard the original bolts and obtain new ones.

19   Lift the cylinder head from the block (photo). If it is stuck, tap it free with a soft-faced mallet. Do not insert a lever into the gasket joint – you may damage the mating surfaces.
20   With the cylinder head removed, recover the gasket (photo).
21   If the cylinder head has been removed for decarbonising or for attention to the valves or springs, reference should be made to Sections 9 and 10.
22   Before refitting the cylinder head ensure that the cylinder block and head mating faces are spotlessly clean and dry with all traces of old gasket removed. Use a scraper and wire brush to do this, but take care to cover the water passages and other openings with masking tape or rag to prevent dirt and carbon falling in. Remove all traces of oil and water from the bolt holes, otherwise hydraulic pressure created by the bolts being screwed in could crack the block or give inaccurate torque settings. Ensure that the bolt threads are clean and dry.
23   When all is clean, screw two guide studs into the cylinder block. These can be made from the two old cylinder head bolts by cutting off their heads and sawing a screwdriver slot in their ends.

Fig. 1.2 Cylinder head guide studs made from old head retaining bolts (Sec 8)

24   Locate a new gasket in position on the block as shown (photo). *Do not use any jointing compound on the gasket.*
25   Lower the cylinder head carefully into position. Screw in the bolts finger tight, remove the guide pins and screw in the two remaining bolts.
26   Tighten the new cylinder head bolts in the order shown (photo) to the first stage specified torque. Follow the same order and tighten the bolts through the angles specified for the second, third and fourth stages. No further tightening is required.
27   Refit the pushrods, making quite sure that each one is located in its tappet (photo).
28   Reposition the rocker arms over the ends of the pushrods and then adjust the valve clearances, as described in Section 7.
29   Place a new gasket in position and refit the inlet manifold and carburettor (photo).
30   Refit the rocker cover, using a new gasket, and secure with the four bolts and spreader washers.
31   Refit the heater hoses and radiator hoses to the outlets on the water pump and cylinder head.
32   Refit the fuel hose to the carburettor, the vacuum advance pipe to the distributor and the breather hoses to the rocker cover.
33   Refit and adjust the accelerator and choke cables, as described in Chapter 3.
34   Refit the brake servo vacuum hose to the inlet manifold.
35   Refit the distributor cap and reconnect the plug leads and coil lead.
36   Slip the drivebelt over the pulleys and adjust its tension, as described in Chapter 2.
37   Reconnect the exhaust front pipe to the manifold and tighten the bolts to compress the tension springs.
38   Refill the cooling system, as describe in Chapter 2, refit the air cleaner, as described in Chapter 3 and connect the battery negative terminal.

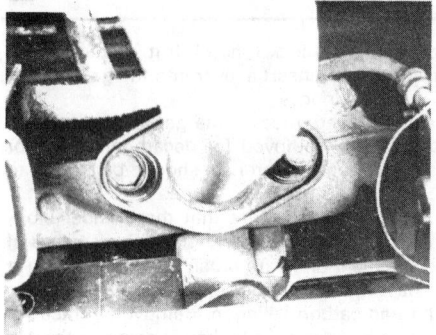

8.4 Exhaust front pipe-to-manifold bolts

8.5 Disconnect the radiator top hose ...

8.7A ... heater hose ...

8.7B ... and radiator bottom hose from the water pump

8.8 Disconnect the other heater hose at the cylinder head outlet

8.9 Disconnect the brake servo vacuum hose union

8.14 Undo the inlet manifold socket-headed bolts (arrowed) ...

8.15 ... and lift off the manifold and carburettor

8.16 Remove the rocker cover

8.17 Slacken the rocker arms and lift out the pushrods

8.19 Remove the cylinder head ...

8.20 ... and the old gasket

8.24 Fit a new cylinder head gasket without jointing compound

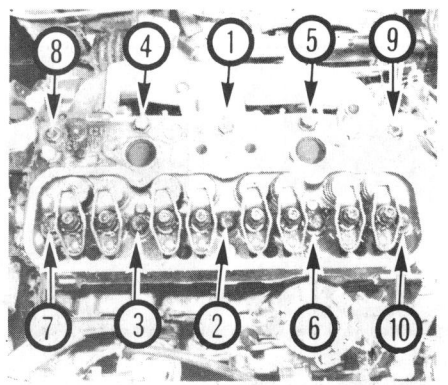
8.26 Cylinder head bolt tightening sequence

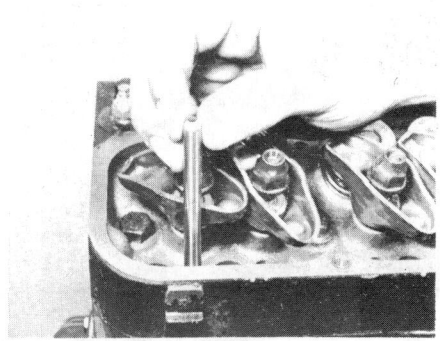
8.27 Refit the pushrods

8.29 Inlet manifold gasket in position

9.5A Remove the valve collar ...

9.5B ... and spring from the valve after releasing the collets

9.18A Refit the valve to the guide ...

9.18B ... then place the spring seat in position ...

9.18C ... followed by the spring

9.19 Place the collar over the spring

9.21 Compress the spring and refit the collets

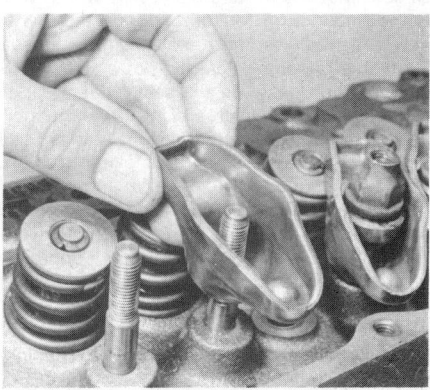
9.24A Place the rocker arm over the stud ...

9.24B ... followed by the pivot ball

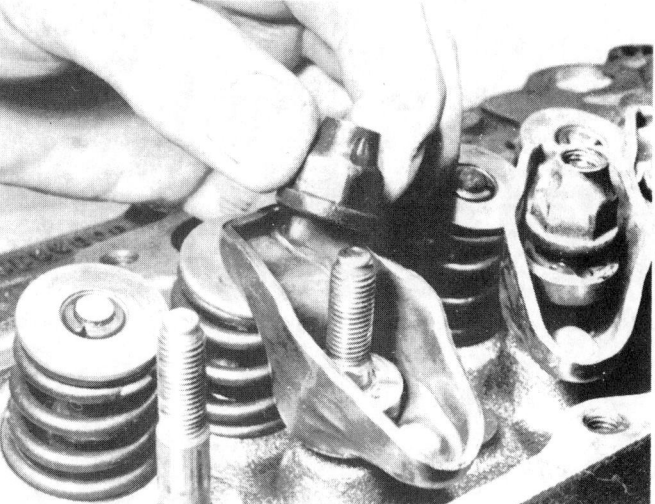

9.25 Refit the nut with the self-locking collar uppermost

9 Cylinder head – overhaul

1 Unscrew the rocker arm retaining/adjustment nuts and withdraw the rocker arms from the studs. Keep them in order as they are removed.
2 To remove the valves, the springs will have to be compressed to allow the split collets to be released from the groove in the upper section the valve stems. A valve spring compressor will therefore be necessary.
3 Locate the compressor to enable the forked end of the arm to be positioned over the valve spring collar whilst the screw part of the clamp is situated squarely on the face of the valve.
4 Screw up the clamp to compress the spring and release the pressure of the collar acting on the collets. If the collar sticks, support the head and clamp frame and give the end of the clamp a light tap with a hammer to help release it.
5 Extract the two collets and then release the tension of the clamp. Remove the clamp, withdraw the collar and spring and extract the valve (photos).
6 As they are released and removed, keep the valves in order so that if they are to be refitted they will be replaced in their original positions in the cylinder head. A piece of stiff card with eight holes punched in it is a sure method of keeping the valves in order.
7 Examine the head of the valves for pitting and burning, especially the heads of the exhaust valves. The valve seatings should be examined at the same time. If the pitting on valve and seat is very slight, the marks can be removed by grinding the seats and valves together with coarse, and then fine, valve grinding paste.
8 Where bad pitting has occurred to the valve seats it will be necessary to recut them and fit new valves. The latter job should be entrusted to the local agent or engineering works. In practice it is very seldom that the seats are so badly worn. Normally it is the valve that is too badly worn for refitting, and the owner can easily purchase a new set of valves and match them to the seats by valve grinding.
9 Valve grinding is carried out as follows. Smear a trace of coarse carborundum paste on the seat face and apply a suction grinder tool to the valve head. With a semi-rotary motion, grind the valve head to its seat, lifting the valve occasionally to redistribute the grinding paste. When a dull matt even surface is produced on both the valve seat and the valve, wipe off the paste and repeat the process with fine carborundum paste, lifting and turning the valve to redistribute the paste as before. A light spring placed under the valve head will greatly ease this operation. When a smooth unbroken ring of light grey matt finish is produced, on both valve and valve seat faces, the grinding operation is complete.
10 Scrape away all carbon from the valve head and the valve stem. Carefully clean away every trace of grinding compound; take great care to leave none in the ports or in the valve guides. Clean the valves and valve seats with a paraffin-soaked rag, then with a clean rag and finally, if an air line is available, blow the valves, valve guides and valve ports clean.
11 Check that all valve springs are intact. If any one is broken, all should be renewed. Check the free height of the springs against new ones. If some springs are not within specification, replace them all. Springs suffer from fatigue and it is a good idea to renew them even if they look serviceable.
12 Check that the oil supply holes in the rocker arm studs are clear.
13 The cylinder head can be checked for warping either by placing it on a piece of plate glass or using a straight-edge and feeler blades. Slight distortion may be corrected by having the head machined to remove metal from the mating face.
14 The renewal of worn valve guides and seats should be left to your dealer.
15 With the cylinder head thoroughly cleaned and all new parts obtained as necessary, reassemble the head as follows:
16 Where a valve stem oil seal has been previously fitted, new seals should be placed over the inlet valve guides.
17 Each valve can now be fitted in turn using the following procedure.
18 Liberally lubricate the valve stem and locate it in its guide (photo). Place the valve spring seat in position followed by the spring (photos).
19 Then place the collar over the spring with the recessed part inside the coil of the spring (photo).
20 Place the end of the spring compressor over the collar and valve stem and, with the screw head of the compressor over the valve head, screw up the clamp until the spring is compressed past the groove in the valve stem. Then put a little grease round the groove.
21 Place the two halves of the split collar (collets) into the groove with the narrow ends pointing towards the spring (photo). The grease will hold them in the groove.
22 Release the clamp slowly and carefully, making sure that the collets are not dislodged from the groove. When the clamp is fully released the top edges of the collets should be in line with each other. Give the top of each spring a smart tap with a soft-faced mallet when assembly is complete to ensure that the collets are properly settled.
23 The rocker gear can be refitted with the head either on or off the engine. The only part of the procedure to watch is that the rocker nuts must not be screwed down too far or it will not be possible to refit the pushrods.
24 Next put the rocker arm over the stud followed by the pivot ball (photos). Make sure that the spring fits snugly round the rocker arm centre section and that the two bearing surfaces of the interior of the arm and the ball face, are clean and lubricated with engine oil.
25 Oil the stud thread and fit the nut with the self-locking collar uppermost (photo). Screw it down until the locking collar is on the stud.

10 Cylinder head and pistons – decarbonising

1 This can be carried out with the engine either in or out of the car. With the cylinder head removed, carefully use a wire brush and blunt scraper to clean all traces of carbon deposits from the combustion

Are your plugs trying to tell you something?

Normal.
Grey-brown deposits, lightly coated core nose. Plugs ideally suited to engine, and engine in good condition.

Heavy Deposits.
A build up of crusty deposits, light-grey sandy colour in appearance.
Fault: Often caused by worn valve guides, excessive use of upper cylinder lubricant, or idling for long periods.

Lead Glazing.
Plug insulator firing tip appears yellow or green/yellow and shiny in appearance.
Fault: Often caused by incorrect carburation, excessive idling followed by sharp acceleration. Also check ignition timing.

Carbon fouling.
Dry, black, sooty deposits.
Fault: over-rich fuel mixture.
Check: carburettor mixture settings, float level, choke operation, air filter.

Oil fouling.
Wet, oily deposits. Fault: worn bores/piston rings or valve guides; sometimes occurs (temporarily) during running-in period.

Overheating.
Electrodes have glazed appearance, core nose very white – few deposits. Fault: plug overheating. Check: plug value, ignition timing, fuel octane rating (too low) and fuel mixture (too weak).

Electrode damage.
Electrodes burned away; core nose has burned, glazed appearance. Fault: pre-ignition. Check: for correct heat range and as for 'overheating'.

Split core nose.
(May appear initially as a crack). Fault: detonation or wrong gap-setting technique. Check: ignition timing, cooling system, fuel mixture (too weak).

WHY DOUBLE COPPER IS BETTER FOR YOUR ENGINE.

Unique Trapezoidal Copper Cored Earth Electrode — 50% Larger Spark Area — Copper Cored Centre Electrode

Champion Double Copper plugs are the first in the world to have copper core in both centre *and* earth electrode. This innovative design means that they run cooler by up to 100°C – giving greater efficiency and longer life. These double copper cores transfer heat away from the tip of the plug faster and more efficiently. Therefore, Double Copper runs at cooler temperatures than conventional plugs giving improved acceleration response and high speed performance with no fear of pre-ignition.

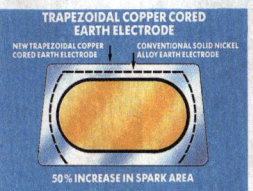
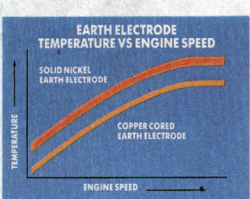

Champion Double Copper plugs also feature a unique trapezoidal earth electrode giving a 50% increase in spark area. This, together with the double copper cores, offers greatly reduced electrode wear, so the spark stays stronger for longer.

 FASTER COLD STARTING

 FOR UNLEADED OR LEADED FUEL

 ELECTRODES UP TO 100°C COOLER

 BETTER ACCELERATION RESPONSE

 LOWER EMISSIONS

 50% BIGGER SPARK AREA

 THE LONGER LIFE PLUG

Plug Tips/Hot and Cold.
Spark plugs must operate within well-defined temperature limits to avoid cold fouling at one extreme and overheating at the other.
Champion and the car manufacturers work out the best plugs for an engine to give optimum performance under all conditions, from freezing cold starts to sustained high speed motorway cruising.
Plugs are often referred to as hot or cold. With Champion, the higher the number on its body, the hotter the plug, and the lower the number the cooler the plug.

Plug Cleaning
Modern plug design and materials mean that Champion no longer recommends periodic plug cleaning. Certainly don't clean your plugs with a wire brush as this can cause metal conductive paths across the nose of the insulator so impairing its performance and resulting in loss of acceleration and reduced m.p.g.
However, if plugs are removed, always carefully clean the area where the plug seats in the cylinder head as grit and dirt can sometimes cause gas leakage.
Also wipe any traces of oil or grease from plug leads as this may lead to arcing.

This photographic sequence shows the steps taken to repair the dent and paintwork damage shown above. In general, the procedure for repairing a hole will be similar; where there are substantial differences, the procedure is clearly described and shown in a separate photograph.

First remove any trim around the dent, then hammer out the dent where access is possible. This will minimise filling. Here, after the large dent has been hammered out, the damaged area is being made slightly concave.

Next, remove all paint from the damaged area by rubbing with coarse abrasive paper or using a power drill fitted with a wire brush or abrasive pad. 'Feather' the edge of the boundary with good paintwork using a finer grade of abrasive paper.

Where there are holes or other damage, the sheet metal should be cut away before proceeding further. The damaged area and any signs of rust should be treated with Turtle Wax Hi-Tech Rust Eater, which will also inhibit further rust formation.

For a large dent or hole mix Holts Body Plus Resin and Hardener according to the manufacturer's instructions and apply around the edge of the repair. Press Glass Fibre Matting over the repair area and leave for 20-30 minutes to harden. Then ...

... brush more Holts Body Plus Resin and Hardener onto the matting and leave to harden. Repeat the sequence with two or three layers of matting, checking that the final layer is lower than the surrounding area. Apply Holts Body Plus Filler Paste as shown in Step 5B.

For a medium dent, mix Holts Body Plus Filler Paste and Hardener according to the manufacturer's instructions and apply it with a flexible applicator. Apply thin layers of filler at 20-minute intervals, until the filler surface is slightly proud of the surrounding bodywork.

For small dents and scratches use Holts No Mix Filler Paste straight from the tube. Apply it according to the instructions in thin layers, using the spatula provided. It will harden in minutes if applied outdoors and may then be used as its own knifing putty.

Use a plane or file for initial shaping. Then, using progressively finer grades of wet-and-dry paper, wrapped round a sanding block, and copious amounts of clean water, rub down the filler until glass smooth. 'Feather' the edges of adjoining paintwork.

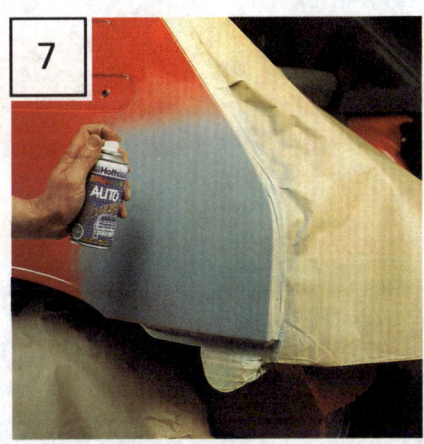

7 Protect adjoining areas before spraying the whole repair area and at least one inch of the surrounding sound paintwork with Holts Dupli-Color primer.

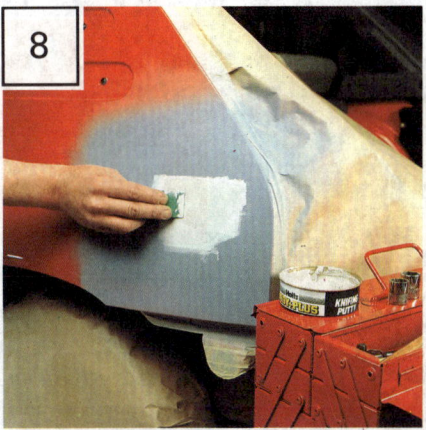

8 Fill any imperfections in the filler surface with a small amount of Holts Body Plus Knifing Putty. Using plenty of clean water, rub down the surface with a fine grade wet-and-dry paper – 400 grade is recommended – until it is really smooth.

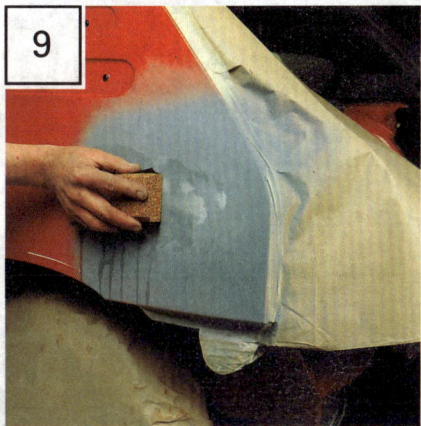

9 Carefully fill any remaining imperfections with knifing putty before applying the last coat of primer. Then rub down the surface with Holts Body Plus Rubbing Compound to ensure a really smooth surface.

10 Protect surrounding areas from overspray before applying the topcoat in several thin layers. Agitate Holts Dupli-Color aerosol thoroughly. Start at the repair centre, spraying outwards with a side-to-side motion.

10A If the exact colour is not available off the shelf, local Holts Professional Spraymatch Centres will custom fill an aerosol to match perfectly.

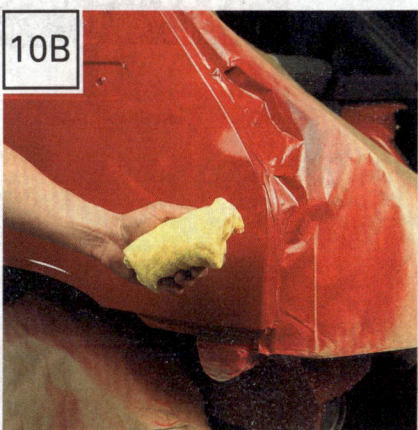

10B To identify whether a lacquer finish is required, rub a painted unrepaired part of the body with wax and a clean cloth.

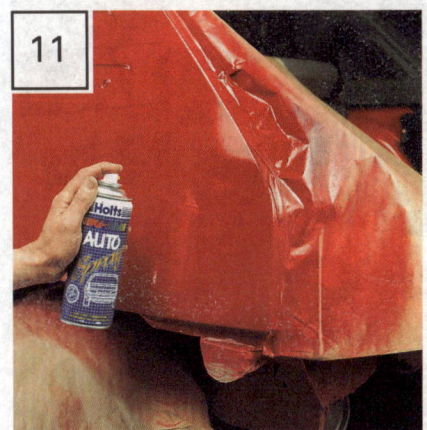

11 If *no* traces of paint appear on the cloth, spray Holts Dupli-Color clear lacquer over the repaired area to achieve the correct gloss level.

12 The paint will take about two weeks to harden fully. After this time it can be 'cut' with a mild cutting compound such as Turtle Wax Minute Cut prior to polishing with a final coating of Turtle Wax Extra.

14 When carrying out bodywork repairs, remember that the quality of the finished job is proportional to the time and effort expended.

HAYNES No1 for DIY

Haynes publish a wide variety of books besides the world famous range of **Haynes Owners Workshop Manuals**. They cover all sorts of DIY jobs. Specialist books such as the **Improve and Modify** series and the **Purchase and DIY Restoration Guides** give you all the information you require to carry out everything from minor modifications to complete restoration on a number of popular cars. In addition there are the publications dealing with specific tasks, such as the **Car Bodywork Repair Manual** and the **In-Car Entertainment Manual**. The **Household DIY** series gives clear step-by-step instructions on how to repair everyday household objects ranging from toasters to washing machines.

Whether it is under the bonnet or around the home there is a Haynes Manual that can help you save money. Available from motor accessory stores and bookshops or direct from the publisher.

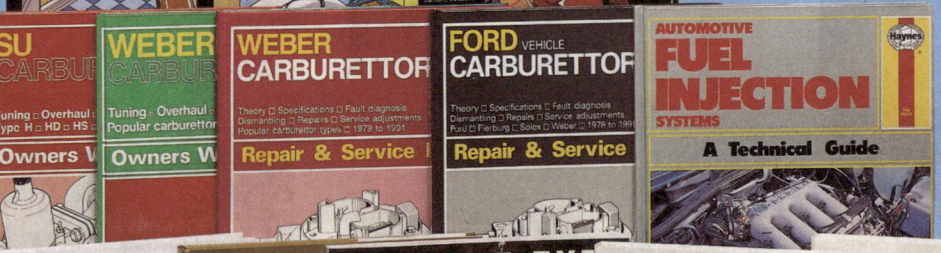

spaces and the ports. The valve head stems and valve guides should also be freed from any carbon deposits. Wash the combustion spaces and ports down with petrol and scrape the cylinder head surface free of any foreign matter with the side of a steel rule or a similar article.

2 If the engine is installed in the car, clean the pistons and the top of the cylinder bores. If the pistons are still in the block, then it is essential that great care is taken to ensure that no carbon gets into the cylinder bores as this could scratch the cylinder walls or cause damage to the piston and rings. To ensure this does not happen, first turn the crankshaft so that two of the pistons are at the bottom of their bores. Stuff rag into the other two bores or seal them off with paper and masking tape. The waterways should also be covered with small pieces of masking tape to prevent particles of carbon entering the cooling system and damaging the water pump.

3 Press a little grease into the gap between the cylinder walls and the two pistons which are to be worked on. With a blunt scraper carefully scrape away the carbon from the piston crown, taking great care not to scratch the aluminium. Also scrape away the carbon from the surrounding lip of the cylinder wall. When all carbon has been removed, scrape away the grease which will now be contaminated with carbon particles, taking care not to press any into the bores. To assist prevention of carbon build-up the piston crown can be polished with a metal polish. Remove the rags or masking tape from the other two cylinders and turn the crankshaft so that the two pistons which were at the bottom are now at the top. Place rag or masking tape in the cylinders which have been decarbonised and proceed as just described. Decarbonising is now complete.

Fig. 1.3 Apply jointing compound to the gasket joints at the main bearing caps before fitting the sump (Sec 11)

11 Sump – removal and refitting

1 Jack up the front of the car and securely support it on axle stands.
2 Drain the engine oil into a suitable container (Section 2) and refit the plug after draining
3 Undo the bolts securing the flywheel cover plate and side support braces and remove the cover (photo).
4 Undo the retaining bolts and lift away the sump. It will probably be necessary to tap the sump from side to side with a hide or plastic mallet to release the joint face.
5 Thoroughly clean the sump in paraffin or a suitable solvent and remove all traces of external dirt and internal sludge. Scrape away all traces of old gasket from the sump and crankcase faces and ensure that they are clean and dry. Also clean the bearing cap grooves.
6 Apply a thick bead of jointing compound to the crankcase flange and at the joints of the front and rear main bearing caps.
7 Position the cork side gaskets on the crankcase flanges (photo) and then insert the cork end sealing strips to the main bearing cap grooves.
8 Apply a further bead of jointing compound to the gasket faces and to the gasket joints at the bearing caps.
9 Refit the sump (photo) and secure it in place with the retaining bolts which should be progressively tightened in a diagonal sequence.

11.3 Removing the flywheel cover plate

11.7 Locate the cork strips into the bearing cap grooves

11.9 Refitting the sump

12 Oil pump – removal and refitting

1 Remove the sump, as described in the previous Section.
2 Undo the two socket-headed bolts (photo) and withdraw the pump from the crankcase (photo).
3 Refitting the pump is the reverse sequence to removal, but engage the pump shaft in the distributor driveshaft slot, and tighten the retaining bolts to the specified torque.

13 Oil pump – overhaul

1 Remove the pump, as described in the previous Section.
2 Undo the two pump cover bolts and lift off the cover and oil pick-up tube. Remove the cover gasket.
3 Take out the driving gear and driven gear (photos).
4 Undo the large nut on the side of the housing and remove the sealing washer and oil pressure relief spring and ball valve (photo).
5 Clean all the parts in paraffin and dry with a lint free cloth.
6 Inspect the pump gears, housing, cover and relief valve ball for scoring, scuff marks or other signs of wear and renew the pump if evident.
7 If the pump condition is satisfactory, check the pump clearances as follows:
8 Using a feeler gauge check the clearance between the gear teeth and between the teeth and housing (photos). Place a straight-edge across the top edge of the pump housing and check the endfloat of the gears (photo). If any of the clearances exceed the tolerances given in the Specifications, renew the pump.
9 If the clearances are satisfactory refit the relief valve assembly and assemble the pump gears. Fill the pump with oil and refit the cover using a new gasket.

14 Timing gear components – removal and refitting

1 For greater access remove the front right-hand wheel trim and slacken the wheel bolts. Jack up the front of the car, support it securely on axle stands and remove the roadwheel.
2 Undo the four retaining bolts and remove the clutch access plate at the base of the bellhousing (photo).
3 Slacken the alternator mounting and adjustment arm bolts, move the alternator towards the engine and slip the drivebelt off the pulleys.
4 Lock the flywheel by wedging a screwdriver between the ring gear teeth and the side of the bellhousing.
5 Using a socket or spanner undo the crankshaft pulley retaining bolt and withdraw the pulley.
6 Undo the bolts securing the timing cover to the front of the engine and lift off the cover.
7 Withdraw the oil slinger from the crankshaft, noting which way round it is fitted (photo).
8 Temporarily refit the pulley and turn the crankshaft until the crankshaft sprocket keyway is uppermost and the timing marks on the two sprockets are in alignment (photo). Remove the pulley.
9 Undo the two retaining bolts and remove the timing chain tensioner.
10 Undo the camshaft sprocket retaining bolt and remove the bolt and washer (photo). Place a screwdriver through one of the sprocket holes and in contact with the camshaft retaining plate behind the sprocket to stop it turning as the bolt is undone.
11 Withdraw the camshaft sprocket and crankshaft sprocket from their respective locations, using a screwdriver as a lever if necessary, then remove the sprockets complete with chain (photo).
12 Thoroughly clean all the components in paraffin and dry them with a lint-free cloth. Remove all traces of old gasket from the faces of the timing cover and engine.
13 To renew the oil seal in the timing cover, place the cover outer face downwards over two blocks of wood and drive out the old seal and holder using a hammer and drift (photo).
14 Place the new seal, which has been soaked in engine oil for 24 hours, in the holder (photo) and then tap the holder into the cover using a block of wood (photo). The seal holder must be fitted flush with the outer edge of the timing cover.
15 Commence reassembly by engaging the chain around the crankshaft sprocket.
16 Engage the camshaft sprocket within the loop of the chain so that it can be fitted to the camshaft and will have its timing mark in alignment with the one on the crankshaft sprocket. Adjust the camshaft sprocket as necessary within the chain loop to achieve this.
17 Fit the sprocket to the camshaft, screw in the bolt and washer and tighten the bolt while holding the sprocket with a screwdriver, as was done during removal.
18 Place the tensioner in position and fit the lower retaining bolt finger tight only at this stage (photo).
19 Move the spring blade away from the tensioner body using a screwdriver, move the tensioner upright, and fit the upper retaining bolt (photo). Now tighten both bolts.
20 Position the oil slinger over the crankshaft and place a new gasket on the front of the engine (photo). Apply jointing compound to both sides of the gasket.
21 Refit the cover (photo) and the retaining bolts, but only screw the bolts in two or three turns. Position the crankshaft pulley on the crankshaft to centralise the cover and then tighten the bolts progressively in a diagonal sequence.
22 Refit the pulley retaining bolt and tighten to the specified torque.
23 Refit the drivebelt and adjust its tension, as described in Chapter 2.
24 Refit the roadwheel and lower the car to the ground.

12.2A Undo the two socket-headed bolts (arrowed) ...

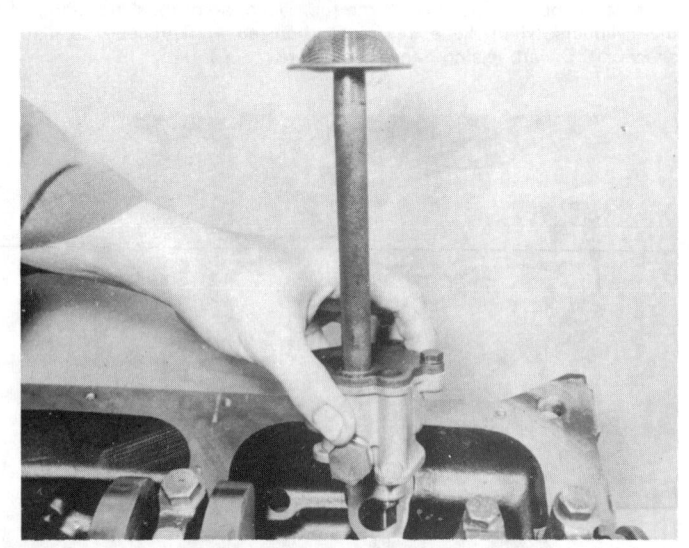

12.2B ... and remove the oil pump

35

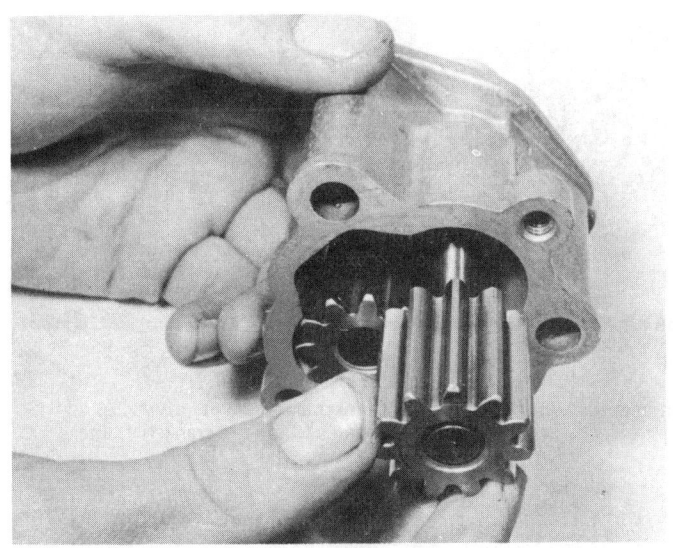

13.3A Remove the oil pump driving gear ...

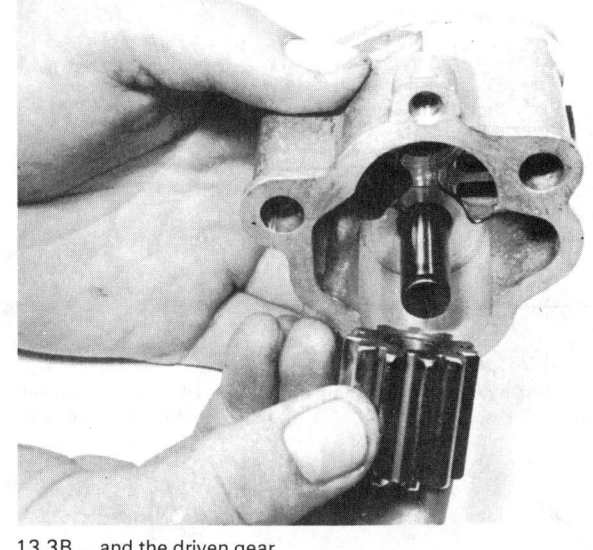

13.3B ... and the driven gear

13.4 Remove the pressure relief valve assembly for inspection

13.8A Check the pump gear teeth clearance ...

13.8B ... gear-to-housing clearance

13.8C ... and the gear endfloat

36

14.2 Clutch and flywheel access plate under transmission bellhousing

14.7 Note the fitted position of the oil slinger before removal

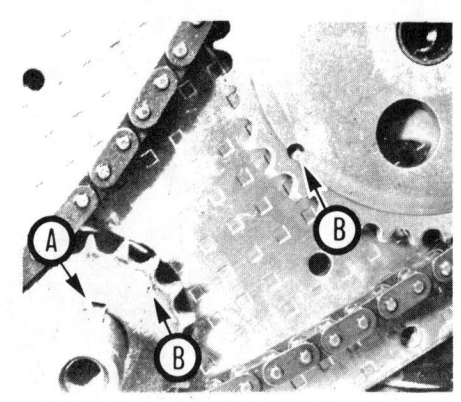

14.8 Crankshaft sprocket keyway (A) and timing marks (B) on the crankshaft and camshaft sprockets

14.10 Camshaft sprocket retaining bolt and washer

14.11 Timing sprocket and chain removal

14.13 Use a drift to remove the timing cover oil seal

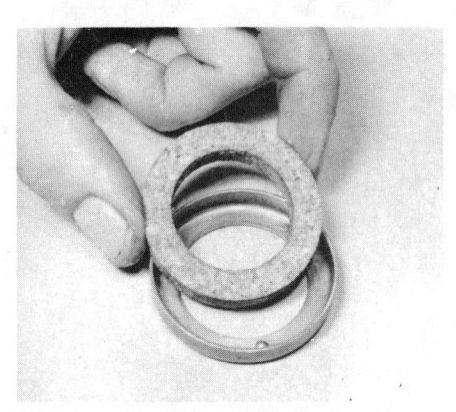

14.14A Position the new seal in the holder ...

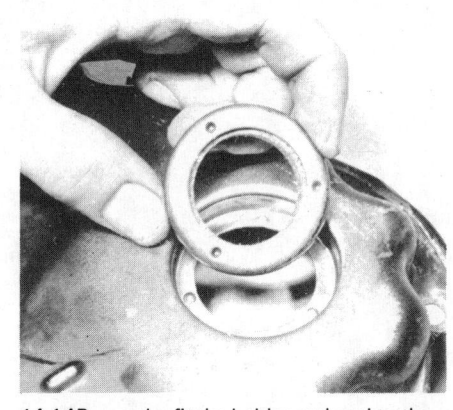

14.14B ... and refit the holder and seal to the cover

14.18 Fit the tensioner lower bolt first ...

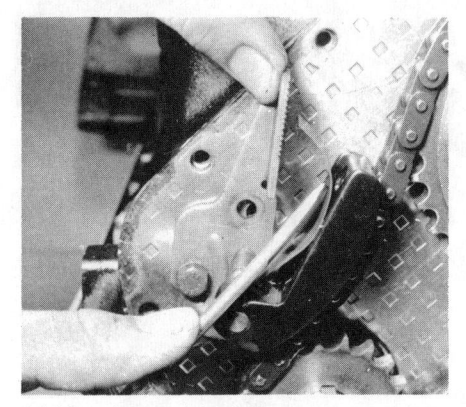

14.19 ... then push back the spring, move the tensioner upright and fit the upper bolt

14.20 With the oil slinger and gasket in place ...

14.21 ... refit the cover

15 Pistons and connecting rods – removal and refitting

1 Remove the cylinder head, the sump and the oil pump, as described in earlier Sections.
2 The connecting rod big-end caps and rods may not be marked numerically for location when new and therefore they must be inspected for identification marks before dismantling. If no marks are evident, punch, scribe or file identification marks on the caps and rods starting with No 1 at the timing cover end. Mark them all on the same side to avoid confusion during reassembly. If they have already been marked then this will not, of course, be necessary.
3 Undo and remove the big-end cap retaining bolts and keep them in order for correct refitting.
4 Detach the big-end bearing caps. If they are stuck, lightly tap them free using a soft-faced mallet.
5 To remove the bearing shells for inspection and/or renewal, press the bearing end opposite the groove in both connecting rod and bearing cap and the shells will slide out. Again keep the shells in order of removal.
6 The piston and rod assemblies are removed through the top of each cylinder bore, being pushed upwards from underneath using a wooden hammer handle which is pushed against the connecting rod. Rotate the crankshaft accordingly to gain suitable access to each rod assembly. Note that if there is a pronounced wear ridge at the top of the cylinder bore, there is a risk of piston ring damage unless the ridge is first removed using a suitable ridge reaming tool, or scraper.
7 The pistons should not be separated from their connecting rods unless they or the gudgeon pins are to be renewed. The gudgeon pin is a press fit and special tools are required for removing and installation. This task should therefore be entrusted to your local agent or automotive machine shop.
8 If for any reason the pistons are separated from their rods, mark them numerically on the same side as the rod markings to ensure correct refitting.
9 If new pistons or piston rings are being fitted to the old bores, it is essential to roughen the cylinder bore walls slightly with medium grit emery cloth to allow the rings to bed in. Do this with a circular up-and-down action to produce a criss-cross pattern on the cylinder bore walls. Make sure that the bearing journal on the crankshaft is protected with marking tape during this operation. Thoroughly clean the bores with a paraffin-soaked rag and dry with a lint-free cloth. Remove the tape from the crankshaft journals and clean them also.
10 Commence reassembly by liberally lubricating the cylinder bores and crankshaft journals. Refit each assembly in turn as follows:
11 Space the piston rings around the pistons so that their gaps are 120° apart and then insert the piston/connecting rod assembly into its respective bore the correct way round as indicated by the markings made during removal (photo).
12 Place a piston ring compressor over the top of the piston and tighten it to compress the rings (photo).
13 Gently tap the piston through the ring compressor and into its bore using the hammer handle. Guide the connecting rod near to its crankshaft journal and then fit the bearing shell upper half.
14 Ease the connecting rod onto the journal, fit the lower shell to the cap and fit the cap to the rod (photo). Refit and tighten the retaining bolts to the specified torque (photo).
15 Repeat the sequence described for the remaining three piston/connecting rod assemblies.
16 Refit the cylinder head, oil pump and sump, as described in earlier Sections.

16 Flywheel – removal and refitting

1 Remove the clutch assembly and the release bearing, as described in Chapter 5.
2 Undo the three bolts and remove the release bearing guide tube.
3 Mark the position of the flywheel in relation to the crankshaft mounting flange or pulley.
4 Wedge a screwdriver between the ring gear teeth and transmission casing and then undo the socket-headed retaining bolts using a multi-tooth key or socket bit (photo). Remove the bolts and withdraw the flywheel.
5 Refitting is the reverse sequence to removal. Tighten the flywheel retaining bolts to the specified torque.

15.11 Insert the piston and connecting rod with the rings spaced at 120° intervals

15.12 Compress the rings and tap the assembly down the bore

15.14A Fit the bearing cap ...

15.14B ... and tighten the bolts to the specified torque

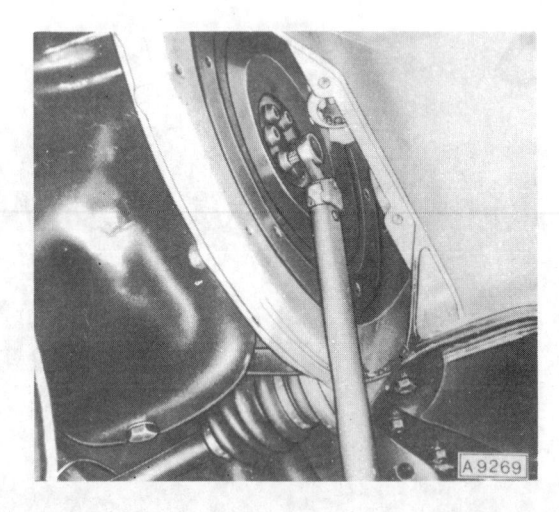

Fig. 1.4 Removing the flywheel retaining bolts (Sec 16)

16.4 A multi-toothed key or socket bit will be needed to undo the flywheel retaining bolts

17 Engine/transmission mountings – removal and refitting

1 The engine/transmission assembly is supported in a triangular arrangement of three mountings. One on the right-hand side supporting the engine, one on the left-hand side supporting the transmission and a third centrally sited mount supporting the complete assembly at the rear.
2 To remove either of the front mountings position a jack under the engine or transmission adjacent to the mounting and just take the weight of the engine or transmission.
3 Undo the bolts securing the support bracket to the engine or gearbox and the bolts securing the mounting to the bodyframe. Lift off the bracket and remove the relevant mounting.
4 To remove the rear mounting jack up the front of the car and support it on axle stands.
5 Support the engine/transmission assembly under the differential cover plate using a jack and interposed block of wood.
6 Undo the two bolts securing the mounting to the underbody and the through-bolt and nut securing the mounting to the support bracket. Slide the mounting rearwards out of the bracket and remove it from under the car.
7 In all cases refitting is the reverse sequence to removal, but tighten the retaining bolts to the specified torque. Note that there is an arrow stamped on the rear mounting which must point to the front when fitted.

18 Camshaft and tappets – removal and refitting

1 Remove the engine from the car, as described in Section 22, and then remove the sump and timing gear components, as described in earlier Sections.
2 If the cylinder head is still in place, slacken the rocker arm nuts, move the rocker arms to one side and lift out the pushrods, keeping them in order.
3 Invert the engine or, if the cylinder head is still in place, lay the engine on its side.
4 Undo the two bolts securing the camshaft retaining plate in position and lift off the plate (photos). The engine front plate should also be removed as all the bolts securing it in place have now been undone.
5 Carefully withdraw the camshaft from the cylinder block, taking care not to scratch the bearing journals with the sharp edges of the cam lobes (photo).
6 From within the crankcase withdraw each tappet from its bore and keep them in order for refitting (photo).
7 Scrape away all traces of old gasket from the engine front plate and cylinder block. Make sure that both mating faces are clean and dry.
8 Lubricate the tappet bores in the crankcase and insert each tappet into its respective bore.
9 Lubricate the camshaft bearing journals and carefully insert the camshaft
10 Place a new gasket in position (photo) and then locate the front plate, patterned side outward, over the gasket. Temporarily refit two or three of the timing cover bolts to act as alignment guides, but only tighten them finger tight.
11 Now position the camshaft retaining plate with its forks located into the groove in the boss on the end of the camshaft. Note that the fork section faces upwards. Secure the retaining plate with the two bolts.
12 Check that the camshaft is free to turn.
13 Refit the timing gear components and the sump, as described in earlier Sections. If the cylinder head is in place, refit the pushrods and adjust the valve clearances.
14 Refit the engine to the car.

19 Crankshaft rear oil seal – removal and refitting

1 Remove the engine from the car, as described in Section 22, and then remove the sump and flywheel, as described in earlier Sections.
2 Slacken the rear main bearing cap bolts slightly and withdraw the oil seal from its location.

Chapter 1 Engine

18.4A Undo the camshaft retaining plate bolts ...

18.4B ... lift off the engine front plate ...

18.5 ... and carefully slide out the camshaft

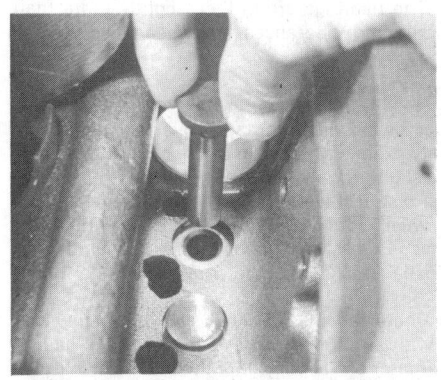

18.6 Remove the tappets and keep them in order

18.10 A new gasket must be used when refitting the front plate

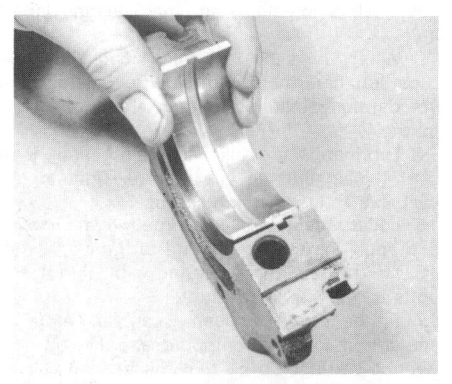

20.9 The tabs on the bearing shells engage with the grooves in the cap and crankcase

20.12 Liberally lubricate the bearing shells (centre shell shown incorporating thrust flanges)

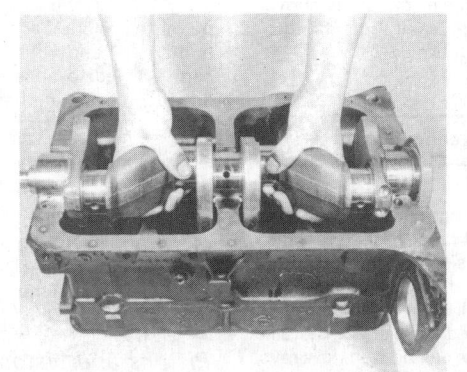

20.13 Place the crankshaft in position in the crankcase

20.14 Refit the centre main bearing cap first

3 Lubricate the lips of a new oil seal and carefully ease it over the crankshaft boss and into position. Make sure that the seal is fully entered into its location so that its outer face is flush with the edge of the bearing cap and cylinder block.
4 Apply jointing compound to the contact edges of the main bearing cap and then tighten the retaining bolts to the specified torque.
5 Refit the sump and flywheel, as described earlier, and then refit the engine to the car, as described in Section 22.

20 Crankshaft and main bearings – removal and refitting

1 With the engine removed from the car, as described in Section 22, and all the components removed from it, as described in earlier Sections, the crankshaft can be removed as follows:
2 Invert the engine. The three main bearing caps are all different so note their locations.
3 Undo the retaining bolts and remove the bearing caps.

4 Lift out the crankshaft and remove the rear oil seal from the crankshaft boss.
5 Remove the main bearing shells from the crankcase and bearing caps and identify them for location.
6 Commence reassembly as follows:
7 Ensure that the crankcase and crankshaft are thoroughly clean and that all oilways are clear. If possible blow the drillings out with compressed air, and then inject clean engine oil through them to ensure they are clear.
8 Avoid using old bearing shells; wipe the shell seats in the crankcase clean and then fit the upper halves of the main bearing shells into their seats.
9 Note that there is a tab on the back of each bearing which engages with a groove in the shell seating (in both crankcase and bearing cap) (photo).
10 Wipe away all traces of protective grease on the new shells.
11 The central bearing shell also takes up the crankshaft endfloat. Note that the half-shells fitted to the cylinder block all have oil duct

Chapter 1 Engine

20.15 With the oil seal in position ...

20.16 ... refit the rear main bearing cap

20.18 Check the crankshaft endfloat at the centre main bearing

holes, while only the centre main bearing cap half-shell has an oil duct hole.
12 When the shells are fully located in the crankcase and bearing caps, lubricate them with clean engine oil (photo).
13 Carefully install the crankshaft into position in the crankcase (photo).
14 Lubricate the crankshaft main bearing journals and then refit the centre main bearing cap (photo). Tighten the retaining bolts to the specified torque wrench setting.
15 Locate the new oil seal onto the rear end of the crankshaft (photo), and apply jointing compound to the block mating flange.
16 Fit the rear main bearing cap (photo) and tighten the retaining bolts to the specified torque.
17 Fit the front main bearing cap, but before fitting the retaining bolts smear with jointing compound and then tighten to the specified torque wrench setting. Check that the bearing cap is exactly flush with the end face of the crankcase as it is tightened.
18 Now rotate the crankshaft and check that it turns freely, and shows no signs of binding or tight spots. Check that the crankshaft endfloat is within the limits specified using a feeler gauge as shown (photo). If it is not, check the centre main bearing shell thrust flanges, or for a fault introduced during grinding of the journals.

21 Engine components – examination and renovation

Crankshaft
1 Examine the crankpin and main journal surfaces for signs of scoring or scratches, and check the ovailty and taper of the crankpins and main journals. If the bearing surface dimensions do not fall within the tolerance ranges given in the Specifications at the beginning of this Chapter, the crankpins and/or main journals will have to be reground.
2 Big-end and crankpin wear is accompanied by distinct metallic knocking, particularly noticed when the engine is pulling from low revs, and some loss of oil pressure.
3 Main bearing and main journal wear is accompanied by severe engine vibration rumble – getting progressively worse as engine revs increase – and again by loss of oil pressure.
4 If the crankshaft requires regrinding take it to an engine reconditioning specialist, who will machine it for you and supply the correct undersize bearing shells.
5 Note: *On some engines, the crankshaft journal diameters are machined undersize in production to allow for greater manufacturing tolerances*

Big-end and main bearing shells
6 Inspect the big-end and main bearing shells for signs of general wear, scoring, pitting and scratches. The bearings should be matt grey in colour. With lead-indium bearings, should a trace of copper colour be noticed, the bearings are badly worn as the lead bearing material has worn away to expose the indium underlay. Renew the bearings if they are in this condition or if there are any signs of scoring or pitting. **You are strongly advised to renew the bearings – regardless of their condition – at time of major overhaul. Refitting used bearings is a false economy.**
7 The undersizes available are designed to correspond with crankshaft regrind sizes. The bearings are in fact, slightly more than the stated undersize as running clearances have been allowed for during their manufacture.

Cylinder bores
8 The cylinder bores must be examined for taper, ovality, scoring and scratches. Start by carefully examining the top of the cylinder bores. If they are at all worn a very slight ridge will be found on the thrust side. This marks the top of the piston travel. The owner will have a good indication of the bore wear prior to dismantling the engine, or removing the cylinder head. Excessive oil consumption accompanied by blue smoke from the exhaust is a sure sign of worn cylinder bores and piston rings.
9 Measure the bore diameter across the block and just below any ridge. This can be done with an internal micrometer or a dial gauge. Compare this with the diameter of the bottom of the bore, which is not subject to wear. If no measuring instruments are available, use a piston from which the rings have been removed and measure the gap between it and the cylinder wall with a feeler gauge.
10 Refer to the Specifications. If the cylinder wear exceeds the permitted tolerances then the cylinders will need reboring. If the wear is marginal and within the tolerances given, new special piston rings can be fitted to offset the wear.
11 If the cylinders have already been bored out to their maximum it may be possible to have liners fitted. This situation will not often be encountered.

Connecting rods
12 Examine the mating faces of the big-end caps to see if they have ever been filed in a mistaken attempt to take up wear. If so, the offending rods must be renewed.
13 Check the alignment of the rods visually, and if all is not well, take the rods to your local agent for checking on a special jig.

Pistons and piston rings
14 If the pistons and/or rings are to be re-used, remove the rings from the pistons. Three strips of tin or 0.015 in (0.38 mm) feeler gauges should be prepared and the top ring then sprung open just sufficiently to allow them to be slipped behind the ring. The ring can then be slid off the piston upwards without scoring or scratching the piston lands.
15 Repeat the process for the second and third rings.
16 Mark the rings or keep them in order so they may be refitted in their original locations.
17 Inspect the pistons to ensure that they are suitable for re-use. Check for cracks, damage to the piston ring grooves and lands, and scores or signs of picking-up the piston walls.
18 Clean the ring grooves using a piece of old piston ring ground to a suitable width and scrape the deposits out of the grooves, taking care not to remove any metal or score the piston lands. Protect your fingers – piston rings are sharp.
19 Check the rings in their respective bores. Press the ring down to the unworn lower section of the bore (use a piston to do this, and keep the ring square in the bore). Measure the ring end gap and check that it is within the tolerance allowed (see Specifications). Also check the ring's side clearance in its groove. If these measurements exceed the specified tolerances the rings will have to be renewed, and if the ring grooves in the pistons are worn new pistons may be needed.

20 If genuine spares are used new pistons and rings are not supplied separately; however if the pistons are in good condition, new rings can be obtained from specialist suppliers who will also undertake any machining work necessary to modify the pistons to suit the new rings.
21 If new rings (or pistons and rings) are to be fitted to an existing bore the top ring must be stepped to clear the wear ridge at the top of the bore, or the bore must be de-ridged.
22 Check the clearance and end gap of any new rings, as described in paragraph 19. If a ring is slightly tight in its groove it may be rubbed down using an oilstone or a sheet of carborundum paper laid on a sheet of glass. If the end gap is inadequate the ring can be carefully ground until the specified clearance is achieved.
23 If new pistons are to be installed they will be selected from the grades available (see Specifications), after measuring the bores as described in paragraph 9. Normally the appropriate oversize pistons are supplied by the repairer when the block is rebored.
24 Removing and refitting pistons on the connecting rod is a job for your dealer or specialist repairer. Press equipment and a means of accurately heating the connecting rod will be required for removal and insertion of the gudgeon pin.

Camshaft and bearings
25 With the camshaft removed, examine the bearings for signs of obvious wear and pitting. If there are signs, then the three bearings will need renewal. This is not a common requirement and to have to do so is indicative of severe engine neglect at some time. As special tools are necessary to do this work properly, it is recommended that it is done by your dealer. Check that the bearings are located properly so that the oilways from the bearing housings are not obstructed.
26 The camshaft itself should show no marks on either the bearing journals or the profiles. If it does, it should be renewed.
27 Examine the skew gear for signs of wear or damage. If this is badly worn it will mean renewing the camshaft.
28 The thrust plate (which also acts as the locating plate) should not be ridged or worn in any way. If it is, renew it.

Timing chain, sprockets and tensioner
29 Examine the teeth of both sprockets for wear. Each tooth is the shape of an inverted V and if the driving (or driven) side is concave in shape the tooth is worn and the sprocket should be renewed. The chain should also be renewed if the sprocket teeth are worn. It is sensible practice to renew the chain anyway.
30 Inspect the chain tensioner, which is automatic in operation. The most important item to check is the shoe which wears against the chain. If it is obviously worn, scratched or damaged in any way, then it must be renewed. Check the spring for signs of wear and renew the unit if generally worn or defective, or when a new chain is being fitted.

Valve rocker arms, pushrods and tappets
31 Each rocker arm has three wearing surfaces, namely the pushrod recess, the valve stem contact, and the centre pivot recess. If any of these surfaces appears severely grooved or worn the arm should be renewed. If only the valve stem contact area is worn it is possible to clean it up with a fine file.
32 If the rocker ball is pitted, or has flats worn in it, this should also be renewed.
33 The nut on the rocker stud is a self-locking type. If it has been removed or adjusted many times, the self-locking ring may have become ineffective and the nut may be slack enough to turn involuntarily and alter the tappet clearance.
34 The rocker studs should be examined to ensure that the threads are undamaged and that the oil delivery hole in the side of the stud at the base of the thread is clear. Place a straight-edge along the top of all the studs to ensure that none is standing higher than the rest. If any are, it means that they have pulled out of the head some distance. They should be removed and replaced with an oversize stud. As this involves reaming out the stud hole to an exact size to provide an interference fit for the replacement stud, you should seek professional advice and assistance to ensure that the new oversize stud is securely fitted at the correct angle.
35 Any pushrods which are bent should be renewed. On no account attempt to straighten them. They are easily checked by rolling over a perfectly flat surface such as a sheet of glass.
36 Examine the bearing surfaces of the tappets which lie on the camshaft. Any indentation in these surfaces or any cracks indicate serious wear and the tappets should be renewed. Thoroughly clean them out, removing all traces of sludge. It is most unlikely that the sides of the tappets will prove worn but, if they are a very loose fit in their bores and can readily be rocked, they should be exchanged for new ones. It is very unusual to find any wear in the tappets, and any wear present is likely to occur only at very high mileages, or in cases of neglect. If the tappets are worn, examine the camshaft carefully as well.

Flywheel
37 If the teeth on the flywheel starter ring are badly worn, or if some are missing, then it will be necessary to remove the ring and fit a new one.
38 Either split the ring with a cold chisel after making a cut with a hacksaw blade between two teeth, or use a soft-headed hammer (not steel) to knock the ring off, striking it evenly and alternately at equally spaced points. Take great care not to damage the flywheel during this process, and protect your eyes from flying fragments.
39 Clean and polish with emery cloth four evenly spaced areas on the outside face of the new starter ring.
40 Heat the ring evenly with a flame until the polished portions turn dark blue. Alternatively heat the ring in a bath of oil to a temperature of 350°C. (If a naked flame is used take careful fire precautions.) Hold the ring at this temperature for five minutes and then quickly fit it to the flywheel so the chamfered portion of the teeth faces the gearbox side of the flywheel. Wipe all oil off the ring before fitting it.
41 The ring should be tapped gently down onto its register and left to cool naturally when the contraction of the metal on cooling will ensure that it is a secure and permanent fit. Great care must be taken not to overheat the ring, indicated by it turning light metallic blue, as if this happens the temper of the ring will be lost.
42 If the driven plate contact surface of the flywheel is scored or on close inspection shows evidence of small hair cracks, caused by overheating, it may be possible to have the flywheel surface ground provided the overall thickness of the flywheel is not reduced too much. Consult your specialist engine repairer and if it is not possible, renew the flywheel complete.
43 If the needle bearing in the centre of the crankshaft flange is worn, fill it with grease and tap in a close-fitting rod. Hydraulic pressure will remove it. Tap the new bearing into position and apply a little grease.

22 Engine – removal and refitting

1 Due to the layout of the engine and transmission assembly and the available clearances the engine must be removed on its own leaving the transmission in the car. The engine is removed and refitted from above as follows:
2 Disconnect the battery negative terminal.
3 Remove the air cleaner, as described in Chapter 3, and the bonnet, as described in Chapter 12.
4 Drain the cooling system, as described in Chapter 2.
5 Refer to the procedure described in Chapter 5, Section 5 and withdraw the transmission input shaft.
6 Slacken the hose clips and disconnect the radiator top and bottom hoses and the heater hose from the water pump.
7 Slacken the hose clip and disconnect the remaining heater hose from the outlet on the cylinder head.
8 Undo the union nut and detach the brake servo vacuum hose from the inlet manifold.
9 Refer to Chapter 3 if necessary and disconnect the accelerator cable and choke cable from the carburettor.
10 Mark the spark plug leads to aid refitting and pull the leads off the spark plugs. Undo the two screws, lift off the distributor cap and place the cap and leads to one side.
11 Make a careful note of their positions and disconnect the electrical leads at the alternator and starter motor solenoid.
12 Slacken the alternator mounting and adjustment arm bolts and move the alternator in towards the engine as far as it will go.
13 Disconnect the temperature transmitter, oil pressure switch and distributor LT electrical leads.
14 Disconnect the fuel inlet hose at the fuel pump and plug its end after removal.

Fig. 1.5 Engine to transmission upper retaining bolts – arrowed (Sec 22)

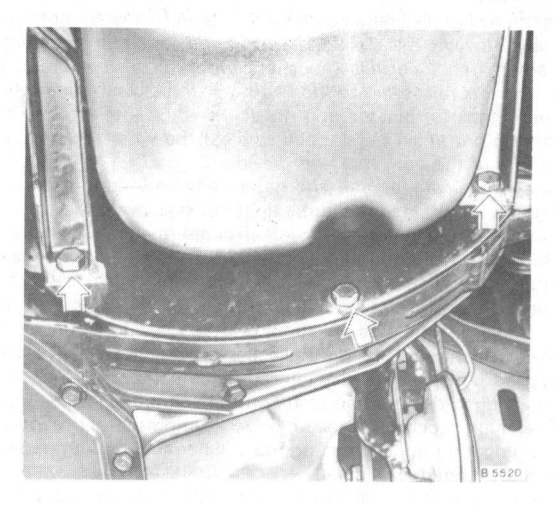

Fig. 1.6 Flywheel cover plate retaining bolts – arrowed (Sec 22)

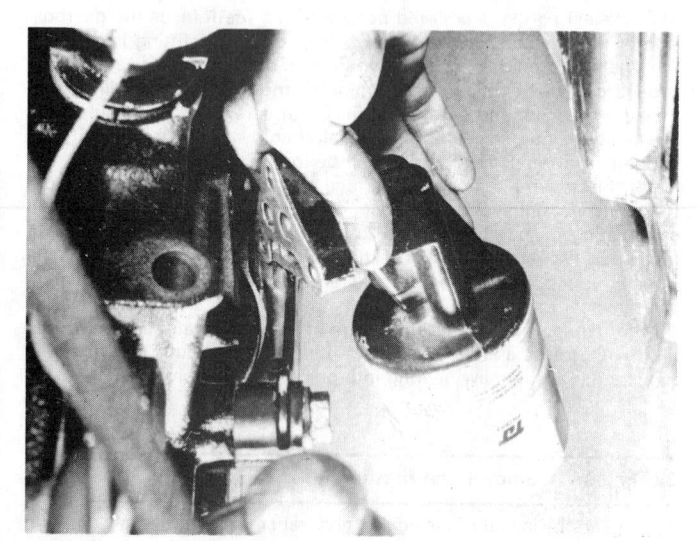

22.15 The oil filter and housing must be removed to provide sufficient clearance for engine removal

22.19 Right-hand engine mounting bracket retaining bolts (arrowed)

22.23 Removing the engine from the car

15 Undo the three bolts securing the oil filter housing to the cylinder block and lift off the housing (photo).
16 Undo the two bolts securing the exhaust front pipe to the manifold. Remove the bolts and tension springs and then separate the joint.
17 Place a jack beneath the transmission and just take its weight.
18 Attach chains or rope slings from a suitable hoist to the engine and just take its weight.
19 Undo the bolts securing the right-hand engine mounting bracket to the engine and to the bodyframe and remove the bracket (photo).
20 Undo the four bolts, accessible from above, securing the transmission bellhousing to the engine.
21 From below undo the three bolts securing the flywheel cover plate and side braces to the transmission.
22 Make a final check that all cables, connections and hoses have been disconnected and are clear of the engine.
23 Raise the engine slightly and separate it from the transmission. Continue raising the engine and turn it as shown (photo) until it can be swung over the front body panel and lowered to the ground.
24 Refitting the engine is the reverse sequence to removal, bearing in mind the following points:

  (a) Tighten all retaining bolts to the specified torque
  (b) Use a new gasket when refitting the oil filter housing and ensure that the mating surfaces are clean

Chapter 1 Engine 43

(c) Adjust the drivebelt, as described in Chapter 2
(d) Refit and adjust the accelerator and choke cables, as described in Chapter 3
(e) Refill the cooling system using the procedure described in Chapter 2
(f) Top up or refill the engine with oil, as described in Section 2

23 Engine lubrication system – general description

The engine lubrication system is quite conventional. A gear type oil pump draws oil up from the sump, via the suction pipe and strainer, and pumps the oil under pressure into the cartridge oil filter. From the oil filter the oil flows into galleries drilled in the engine block to feed the main bearings on the crankshaft and the moving components of the cylinder head. Oil is bled from the main bearing journals in the crankshaft to supply the big-end bearings.

Therefore, the bearings which receive pressure lubrication are the main crankshaft bearings, the big-end bearings, the camshaft bearings, and the rocker arms.

The remaining moving parts receive oil by splash or drip feed and these include the timing chain and associated items, the distributor and fuel pump drive, the tappets, the valve stems and to a certain extent the pistons.

The lubrication system incorporates two safeguards. The first is a pressure operated ball valve situated in the gallery between the oil pump and oil filter. This is in effect a filter bypass valve and allows oil to pass directly into the engine block gallery, downstream of the filter, when the filter is clogged up and resists the flow of oil.

The second system is an oil pressure relief valve, located in the oil pump casing, which controls the oil pressure to the specified maximum.

PART B: 1.2 AND 1.3 LITRE ENGINE

24 General description

The engine is of four cylinder, in-line overhead camshaft type, mounted transversely at the front of the car.

The crankshaft is supported in five shell type main bearings. Thrust flanges are incorporated in the centre main bearing to control crankshaft endfloat.

The connecting rods are attached to the crankshaft by horizontally split shell type main bearings, and to the pistons by gudgeon pins which are an interference fit in the connecting rod small-end bore. The aluminium alloy pistons are fitted with three piston rings: two compression rings and an oil control ring.

The camshaft is driven by a toothed rubber belt from the crankshaft and operates the valve via rocker arms. The rocker arms are supported at their pivot end by hydraulic self-adjusting valve lifters (tappets) which automatically take up any clearance between the camshaft, rocker arm and valve stem. The inlet and exhaust valves are each closed by a single spring and operate in guides pressed into the cylinder head.

Engine lubrication is by a gear type pump located in a housing attached to the front of the cylinder block. The oil pump is driven by the crankshaft, while the fuel pump and distributor are driven by the camshaft.

25 Maintenance and inspection

Refer to Section 2, but disregard the reference to valve clearance adjustment, as on these engines no adjustment is necessary.

Note that although a maximum service life for the camshaft toothed belt is not specified by the manufacturers, it is recommended that renewal should be carried out at 36 000 miles (60 000 km). Renewal procedures are contained in Section 31.

26 Operations possible with the engine in the car

The following operations may be carried out without having to remove the engine from the car:

(a) Removal and refitting of oil pressure regulator valve
(b) Removal and refitting of camshaft toothed belt
(c) Removal and refitting of camshaft housing
(d) Removal and refitting of camshaft
(e) Removal and refitting of cylinder head
(f) Removal and refitting of sump
(g) Removal and refitting of oil pump
(h) Removal and refitting of piston/connecting rod assemblies
(j) Removal and refitting of flywheel
(k) Renewal of crankshaft front oil seal
(l) Removal and refitting of engine/transmission mountings

27 Operations requiring engine removal

The design of the engine is such that great accessibility is afforded and it is only necessary to remove the engine for attention to the crankshaft and main bearings. It is possible to renew the crankshaft rear oil seal with the engine in the car, but this entails the use of the manufacturer's special tools and is quite a difficult operation due to lack of working clearance. For this reason this operation is described with the engine removed.

28 Engine dismantling and reassembly – general

Refer to Section 5. When starting the engine after overhaul, expect some initial noise from the hydraulic valve lifters, until they are properly pressurized with oil.

29 Ancillary components – removal and refitting

Refer to Section 6.

30 Oil pressure regulator valve – removal and refitting

1   From just to the rear of the crankshaft pulley, unscrew the pressure regulator valve plug and extract the spring and plunger.
2   Renew the spring if it is distorted or weak (compare it with a new one if possible).
3   If the plunger is scored, renew it.
4   Clean out the plunger hole and reassemble using a new plug sealing washer.

31 Camshaft toothed belt – removal, refitting and adjustment

1   Undo the five retaining bolts and remove the belt cover (photo).
2   Using a socket or spanner on the crankshaft until No 1 piston is on the firing stroke. The notch on the pulley should be in alignment with the ignition timing pointer (photo) but to ensure that it is No 1 piston that is on the firing stroke and not No 4, either remove the No 1 spark plug and feel the compression being generated as the crankshaft is turned, or remove the distributor cap and check that the rotor is in alignment with No 1 spark plug contact in the cap.
3   Release the alternator adjustment arm and mounting bolts, push the alternator in towards the engine and slip the drivebelt from the pulleys.
4   Unscrew the crankshaft pulley bolt without disturbing the set position of the crankshaft. To prevent the crankshaft rotating as the bolt is unscrewed either engage a gear and apply the handbrake fully, or remove the flywheel cover plate and lock the flywheel ring gear with a suitable tool. Withdraw the pulley from the crankshaft.
5   Refer to Chapter 2 and drain the cooling system.
6   Using a socket inserted through the access holes in the belt cover backplate (photo), slacken the three water pump retaining bolts just enough to be able to swivel the pump and release the tension of the toothed belt.
7   If the toothed belt is to be used again, note its running direction before removing it.

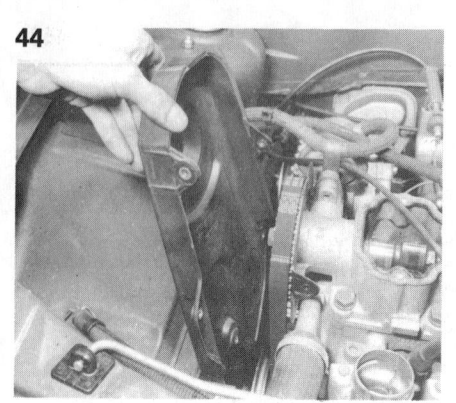

31.1 Remove the camshaft belt cover

31.2 Timing mark on pulley (arrowed) in line with pointer

31.6 Slacken the water pump retaining bolts to release the belt tension

31.9 A spanner engaged with the flats on the pump can be used to tension the toothed belt

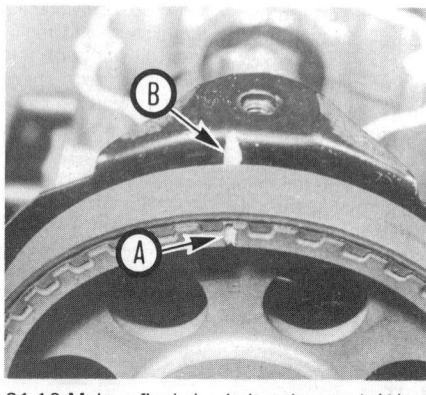

31.10 Make a final check that the notch (A) on the sprocket is aligned with the groove (B) in the backplate ...

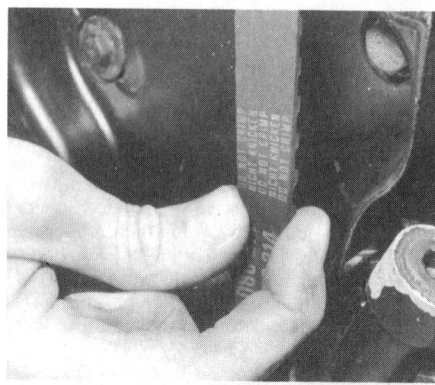

31.11 ... before finally tensioning the belt

32.4 Remove the camshaft housing cover

32.5A Disconnect the fuel inlet hose at the pump (outlet hose shown disconnected) ...

32.5B ... and the return hose at the carburettor

32.6 Remove the fuel hose from its clip

32.7 Disconnect the heater hose from the inlet manifold

32.8 Release the water pipe bracket from the inlet manifold

Chapter 1 Engine 45

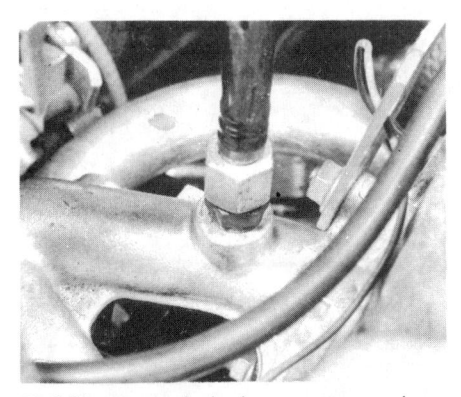

32.9 Disconnect the brake servo vacuum hose from the manifold

32.13 Align the camshaft sprocket and backplate notches (arrowed)

32.14 Remove the toothed belt from the sprocket

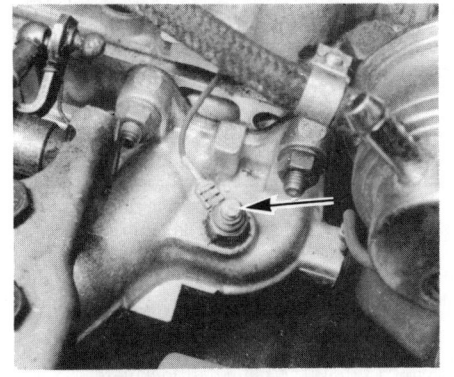

32.15 Disconnect the temperature gauge transmitter lead (arrowed)

32.16 Undo the bolts and separate the exhaust front pipe-to-manifold joint

32.19 Lift off the camshaft housing ...

8 Take the belt off the sprockets and fit the new one without moving the set position of the camshaft or crankshaft.
9 Engage the new belt over the sprockets and apply some tension by moving the water pump. Flats are provided on the pump body behind the sprocket and if necessary a large thin spanner can be used to turn the eccentrically mounted pump (photo).
10 Refit the crankshaft pulley and then check that the pulley notch is still in alignment with the timing pointer and that the camshaft sprocket mark is aligned with the groove in the plate behind it (photo). If not, release the belt tension and readjust the position of the sprockets as necessary.
11 The belt tension should now be adjusted in the following way if the official tool (KM 510A) is not available. Tighten the screws on the water pump and using moderate thumb pressure at the midpoint of the longest run of the belt between the crankshaft and camshaft sprockets, deflect the belt towards the coolant pump sprocket (photo). This deflection should be approximately 6.35 mm ($\frac{1}{4}$ in). Adjust the tension as necessary by moving the water pump. If the belt is overtight, it will be heard to hum when the engine is running. Where the official tension gauging tool is being used, check tension with the engine cold. Gauge readings are 5.5 (1.2 litre)/6.0 (1.3 litre) for a new belt, or 4.0 (1.2 litre)/5.0 (1.3 litre) for a used belt.
12 Refit the belt cover and the alternator drivebelt. Adjust the drivebelt and refill the cooling system, as described in Chapter 2.

32 Cylinder head – removal and refitting

1 Make sure that the engine is cold before commencing operations to avoid any chance of the head distorting.
2 Disconnect the battery negative terminal.
3 Drain the cooling system, as described in Chapter 2, and remove the air cleaner, as described in Chapter 3.
4 Undo the retaining bolts and lift off the camshaft housing cover (photo). Note the position of the fuel hose and spark plug lead clips. Recover the cover gasket.
5 Slacken the clips and disconnect the fuel inlet and outlet hoses from the fuel pump (photo) and the fuel return hose at the carburettor (photo). Plug the hose ends after removal.

6 Release the fuel inlet hose from the clip adjacent to the pump (photo).
7 Slacken the retaining clip and disconnect the heater hose from the inlet manifold (photo).
8 Undo the retaining bolt and release the water pipe bracket from the inlet manifold (photo). Remove the radiator top hose at the thermostat housing.
9 Undo the nut securing the brake servo vacuum hose at the inlet manifold (photo).
10 Refer to Chapter 3 and disconnect the accelerator cable and choke cable from the carburettor and support bracket.
11 Slacken the alternator mounting and adjustment arm bolts, move the alternator towards the engine and slip the drivebelt off the pulleys.
12 Undo the five retaining bolts and lift off the camshaft toothed belt cover.
13 Using a socket or spanner on the crankshaft pulley bolt, turn the crankshaft until the notch on the pulley is in line with the ignition pointer and the camshaft sprocket notch is aligned with the notch in the plate behind it (photo).
14 Using a socket inserted through the access hole in the belt cover backplate, slacken the three water pump retaining bolts just enough to be able to swivel the pump and release the tension of the toothed belt. Slip the belt off the camshaft sprocket (photo).
15 Disconnect the electrical lead at the temperature gauge transmitter on the inlet manifold (photo).
16 Undo the two bolts securing the exhaust front pipe to the manifold. Remove the bolts and, where fitted, the tension springs and separate the joint (photo).
17 Detach the spark plug leads from the spark plugs, ensuring they are labelled for correct refitment, and then remove the distributor, as described in Chapter 4.
18 Working in a spiral pattern starting at the ends and working toward the centre, slacken each cylinder head retaining bolt a quarter of a turn. Now slacken them a further half a turn in the same spiral pattern. Finally undo the bolts fully and remove them from the cylinder head. Discard the original bolts and obtain new ones.
19 Lift the camshaft housing off the locating dowels on the cylinder head (photo).

32.20 ... and then remove the cylinder head

32.24 The new head gasket must be fitted with the work OBEN uppermost

32.28 Using a marked card to measure the torque angle

20 Now lift off the cylinder head (photo), tapping it gently with a plastic mallet if it is stuck. Remove the cylinder head gasket.
21 Remove the rocker arms and thrust pads from the cylinder head. Withdraw the hydraulic valve lifters and immerse them in a container of clean engine oil to avoid any possibility of them draining. Keep all components in their original order if they are to be refitted.
22 If the cylinder head has been removed for decarbonising or for attention to the valves and springs, reference should be made to Sections 34 and 35.
23 Before refitting the cylinder head ensure that the block and head mating faces are spotlessly clean and dry with all traces of old gasket removed. Use a scraper to do this, but take care to cover the water passages and other openings with masking tape or rag to prevent dirt and carbon falling in. Remove all traces of oil and water from the bolt holes otherwise hydraulic pressure created by the bolts being screwed in could crack the block or give inaccurate torque settings. Ensure that the bolt threads are clean and dry.
24 When all is clean locate a new gasket on the block so that the word OBEN can be read from above (photo). Do not use any jointing compound on the gasket.
25 Refit the hydraulic valve lifters, thrust pads and rocker arms to the cylinder head in their original positions. If new hydraulic valve lifters are being used, initially immerse each one in a container of clean engine oil and compress it (by hand) several times to charge it.
26 Locate the cylinder head on the block so that the positioning dowels engage in their holes.
27 Apply a uniform bead of jointing compound to the mating face of the cylinder head and lower the camshaft housing into place. Position the sprocket with the timing marks aligned.
28 Fit the new cylinder head bolts and tighten them in a spiral pattern starting at the centre and working outwards, in the stages given in the Specifications. The required angular measurement can be marked on a card and then placed over the bolt as a guide to the movement of the bolt (photo).
29 Refit the distributor, as described in Chapter 4.
30 Refit the exhaust front pipe and tighten the bolts to compress the tension springs, where fitted.
31 Refit the spark plug leads and the temperature transmitter wire.
32 With the marks on the crankshaft pulley and camshaft sprocket aligned with their respective pointers or notches, fit the toothed belt to the sprockets. Adjust the belt tension, as described in Section 31, and refit the belt cover.
33 Refit the alternator drivebelt and adjust its tension, as described in Chapter 2.
34 Refit the accelerator and choke cable, as described in Chapter 3.
35 Refit the brake servo vacuum hose, the water hoses and the water pipe bracket bolt.
36 Refit the fuel hoses to the pump and carburettor.
37 Using a new gasket refit the camshaft housing cover and tighten the bolts in a diagonal sequence.
38 Refit the air cleaner, as described in Chapter 3 and refill the cooling system using the procedure in Chapter 2.
39 Refit the battery negative terminal.
40 Start the engine and allow it to reach its normal operating temperature. There may be some initial noise when the engine is started until the hydraulic valve lifters pressurize with oil.
41 Adjust the ignition timing, as described in Chapter 4.

42 Switch off the engine and tighten the cylinder head bolts by a further 30° (Stage 5) in the previously described spiral pattern. Further tightening of the bolts is not necessary.

33 Camshaft housing and camshaft – dismantling and reassembly

1 With the camshaft housing removed from the cylinder head, as described in the previous Section, undo the two bolts and remove the fuel pump.
2 With an open-ended spanner engaged with the flats on the camshaft to stop it turning, undo the sprocket retaining bolt (photo). Withdraw the sprocket.
3 At the opposite end of the camshaft housing, use an Allen key to unscrew the two screws which retain the camshaft lockplate.
4 Withdraw the lockplate (photo).
5 Remove the camshaft carefully out of the distributor end of the camshaft housing taking care not to damage the camshaft bearing surfaces (photo).
6 Undo the two bolts securing the toothed belt cover backplate and remove the plate (photo).
7 Prise out the oil seal using a screwdriver and drive in a new seal until it is flush with the housing, using a hammer and block of wood (photo).
8 Liberally lubricate the camshaft bearings and oil seal lip and carefully insert the camshaft.
9 Refit the lockplate and check the camshaft endfloat using a feeler blade (photo). If the endfloat is excessive, renew the lockplate.
10 Refit the belt cover backplate and sprocket, tightening the bolts to the specified torque.
11 Refit the fuel pump using new gaskets on either side of the insulator block.

34 Cylinder head – overhaul

1 With the cylinder head removed, clean away external dirt.
2 To remove the valves the springs will have to be compressed to allow the split collets to be released from the groove in the upper section of the valve stems. A valve spring compressor will therefore be necessary.
3 Locate the compressor to enable the forked end of the arm to be positioned over the valve spring collar whilst the screw part of the clamp is situated squarely on the face of the valve.
4 Screw up the clamp to compress the spring and release the pressure of the collar acting on the collets. If the collar sticks, support the head and clamp frame and give the end of the clamp a light tap with a hammer to help release it.
5 Extract the two collets and then release the tension of the clamp. Remove the clamp, withdraw the collar and spring and extract the valve.
6 As they are released and removed, keep the valve assemblies together and in order (photo) so that if they are to be refitted they will be replaced in their original positions in the cylinder head. A piece of stiff card with eight holes punched in it is a sure method of keeping the valves in order.

33.2 Undo the camshaft sprocket retaining nut

33.4 Remove the lockplate ...

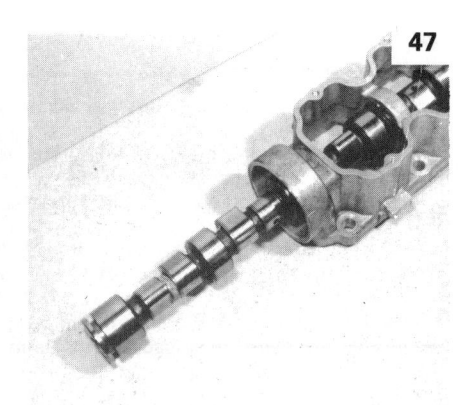

33.5 ... and withdraw the camshaft

33.6 Remove the belt cover backplate ...

33.7 ... to gain access to the oil seal

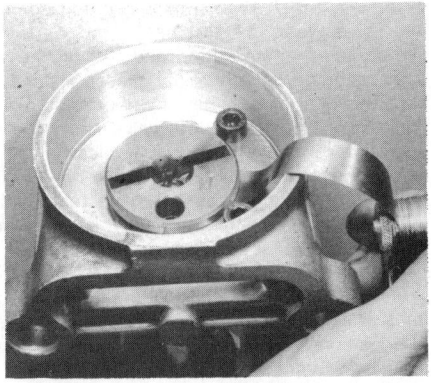

33.9 Check the camshaft endfloat with a feeler blade

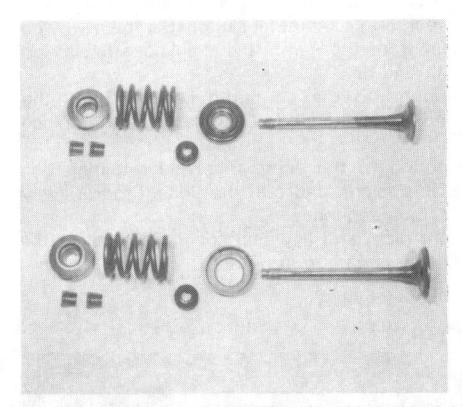

34.6 Keep the valve assemblies together and in order after removal

34.14 Fit new valve stem oil seals to the guides

34.15 Liberally lubricate the valve stem and insert it into the guide

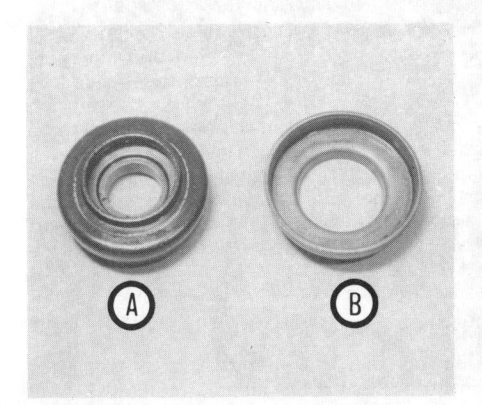

34.16 Exhaust valve spring rotator (A) and inlet valve spring seat (B)

34.17 Refit the spring and collar

34.19 Compress the spring and fit the collets

7 Remove the valve stem oil seals, inlet valve spring seats and exhaust valve spring rotators.
8 Examine the head of the valves for pitting and burning, especially the heads of the exhaust valves. The valve seatings should be examined at the same time. If the pitting on the valve and seat is very slight, the marks can be removed by grinding the seats and valves together with coarse, and then fine, valve grinding paste.
9 Where bad pitting has occurred to the valve seats it will be necessary to recut them and fit new valves. This latter job should be entrusted to the local agent or engineering works. In practice it is very seldom that the seats are so badly worn. Normally it is the valve that is too badly worn for refitting, and the owner can easily purchase a new set of valves and match them to the seats by valve grinding.
10 Valve grinding is carried out as follows. Smear a trace of coarse carborundum paste on the seat face and apply a suction grinder tool to the valve head. With a semi-rotary motion, grind the valve head to its seat, lifting the valve occasionally to redistribute the grinding paste. When a dull matt even surface is produced on both the valve seat and the valve, wipe off the paste and repeat the process with fine carborundum paste, lifting and turning the valve to redistribute the paste as before. A light spring placed under the valve head will greatly ease this operation. When a smooth unbroken ring of light grey matt finish is produced, on both valve and valve seat faces, the grinding operation is complete. Carefully clean away every trace of grinding compound, take great care to leave none in the ports or in the valve guides. Clean the valves and valve seats with a paraffin-soaked rag, then with a clean rag, and finally, if an air line is available, blow the valves, valve guides and valve ports clean.
11 Check that all valve springs are intact. If any one is broken, all should be renewed. Check the free height of the springs against new ones. If some springs are not within specifications, replace them all. Springs suffer from fatigue and it is a good idea to renew them even if they look serviceable.
12 The cylinder head can be checked for warping either by placing it on a piece of plate glass or using a straight-edge and feeler blades. If there is any doubt or if its block face is corroded, have it re-faced by your dealer or motor engineering works.
13 Ensure that all valves and springs are clean and free from carbon deposits and that the ports and valve guides in the cylinder head have no carbon dust or valve grinding paste left in them.
14 Starting at one end of the cylinder head, fit new valve stem oil seals to the ends of the valve guides (photo).
15 Oil the stem of the first valve and insert it into its guide (photo). The valves must be installed into the seats into which they have been ground, which in the case of the original valves will mean that their original sequence of fitting is retained.
16 With inlet valves, fit the spring seat. With exhaust valves fit the valve rotator. Fit the valve spring (photo).
17 Then place the collar over the spring with the recessed part inside the coil of the spring (photo).
18 Place the end of the spring compressor over the collar and valve stem and with the screw head of the compressor over the valve head, screw up the clamp until the spring is compressed past the groove in the valve stem. Then put a little grease round the groove.
19 Place the two halves of the split collar (collets) into the groove with the narrow ends pointing towards the spring (photo). The grease will hold them in the groove.
20 Release the clamp slowly and carefully, making sure that the collets are not dislodged from the groove. When the clamp is fully released the top edges of the collets should be in line with each other. Give the top of each spring a smart tap with a soft-faced mallet when assembly is complete to ensure that the collets are properly settled.

35 Cylinder head and pistons – decarbonising

1 Bearing in mind that the cylinder head is of light alloy construction and is easily damaged, use a blunt scraper or rotary wire brush to clean all traces of carbon deposits from the combustion spaces and the ports. The valve head stems and valve guides should also be freed from any carbon deposits. Wash the combustion spaces and ports down with paraffin and scrape the cylinder head surface free of any foreign matter with the side of a steel rule, or a similar article.
2 If the engine is installed in the car, clean the pistons and the top of the cylinder bores. If the pistons are still in the block, then it is essential that great care is taken to ensure that no carbon gets into the cylinder bores as this could scratch the cylinder walls or cause damage to the piston and rings. To ensure this does not happen, first turn the crankshaft so that two of the pistons are at the top of their bores. Stuff rag into the other two bores or seal them off with paper and masking tape. The waterways should also be covered with small pieces of masking tape to prevent particles of carbon entering the cooling system and damaging the coolant pump.
3 Press a little grease into the gap between the cylinder walls and the two pistons which are to be worked on. With a blunt scraper carefully scrape away the carbon from the piston crown, taking great care not to scratch the aluminium. Also scrape away the carbon from the surrounding lip of the cylinder wall. When all carbon has been removed, scrape away the grease which will now be contaminated with carbon particles, taking care not to press any into the bores. To assist prevention of carbon build-up the piston crown can be polished with a metal polish. Remove the rags or masking tape from the other two cylinders and turn the crankshaft so that the two pistons which were at the bottom are now at the top. Place rag or masking tape in the cylinders which have been decarbonised, and proceed as just described.

36 Sump – removal and refitting

1 Jack up the front of the car and support it on axle stands.
2 Drain the engine oil into a suitable container and refit the plug after draining.
3 Undo the two bolts securing the exhaust front pipe to the manifold and, where fitted, remove the two bolt tension springs.
4 Undo the bolts securing the flywheel cover plate to the transmission bellhousing and remove the plate.
5 Undo the retaining bolts and lift away the sump (photo). It will probably be necessary to tap the sump from side to side with a hide or plastic mallet to release the joint face.
6 Thoroughly clean the sump in paraffin or a suitable solvent and remove all traces of external dirt and internal sludge. Scrape away the remains of the old gasket from the sump and crankcase faces and ensure that they are clean and dry.
7 Apply jointing compound to the oil pump housing joint, the crankcase mating face and the rear main bearing cap joint then place a new gasket in position.
8 Apply jointing compound to the sump face and retaining bolt threads, place the sump in position and refit the bolts. Progressively tighten the bolts in a diagonal sequence.
9 Refit the exhaust front pipe, and flywheel cover plate, lower the car to the ground and fill the engine with oil.

36.5 Removing the sump

37 Oil pump – removal and refitting

1 Remove the camshaft toothed belt and the sump, as described in earlier Sections of this Chapter.
2 Undo the retaining bolts and remove the toothed belt cover backplate.
3 Using two screwdrivers, lever off the crankshaft sprocket and remove the Woodruff key (photos).
4 Undo the two bolts securing the oil pick-up pipe to the oil pump housing and the bolt securing the support bracket to the centre main bearing cap (photo). Remove the pick-up pipe.
5 Undo the retaining bolts and withdraw the oil pump housing from the front of the engine.
6 To refit the pump housing assembly ensure that all mating faces are clean and place a new gasket which is smeared with jointing compound on both sides in position.
7 Before refitting the oil pump, steps must be taken to protect the seal lips from damage or turning back on the shoulder at the front end of the crankshaft. To do this, grease the seal lips and then bind tape around the crankshaft to form a gentle taper (photo).
8 Refit the oil pump and unwind and remove the tape.
9 Tighten the bolts to the specified torque and fit the belt sprocket.
10 Refit the oil pick-up pipe using a new gasket.
11 Refit the sump, then refit and adjust the camshaft toothed belt, as described in previous Sections.

38 Oil pump – overhaul

1 With the oil pump removed from the vehicle, withdraw the rear cover. The cross-head fixing screws are very tight and an impact driver will be required to remove them (photo).
2 Check the clearance between the inner and outer gear teeth and the outer gear and the pump body (photos).
3 Using a straight-edge across the pump cover flange, measure the gear endfloat.
4 If any of the clearances are outside the specified tolerances, renew the components as necessary. Note that the outer gear face is marked for position (photo).
5 The pressure regulator valve can be unscrewed from the oil pump housing and the components cleaned and examined (photo).
6 Always renew the oil seal; a socket is useful to remove and install it (photo).

39 Pistons and connecting rods – removal and refitting

1 Remove the cylinder head, the sump and the oil pick-up pipe, as described in earlier Sections of this Chapter.
2 Check that the rods and caps are marked with their position in the crankcase. If they are not, centre-punch them at adjacent points either side of the cap/rod joint. Note to which side of the engine the marks face.
3 Unscrew the big-end cap bolts from the first rod and remove the cap. If the bearing shells are to be used again, tape the cap and shell together.
4 Check the top of the cylinder bores for a wear ridge. If evident, carefully scrape it away with a ridge reaming tool, otherwise as the piston is pushed out of the block the piston top ring may jam against it.
5 Place the wooden handle of a hammer against the bottom of the connecting rod and push the piston/rod assembly up and out of the cylinder bore.
6 Remove the remaining three assemblies in a similar way. Rotate the crankshaft as necessary to bring the big-end bolts to the most accessible position.
7 If the piston must be separated from its rod, leave this job to your dealer as special tools and a press will be required.
8 If new pistons or piston rings are being fitted to old bores, it is essential to roughen the cylinder bore walls slightly with medium grit emery cloth to allow the rings to bed in. Do this with a circular up-and-down action to produce a criss-cross pattern on the cylinder bore walls. Make sure that the bearing journal on the crankshaft is protected with masking tape during this operation. Thoroughly clean the bores with a paraffin-soaked rag and dry with a lint-free cloth.

37.3A Remove the crankshaft sprocket

37.3B ... and the Woodruff key (arrowed)

Remove the tape from the crankshaft journals and clean them also.
9 Before refitting the pistons space the rings so that their grooves are 120° apart.
10 Commence reassembly by laying the piston/connecting rod assemblies out in their correct order, complete with bearing shells, ready for refitting into their respective bores in the cylinder block (photo). Make sure that the seats for the shells are absolutely clean and then fit the shells into the seats.
11 Wipe out the bores and oil them. Oil the piston rings liberally.
12 Fit a piston ring compressor to the first assembly to be installed.
13 Insert the rod and piston into the top of the bore so that the base of the compressor stands on the block. Check that the rod markings are towards the side of the engine as noted before dismantling. This is very important as the piston crowns do not have front directional marks. The underside of the piston does indicate the flywheel end of the engine (photo).
14 Apply the wooden handle of a hammer to the piston crown and tap the assembly into the bore, at the same time releasing the compressor (photo).
15 Guide the big-end of the connecting rod near to the crankpin and then pull it firmly onto the crankpin which should have been oiled liberally.
16 Fit the cap and bolts and tighten to the specified torque (photo).
17 Repeat the operations on the other pistons/rods.
18 Refit the cylinder head, the pick-up pipe and the sump, all as described in earlier Sections of this Chapter.

37.4 Oil pick-up pipe retaining bolts (arrowed)

37.7 Protect the oil seal lips by wrapping tape around the crankshaft

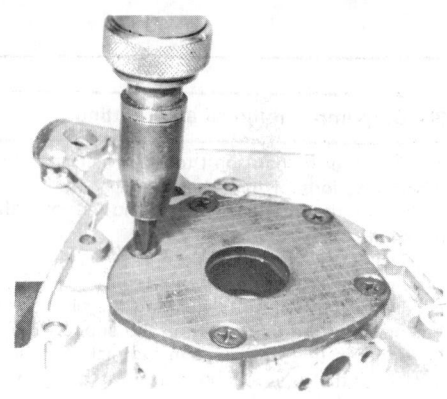

38.1 Using an impact driver to remove the oil pump rear cover

38.2a Check the clearance between the oil pump gear teeth ...

38.2b ... and between the outer gear and pump body

38.4 Outer gear outer face identification mark (arrowed)

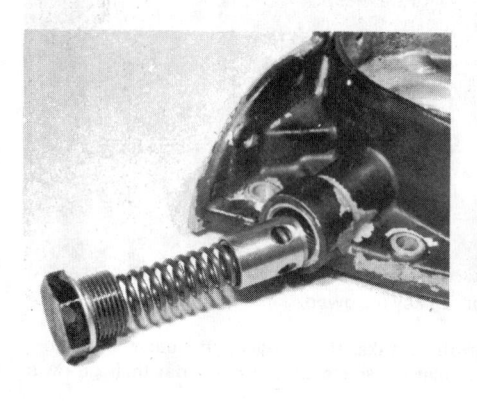

38.5 Remove the pressure relief valve components for inspection

38.6 Using a socket to install the front oil seal

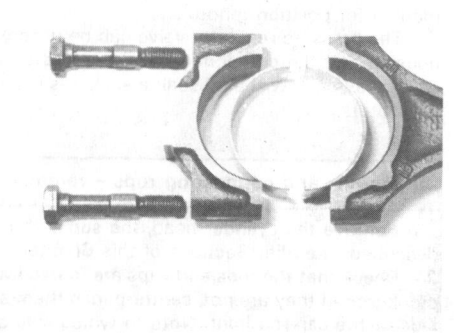

39.10 Connecting rod big-end bearing assembly prior to fitting

39.13 The large land on the underside of the piston must face the flywheel end of the engine

39.14 With the piston rings compressed, tap the piston into the cylinder bore

39.16 Refit the big-end bearing cap

40 Flywheel – removal and refitting

Refer to Section 16, but note that conventional hexagon-headed retaining bolts are used and these should have thread locking compound applied to their threads before refitting.

41 Crankshaft front oil seal – removal and refitting

1 Remove the camshaft toothed belt, as described in earlier Sections of this Chapter.
2 Using two screwdrivers as levers, prise the sprocket from the crankshaft. Remove the Woodruff key.
3 Punch or drill a small hole in the metal face of the oil seal and screw in a self-tapping screw. Use the head of the screw to lever out the seal.
4 Apply masking tape to the flange of the crankshaft to protect the seal lip as it is fitted.
5 Lubricate the lip of the seal and using a suitable tube, tap the seal into its location. Remove the masking tape.
6 Refit the Woodruff key and sprocket.
7 Refit and adjust the toothed belt, as previously described.

44.3 Identification number on No 3 main bearing cap

Fig. 1.7 Screw inserted into crankshaft front oil seal for removal (Sec 41)

44.5 Centre main bearing shell and integral thrust flanges

42 Engine/transmission mountings – removal and refiiitting

Refer to Section 17.

43 Crankshaft rear oil seal – removal and refitting

1 Remove the engine from the car, as described in Section 46, and then remove the seal, as described in Section 19.

44 Crankshaft and main bearings – removal and refitting

1 With the engine removed from the car, as described in Section 46, and all the components removed from it, as described in earlier Sections, the crankshaft can be removed as follows:
2 Invert the engine so that it is standing on the top surface of the cylinder block.
3 The main bearing caps are numbered 1 to 4 from the toothed belt end of the engine. The rear cap is not marked. To ensure that the caps are refitted the correct way round, note that the numbers are read from the water pump side when the crankcase is inverted (photo).
4 Unscrew and remove the main bearing cap bolts and tap off the caps. If the bearing shells are to be used again, keep them with their respective caps. The original shells are colour coded and if used again must be returned to their original locations. Renew the main bearing cap bolts.
5 Note that the centre bearing shell incorporates thrust flanges to control crankshaft endfloat (photo).
6 Lift the crankshaft from the crankcase. Extract the upper half shells and again identify their position in the crankcase if they are to be used again.
7 The rubber plug location adjacent to the bellhousing flange on the crankcase covers the aperture for installation of a TDC sensor. This sensor when connected to a suitable monitoring unit, indicates TDC from the position of the contact pins set in the crankshaft counter-balance weight (photo).
8 Ensure that the crankcase and crankshaft are thoroughly clean and that all oilways are clear. If possible blow the drillings out with compressed air, and then inject clean engine oil through them to ensure they are clear.
9 Wipe the shell seats in the crankcase and bearing caps clean and then fit the upper halves of the main bearing shells into their seats.
10 Note that there is a tag on the back of each bearing which engages with a groove in the shell seating in both crankcase and bearing cap (photo).
11 Wipe away all traces of protective grease on the new shells.
12 Note that the half shells fitted to the cylinder block all have oil duct holes, while only the centre main bearing cap half shell has an oil duct hole.

Chapter 1 Engine

13 When the shells are fully located in the crankcase and bearing caps, lubricate them with clean engine oil.
14 Fill the lips of a new crankshaft oil seal with grease and fit it to the end of the crankshaft (photo).
15 Carefully install the crankshaft into position in the crankcase (photo).
16 Lubricate the crankshaft main bearing journals and then refit the centre and intermediate main bearing caps (photo). Tighten the retaining bolts to the specified torque wrench setting (photo).
17 Clean the grooves of the rear main bearing cap free from old sealant. Fit the cap and tighten the bolts to the specified torque (photo). Now fill the grooves in the bearing cap with RTV type jointing compound (photo). Inject straight from the tube until the material is seen to exude from the cap joints to prove that any trapped air has been expelled.
18 Fit the front main bearing cap, but before fitting the retaining bolts, smear them with jointing compound, and then tighten to the specified torque wrench setting. Check that the bearing cap is exactly flush with the end face of the crankcase as it is tightened.
19 Now rotate the crankshaft and check that it turns freely, and shows no signs of binding or tight spots. Check that the crankshaft endfloat is within the limits specified. The endfloat can be checked using a dial gauge or with feeler blades inserted between the flange of the centre bearing shell and the machined surface of the crankshaft (photo). Before measuring, make sure that the crankshaft has been forced fully towards one end of the crankcase to give the widest gap at the measuring location. If the endfloat is not within specified tolerance, check the centre main bearing shell thrust flanges, or for a fault introduced during grinding of the journals.

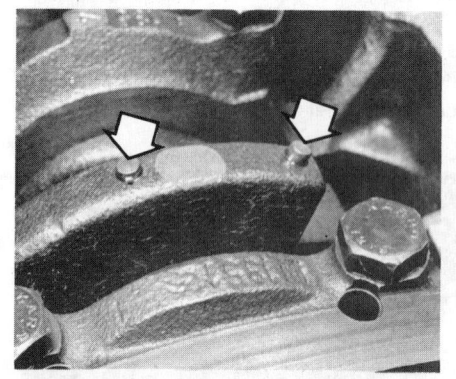

44.7 TDC sensor contact pins (arrowed) in crankshaft web

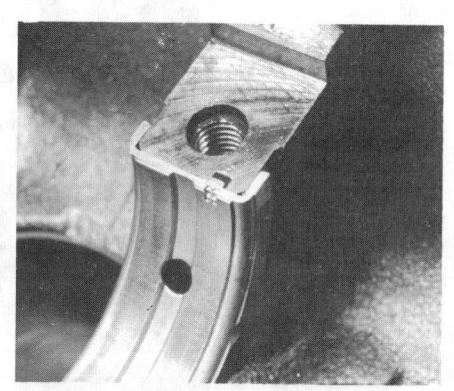

44.10 Centre main bearing shell correctly located with its tag engaged with the notch

44.14 Fit the oil seal to the crankshaft ...

44.15 ... and position the crankshaft in its bearings

44.16A Refit the centre and intermediate main bearing caps ...

44.16B ... and tighten the bolts to the specified torque

44.17A Refit the rear main bearing cap ...

44.17B ... and fill the grooves in the cap with jointing compound

44.19 Check the crankshaft endfloat with a feeler blade

45 Engine components – examination and renovation

1 Refer to Section 21, but note the following additional information.

Camshaft
2 With the camshaft removed, examine the bearings for signs of obvious wear and pitting. If evident, a new camshaft housing will probably be required.
3 The camshaft itself should show no marks or scoring on the journal or cam lobe surfaces. If evident, renew the camshaft.
4 The retaining plate should appear unworn and without grooves. In any event, check the camshaft endfloat and fit a new plate where necessary.
5 The housing front oil seal should always be renewed at major overhaul.

Camshaft toothed belt
6 Closely inspect the belt for cracking, fraying or tooth deformation. Where evident, renew the belt.
7 If the belt has been in use for 36 000 miles (60 000 km) or more, it is recommended that it is renewed even if it appears in good condition.
8 Whenever the original belt is to be removed, but is going to be used again, always note its running direction before removing it. It is even worthwhile marking the tooth engagement points on each sprocket. As the belt will have worn in a set position, refitting it in exactly the same way will prevent any increase in noise which might otherwise occur when the engine is running.

Valve lifters, rockers and thrust pads
9 Any signs of wear in a hydraulic valve lifter can only be rectified by renewal, the unit cannot be dismantled.
10 Inspect the rockers and thrust pads for wear or grooving. Again, renew if evident.

46 Engine – removal and refitting

1 Due to the layout of the engine and transmission assembly and the available clearances the engine must be removed on its own, leaving the transmission in the car. The engine is removed and refitted from above as follows:
2 Disconnect the battery negative terminal.
3 Remove the air cleaner, as described in Chapter 3, and the bonnet, as described in Chapter 12.
4 Drain the cooling system, as described in Chapter 2.
5 Refer to the procedure described in Chapter 5, Section 5 and withdraw the transmission input shaft.
6 Slacken the radiator top hose clips and remove the hose from the radiator and thermostat housing.
7 Slacken the radiator bottom hoses at the radiator and at the connection at the rear of the water pump. Slacken the left-hand heater hose clip at its heater outlet connection. Slacken the expansion tank supply hose at the expansion tank. Detach the hoses.
8 Undo the bolt at the rear of the inlet manifold securing the water pipe bracket in place and then remove the nut securing the pipe to the transmission bellhousing. Remove the pipe complete with hoses (photo).
10 Undo the nut and remove the brake servo vacuum hose from the inlet manifold (photo).
11 Refer to Chapter 10 and remove the alternator.
12 Make a note of the electrical connections at the starter motor solenoid and disconnect them.
13 Slacken the clips and remove the fuel inlet hose from the fuel pump and the return hose from the carburettor. Release the hoses from their retaining clips and plug their ends (photo).
14 Refer to Chapter 3 and disconnect the accelerator and choke cables from the carburettor and support brackets.
15 Disconnect the oil pressure switch and temperature gauge transmitter leads.
16 Disconnect the distributor HT lead and LT plug at the ignition coil.
17 Undo the two bolts securing the exhaust front pipe to the manifold, remove the bolts and, where fitted, the tension springs and separate the joint.
18 Place a jack beneath the transmission and just take its weight.

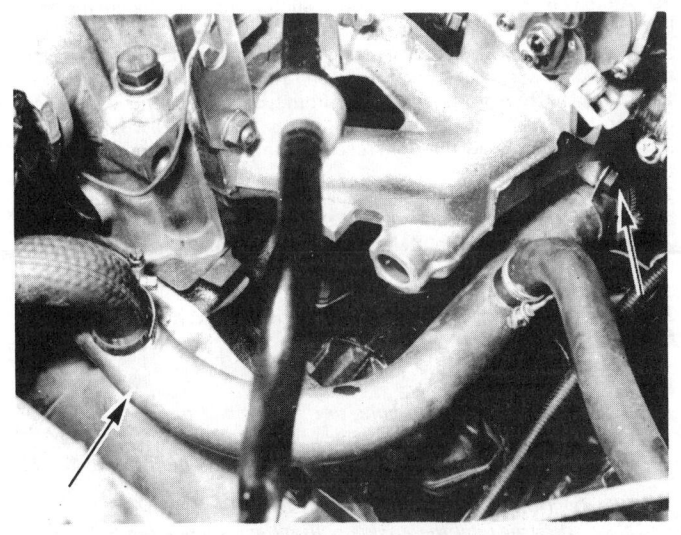

46.8 Water pipe bracket retaining bolt locations (arrowed)

46.9 Remove the remaining heater hose from the inlet manifold

46.10 Undo the nut and remove the brake servo vacuum hose

19 Attach chains or rope slings from a suitable hoist to the engine and just take its weight.
20 Undo the bolts securing the right-hand engine mounting bracket to the engine and to the bodyframe and remove the bracket (photo).
21 Undo the four bolts accessible from above, securing the transmission bellhousing to the engine.
22 From below undo the bolts securing the flywheel cover plate to the transmission (photo).
23 Make a final check that all cables, connections and hoses have been disconnected and are clear of the engine.
24 Raise the engine slightly and separate it from the transmission. Continue raising the engine (photo) and turn it until it can be swung over the front body panel and lowered to the ground.
25 Refitting the engine is the reverse sequence to removal, bearing in mind the following points:

(a) Tighten all retaining bolts to the specified torque
(b) Adjust the alternator drivebelt, as described in Chapter 2
(c) Refit and adjust the accelerator and choke cables, as described in Chapter 3
(d) Refill the cooling system, as described in Chapter 2
(e) Top up or refill the engine with oil as described in Section 2 of this Chapter

47 Engine lubrication and crankcase ventilation systems — general description

1 Oil pressure for all moving components is provided by a gear type oil pump which is driven from the front end of the crankshaft. The crankshaft has flats for this purpose.
2 The pump draws oil from the sump through a pick-up pipe and strainer and pumps it through the oil filter and oil galleries to the engine friction surfaces.
3 A pressure regulator valve is screwed into the body of the oil pump. A relief valve, located in the oil filter mounting base opens should the filter block due to clogging caused by neglected servicing. An oil pressure switch is screwed into the pump casing.
4 The cylinder bores are lubricated by oil splash from the sump.
5 The hydraulic, valve lifters are pressurised with oil to maintain optimum valve clearance at all times.
6 The crankcase ventilation system is designed to draw oil fumes and blow-by gas (combustion gas which has passed the piston rings) from the crankcase into the air cleaner, whence they are drawn into the engine and burnt during the normal combustion cycle.
7 Although not a specified maintenance task, the oil separator can be unbolted from the crankcase (photo), washed out with paraffin and shaken dry. Clean out the breather hoses at the same time.

46.13 Disconnect the fuel inlet and fuel return hoses (arrowed)

46.20 Undo the engine mounting bracket retaining bolts (arrowed)

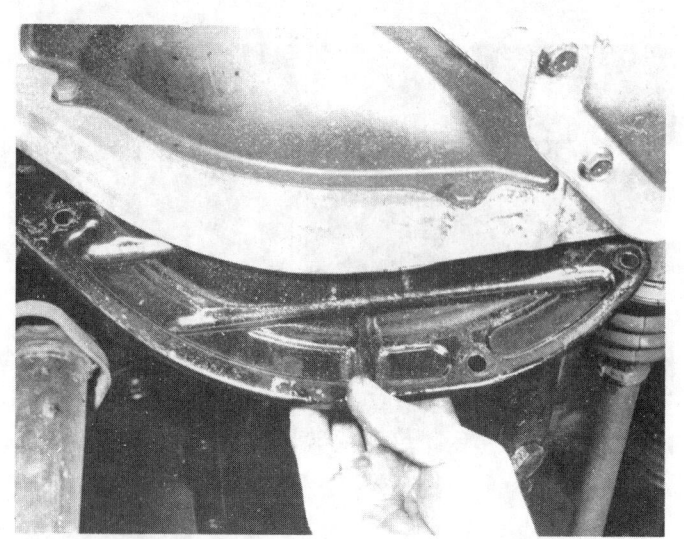

46.22 Remove the flywheel cover plate

46.24 Removing the engine from the car

47.7 Crankcase ventilation system oil separator removal

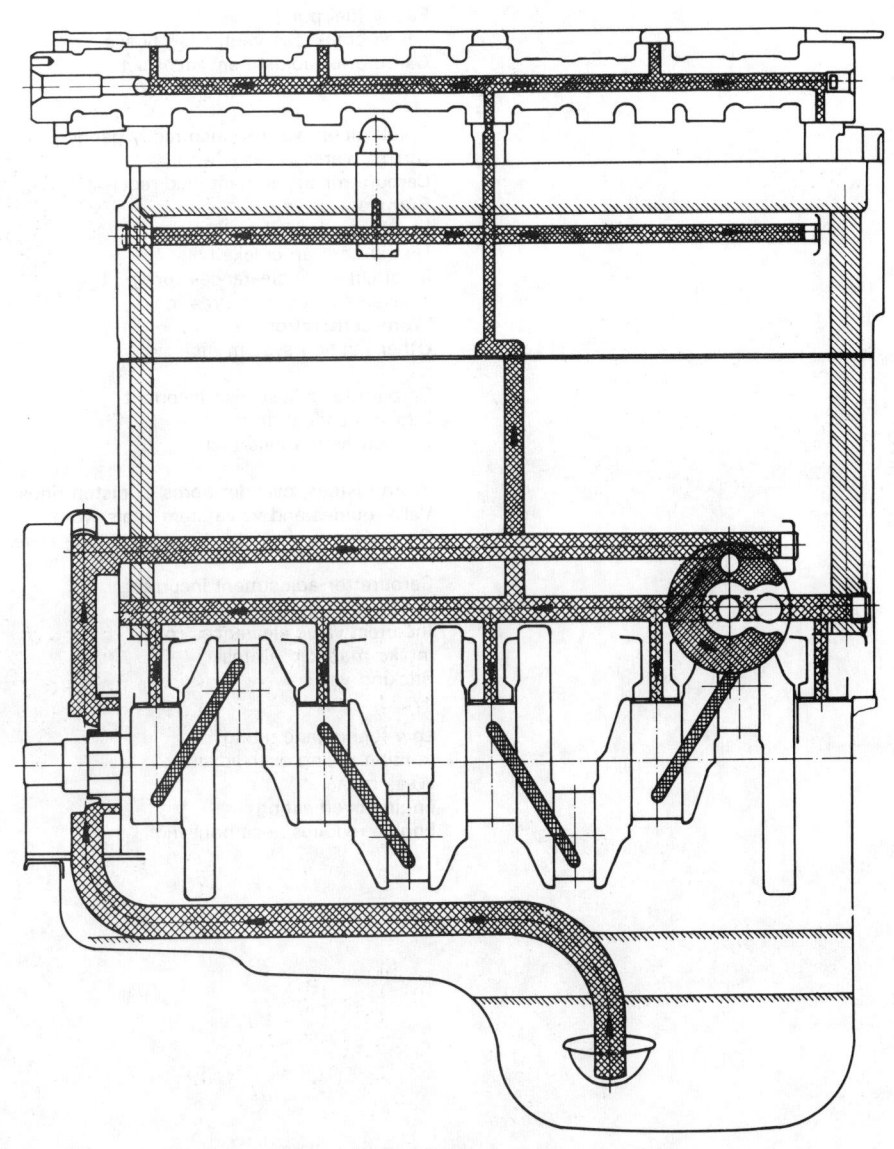

Fig. 1.8 Engine lubrication circuit (Sec 47)

PART C: ALL ENGINES

48 Fault diagnosis – engine

Symptom	Reason(s)
Engine fails to start	Discharged battery
	Loose battery connection
	Loose or broken ignition leads
	Moisture on spark plugs, distributor cap, or HT leads
	Incorrect spark plug gaps
	Cracked distributor cap or rotor
	Other ignition system fault
	Dirt or water in carburettor
	Empty fuel tank
	Faulty fuel pump
	Other fuel system fault
	Faulty starter motor
	Low cylinder compressions
Engine idles erratically	Intake manifold air leak
	Leaking cylinder head gasket
	Worn rocker arms, timing chain, and gears (where applicable)
	Worn camshaft lobes
	Faulty fuel pump
	Loose crankcase ventilation hoses
	Carburettor adjustment incorrect
	Uneven cylinder compressions
Engine misfires	Spark plugs worn or incorrectly gapped
	Dirt or water in carburettor
	Carburettor adjustment incorrect
	Burnt out valve
	Leaking cylinder head gasket
	Distributor cap cracked
	Incorrect valve clearances (ohv)
	Uneven cylinder compressions
	Worn carburettor
	Other ignition system fault
Engine stalls	Carburettor adjustment incorrect
	Intake manifold air leak
	Ignition timing incorrect
Excessive oil consumption	Worn pistons, cylinder bores or piston rings
	Valve guides and valve stem seals worn
	Oil leaking from rocker cover or timing cover oil seal
Engine backfires	Carburettor adjustment incorrect
	Ignition timing incorrect
	Incorrect valve clearances (ohv)
	Intake manifold air leak
	Sticking valve
Engine 'pinks'	Low fuel octane rating
	Ignition timing over-advanced
	Weak mixture
	Engine overheating
	Engine requires decarbonising

Chapter 2 Cooling system

For modifications, and information applicable to later models, see Supplement at end of manual

Contents

Antifreeze mixture 13	General description 1
Cooling fan thermal switch – testing, removal and refitting 11	Maintenance and inspection 2
Cooling system – draining 3	Radiator – removal, inspection and refitting 6
Cooling system – filling 5	Radiator electric cooling fan – removal and refitting 10
Cooling system – flushing 4	Temperature gauge transmitter – removal and refitting 12
Drivebelt – removal, refitting and adjustment 9	Thermostat – removal, testing and refitting 7
Fault diagnosis – cooling system 14	Water pump – removal and refitting 8

Specifications

System type .... Pressurised sealed system with remote expansion tank

Thermostat
Identification:
- 1.0 models 102
- 1.2 models 92
- 1.3 models .... 91 G 195 F

Starts to open:
- 1.0 models .... 91°C (195°F)
- 1.2 models .... 92°C (197°F)
- 1.3 models .... 91°C (195°F)

Fully open:
- 1.0 models .... 107°C (224°F)
- 1.2 models .... 107°C (224°F)
- 1.3 models .... 103°C (217°F)

Expansion tank filler cap
Identification:
- 1.0 models .... Blue 90 096 561
- 1.2 and 1.3 models .... Blue 90 108 850

Opening pressure:
- 1.0 models .... 1.25 to 1.30 bar (18.13 to 18.85 lbf/in^2)
- 1.2 and 1.3 models .... 1.30 to 1.35 bar (18.85 to 19.57 lbf/in^2)

Temperature at opening pressure (boiling point) .... 125°C (257°F)

Cooling system capacity
- 1.0 models .... 5.5 litres (9.7 pints)
- 1.2 and 1.3 models .... 6.3 litres (11.1 pints)

Coolant type/specification .... Soft water and antifreeze to GME L 6368 (Duckhams Universal Antifreeze and Summer Coolant)

Torque wrench settings
	Nm	lbf ft
Water pump bolts	8	6
Thermostat housing bolts	10	7
Temperature gauge transmitter	10	7

Chapter 2 Cooling system

1 General description

Engine cooling is achieved by a conventional thermo-syphon, pump-assisted system in which the cooling water is pressurised. The system consists of a radiator, water pump, electric fan, thermostat expansion tank and associated hoses.

The system functions as follows. Cold water from one side of the radiator, which is mounted at the front of the engine compartment, is directed to the inlet side of the water pump where it is then forced round the cooling passages in the engine cylinder block and cylinder head. The cooling water, now hot, is returned to the other side of the radiator where it flows across and cools before repeating the cycle.

To enable the engine to warm up quickly when starting from cold the thermostat prevents cooling water returning to the radiator until a predetermined temperature is reached. Instead the same water is recirculated around the passages in the cylinder block and cylinder head. When hot the thermostat opens, allowing the water to return to the radiator.

Air flows through the radiator to cool the water as a result of the car's forward motion. However, if the temperature of the cooling water exceeds a given figure, a temperature switch in the radiator switches on an electrical fan to assist and increase the airflow through the radiator.

An expansion tank is incorporated in the system to accommodate expansion of the cooling water. The system is topped up through a filler cap on this tank.

2 Maintenance and inspection

1 Weekly or before a long journey check the level of cooling water in the expansion tank, with the engine cold, and top up to just above the KALT mark if necessary. Under normal operating conditions coolant loss should be negligible and any need for frequent topping-up should be investigated.
2 Every 9000 miles (15 000 km) or 6 months, whichever comes first, carefully inspect the drivebelt for signs of cracks, fraying or incorrect tension. Renew and/or adjust the belt tension, as described in Section 9.
3 At the same service intervals check the condition and security of all hoses and hose clips. Also inspect the joint faces of the water pump, thermostat housing, cylinder head etc for any sign of coolant seepage.
4 Every 18 000 miles (30 000 km) or 12 months, whichever comes first, have the concentration of antifreeze in the cooling system checked and if necessary add fresh antifreeze to maintain the required strength, as described in Section 13. To provide adequate protection against corrosion in the cooling system renew the coolant completely every two years.

3 Cooling system – draining

Note: *Take care to protect the hands from escaping steam when removing the expansion tank filler cap if the system is hot.*
1 With the car parked on level ground, unscrew the expansion tank filler cap. If the system is hot unscrew the cap slowly and allow the pressure in the system to be released before completely removing the cap.
2 Position a clean container beneath the bottom hose connection on the left-hand side of the radiator. Slacken the hose clip, ease the hose off the outlet and allow the water to drain into the container.
3 On 1.0 models a cylinder block drain plug is provided and is located on the side of the engine beneath the exhaust manifold. Reposition the container beneath the engine and unscrew the plug using a suitable Allen key. Refit and tighten the plug after draining.
4 On completion, remove the container to a safe place and cover it to prevent contamination of the coolant if it is to be re-used.
5 If renewing the coolant, flush the system.

4 Cooling system – flushing

1 If coolant renewal has been neglected, or if the antifreeze mixture has become diluted, then in time the cooling system will gradually lose efficiency as the cooling passages become choked with rust, scale deposits and other sediment. To restore cooling system efficiency it is necessary to flush the system clean. This is particularly necessary if the engine has no cylinder block drain plug, as large amounts of old coolant will remain in the system.
2 First drain the system, as described in the previous Section, and then remove the thermostat, as described in Section 7. Temporarily refit the thermostat housing and reconnect the hose.
3 Insert a garden hose into the disconnected radiator bottom hose and secure it in place with rags. Turn on the supply and allow clean water to flow through the system and out of the radiator bottom outlet. Continue flushing for ten to fifteen minutes or until clean rust-free water emerges from the radiator.
4 If the contamination is particularly bad, reverse flush the system by inserting the garden hose in the radiator bottom outlet and allow the water to flow through the system and out of the radiator bottom hose. This should dislodge deposits that were not moved by conventional flushing. If any doubt exists about the cleanliness of the radiator after flushing, it should be removed, as described in Section 6, so that it can be flushed and agitated at the same time. After severe flushing carry out a normal flow flush before refitting the thermostat and reconnecting the system hoses.
5 If after a reasonable period the water still does not run clear, the radiator can be flushed with a good proprietary cleaning agent such as Holts Radflush or Holts Speedflush. It is important that the manufacturer's instructions are followed carefully.

5.3A Bleed the cooling system through the heater hose outlet on 1.0 models ...

5.3B ... and through the temperature gauge transmitter orifice on 1.2 and 1.3 models

Chapter 2 Cooling system

5 Cooling system – filling

1 Before filling the cooling system make sure that all the hoses and hose clips are in good condition and that the clips are tight. On 1.0 models ensure that the cylinder block drain plug is fully tightened.

2 To ensure adequate protection against corrosion, as well as the effects of winter weather, a proportion of antifreeze must be maintained in the cooling system at all times. *Never fill the system with water only, even in summer.* Refer to Section 13 for details of the required antifreeze strength. In areas where the lime content of the mains water supply is high it is advantageous to use rainwater in the system, as this will reduce the build-up of scale in the radiator.

3 Remove the filler cap from the expansion tank. To allow air to escape from the system as it is being filled, slacken the clip and remove the heater hose at the cylinder head on 1.0 models (photo), or disconnect the lead and unscrew the temperature gauge transmitter from the inlet manifold on 1.2 and 1.3 models (photo).

4 Pour the required quantity of antifreeze into the expansion tank and then add water. Do not use a prepared mixture of antifreeze and water as any water remaining in the system will reduce the concentration. When coolant emerges from the heater hose outlet or temperature gauge transmitter orifice, refit the hose or transmitter and ensure that they are secure.

5 Continue adding water until the level in the expansion tank is just above the KALT mark. Repeated squeezing of the large coolant hoses will induce surging of the mixture in the system which will help to dislodge any air bubbles. Refit the expansion tank filler cap and wipe up any spilt coolant.

6 Run the engine at a fast tickover until the cooling fan motor engages and, particularly if the system has been disturbed in any way, examine carefully for leaks. Stop the engine and allow it to cool before topping-up the level in the expansion tank if necessary. Remember that the system must be cold before an accurate level is indicated in the expansion tank.

6 Radiator – removal, inspection and refitting

Note: *The radiator can be removed with the cooling fan still attached, but due to the limited space available care must be taken to avoid damaging the radiator. If preferred, the cooling fan may be removed first, as described in Section 10, to provide greater access.*

1 Drain the cooling system, as described in Section 3, and disconnect the battery earth terminal.

2 Slacken the retaining clips and detach the radiator top and bottom hoses and also the smaller diameter expansion tank vent hose.

3 Disconnect the two electrical leads at the thermal switch on the lower right-hand side of the radiator (photo).

4 If the cooling fan is still in position, disconnect the electrical leads at the multi-plug adjacent to the fan motor. Release the cable clips securing the wiring harness to the fan cowl bracket and move the harness to one side.

5 Using pliers, compress the two retaining clips, one each side of the radiator, and release them from their locations (photo).

6 Move the top of the radiator towards the engine to free the upper mountings and then lift the radiator out of the lower mountings and clear of the car.

7 When the radiator assembly is removed it is easier to examine for leaks which will show up as corroded or stained areas. If a leak is detected it is better to renew the complete radiator. In an emergency however, minor leaks from the radiator can be cured by using a radiator sealant such as Holts Radweld with the radiator *in situ*.

8 Clean out the inside of the radiator by flushing, as described in Section 4, and also clean the matrix, removing all the dead flies and bugs which reduce the radiator's efficiency. Take this opportunity to inspect the hoses and clips, making sure that all are fit for further use.

9 Refitting the radiator is the reverse of the removal procedure. Check that the rubber mountings are in good condition (photo) and ensure that the bottom location pegs fit correctly on installation. Refer to Section 5 for refilling the system.

7 Thermostat – removal, testing and refitting

1.0 models

1 Drain the cooling system, as described in Section 3, and

6.3 Radiator thermal switch location

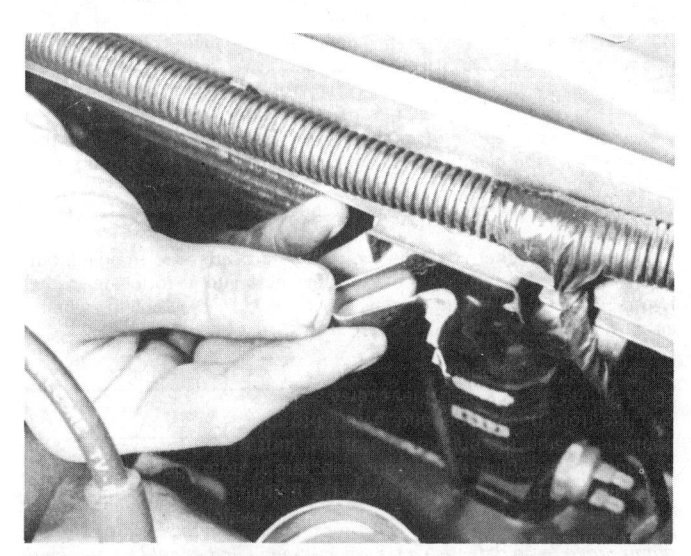

6.5 Release the radiator retaining clips

6.9 Radiator rubber mounting block in position over peg

disconnect the battery negative terminal.
2 After draining the coolant, disconnect the radiator top hose from the outlet connection at the top of the water pump. This will expose the thermostat and it will be seen that it is retained in position by a snap-ring.
3 Prise the snap-ring free using a suitable screwdriver blade and then remove the thermostat from the pump outlet (photos).
4 The thermostat can be tested easily for correct functioning if this should be in doubt. Boil a pan of water and suspend the thermostat on a piece of cord. Lower the thermostat into the hot water and it should be seen to open on immersion. Remove the thermostat from the water and it should be seen to close. This is only a simple functional test but it will identify a failed thermostat. With a thermometer you can check the correct opening temperature, see Specifications, but the full open temperature will be difficult to check as it is above the boiling point of water. When renewing this component make sure that the replacement item is the correct one for your car, as a wide range of thermostats are made for different models and conditions.
5 Refitting the thermostat is the reverse sequence to removal, but use a new rubber seal and install the thermostat with the arrow on the web pointing upwards. Refill the cooling system, as described in Section 5.

1.2 and 1.3 models

6 Drain the cooling system, as described in Section 3, and disconnect the battery negative terminal.
7 After draining the system, undo and remove the five small retaining bolts and lift off the belt cover.
8 Slacken the clip and detach the radiator top hose at its connection on the engine which is the thermostat housing.
9 Undo the two securing bolts and remove the housing to reveal the thermostat in the cylinder head. Remove the thermostat, noting how it fits in the recesses in the aperture.
10 Testing this thermostat is exactly the same as already described in paragraph 4.
11 Refitting the thermostat is the reverse procedure to removal, but fit a new rubber seal to the thermostat and install it to locate in the two recesses noted during removal (photo). Refill the cooling system, as described in Section 5.

8 Water pump – removal and refitting

Note: *Water pump failure is indicated by water leaking from the front of the pump, or by rough and noisy operation. This is usually accompanied by excessive play of the pump spindle which can be checked by moving the pulley or sprocket from side to side. Should the pump prove to be defective a factory exchange unit should be obtained as repair of the old unit is not a practical proposition.*

1.0 models

1 Drain the cooling system, as described in Section 3, and disconnect the battery earth terminal.
2 Slacken the alternator mounting nuts and bolts, swing the alternator in towards the engine and remove the drivebelt.
3 Slacken the retaining clips and detach the three hoses connected to the water pump.
4 Undo and remove the six water pump retaining bolts and carefully separate the pump from the cylinder head (photo). Recover the gasket.
5 If a new pump is to be fitted, undo and remove the three retaining bolts and transfer the pulley to the new unit. Transfer the thermostat also, referring to Section 7 if necessary.
6 Before refitting the water pump clean away all traces of old gasket from the pump and cylinder head mating faces.
7 Apply a little grease to a new gasket and place it in position on the pump. Refit the pump and secure the unit with the six bolts tightened progressively to the specified torque.
8 Refit the three hoses and tighten the clips securely. Refit the drivebelt and adjust the tension, as described in Section 9. Finally, refit the cooling system, as described in Section 5, and reconnect the battery.

1.2 and 1.3 models

9 Drain the cooling system, as described in Section 3, and disconnect the battery negative terminal.

7.3A Remove the thermostat retaining snap-ring ...

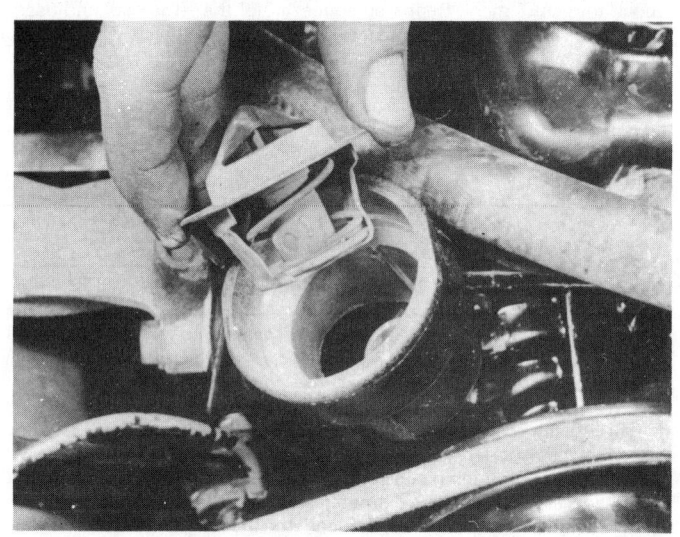

7.3B ... and lift the thermostat from the pump outlet

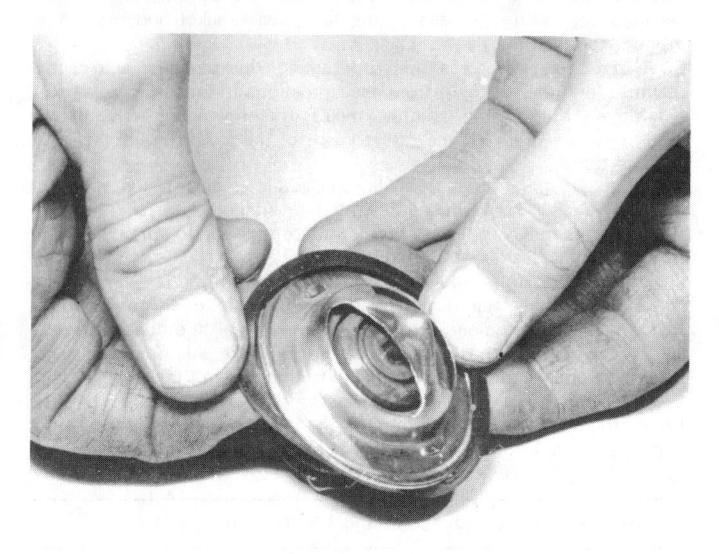

7.11 Fit a new seal ring to the thermostat on 1.2 and 1.3 models

Chapter 2 Cooling system

8.4 Removing the water pump on 1.0 models

8.11 On 1.2 and 1.3 models first remove the toothed belt cover

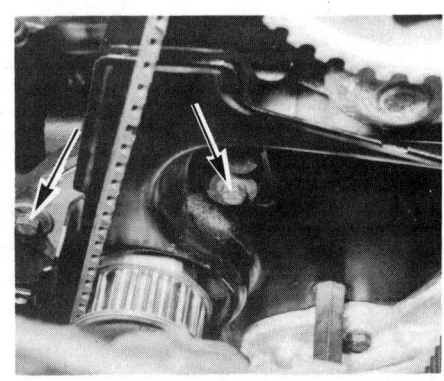

8.13A Slacken the water pump retaining bolts (arrowed) ...

8.13B ... and slip the toothed belt off the sprockets

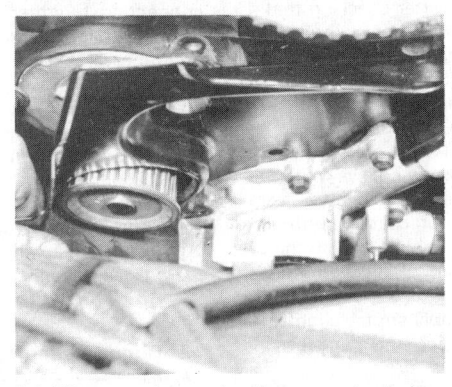

8.14 Remove the toothed belt cover backplate for access to the pump

8.15 Remove the bolts and withdraw the pump

10 Slacken the alternator mounting nuts and bolts, swing the alternator in towards the engine and remove the drivebelt.
11 Undo and remove the five small retaining bolts and lift off the toothed belt cover (photo).
12 Using a spanner or socket on the crankshaft pulley retaining bolt, turn the crankshaft until the notch on the outer edge of the camshaft sprocket is aligned with the groove in the plate behind the sprocket, and at the same time the notch in the crankshaft pulley is in line with the timing pointer. Turn the crankshaft in the normal direction of rotation only, ie clockwise when viewed from the belt cover end.
13 Slacken the three bolts securing the water pump to the cylinder block (photo). The pump shaft is eccentric in the pump body so that by rotating the pump body, the tension in the toothed belt can be released. Turn the pump inwards to slacken the toothed belt and slip the belt off the pump sprocket (photo). Flats are provided on the pump body behind the sprocket and if necessary a large thin spanner can be used to turn the pump.
14 Undo and remove the bolts securing the belt cover backplate to the cylinder block, noting the location of the longer stud bolt. Withdraw the backplate from the engine (photo).
15 The three water pump retaining bolts can now be removed and the pump withdrawn from its location (photo).
16 Before refitting the pump, clean its mounting in the cylinder block. Fit a new O-ring seal to the pump body and apply silicone grease to the seal and to the sealing surface in the block (photo). Install the pump and refit the three retaining bolts and washers, but only hand tighten them at this stage.
17 Refit the toothed belt cover backplate and secure it with the retaining bolts and stud bolt.
18 Check that the timing mark on the camshaft sprocket is still aligned with the notch on the backplate and that the notch on the crankshaft pulley is in line with the timing pointer. Now slip the toothed belt over the camshaft and water pump sprockets.
19 Move the water pump body away from the engine until it is just possible to deflect the timing belt approximately 6.35 mm (0.25 in) at a point midway between the camshaft and crankshaft sprockets.

Alternatively, twist the belt between thumb and forefinger. When the tension is correct it should be just possible to twist the belt through 90° using moderate pressure. When the correct tension on the timing belt has been achieved, tighten the three water pump retaining bolts.
20 Refit the toothed belt cover and secure it with the five small bolts. Refit the alternator drivebelt and adjust its tension, as described in Section 9. Refer to Section 5 and refill the cooling system, then reconnect the battery.

8.16 Renew the O-ring seal

9 Drivebelt – removal, refitting and adjustment

Note: On 1.0 models the drivebelt runs in the crankshaft, water pump and alternator pulleys. On 1.2 and 1.3 models the belt only runs in the crankshaft and alternator pulleys as the water pump is driven by the toothed camshaft belt. However, the removal, refitting and adjustment procedures on all models are similar.

1 Correct tensioning of the drivebelt will ensure that it has a long and useful life. Beware, however, of overtightening as this can cause excessive wear in the alternator and/or water pump bearings.
2 A regular inspection of the belt should be made and if it is found to be overstretched, worn, frayed or cracked it should be renewed before it breaks in service. To insure against such an event arising it is a good idea to carry a spare belt, of the correct type, in the car at all times.
3 To remove an old belt, loosen the alternator mounting bolts and nuts just sufficiently to allow the unit to be pivoted in towards the engine. This will release all tension from the belt which can now be slipped off the respective pulleys. Fit a new belt after checking that it is of the correct type and take up the slack in the belt by swinging the alternator away from the engine and lightly tightening the bolts just to hold it in that position.
4 Although special tools are available for measuring the belt tension a good approximation can be achieved if the belt is tensioned so that there is 13 mm (0.5 in) of movement under firm thumb pressure at the mid-point position on the longest run of belt between pulleys. With the alternator bolts just holding the unit firm, lever the alternator away from the engine using a wooden lever at the mounting bracket end until the correct tension in the belt is reached and then tighten the alternator bolts. On no account apply any loads at the free end of the alternator as serious damage can be caused internally.
5 When a new belt has been fitted it will probably stretch slightly to start with and the tension should be rechecked, and if necessary adjusted, after about 250 miles (400 km).

10 Radiator electric cooling fan – removal and refitting

1 Disconnect the battery negative terminal.
2 Disconnect the fan motor electrical leads at the multi-plug adjacent to the motor. Release the wiring harness cable ties at the fan cowl bracket and move the harness to one side.
3 Undo and remove the two small bolts securing the fan cowl to the top of the radiator. Lift the fan and cowl assembly upwards to release the lower mounting lugs and remove the unit from the car (photo).

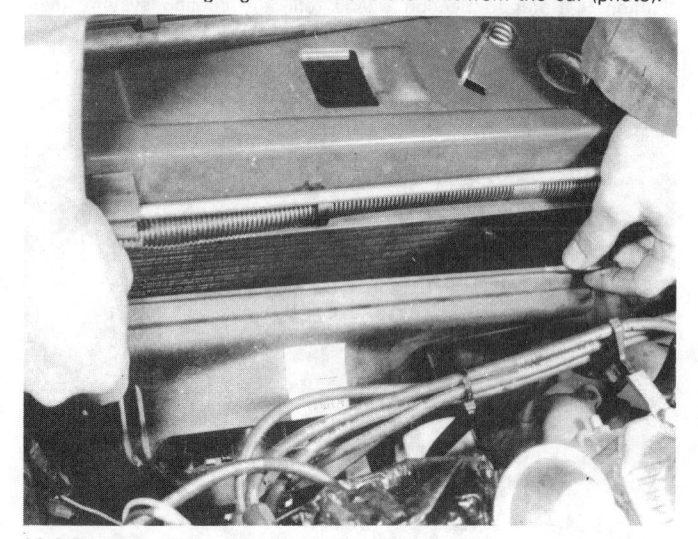

10.3 Removing the cooling fan and cowl assembly

4 To separate the fan motor from the cowl unscrew the three nuts. The fan blades may be withdrawn from the motor spindle after removal of the retaining clip.
5 Further dismantling of the assembly depends on the extent of the problem. If the motor is defective it would be better to have it overhauled by a specialist, as spare parts may be difficult to obtain. The alternative is to renew the motor which may prove cheaper and quicker in the long run.
6 Reassembly, if the unit was dismantled, and refitting to the car are the reverse of the dismantling and removal sequences. On completion run the engine up to normal operating temperature and check the fan for correct functioning.

11 Cooling fan thermal switch – testing, removal and refitting

1 The cooling fan is controlled by a thermal switch located at the lower right-hand side of the radiator (photo 6.3). If the cooling fan fails to operate at high engine temperatures the circuit can be tested by connecting together the two electrical leads serving the switch and turning the ignition on. If the fan now works then the switch is at fault and renewal is necessary. If the motor still fails to function then either a fuse has blown (fuse No 10), the motor itself is faulty, or there is a break or bad connection in the electrical supply to the motor.
2 To remove the thermal switch, first drain the cooling system, as described in Section 3, and disconnect the battery earth terminal.
3 Disconnect the two electrical leads at the switch and then unscrew the switch from its location in the radiator.
4 Refitting the switch is the reverse sequence to removal, but apply a little sealant to the switch threads before fitting. Refill the cooling system, as described in Section 5, and reconnect the battery.

12 Temperature gauge transmitter – removal and refitting

1 The transmitter unit is located in the cylinder head behind the water pump on 1.0 models and on the top right-hand side of the inlet manifold on 1.2 and 1.3 models.
2 Before removing the transmitter, refer to Section 3 and drain the cooling system sufficiently to avoid spillage. Also disconnect the battery earth terminal.
3 Disconnect the electrical lead from the transmitter and unscrew the unit from its location.
4 Refitting is the reverse sequence to removal. Apply a little sealant to the transmitter threads and tighten the unit to the specified torque. Refill the cooling system, as described in Section 5, and reconnect the battery.

13 Antifreeze mixture

1 It is essential that an antifreeze mixture is retained in the cooling system at all times to act as a corrosion inhibitor and to protect the engine against freezing in winter months. The mixture should be made up from clean water with a low lime content (preferably rainwater) and a good quality ethylene glycol based antifreeze which contains a corrosion inhibitor and is suitable for use in aluminium engines.
2 The proportions of antifreeze to water required will depend on the maker's recommendations, but the mixture must be adequate to give protection down to approximately -30°C (-22°F).
3 Before filling with fresh antifreeze drain and flush the cooling system, as described in Sections 3 and 4. Check that all hoses are in good condition and that all clips are secure, then fill the system, as described in Section 5.
4 The antifreeze should be renewed every two years to maintain adequate corrosion protection. Do not use engine coolant antifreeze in the windscreen or tailgate wash systems; it will damage the car's paintwork and smear the glass. *Finally remember that antifreeze is poisonous and must be handled with due care.*

14 Fault diagnosis – cooling system

Symptom	Reason(s)
Overheating	Insufficient coolant in system
	Pump ineffective due to slack drivebelt (ohv)
	Radiator blocked either internally or externally
	Kinked or collapsed hose causing coolant flow restriction
	Thermostat not working properly
	Faulty electric fan
	Faulty fan thermal switch
	Engine out of tune
	Ignition timing retarded or auto advance malfunction
	Cylinder head gasket blown
	Engine not yet run-in
	Exhaust system partially blocked
	Engine oil level too low
	Brakes binding
Engine running too cool	Faulty, incorrect or missing thermostat
Loss of coolant	Loose hose clips
	Hoses perished or leaking
	Radiator leaking
	Filler/pressure cap defective
	Blown cylinder head gasket
	Cracked cylinder block or head

Chapter 3 Fuel and exhaust systems

For modifications, and information applicable to later models, see Supplement at end of manual

Contents

Accelerator cable – removal, refitting and adjustment	9
Accelerator pedal – removal and refitting	10
Air cleaner – description	5
Air cleaner – servicing, removal and refitting	6
Carburettor – description	12
Carburettor – removal and refitting	13
Carburettor (1.0 models) – idle speed and mixture adjustment	14
Carburettor (1.0 models) – overhaul	16
Carburettor (1.0 models) – setting and adjustment of components	15
Carburettor (1.2 and 1.3 models) – idle speed and mixture adjustment	17
Carburettor (1.2 and 1.3 models) – overhaul	19
Carburettor (1.2 and 1.3 models) – setting and adjustment of components	18
Choke cable – removal, refitting and adjustment	11
Exhaust system – inspection, removal and refitting	22
Fault diagnosis – fuel and exhaust systems	23
Fuel gauge sender unit – removal and refitting	8
Fuel pump – description and maintenance	3
Fuel pump – testing, removal and refitting	4
Fuel tank – removal and refitting	7
General description	1
Maintenance and inspection	2
Manifolds (1,0 models) – removal and refitting	20
Manifolds (1.2 and 1.3 models) – removal and refitting	21

Specifications

General
System type	Rear mounted fuel tank, mechanically operated fuel pump, downdraught single barrel carburettor
Fuel tank capacity	42 litres (9.2 gals)
Fuel octane rating	98 RON (4-star)

Fuel pump
Type	Mechanical, operated by camshaft
Delivery pressure	0.19 to 0.25 bar (2.8 to 3.7 lbf/in^2) at 2000 rpm

Air filter element
Type:
1.0 litre model	Champion W101
1.2 litre model	Champion W101 to August 1985
1.3 litre model	Champion W103

Carburettor
Type:
1.0 models	Weber 32 TL
1.2 and 1.3 models	Pierburg (Solex) 1B1

Weber carburettor specification
Venturi diameter	25 mm (1.0 in)
Main jet	117
Main mixture outlet diameter	2.5 mm (1.0 in)
Air correction jet	75
Mixture tube identification	F96
Idle fuel jet	47
Idle jet	90
Idle mixture jet	210
Auxiliary mixture fuel jet	35
Auxiliary mixture air jet	170
Float weight	9.75 to 10.75 g (0.34 to 0.38 oz)
Float needle valve diameter	1.75 mm (0.07 in)

Chapter 3 Fuel and exhaust systems

Auxiliary mixture jet	100
Full load enrichment jet	65
Partial load enrichment idle jet	40
Partial load enrichment main jet	40
Accelerator pump jet	45
Accelerator pump return jet	35
Accelerator pump injection quantity	0.75 to 1.35 cc/stroke
Vacuum pull-down reduction jet	35
Fast idle throttle valve gap	0.60 to 0.70 mm (0.02 to 0.03 in)
Choke valve gap	4.30 to 4.80 mm (0.17 to 0.18 in)
Vacuum at idle speed	1 to 20 mbar (0.015 to 0.29 lbf/in^2)
Idle speed	850 to 900 rpm
Fast idle speed	3600 to 4000 rpm
Exhaust gas CO content at idle	0.5 to 1.5%

Pierburg (Solex) carburettor specification

Venturi diameter:	
1.2 models	23 mm (0.90 in)
1.3 models	25 mm (1.0 in)
Main jet:	
1.2 models	x105
1.3 models	x120
Air correction jet/mixture tube identification	75/17
Auxiliary fuel jet	42.5/147.5
Idle fuel jet	47.5/140
Idle mixture outlet diameter:	
1.2 models	1.3 mm (0.05 in)
1.3 models	1.5 mm (0.06 in)
Accelerator pump injection quantity	0.45 to 0.75 cc/stroke
Choke valve gap:	
1.2 models	2.3 to 2.7 mm (0.09 to 0.11 in)
1.3 models	3.7 to 4.1 mm (0.14 to 0.16 in)
Vacuum at idle speed	1 to 20 mbar (0.015 to 0.29 lbf/in^2)
Idle speed:	
1.2 models	900 to 950 rpm
1.3 models	900 to 950 rpm
Fast idle speed:	
1.2 models	3600 to 4000 rpm
1.3 models	3800 to 4200 rpm
Exhaust gas CO content at idle	0.5 to 1.5%

Torque wrench settings

	Nm	lbf ft
Fuel pump retaining bolts	20	15
Carburettor-to-inlet manifold nuts:		
1.0 models	18	13
1.2 and 1.3 models	20	15
Inlet and exhaust manifolds to cylinder head:		
1.0 models	23	17
1.2 and 1.3 models	20	15

1 General description

The fuel system consists of a rear mounted fuel tank, camshaft operated mechanical fuel pump and a downdraught fixed jet carburettor.

The air cleaner on 1.3 models features automatic intake air temperature control. The unit fitted to 1.0 models incorporates a similar device, but with manual control. On 1.2 models, a blended supply of hot and cold intake air is supplied, and no control is provided. A disposable paper filter element is used on all models.

The exhaust system is in three sections. The front section incorporates a flexible joint to allow a degree of engine movement without placing undue strain on the system. A single front downpipe is used on 1.0 and 1.2 models with twin front downpipes being used on 1.3 models. The intermediate section incorporates the front silencer and the rear section incorporates the rear silencer and tailpipe. The complete system is suspended from the underbody on rubber ring type mountings at the centre and rear.

Warning: *Many of the procedures in this Chapter entail the removal of fuel pipes and connections which may result in some fuel spillage. Before carrying out any operation on the fuel system refer to the precautions given in Safety First! at the beginning of this manual and follow them implicitly. Petrol is a highly dangerous and volatile liquid and the precautions necessary when handling it cannot be overstressed.*

2 Maintenance and inspection

1 Every 9000 miles (15 000 km) or 6 months, whichever comes first, the following checks and adjustments should be carried out on the fuel and exhaust system components.

2 Remove the air cleaner, as described in Section 6, wipe it clean inside and out and renew the paper element. On 1.0 models set the temperature control according to expected seasonal weather conditions. On 1.3 models check the condition of the vacuum hose and the operation of the flap valve in the air cleaner spout.

3 Remove and clean the gauze filter in the fuel pump, as described in Section 3 (where applicable). On 1.0 models there is an additional filter located in the carburettor which is accessible after unscrewing the hexagon-headed plug adjacent to the fuel inlet hose connection.

4 Check the exhaust system for condition, leaks and security of the rubber mountings. Repair or renew the sections of the exhaust system where necessary, as described in Section 22.

5 Check the carburettor idle settings, as described in Section 14 or 17, and adjust as necessary.

6 Every 18 000 miles (30 000 km) or 12 months, whichever comes first the following additional checks should be carried out.

7 With the car over a pit, raised on a vehicle hoist or securely supported on axle stands carefully inspect the fuel pipes, hoses and unions for chafing, leaks and corrosion. Renew any pipes that are

severely pitted with corrosion or in any way damaged. Renew any hoses that show signs of cracking or other deterioration.
8 Examine the fuel tank for leaks, particularly around the fuel gauge sender unit, and for signs of corrosion or damage.
9 From within the engine compartment check the security of all fuel hose attachments and inspect the fuel hoses and vacuum hoses for kinks, chafing or deterioration.
10 Check the operation of the accelerator and choke controls and lubricate the linkage, cables and pivots with a few drops of engine oil.

3 Fuel pump – description and maintenance

1 On 1.0 models the fuel pump is located on the side of the engine facing the front of the car, and is driven by a lobe on the camshaft which is in contact with the rocker arm of the pump.
2 On 1.2 and 1.3 models the fuel pump is mounted at the front end of the camshaft housing and is again driven by a lobe on the camshaft, this time with a pushrod.
3 One of two types of pump may be fitted; a semi-sealed unit or a completely sealed type. The completely sealed type of pump is easily identifiable by the position of the hose nozzles, one of which is horizontal with respect to the pump body and the other vertical. This type of pump cannot be dismantled for overhaul or repair and no maintenance is required.
4 The semi-sealed pump can also be identified by the position of the hose nozzles which are both horizontal in relation to the pump body. Also, on these pumps a screw will be found in the centre of the top cover. As with the completely sealed type, dismantling and overhaul is not possible; however, the top cover can be removed to allow access to the gauze filter.
5 To clean the filter, undo the retaining screw and lift off the top cover complete with hose. Be prepared for some fuel spillage and ensure adequate ventilation.
6 Lift off the filter gauze and clean it by blowing through it or by using an air line (photo).
7 Check the condition of the rubber seal in the top cover and renew it if it is in any way damaged, cracked or deformed.
8 Refit the filter, cover and retaining screw, but take care not to overtighten the screw.

4 Fuel pump – testing, removal and refitting

1 Note the location of the fuel inlet hose and outlet hose and then disconnect them from the pump nozzles after slackening the clips (photo). Plug the hose ends with a metal rod or old bolt after removal.
2 Undo the two pump retaining bolts and washers and withdraw the pump from the engine. On 1.2 and 1.3 models note the position of the hose clip. Recover the insulating block and gaskets.

3.6 Fuel pump cover, filter gauze and rubber seal

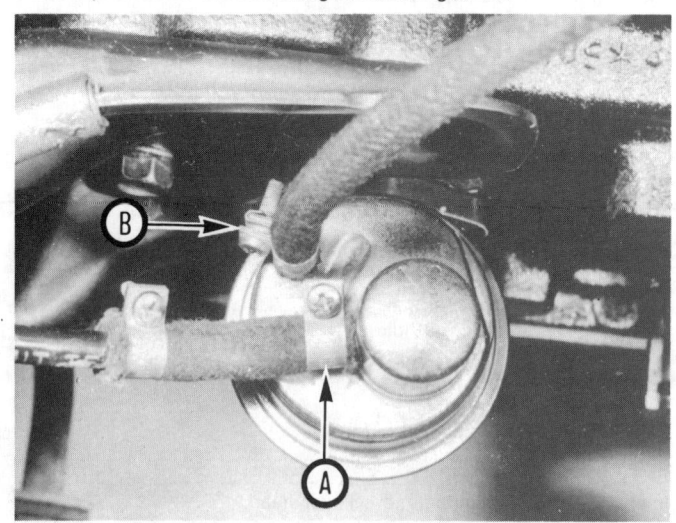

4.1A Fuel inlet (A) and outlet (B) hose connections at the fuel pump on 1.0 models

4.1B Removing the fuel inlet hose from the fuel pump on 1.2 and 1.3 models

5.2 Air cleaner flap valve on 1.0 models (arrowed)

Chapter 3 Fuel and exhaust systems

3 To test the pump operation refit the fuel inlet hose to the pump inlet nozzle and hold a rag near the outlet. Operate the pump lever by hand and if the pump is in a satisfactory condition a strong jet of fuel should be ejected from the outlet as the lever is released. If this is not the case, check that fuel will flow from the inlet hose when it is held below tank level; if so the pump is faulty and renewal is necessary.

4 Before refitting the pump, clean all traces of old gasket from the pump and engine mating faces. If necessary use a new gasket on each side of the insulating block and refit the pump using the reverse sequence to removal.

5 Air cleaner – description

1 The provision of intake air to the carburettor at the correct temperature for optimum combustion is carried out as follows:

1.0 models

2 Intake air pre-heating is controlled manually by a flap valve located in the side of air cleaner casing (photo). The valve can be set in any one of three positions according to seasonal operating temperature as shown in the following table.

Summer position	–	above 10°C (50°F)
Intermediate position	–	10°C to -5° (50°F to 23°F)
Winter position	–	below -5°C (23°F)

3 In terms of fuel economy the engine will run most efficiently with the valve set in the summer position and least efficiently in the winter position. Providing the engine is running smoothly, and accelerates evenly, the summer position may be retained down to 0°C (32°F). If roughness or hesitation occurs, move the flap valve to the next position.

4 The three positions are shown on the air cleaner cover and the valve can be repositioned after slackening the wing nut. Tighten the nut after repositioning. In the winter position only hot air from the hot air box on the exhaust manifold enters the air cleaner. In the summer position only cold air from the air cleaner intake spout enters. In the intermediate position a blended supply from both sources enters the air cleaner.

1.2 models

5 The air cleaner fitted to 1.2 models provides a supply of warm air to the carburettor blended from the hot air box on the exhaust manifold and from the cold air intake spout. The air supply is continuous from both sources and, as no control is provided, the temperature of the air entering the air cleaner will be largely dependent on ambient temperatures.

1.3 models

6 A thermostatically controlled air cleaner is used to regulate the temperature of the air entering the carburettor according to ambient temperatures and engine load. The air cleaner has two sources of supply, through the normal intake spout (cold air) or from a hot air box mounted on the exhaust manifold (hot air).

7 The air flow through the air cleaner is controlled by a flap valve in the air cleaner spout, which covers or exposes the hot or cold air ports according to temperature and manifold vacuum.

8 A vacuum motor operates the flap valve and holds it fully open when the temperature in the air cleaner is below a predetermined level. As the air intake temperature rises the vacuum motor opens or closes the flap valve dependent entirely on manifold vacuum. Thus, during light or constant throttle applications, the flap valve will remain open, supplying the carburettor with hot air, and will close under heavy throttle application so that only cold air enters the carburettor.

9 As the temperature in the air cleaner rises further the vacuum motor closes the flap valve therefore allowing only cold air to enter the carburettor under all operating conditions.

10 The vacuum motor is operated by vacuum created in the inlet manifold and is controlled by a temperature sensing unit located inside the air cleaner.

6 Air cleaner – servicing, removal and refitting

Servicing

1 To remove the paper filter element, undo the three screws on the

6.1A Undo the three air cleaner retaining screws ...

6.1B ... spring back the retaining clips ...

6.2 ... remove the cover and lift out the filter element

6.6 Detach the air cleaner spout from the hot air box supply tube

6.7 Detach the vacuum hose and breather hose from the base of the air cleaner

6.8 Ensure that the sealing ring is in place before refitting the air cleaner

Chapter 3 Fuel and exhaust systems

top of the cover (photo) spring back the retaining clips (photo) and lift off the cover.
2   Remove the filter element from the air cleaner (photo) and, if dirty, discard it and obtain a new element. The filter element should be renewed regardless of its condition at the intervals given in Routine Maintenance.
3   Clean the inside of the air cleaner body and place a new element in position.
4   Refit the cover and secure with the clips and retaining screws.

Removal and refitting

5   Undo and remove the three screws on the top of the air cleaner cover.
6   Lift the air cleaner off the carburettor and detach the spout from the hot air box supply tube (photo).
7   Detach the vacuum hose and breather hose from the base of the air cleaner (photo) and remove the unit from the engine.
8   Refitting is the reverse sequence to removal, but ensure that the sealing ring is in place on the carburettor (photo).

7 Fuel tank – removal and refitting

Note: *Refer to the warning note in Section 1 before proceeding*
1   Disconnect the battery negative lead. Remove the fuel tank filler cap.
2   A drain plug is not provided and it will therefore be necessary to syphon or hand pump all the fuel from the tank before removal.
3   Having emptied the tank, jack up the rear of the car and support it on axle stands.
4   Refer to Section 22 and remove the exhaust system.
5   Measure and record the length of exposed thread protruding through the handbrake cable adjusting locknut at the compensating yoke on the rear axle.
6   Hold the cable with pliers or a spanner, unscrew the adjusting nut and remove the cable end from the yoke.
7   Remove the retainer and detach the cable from the connecting link located just to the rear of the handbrake lever rod.
8   Detach the cable from its retainers on the fuel tank and underbody and move it clear of the tank.
9   Disconnect the two electrical leads from the fuel gauge sender unit.
10 Slacken the retaining clips and disconnect the fuel filler hose and the feed and return hoses from the tank (photos).
11 Support the tank with a jack and suitable blocks of wood and then undo the two tank mounting strap retaining nuts (photos). Pivot the straps down and out of the way of the tank.
12 Lower the jack slowly and, when sufficient clearance exists, disconnect the vent hose from the top of the tank.
13 Lower the jack fully and slide the tank out from under the car.
14 If the tank is contaminated with sediment or water, remove the sender unit, as described in Section 8, and swill the tank out with clean

7.10A Fuel tank filler hose connections ...

7.10B ... and fuel feed and return hose connections

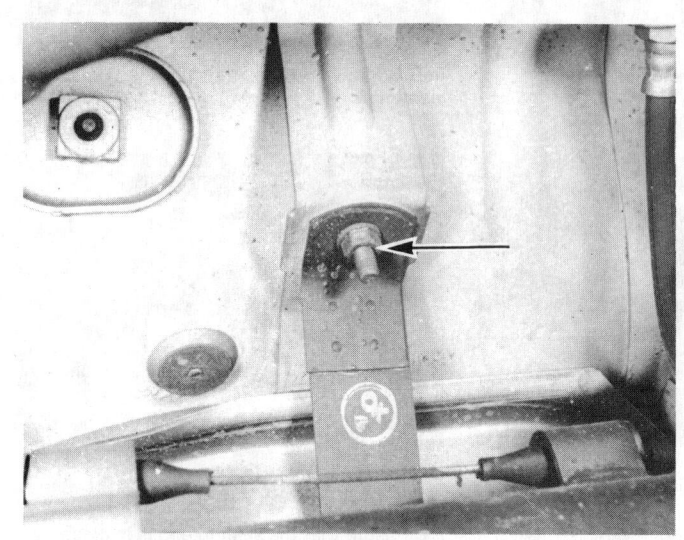

7.11A Fuel tank strap retaining nut (arrowed)

7.11B ... and retaining strap front pivot mounting (arrowed)

Chapter 3 Fuel and exhaust systems

fuel. If the tank is damaged, or leaks, it should be repaired by a specialist or, alternatively, renewed. **Do not** under any circumstances solder or weld the tank.
15 Refitting is the reverse sequence to removal. Refit the handbrake cable adjusting locknut to the position noted during removal and if necessary adjust the handbrake, as described in Chapter 9.

8 Fuel gauge sender unit – removal and refitting

Note: *Refer to the warning note in Section 1 before proceeding*
1 Follow the procedure given in Section 7, paragraphs 1 to 3 inclusive.
2 Disconnect the two electrical leads from the sender unit terminals (photo).
3 Engage a flat bar between two of the raised tabs on the sender unit and turn it anti-clockwise to release it.
4 Withdraw the unit carefully from the tank to avoid bending the float arm. Recover the sealing ring.
5 Refitting is the reverse sequence to removal, but make sure that the sealing ring is in good condition and renew it if necessary.

9 Accelerator cable – removal, refitting and adjustment

1 Remove the air cleaner, as described in Section 6.
2 For greater access to the cable fitting at the accelerator pedal remove the parcel shelf, as described in Chapter 12.
3 Working inside the car, release the cable from the accelerator pedal by easing back the spring on the end fitting and prising out the cable retainer from the pedal slot.
4 Release the grommet from the bulkhead and pull the cable through into the engine compartment.
5 Extract the small retaining clip (photo) and disconnect the ball end fitting from the carburettor linkage (photo).
6 Slide the cable bushing out of the support bracket on the carburettor and remove the cable from the car.

7 Refitting is the reverse sequence to removal. The cable should be adjusted to give a small amount of free play at the carburettor linkage with the accelerator pedal released. To adjust the cable, release the retaining clip located behind the cable bushing on the carburettor bracket and reposition the bushing as necessary. Refit the clip to retain the bushing in position.

10 Accelerator pedal – removal and refitting

1 Remove the parcel shelf under the facia, as described in Chapter 12.
2 Release the accelerator cable from the pedal by easing back the spring on the end fitting and prising out the cable retainer from the pedal slot.
3 Disengage the pedal return spring, extract the pedal pivot retaining clip and slide the pedal off the pivot (photo).
4 Refitting is the reverse sequence to removal. Lubricate the pedal pivot with multi-purpose grease and if necessary adjust the accelerator cable, as described in the previous Section, after fitting.

11 Choke cable – removal, refitting and adjustment

1 Disconnect the battery negative terminal.
2 Tap out the small retaining pin securing the choke control knob to the cable end fitting and unscrew the knob.
3 Undo the retaining ring securing the choke cable to the facia. Release the bulkhead grommet and pull the cable out of the facia, through the bulkhead and into the engine compartment.
4 Slacken the retaining screw and release the outer cable from the clamp on the carburettor bracket.
5 On 1.0 models slacken the retaining screw and slide the inner cable out of the operating lever on the carburettor (photo).
6 On 1.2 and 1.3 models slacken the grub screw securing the inner cable to the carburettor operating lever using an Allen key and withdraw the cable (photo).
7 Remove the cable from the car.

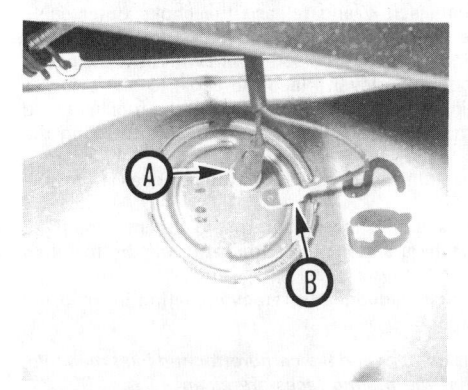

8.2 Electrical feed connection (A) and earth lead (B) at the fuel gauge sender unit

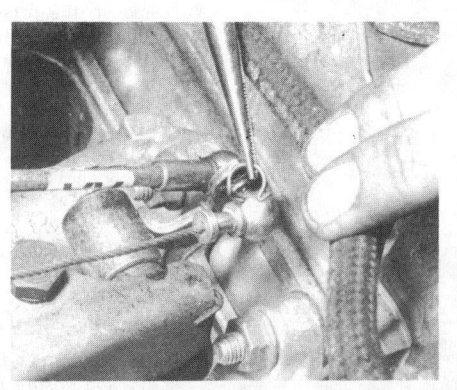

9.5A Extract the small retaining clip ...

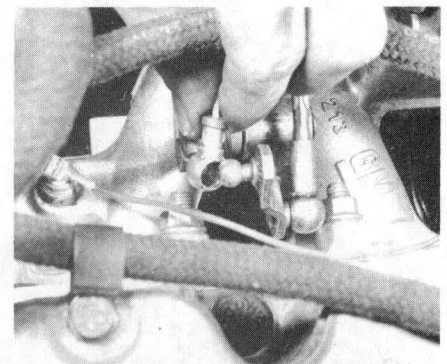

9.5B ... and disconnect the accelerator cable ball end fitting from the carburettor linkage

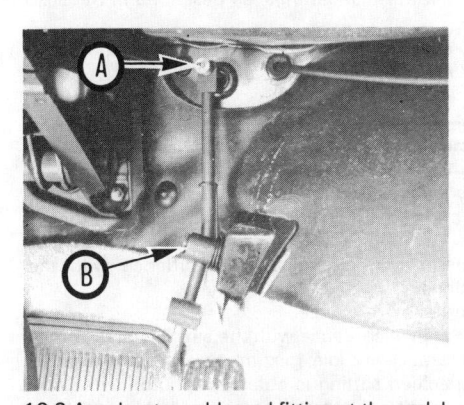

10.3 Accelerator cable end fitting at the pedal (A) and pedal pivot retaining clip (B)

11.5 Choke cable attachments at the carburettor on 1.0 models (arrowed)

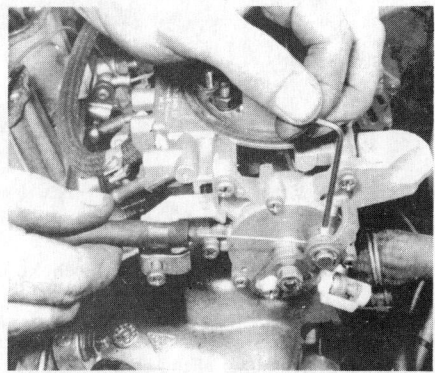

11.6 An Allen key is needed to release the cable on 1.2 and 1.3 models

Chapter 3 Fuel and exhaust systems

Fig. 3.1 Choke cable positioning dimension – 1.0 models (Sec 11)

X = 25 to 30 mm (1.0 to 1.2 in)

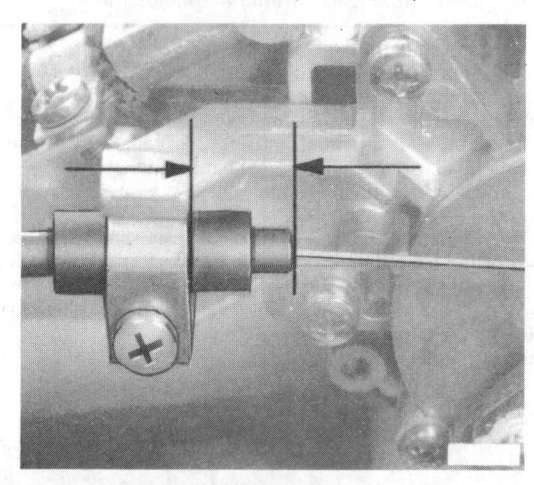

Fig. 3.2 Choke cable positioning dimension – 1.2 and 1.3 models (Sec 11)

Projecting cable length = 5 to 10 mm (0.20 to 0.39 in)

Fig. 3.3 Using a drill bit to hold the choke knob out during cable adjustment (Sec 11)

8 Refitting is the reverse sequence to removal, but before tightening the cable attachments at the carburettor, adjust the cable as follows:
9 Position the outer cable in its clamp on the carburettor bracket so that the amount of cable protruding through the clamp is as shown in the accompanying illustrations (see Figs. 3.1 and 3.2).
10 Pull the choke knob out very slightly and clamp a 4 mm (0.16 in) twist drill bit between the knob and the threaded part of the outer cable (see Fig. 3.3).
11 With the choke linkage on the carburettor fully open, ie at rest, tighten the inner cable retaining screw or grub screw.
12 Remove the drill bit, pull the choke knob fully out and then push it fully in. If the cable has been correctly adjusted the knob should spring back by 0.5 to 1.5 mm (0.02 to 0.06 in) after being pushed fully in.
13 Refit the air cleaner and reconnect the battery after making the adjustment.

12 Carburettor – description

1 One of two types of carburettor may be fitted according to engine type.
2 On 1.0 models a Weber 32 TL carburettor is used and on 1.2 and 1.3 models a Pierburg 1B1 carburettor is fitted. The Pierburg carburettor is essentially a Solex unit manufactured under licence in West Germany.
3 Both types of carburettor are of single barrel, downdraught configuration with manually operated choke control.
4 The carburettors are of the fixed jet type with two enrichment systems supporting the main jet system, and a separate idle circuit.
5 Both carburettors are of straightforward design and layout and all adjustments and settings can be carried out using instruments readily available to the home mechanic.

13 Carburettor – removal and refitting

1 Remove the air cleaner, as described in Section 6.
2 Extract the retaining clip and disconnect the ball end fitting from the carburettor linkage.
3 On 1.0 models slide the accelerator cable bushing out of the support bracket on the carburettor and move the cable to one side.
4 Slacken the retaining screw and release the choke outer cable from the clamp on the carburettor bracket.
5 Slacken the retaining screw or grub screw and withdraw the choke inner cable from the operating lever. Place the cable to one side.
6 Slacken the retaining clip screws and disconnect the fuel inlet, and where fitted the return hoses from the carburettor (photo). Plug the hoses after removal.
7 Disconnect the vacuum and engine breather hoses from the carburettor.
8 Undo the two nuts securing the carburettor to the inlet manifold (photo), remove the washers and lift the carburettor off the manifold studs (photo).
9 Refitting is the reverse sequence to removal bearing in mind the following points:

   (a) Use a new gasket between the carburettor and inlet manifold, and ensure that the mating faces are clean
   (b) Adjust the accelerator and choke cables, as described in Sections 9 and 11 respectively
   (c) Adjust the carburettor idle settings, as described in Sections 14 or 17
   (d) On 1.0 models (engine numbers 10S 3 260 000 to 10S 3 300 087) change the fixing nuts to self-locking type

14 Carburettor (1.0 models) – idle speed and mixture adjustment

Note: For the following adjustments the engine must be at its normal operating temperature and the air cleaner must be in position on the carburettor.

1 Connect a tachometer to the engine in accordance with the manufacturer's instructions.
2 Start the engine and allow it to idle.
3 If the idle speed is not in accordance with the settings given in the Specifications, adjust the auxiliary idle mixture screw, using a small screwdriver, until the specified setting is obtained (photo).
4 Normally this is the only adjustment necessary. However if the engine tickover is uneven or erratic, or if the carburettor has been

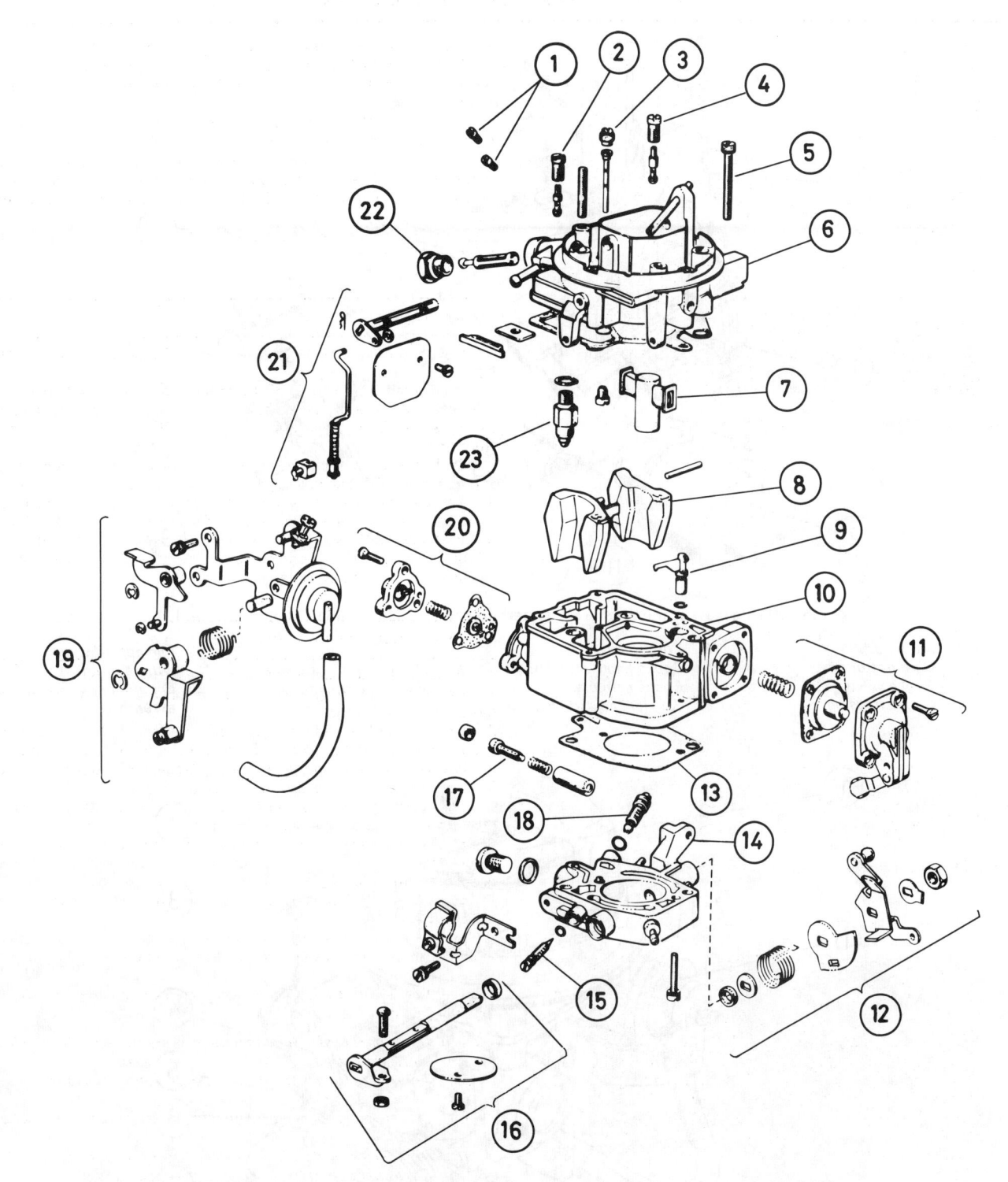

Fig. 3.4 Exploded view of the Weber 32 TL carburettor fitted to 1.0 models (Sec 12)

1 Screw plugs
2 Idle fuel/air jet
3 Air correction jet and mixture tube
4 Auxiliary fuel/air jet
5 Carburettor cover retaining screw
6 Carburettor cover
7 Pre-atomizer
8 Floats
9 Fuel discharge nozzle
10 Float chamber housing
11 Accelerator pump assembly
12 Throttle linkage assembly
13 Gasket
14 Throttle valve housing
15 Basic idle mixture screw
16 Throttle valve assembly
17 Throttle valve stop screw
18 Auxiliary idle mixture screw
19 Choke linkage and vacuum unit assembly
20 Enrichment valve assembly
21 Choke valve assembly
22 Fuel inlet filter and plug
23 Float needle valve

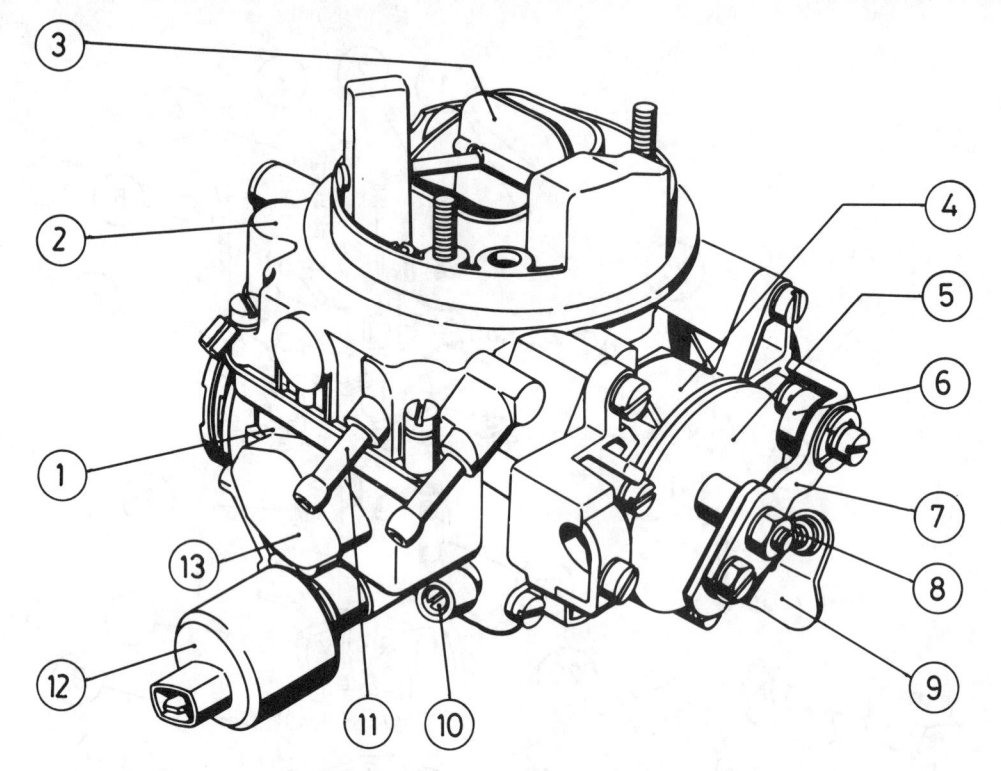

Fig. 3.5 Left side view of the Pierburg 1B1 carburettor fitted to 1.2 and 1.3 models (Sec 12)

1. Main body
2. Carburettor cover
3. Choke valve
4. Choke housing
5. Choke cover
6. Choke cable attachment
7. Choke cam plate
8. Fast idle adjusting screw
9. Fast idle lever
10. Basic idle mixture screw
11. Fuel hose connection
12. Idle cut-off valve (where fitted)
13. Enrichment valve assembly

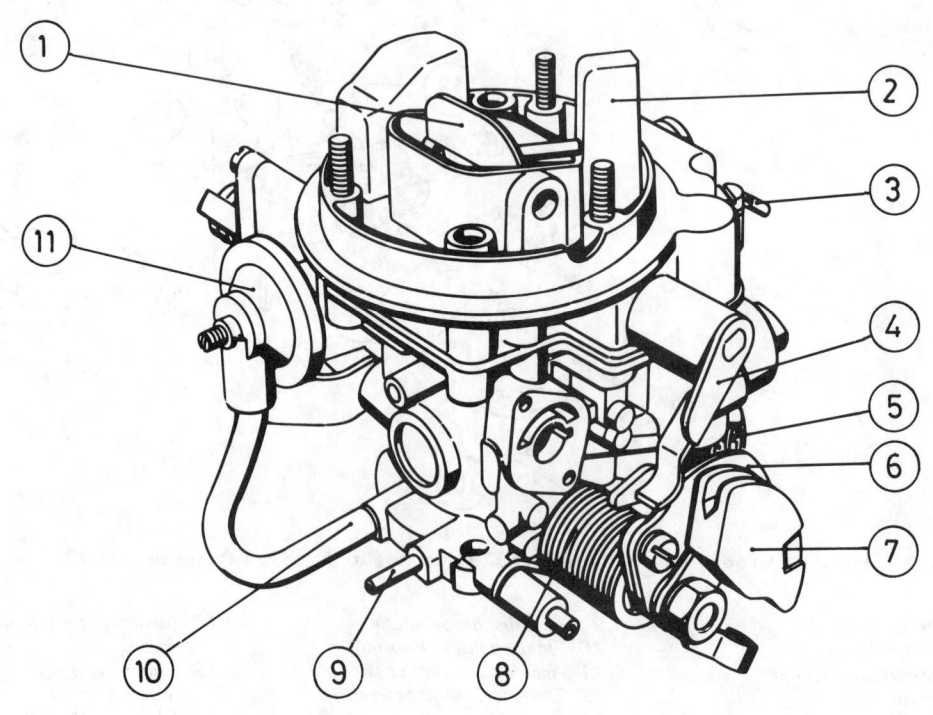

Fig. 3.6 Right side view of the Pierburg 1B1 carburettor (Sec 12)

1. Choke valve
2. Full load enrichment valve (where fitted)
3. Identification tag
4. Accelerator pump shaft
5. Auxiliary idle mixture screw
6. Accelerator pump cam plate
7. Throttle linkage
8. Linkage return spring
9. Air cleaner vacuum connection (where applicable)
10. Vacuum hose
11. Vacuum unit

13.6 Fuel inlet hose connection at the carburettor on 1.0 models

13.8A Carburettor retaining nuts (arrowed)

13.8B Removing the carburettor from the inlet manifold

14.3 Auxiliary idle mixture screw location (arrowed) on 1.0 models

14.7 Basic idle mixture screw location (arrowed) on 1.0 models

15.3 Using a drill bit to check the fast idle gap on 1.0 models

dismantled or renewed, the basic idle mixture setting should be checked as follows.

5 Connect an exhaust gas analyser (CO meter) to the car, in accordance with the manufacturer's instructions.

6 With the engine idling at the specified speed, check that the CO percentage is as given in the Specifications.

7 If adjustment is required, turn the basic idle mixture screw (photo), clockwise to weaken or anti-clockwise to richen the mixture until the specified setting is obtained. After adjustment, correct the idle speed if necessary, as previously described, switch off the engine and disconnect the instruments.

8 If it was found impossible to obtain the specified idle CO percentage during the previous operations, check the settings and adjustments described in Section 15, particularly the throttle valve basic setting.

15 Carburettor (1.0 models) – setting and adjustment of components

Note: *Under normal operating conditions only the carburettor idle adjustments described in Section 14 will need attention. Checking and adjustment of the following settings is not a routine operation and should only be necessary after carburettor overhaul or if the operation of the carburettor is suspect.*

Fast idle

1 This operation may be carried out with the carburettor installed or removed.

2 If the carburettor is removed, rotate the choke linkage on the side of the carburettor until the linkage arm is against its stop and the choke valve is fully closed.

3 With the linkage held in this position a small drill bit of diameter equal to the fast idle throttle valve gap given in the Specifications, should just slide between the throttle valve and the carburettor barrel (photo).

4 If adjustment is necessary slacken the locknut on the fast idle adjusting screw (photo) and turn the screw as necessary to achieve the specified setting. Tighten the locknut after adjustment.

5 If the carburettor is in the car first allow the engine to reach normal operating temperature and then if necessary adjust the idle speed, as described in Section 14. With the engine stopped, remove the air cleaner (Section 6), and also fully pull out the choke control knob, so that the choke linkage on the carburettor is against its stop.

6 Connect a tachometer to the engine in accordance with the equipment manufacturer's instructions.

7 Start the engine, then fully open the choke **valve** by pushing the linkage downwards against the pull-off spring tension. Leave the choke **lever** fully applied, so that the fast idle cam is opening the throttle valve. Compare the engine speed with the fast idle speed setting given in the Specifications. If adjustment is necessary, slacken

15.4 Fast idle adjusting screw location on 1.0 models (arrowed)

15.11A Using a drill bit to check the choke valve gap on 1.0 models

15.11B Choke valve gap adjusting screw on 1.0 models (arrowed)

15.21 Throttle valve stop screw (arrowed) on 1.0 models

Chapter 3 Fuel and exhaust systems

the locknut and turn the fast idle adjusting screw to achieve the specified speed. Tighten the locknut after adjustment.
8 Switch off the engine, and disconnect the tachometer.

Choke valve gap
9 Run the engine until normal operating temperature is reached and then switch off and remove the air cleaner, as described in Section 6.
10 Pull the choke knob fully out and check that the linkage rotates to the fully closed position with the linkage arm against its stop. If necessary adjust the choke cable, as described in Section 11.
11 With the choke knob still pulled out, start the engine and check that a drill bit of diameter equal to the choke valve gap dimension will just slide between the valve and choke barrel (photo). If necessary slacken the locknut and turn the adjusting screw above the vacuum unit until the correct gap is achieved (photo).
12 Switch off the engine, tighten the locknut and refit the air cleaner.

Throttle valve basic setting
13 Adjustment of the throttle valve will only be necessary if it was found impossible to obtain the correct idle mixture setting during the adjustments described in Section 14. For the following operations a tachometer, CO meter and vacuum gauge will be required.
14 Start the engine, allow it to reach normal operating temperature and then switch off.
15 Operate the choke control and ensure that it opens and closes fully without binding or sticking. Check that with the knob pushed fully in, the fast idle adjusting screw is not in contact with the cam on the linkage arm. Adjust the fast idle and choke valve gap, as described earlier in this Section, if the choke operation is not satisfactory.
16 Connect the tachometer and CO meter in accordance with the manufacturers' instructions. Connect the vacuum gauge to the distributor vacuum advance connection on the carburettor.
17 Start the engine and compare the readings on the instruments with the figures given in the Specifications.
18 If the instrument readings are not as specified proceed as follows:
19 Close the auxiliary idle mixture screw (photo 14.3) by screwing it fully in. Do not overtighten it, or the seating will be damaged.
20 Adjust the basic idle mixture screw (photo 14.7) to give a CO reading of 1 to 2%
21 Turn the throttle valve stop screw (photo) to give an idle speed of 550 to 650 rpm with a corresponding vacuum of 1 to 20 mbar (0.015 to 0.29 lbf/in^2).
22 Now adjust the auxiliary idle mixture screw and basic idle mixture screw, as described in Section 14, to obtain the idle speed and mixture settings as given in the Specifications.
23 After adjustment, switch off the engine and disconnect the instruments.

16 Carburettor (1.0 models) – overhaul

1 Remove the carburettor from the engine, as described in Section 13.
2 Clean the carburettor externally using a suitable cleaning solvent, or petrol in a well ventilated area. Wipe the carburettor dry with a lint-free cloth and prepare a clean uncluttered working area.
3 Disconnect the throttle return spring from the linkage and the support bracket on the side of the carburettor (photo).
4 Disconnect the vacuum unit hose from the outlet on the throttle valve housing (photo).
5 Undo the four retaining screws and separate the carburettor cover from the float chamber housing (photos).
6 At the base of the carburettor undo the single screw securing the throttle valve housing to the float chamber housing (photo). Separate the two housings.
7 Undo the screw securing the choke cable support bracket to the throttle valve housing and lift off the bracket. Undo the blanking plug and remove the seal ring from the housing (photo).
8 As a guide to refitting, count and record the number of turns necessary to screw the auxiliary idle mixture screw and the basic idle mixture screw fully into the housing. Now remove the two screws.
9 Undo the four screws and remove the accelerator pump cover, diaphragm, and spring from the float chamber housing (photo).
10 From the other side of the float chamber housing, undo the three screws and remove the enrichment valve cover, diaphragm and spring (photo).

16.3 Begin dismantling of the carburettor fitted to 1.0 models by disconnecting the throttle return spring

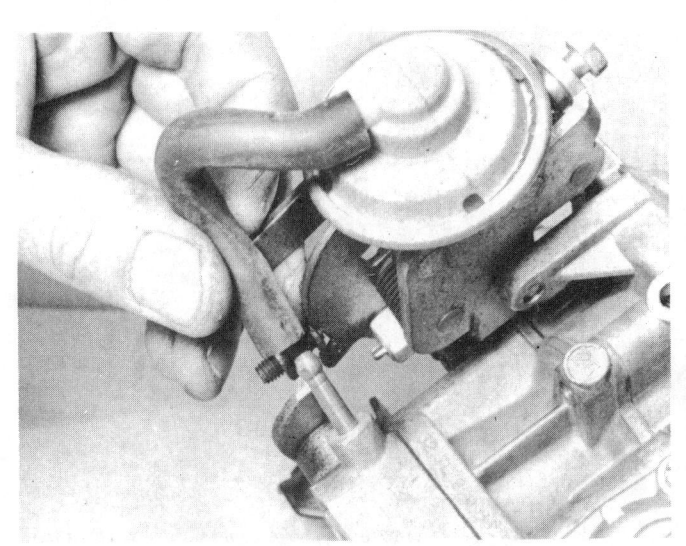

16.4 Disconnect the vacuum unit hose from the outlet

16.5A Undo the screws securing the carburettor cover to the float chamber housing ...

16.5B ... then separate the two housings

16.6 Undo the single screw and remove the throttle valve housing

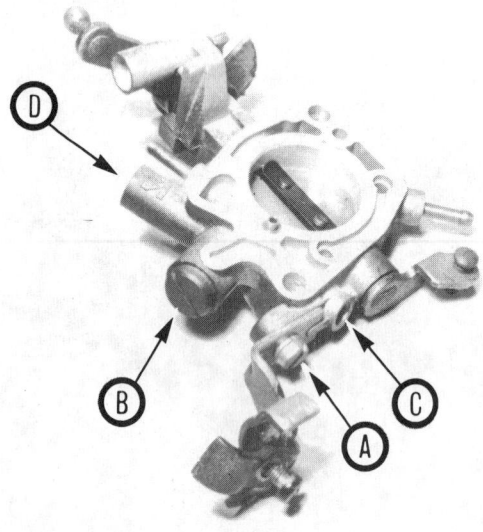

16.7 Undo the screw and remove the choke cable support bracket (A), undo the blanking plug and seal (B) then remove the auxiliary idle mixture screw (D) and basic idle mixture screw (C) after noting their set positions

16.9 Accelerator pump assembly removed from float chamber housing

16.10 Enrichment valve assembly removed from float chamber housing

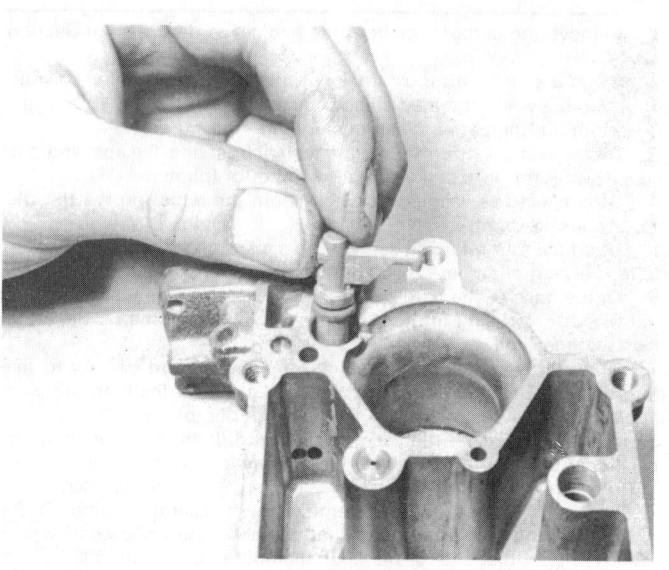

16.11 Withdraw the fuel discharge nozzle

77

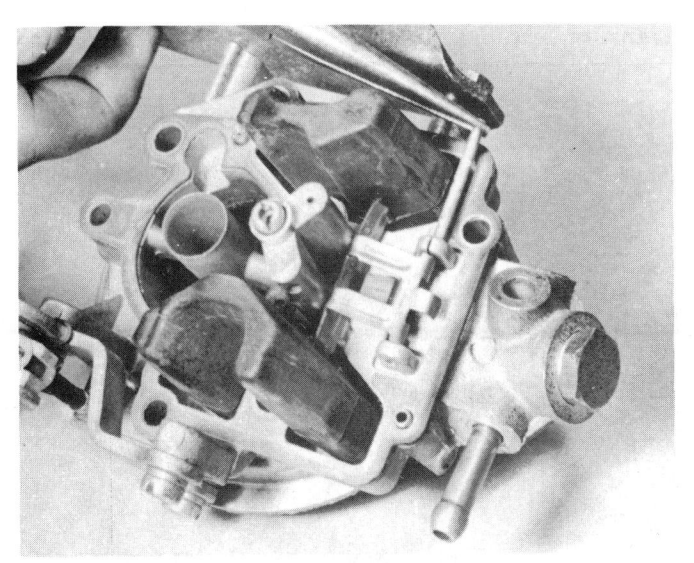

16.12 Extract the float pivot pin and withdraw the floats

16.13 Lift off the gasket

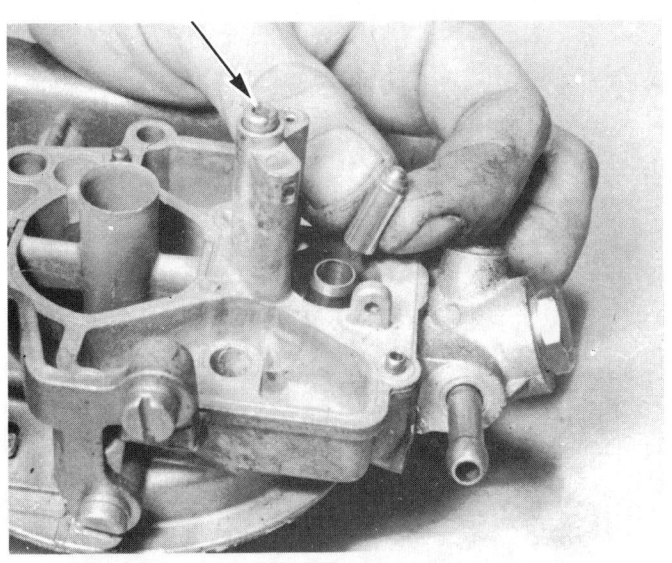

16.14 Remove the float needle valve (main jet arrowed)

16.15 Unscrew the screw plugs (A and B), the idle fuel/air jet (C), air correction jet (D) and the auxiliary fuel/air jet (E) from the carburettor cover

16.16 Withdraw the pre-atomizer

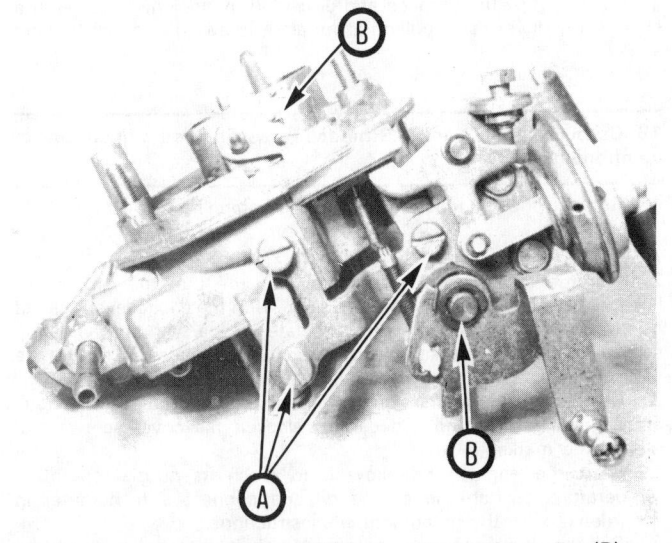

16.18 Unscrew the three screws (A), remove the retaining clips (B) and remove the vacuum unit and linkage assembly from the carburettor cover

Chapter 3 Fuel and exhaust systems

11 Carefully withdraw the fuel discharge nozzle from the housing (photo).
12 Tap the float pivot pin out of the pivot posts and withdraw the pin using long-nosed pliers (photo).
13 Lift out the float and then remove the gasket from the carburettor top cover (photo).
14 Lift out the float needle valve and then unscrew the main jet (photo).
15 Unscrew all the jets and plugs from the carburettor cover making a careful note of their locations (photo). Remove the mixture tube from the air correction jet bore.
16 Withdraw the pre-atomizer from the top cover venturi (photo).
17 Undo the retaining plug and withdraw the fuel filter adjacent to the inlet hose connection on the top cover.
18 If necessary the choke valve operating linkage and vacuum unit can be removed from the top cover. Undo the three retaining screws and the retaining clips for the operating cam and choke valve rod. Remove the cam and spring, disengage the operating rod from the cam and choke valve lever and withdraw the assembly (photo).
19 With the carburettor now dismantled, clean the components in petrol, in a well ventilated area. Allow the parts to air dry.
20 Blow out all the jets and the passages in the housings using compressed air or a tyre foot pump. Never probe with wire.
21 Examine the choke and throttle valve spindles and linkages for wear or excessive side-play. If wear is apparent in these areas it is advisable to obtain an exchange carburettor.
22 Check the diaphragms and renew them if they are punctured or show signs of deterioration.
23 Examine the float for signs of deterioration and shake it, listening for fuel inside. If so renew it, as it is leaking and will give an incorrect float level height causing flooding.
24 Blow through the float needle valve assembly while holding the needle valve closed, then open. Renew the valve if faulty, or as a matter of course if high mileages have been covered.
25 Obtain the new parts as necessary and also a carburettor repair kit which will contain a complete set of gaskets, washers and seals.
26 Reassemble the carburettor using the reverse of the dismantling procedure, but carry out the settings and adjustments described in Section 15 as the work progresses.
27 After refitting the carburettor to the car adjust the idle settings, as described in Section 14.

17 Carburettor (1.2 and 1.3 models) – idle speed and mixture adjustment

1 The procedure is identical to that described in Section 14 for the carburettor fitted to 1.0 models, but note the different locations of the adjustment screws (photos).
2 If the basic idle mixture screw is covered with a tamperproof cap (photo) this may be removed and discarded by hooking it out with a small screwdriver or by pulling it out after screwing in a self-tapping screw.

18 Carburettor (1.2 and 1.3 models) – setting and adjustment of components

1 Refer to the introductory note at the beginning of Section 15.

Fast idle

2 Pull the choke knob fully out and ensure that the carburettor linkage rotates to the fully closed position, with the choke lever against its stop and the notch on the cam plate in line with the adjusting screw. If necessary, slacken the screw on the cam plate and realign the plate (photo).
3 Check that the notches on the choke cover and carburettor body are aligned (photo) and if necessary slacken the cover screws and realign the marks.
4 Start the engine and allow it to reach its normal operating temperature. Switch the engine off and connect a tachometer in accordance with the manufacturer's instructions.
5 Pull the choke knob out and start the engine without touching the accelerator.
6 Compare the fast idle speed on the tachometer with the figure

17.1A Auxiliary idle mixture screw (A) ...

17.1B ... and basic idle mixture screw (B) on 1.2 and 1.3 models

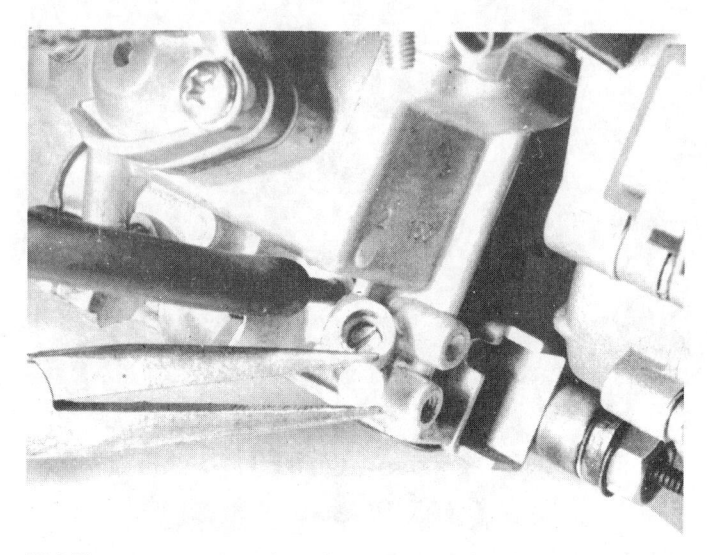

17.2 The tamperproof cap from the basic idle mixture screw

79

18.2 On 1.2 and 1.3 models the notch (B) must be in line with the adjusting screw when the choke lever (A) is against its stop. Slacken screw (C) to adjust

18.3 The notches on the choke cover and carburettor body (arrowed) on 1.2 and 1.3 models must be aligned

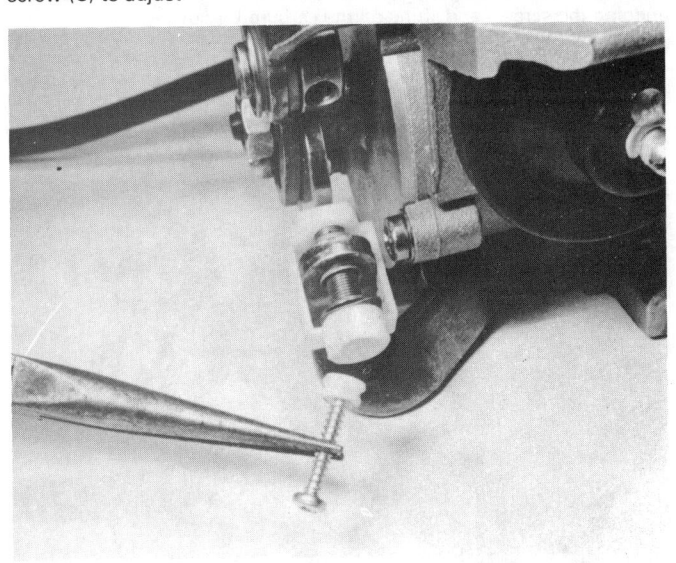

18.6 Remove the tamperproof cap over the fast idle adjusting screw on 1.2 and 1.3 models

18.10A Adjust the choke valve gap on 1.2 and 1.3 models at the adjusting screw (arrowed) ...

18.10B ... until a drill bit of appropriate diameter will slide in the gap with the engine running

Fig. 3.7 Throttle stop screw location (arrowed) – 1.2 and 1.3 models (Sec 18)

given in the Specifications. If necessary turn the fast idle adjusting screw until the specified setting is obtained. If the adjusting screw is covered with a tamperproof cap, screw in a self-tapping screw and withdraw the screw and cap (photo).
7 After adjustment switch off the engine and disconnect the tachometer.

Choke valve gap
8 Start the engine, allow it to reach normal operating temperature then switch it off and remove the air cleaner, as described in Section 6.
9 Pull the choke knob fully out and start the engine without touching the accelerator.
10 Turn the choke valve adjusting screw on the vacuum unit as necessary until a drill bit of diameter equal to the choke valve gap dimension given in the Specifications will just slide between the valve and choke barrel (photos).
11 After adjustment switch off the engine, seal the adjusting screw with a dab of paint and refit the air cleaner.

Throttle valve basic setting
12 Adjustment of the throttle valve will only be necessary if it was found impossible to obtain the correct idle mixture setting during the adjustments described in Section 17. For the following operations a tachometer, CO meter and vacuum gauge will be required.
13 Start the engine, allow it to reach normal operating temperature and then switch off.
14 Operate the choke control and ensure that it opens and closes fully without binding or sticking. Check that with the knob pushed fully in a small clearance exists between the fast idle adjusting screw and the cam plate. If necessary adjust the choke cable, as described in Section 11, and the fast idle setting, as described earlier in this Section.
15 Connect the tachometer and CO meter in accordance with the manufacturers' instructions. Connect the vacuum gauge to the distributor vacuum advance connection on the carburettor.
16 Start the engine and compare the readings on the instruments with the figures given in the Specifications.
17 If the instrument readings are not as specified, proceed as follows:
18 Close the auxiliary idle mixture screw (photo 17.1a) by screwing it fully in. Do not overtighten it, or the seating will be damaged.
19 Adjust the basic idle mixture screw (photo 17.1b) to give a CO reading of 1 to 2%.
20 Turn the throttle valve stop screw, after removing the tamperproof plug if fitted, to give an idle speed of 550 to 650 rpm with a corresponding vacuum of 1 to 20 mbar (0.015 to 0.29 lbf/in^2).
21 Now adjust the auxiliary idle mixture screw and basic idle mixture screw, as described in Section 17, to obtain the idle speed and mixture settings as given in the Specifications.
22 After adjustment switch off the engine and disconnect the instruments.

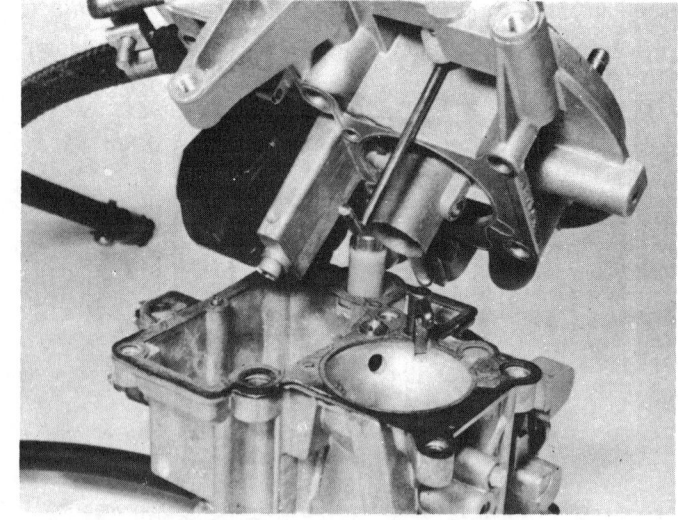

19.4 Begin dismantling the carburettor fitted to 1.2 and 1.3 models by undoing the screws and lifting off the carburettor cover

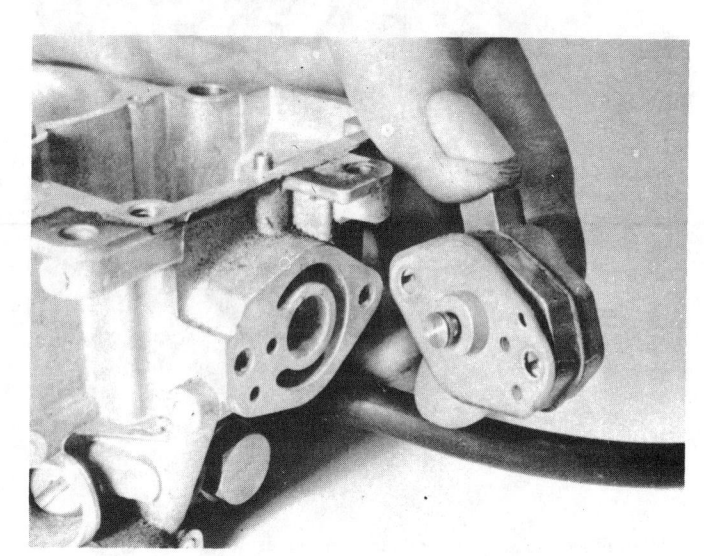

19.7 Remove the partial load enrichment valve assembly

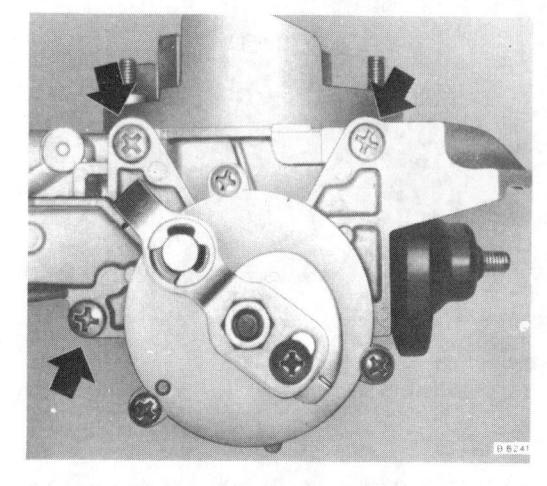

Fig. 3.8 Choke and vacuum housing retaining screws (arrowed) – 1.2 and 1.3

19 Carburettor (1.2 and 1.3 models) – overhaul

1 Remove the carburettor from the engine, as described in Section 13.
2 Clean the carburettor externally using a suitable cleaning solvent, or petrol in a well ventilated area. Wipe the carburettor dry with a lint-free cloth and prepare a clean uncluttered working area.
3 Disconnect the vacuum unit hose from the outlet on the carburettor main body.
4 Undo the four retaining screws and separate the carburettor cover from the main body (photo).
5 As a guide to refitting, count and record the number of turns necessary to screw the auxiliary idle mixture screw and the basic idle mixture screw fully into the main body. Now remove the two adjusting screws.
6 Undo the blanking plug and seal ring located under the partial load enrichment valve on the main body.
7 Undo the two screws and remove the partial load enrichment valve, diaphragm and spring from the main body (photo).
8 Remove the accelerator pump collar, piston and spring (photos).
9 Carefully withdraw the fuel discharge nozzle from the main body (photo).

Chapter 3 Fuel and exhaust systems

10 Tap the float pivot pin out of the pivot posts and lift out the float, followed by the float needle valve (photos).
11 Carefully unscrew the main jet (photo).
12 Unscrew the idle fuel/air jet and auxiliary fuel/air jet and remove them from the carburettor cover (photo).
13 Undo the three screws securing the choke and vacuum housing to the carburettor cover and withdraw the housing.
14 Note the notches on the choke cover plate and cover which must be aligned on reassembly, undo the three screws and remove the cover. As the cover is removed, note the position of the internal opening lever which must be to the left of the driving lever when reassembling.
15 With the carburettor now dismantled, clean the components in petrol in a well ventilated area and allow the parts to air dry. Carry out a careful inspection of the carburettor, as described in Section 16, paragraphs 20 to 25 inclusive.
16 Reassemble the carburettor using the reverse of the dismantling procedure, but carry out the setting and adjustments described in Section 18 during, and after, reassembly.
17 With the carburettor refitted, adjust the idle settings, as described in Section 17.

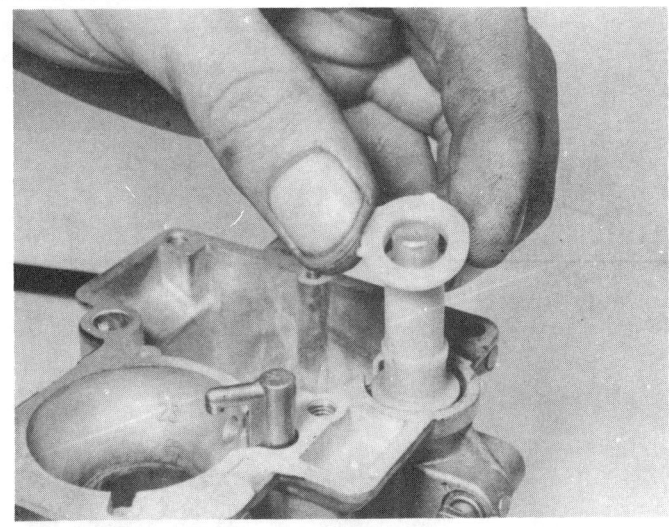

19.8A Withdraw the accelerator pump collar ...

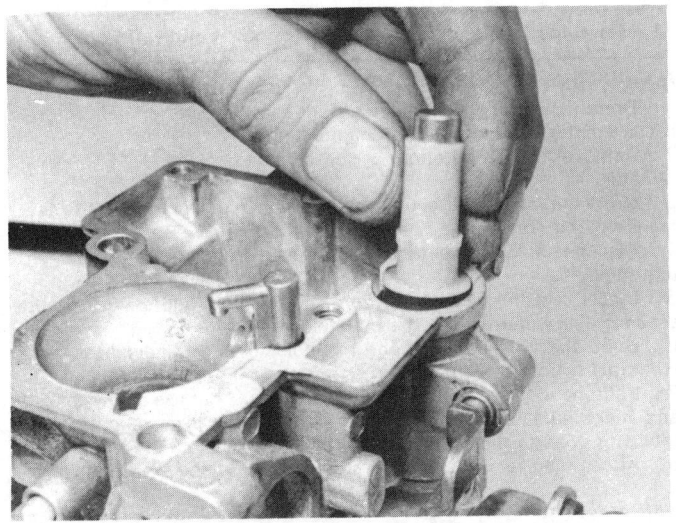

19.8B ... piston assembly ...

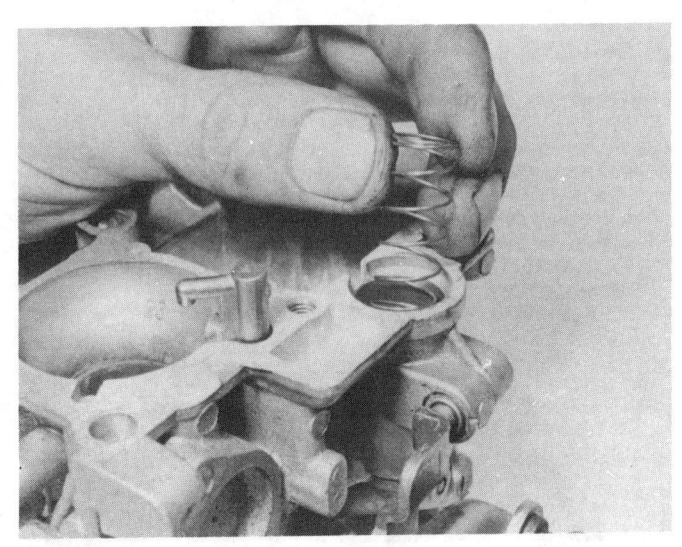

19.8C ... and spring

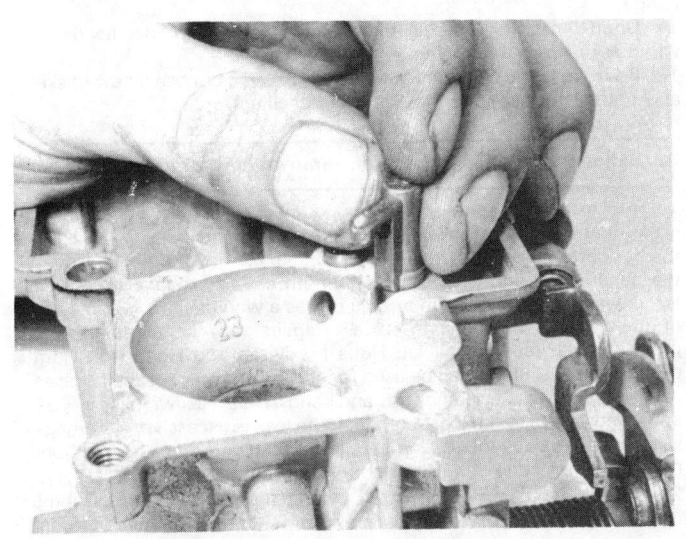

19.9 Carefully withdraw the fuel discharge nozzle

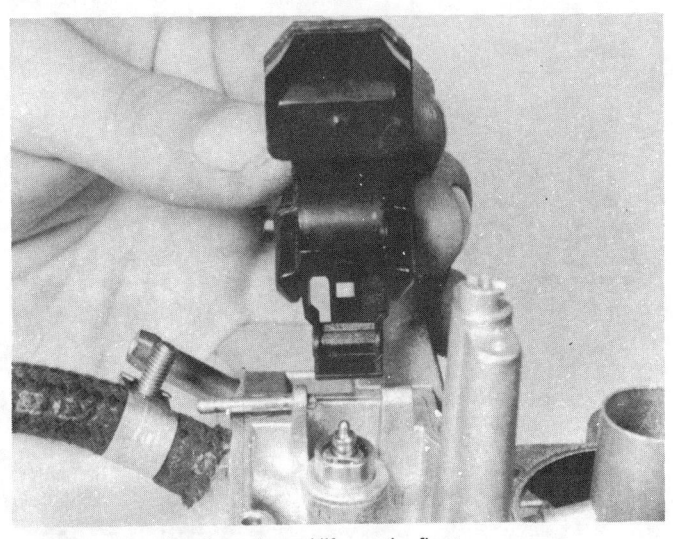

19.10A Tap out the pivot pin and lift out the float ...

19.10B ... followed by the float needle valve

19.11 Unscrew the main jet ...

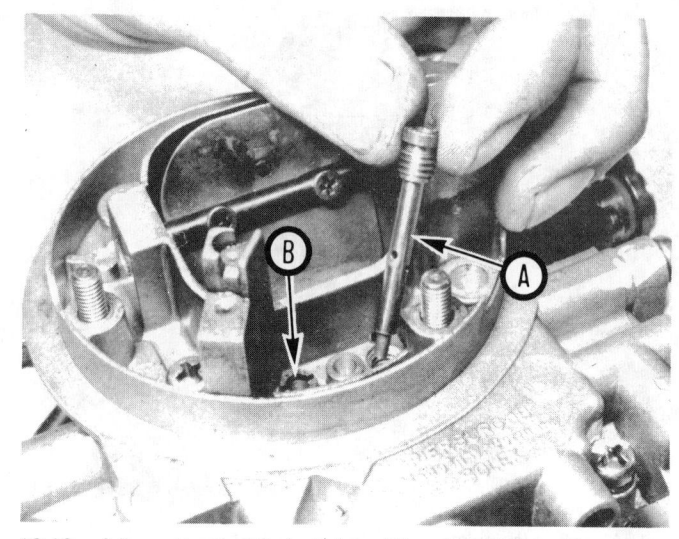

19.12 ... followed by the idle fuel/air jet (A) and auxiliary fuel/air jet (B)

20 Manifolds (1.0 models) – removal and refitting

Inlet manifold
1 Remove the carburettor, as described in Section 13.
2 Undo the union nut securing the brake servo vacuum hose to the manifold and withdraw the hose.
3 Using an Allen key undo the three bolts securing the manifold to the cylinder head and the additional bolt securing the support bracket to the manifold at the rear.
4 Remove the manifold from the cylinder head.
5 Refitting is the reverse sequence to removal, but use a new gasket and tighten the retaining bolts to the specified torque.

Exhaust manifold
6 Jack up the front of the car and support it on axle stands.
7 From under the car undo the two bolts and remove the tension springs, then separate the exhaust downpipe from the manifold flange.
8 Remove the air cleaner, as described in Section 6.
9 Undo the bolts securing the manifold to the cylinder head and remove it from the engine.
10 Remove the gasket and scrape off any traces of remaining gasket from the manifold and cylinder head faces.
11 Refitting is the reverse sequence to removal, but use a new gasket and tighten the retaining bolts to the specified torque.

21 Manifolds (1.2 and 1.3 models) – removal and refitting

Inlet manifold
1 Drain the cooling system, as described in Chapter 2, and remove the alternator, as described in Chapter 10.
2 Remove the carburettor, as described in Section 13 of this Chapter.
3 Undo the union nut securing the brake servo vacuum hose to the manifold and remove the hose.
4 Disconnect the temperature gauge lead from the transmitter on the manifold.
5 Disconnect the water hose from the manifold and undo the bolt securing the water pipe to the manifold.
6 Undo the bolts securing the manifold to the cylinder head and withdraw it from the engine.
7 Refitting is the reverse sequence to removal, but use a new gasket and make sure both mating faces are perfectly clean. Tighten the retaining bolts to the specified torque.
8 After fitting refill the cooling system, as described in Chapter 2.

Exhaust manifold
9 Remove the air cleaner, as described in Section 6.
10 Undo the bolts securing the exhaust downpipe to the manifold, remove the tension springs (where fitted) and separate the downpipe flange.
11 Undo the bolts securing the manifold to the cylinder head and withdraw it from the engine.
12 Refitting is the reverse sequence to removal, but use a new gasket and tighten the retaining bolts to the specified torque.

22 Exhaust system – inspection, removal and refitting

1 The exhaust system should be examined for leaks, damage and security at the intervals given in Routine Maintenance. To do this apply the handbrake and allow the engine to idle. Lie down on each side of the car in turn, and check the full length of the exhaust system for leaks while an assistant temporarily places a wad of cloth over the end of the tailpipe. If a leak is evident repairs may be made using a proprietary exhaust repair kit. Holts Flexiwrap and Holts Gun Gum exhaust repair systems can be used for effective repairs to exhaust pipes and silencer boxes, including ends and bends. Holts Flexiwrap is an MOT approved permanent exhaust repair. If the leak is excessive, or damage is evident the relevant section should be renewed. Check the rubber mountings for condition and security and renew them if necessary.
2 To remove the exhaust system, jack up the front and/or rear of the car and support it securely on axle stands. Alternatively drive the front or rear wheels up on ramps, over a pit, or on a hoist.
3 The system consists of three sections which can be individually removed. If the intermediate section is to be removed it will, however,

Chapter 3 Fuel and exhaust systems

be necessary to remove the front or rear section first. To remove the front or rear sections, unscrew the retaining nuts and remove the U-shaped retaining clamps (photo). At the front undo the bolts and remove the tension springs at the flexible joint or manifold flange (photo). Release the rubber mounting rings (photos) and twist the section free. If the joint is stubborn, liberally apply penetrating oil and leave it to soak. Tap the joint with a hammer and it should now be possible to twist it free. If necessary, carefully heat the joint with a blowlamp to assist removal, but shield the fuel tank, fuel lines and underbody adequately from heat.

4 Refitting is the reverse sequence to removal. Position the joints so that there is adequate clearance between all parts of the system and the underbody, and ensure that there is equal load on all the mountings. Tighten the flexible joint retaining bolts to compress the springs and, on 1.3 models, use a new gasket between the downpipe and manifold flange.

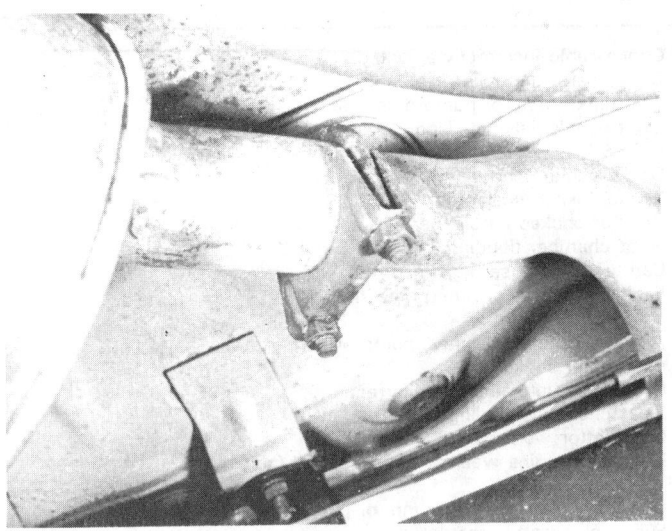

22.3A Exhaust section U-shaped retaining clamp ...

22.3B ... downpipe to manifold flexible joint (1.0 model shown) ...

22.3C ... tailpipe rubber mounting ring

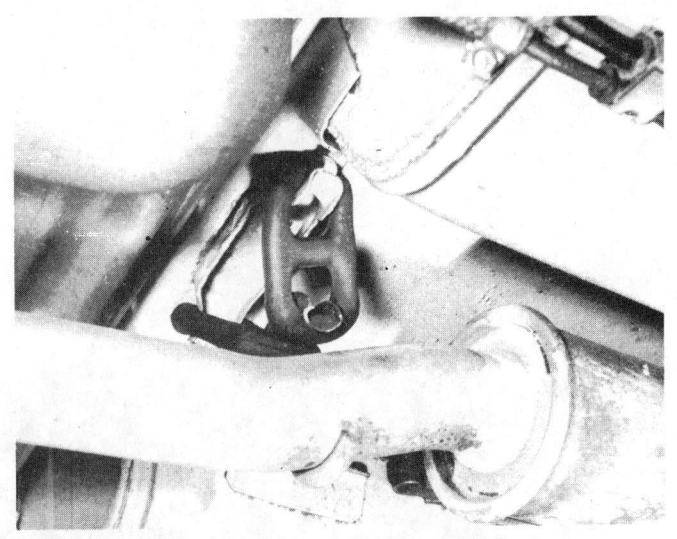

22.3D ... and intermediate section rubber mounting rings

Fault diagnosis overleaf

… Chapter 3 Fuel and exhaust systems

23 Fault diagnosis – fuel and exhaust systems

Unsatisfactory engine performance, bad starting and excessive fuel consumption are not necessarily the fault of the fuel system or carburettor. In fact they more commonly occur as a result of ignition system faults. Before acting on the following it is necessary to check the ignition system first. Even though a fault may lie in the fuel system, it will be difficult to trace unless the ignition system is correct. The faults below therefore assume that, where applicable, this has been attended to first.

Symptom	Reason(s)
Difficult starting when cold	Choke cable incorrectly adjusted Choke valve gap incorrect Fuel tank empty or pump defective Float needle valve sticking
Difficult starting when hot	Choke cable incorrectly adjusted Choke linkage sticking Dirty or choked air cleaner Float chamber flooding Carburettor idle settings incorrect Fuel tank empty or pump defective
Excessive fuel consumption	Leakage from tank, pipes, pump or carburettor Air cleaner choked Carburettor idle settings incorrect Float chamber flooding Carburettor worn Excessive engine wear or other internal fault
Fuel starvation	Leak or suction side of pump, or pump faulty Fuel tank breather restricted Fuel pipes or hoses restricted Incorrect float level Blocked carburettor jets Blocked fuel pump or carburettor filter (where applicable)
Poor performance, hesitation or erratic running	Carburettor idle settings incorrect Carburettor component settings incorrect Faulty carburettor accelerator pump or enrichment valve Air cleaner temperature control inoperative or incorrectly set Leaking carburettor or manifold gasket Blocked carburettor jets Fuel starvation

Chapter 4 Ignition system

For modifications, and information applicable to later models, see Supplement at end of manual

Contents

Coil – general 12	Distributor (1.2 and 1.3 models) – removal and refitting 8
Condenser (1.0 models) – removal and refitting 5	Fault diagnosis – conventional ignition system 14
Contact breaker points (1.0 models) – adjustment 3	Fault diagnosis – electronic breakerless ignition system 15
Contact breaker points (1.0 models) – removal and refitting 4	General description 1
Distributor (1.0 models) – dismantling, inspection and reassembly 7	Ignition timing – adjustment 10
Distributor (1.0 models) – removal and refitting 6	Maintenance and inspection 2
Distributor (1.2 and 1.3 models) – dismantling, inspection and reassembly 9	Spark plugs and HT leads – general 11
	T.D.C sensor – general 13

Specifications

General
System type:
- 1.0 models .... Conventional contact breaker and coil ignition
- 1.2 and 1.3 models .... Electronic breakerless ignition

Firing order .... 1-3-4-2 (No 1 cylinder at crankshaft pulley end of engine)

Distributor – 1.0 models
- Make .... Delco-Remy
- Type 3470269
- Direction of rotation .... Clockwise (viewed from cap)
- Contact breaker points gap .... 0.4 mm (0.016 in)
- Dwell angle .... 47° to 53°

Distributor – 1.2 and 1.3 models
- Make .... Delco-Remy
- Type:
  - 1.2 models 1111398
  - 1.3 models 1111395
- Direction of rotation .... Anti-clockwise (viewed from cap)

Ignition timing
- Timing .... 10° BTDC – engine idling
- Timing mark location .... Notch on crankshaft pulley aligned with mark or pointer on front of engine

Coil – 1.0 models
- Make .... Delco-Remy or Bosch
- Type:
  - Delco-Remy .... 12 VDR 502
  - Bosch .... KW 12
- Primary winding resistance .... 1.2 to 1.6 ohms
- Ignition voltage .... 12 000 to 14 000 volts

Coil – 1.2 and 1.3 models
- Make .... Delco-Remy
- Primary winding resistance .... 0.6 to 0.9 ohms
- Ignition voltage .... 16 000 to 18 000 volts

Spark plugs
Type:
- 1.0 models .... Champion RL82YCC or RL82YC
- 1.2 and 1.3 models .... Champion RN7YCC or RN7YC

Electrode gap:
- RL82YCC and RN7YCC .... 0.8 mm (0.032 in)
- RL82YC and RN7YC .... 0.7 mm (0.028 in)

Torque wrench settings

	Nm	lbf ft
Spark plugs:		
1.0 models	40	30
1.2 and 1.3 models	20	15

1 General description

In order that the engine can run correctly, it is necessary for an electrical spark to ignite the fuel/air mixture in the combustion chamber at exactly the right moment in relation to engine speed and load.

The ignition system is divided into two circuits, low tension and high tension. On 1.0 models with conventional ignition, the low tension circuit consists of the battery, ignition switch, primary coil windings and the contact breaker points and condenser, both located at the distributor. On 1.2 and 1.3 models with electronic ignition, an electronic module controlled by an induction sensor, permanent magnets and a pulse generator wheel performs the same function electronically as the mechanical contact breaker points in the conventional system. In both the conventional and electronic systems the high tension circuit consists of the secondary coil windings, the heavy ignition lead from the centre of the coil to the distributor cap, the rotor arm and the spark plugs and leads. The ignition system is based on feeding low tension voltage from the battery to the coil where it is converted to high tension voltage. The high tension voltage is powerful enough to jump the spark plug gap in the cylinders many times a second under high compression pressures, providing the system is in good order and all adjustments are correct.

The wiring harness on conventional systems includes a resistor wire in the coil feed circuit. During starting this resistor wire is bypassed allowing full battery voltage to be fed to the coil. This ensures that during cold starting when the starter motor current consumption would be high, sufficient voltage is still available at the coil to produce a powerful spark. Under normal running, battery voltage is directed through the resistor cable before reaching the coil.

The ignition advance is controlled both mechanically and by vacuum, to ensure that the spark occurs at just the right instant for the particular engine load and speed. The mechanical governor comprises two weights, which move out from the distributor shaft as the engine speed rises due to centrifugal force. The vacuum control consists of a diaphragm, one side of which is connected via a small bore tube to the carburettor, and the other side to the distributor baseplate. Depression in the inlet manifold and carburettor, which varies with engine speed and throttle opening, causes the diaphragm to move, so moving the baseplate and advancing or retarding the spark.

Warning: The voltages produced by the electronic ignition system are considerably higher than those produced by the conventional system. Extreme care must be used when working on the system with the ignition switched on, particularly by persons fitted with a cardiac pacemaker.

2 Maintenance and inspection

1 Every 9000 miles (15 000 km) or 6 months, whichever comes first, the following service operations should be carried out to the components of the ignition system.
2 Renew the spark plugs. At the same time check the spark plug HT leads for any signs of corrosion of the end fittings which if evident should be carefully cleaned away. Wipe clean the plug leads over their entire length before refitting.
3 On 1.0 models check the condition of the contact breaker points, as described in Section 3, and if unsatisfactory renew them, as described in Section 4. After resetting the contact breaker points gap, refer to Section 10 and adjust the ignition timing. Although not a specific service requirement on electronic ignition systems, it is desirable to check the ignition timing on 1.2 and 1.3 models at this time also.
4 Check the condition and security of all leads and wiring associated with the ignition system. Make sure that no chafing is occurring on any of the wires and that all connections are secure, clean and free of corrosion.

3 Contact breaker points (1.0 models) – adjustment

1 To adjust the contact breaker points so that the correct gap is obtained, first undo the two distributor cap retaining screws, lift off the cap and withdraw the rotor arm from the distributor shaft. At this stage it is a good idea to clean the inside and outside of the cap and inspect its condition. It is unlikely that the four segments inside the cap will be badly burned or corroded, but if they are the cap must be renewed. If only a small deposit is on the segments, it may be scraped away using a small screwdriver.
2 Push in the carbon brush located in the centre of the cap several times to ensure that it moves freely. The brush should protrude by at least 6.3 mm ($\frac{1}{4}$ in).
3 Gently prise the contact breaker points open to examine the condition of their faces. If they are rough, pitted or dirty it will be necessary to remove them to enable new points to be fitted.
4 Assuming that the points are in a satisfactory condition, or that they have been renewed, the gap between the two faces should be measured using feeler gauges as follows:
5 Pull off the plug leads, after marking them to ensure correct refitment, and then remove the spark plugs.
6 With the transmission in gear and the handbrake released, slowly pull the car forward, while at the same time watching the distributor, until the heel of the contact breaker arm is on the peak of one of the four cam lobes. A feeler blade equal to the contact breaker points gap, as given in the Specifications, should now just fit between the contact faces (photo).
7 If the gap varies from this amount slacken the contact breaker plate retaining screw and move the breaker plate in or out to achieve the desired gap. The plate can be easily moved with a screwdriver inserted between the notch in the breaker plate and the two adjacent pips in the distributor baseplate.
8 When the gap is correct, tighten the retaining screw and then recheck the gap. Lightly smear the surfaces of the cam with high melting point grease. Do not over-lubricate as any excess could get onto the surfaces of the points and cause ignition malfunction.
9 Refit the rotor arm, distributor cap, spark plugs and leads.
10 If a dwell meter is available, a far more accurate method of setting the contact breaker points gap is by measuring and setting the distributor dwell angle.
11 The dwell angle is the number of degrees of distributor cam rotation during which the contact breaker points are closed, ie the period from when the points close after being opened by one cam lobe until they are opened again by the next cam lobe. The advantages of setting the points by this method are that any wear of the distributor shaft or cam lobes is taken into account, and also the inaccuracies of using a feeler gauge are eliminated. In general, a dwell meter should be used in accordance with the manufacturer's instructions. However, the use of one type of meter is outlined as follows:
12 To set the dwell angle, remove the distributor cap and rotor arm and connect one lead of the dwell meter to the '+' terminal (15) on the coil and the other lead to the '-' terminal (1) on the coil.
13 Whilst an assistant turns on the ignition and operates the starter,

Chapter 4 Ignition system

3.6 Checking the contact breaker points gap using a feeler gauge

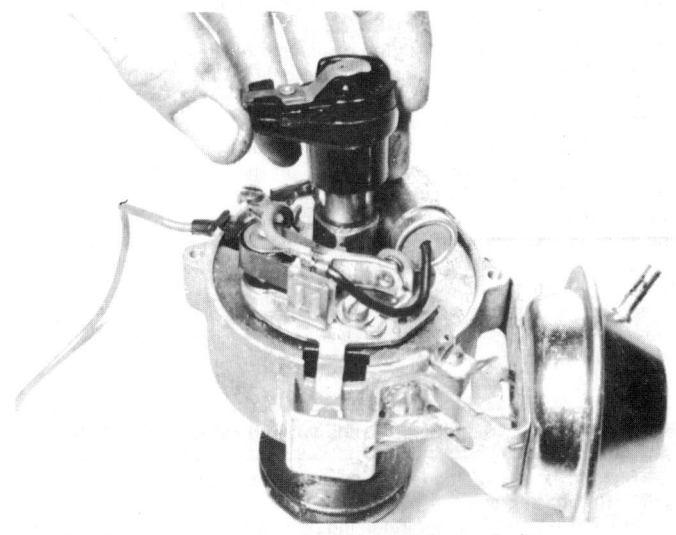

4.2 Withdrawing the rotor arm from the distributor shaft

4.3 Slip the low tension and condenser leads out of the insulator

4.4 The contact breaker points can then be removed after undoing the retaining screw

observe the reading on the dwell meter scale. With the engine turning over on the starter the reading should be as stated in the Specification.
Note: *Fluctuation of the dwell meter needle indicates that the engine is not turning over fast enough to give a steady reading. If this is the case, remove the spark plugs and repeat the checks.*
14 If the dwell angle is too small, the contact breaker point gap is too wide, and if the dwell angle is excessive the gap is too small.
15 Adjust the contact breaker points gap, while the engine is cranking using the method described in paragraph 7, until the correct dwell angle is obtained.
16 When the dwell angle is satisfactory, disconnect the meter and refit the rotor arm, distributor cap and, if removed, the spark plugs and leads.

4 Contact breaker points (1.0 models) – removal and refitting

1 The contact breaker points should be checked and renewed if necessary at the service intervals given in Section 2 or sooner if they become burned, pitted or badly worn.
2 Undo the two distributor cap retaining screws, lift off the cap and withdraw the rotor arm from the distributor shaft (photo).
3 Move the contact breaker arm spring blade away from the plastic insulator and slip the low tension and condenser lead terminals off the insulator (photo).
4 Undo the retaining screw securing the contact breaker plate to the distributor baseplate (photo) and lift off the contact set.
5 Locate the new contact set on the baseplate and refit the retaining screw.
6 Move the contact breaker spring blade away from the insulator, fit the low tension and condenser leads and allow the spring blade to slip back into place. Make sure that the leads and the blade locate squarely in the insulator.
7 Check and adjust the contact breaker points gap or dwell angle, as described in Section 3, then refit the rotor arm and distributor cap.

5 Condenser (1.0 models) – removal and refitting

1 The purpose of the condenser (sometimes known as a capacitor) is to prevent excessive arcing of the contact breaker points, and to ensure that a rapid collapse of the magnetic field, created in the coil, and necessary if a healthy spark is to be produced at the plugs, is allowed to occur.
2 The condenser is fitted in parallel with the contact breaker points. If it becomes faulty it will cause ignition failure, as the points will be prevented from cleanly interrupting the low tension circuit.
3 If the engine becomes very difficult to start, or begins to miss after several miles of running, and the contact breaker points show signs of excessive burning, then the condition of the condenser must be

suspect. A further test can be made by separating the points by hand, with the ignition switched on. If this is accomplished by an excessively strong flash, it indicates that the condenser has failed.

4 Without special test equipment, the only reliable way to diagnose condenser trouble is to renew the suspect unit and note if there is any improvement in performance.

5 To remove the condenser from its location in the distributor, undo the distributor cap retaining screws, lift off the cap and withdraw the rotor arm from the distributor shaft.

6 Move the contact breaker arm spring blade away from the plastic insulator and withdraw the condenser lead.

7 Undo the screw securing the condenser to the distributor base plate and lift the condenser off.

8 Refitting is the reverse sequence to removal, but make sure that the condenser and low tension leads are securely located in the insulator behind the contact breaker arm spring blade.

6 Distributor (1.0 models) – removal and refitting

1 Pull off the spark plug leads, after marking them to ensure correct refitment, and remove the spark plugs.

2 Undo the distributor cap retaining screws, lift off the cap and place it to one side.

3 With the transmission in gear and the handbrake released, pull the car forward until, with a finger over the plug hole, compression can be felt in No 1 cylinder (the cylinder nearest the crankshaft pulley). Continue moving the car forwards until the notch on the crankshaft pulley is in line with the raised mark on the timing cover (photo). The distributor rotor arm should now be pointing to the notch on the rim of the distributor body.

4 Disconnect the distributor low tension lead at the harness connector and detach the vacuum advance pipe from the distributor vacuum unit.

5 Undo the distributor clamp retaining bolt, lift off the clamp plate and withdraw the distributor from its location.

6 Before refitting the distributor check that the engine has not been inadvertently turned whilst the distributor was removed; if it has, return it to the original position, as described in paragraph 3.

7 As the distributor is refitted, the distributor shaft will rotate anti-clockwise slightly due to the meshing action of the skew gears on the distributor shaft and camshaft. To ensure that the distributor shaft is in the correct position after fitting ie with the rotor arm pointing towards the notch in the rim of the distributor body, set the shaft so that the rotor arm is pointing towards the low tension lead grommet in the side of the distributor body (photo), prior to fitting. As the skew gears mesh, the shaft will turn back to the correct position.

8 It is also necessary to position the oil pump driveshaft so that it engages with the slot in the distributor shaft as the distributor is inserted. The shaft should be positioned so that it is at approximately 90° to the crankshaft centreline (photo).

9 Make sure that the O-ring seal is in position at the base of the distributor and, with the shafts set as previously described, insert the distributor into its location. It may take two or three attempts to engage the oil pump driveshaft, and finish with the rotor arm pointing to the notch. If necessary move the distributor shaft very slightly one way or the other, until the correct position is achieved.

10 With the distributor in place, turn the distributor body clockwise a few degrees so that the contact breaker points are closed, and then slowly turn it anti-clockwise until they just open with the rotor arm once more pointing towards the notch in the distributor body rim. Hold the distributor in this position and refit the clamp plate and clamp bolt. Tighten the bolt securely.

11 Reconnect the low tension lead and the vacuum advance pipe. Refit the spark plugs, distributor cap and leads.

12 Refer to Section 10 and adjust the ignition timing.

7 Distributor (1.0 models) – dismantling, inspection and reassembly

1 Remove the distributor from the engine, as described in the previous Section, and then prepare a clean uncluttered working area.

2 Remove the rotor arm, ease the contact breaker arm spring blade away from the plastic insulator and slip the low tension and condenser leads off the insulator.

6.3 Timing marks on 1.0 models in alignment

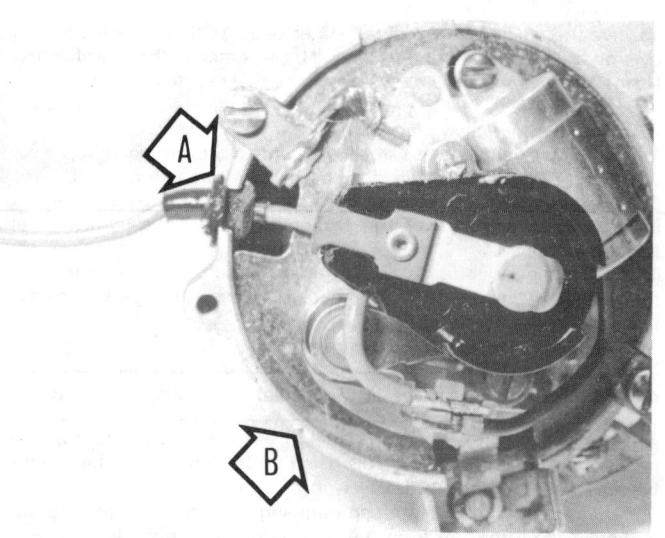

6.7 When refitting the distributor on 1.0 models the rotor arm should point toward the LT lead grommet (A). The rotor arm will move back to the notch (B) on the distributor rim as it is fitted

6.8 Correct position of oil pump driveshaft on 1.0 models prior to fitting distributor

Chapter 4 Ignition system

3 Withdraw the low tension lead grommet from the slot in the side of the distributor body (photo) and remove the lead.
4 Undo the retaining screw and lift off the contact set.
5 Undo the retaining screw and lift off the condenser (photo).
6 On the side of the distributor body, undo the two vacuum unit securing screws (photo). Withdraw the vacuum unit and at the same time disengage the operating arm from the peg on the side of the baseplate (photo).
7 Undo the two baseplate securing screws noting the earth tag under one screw and the spade terminal under the other (photo). Withdraw the baseplate assembly from the distributor body (photo).
8 This is the practical limit of dismantling of these distributors, as none of the components below the baseplate are renewable as separate items. If, however, it is necessary to remove the centrifugal advance springs and weights for cleaning or inspection, this can now be done. Mark each spring and its associated locating post with a dab of paint for identification and then carefully hook off the springs. Ensure that the springs and weights are refitted in the same positions otherwise the ignition advance characteristics of the engine will be altered. The weights can be withdrawn after extracting the small retaining clips (photo).
9 With the distributor dismantled, clean the mechanical components in paraffin and dry with a lint-free cloth.
10 Check the condition of the contact breaker points, as described in Section 3. Check the distributor cap for corrosion of the segments and for signs of tracking, indicated by a thin black line between the segments. Make sure that the carbon brush in the centre of the cap moves freely and stands proud by at least 6.3 mm ($\frac{1}{4}$ in). Renew the cap if necessary.
11 If the metal portion of the rotor arm is badly burned or loose renew the arm. If slightly burnt, clean the arm with a fine file.
12 Check that the plates of the baseplate assembly move freely, but without excessive slackness. If defective the baseplate assembly must be renewed.
13 Suck on the end of the vacuum unit outlet and check that the operating arm moves in as the suction is applied. Release the suction and check that the arm returns to its original position. If this is not the case, renew the vacuum unit.
14 Inspect the distributor body and shaft assembly for excessive side movement of the shaft in its bushes. With the advance weights and springs in position, hold the skew gear at the base of the shaft with one hand, and with the other hand turn the upper shaft clockwise as far as it will go and then release it. Check that as this is done the advance weights move out and then return under the action of the springs. Finally check the drivegear for wear, chips or pitting of the teeth. It will be necessary to renew the complete distributor if the body, shafts, weights, springs or drivegear are worn or in any way unsatisfactory.
15 Reassembly of the distributor is a direct reversal of the removal sequence, but apply a few drops of engine oil to the locating pivot posts of the advance weights and to the felt pad at the top of the distributor shaft. After reassembly adjust the contact breaker points, as described in Section 3, and then refit the distributor to the car, as described in Section 6.

8 Distributor (1.2 and 1.3 models) – removal and refitting

1 Pull off the spark plug leads, after marking them to ensure correct refitment and then remove the spark plugs, referring to Section 11 if necessary.
2 Slide up the distributor cap cover and undo the cap retaining screws. Lift off the distributor cap and place it to one side.
3 With the transmission in gear and the handbrake released, pull the car forward until, with a finger over the plug hole, compression can be felt in No 1 cylinder (the cylinder nearest the crankshaft pulley). Continue moving the car forward until the notch on the crankshaft pulley is aligned with the timing pointer (photo). If the distributor cap is temporarily placed in position, the distributor rotor contact should be pointing towards the No 1 spark plug lead segment in the cap.
4 Disconnect the distributor wiring connector at the ignition coil (photo) and detach the vacuum advance pipe from the distributor vacuum unit.
5 Undo the distributor clamp retaining bolt, lift off the clamp plate and withdraw the distributor from the camshaft housing (photo).
6 Before refitting the distributor check that the engine has not been

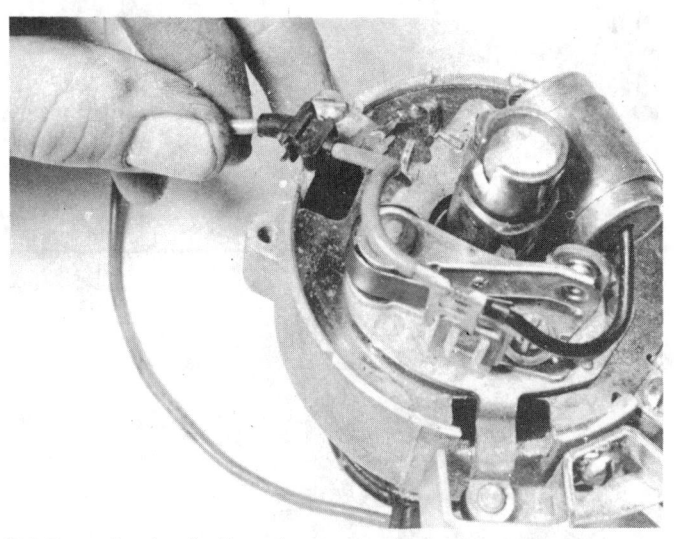

7.3 Removing the distributor low tension lead

inadvertently turned whilst the distributor was removed; if it has, return it to the original position, as described in paragraph 3.
7 Position the distributor rotor so that the rotor contact is in line with the arrow or notch on the distributor body (photo). In this position the offset lug on the distributor drive coupling will be in the correct position to engage the similarly offset slot in the end of the camshaft (photos).
8 Check that the O-ring seal is in place on the distributor body and then insert the distributor into its camshaft housing location. With the rotor contact and arrow on the distributor body still in line, refit the distributor clamp and clamp bolt. Tighten the clamp bolt securely.
9 Refit the distributor cap and cap cover, the spark plugs and plug leads. Reconnect the wiring plug and refit the vacuum advance pipe.
10 Refer to Section 10 and adjust the ignition timing.

9 Distributor (1.2 and 1.3 models) – dismantling, inspection and reassembly

1 Remove the distributor from the engine, as described in the previous Section.
2 Undo the two retaining screws and lift off the rotor (photo).
3 Disconnect the two electrical plugs, one at each end, from the ignition module (photo).
4 Undo the two module retaining screws (photo) and withdraw the unit from the distributor.
5 Undo the two vacuum unit retaining screws (photo), disengage the operating rod and remove the vacuum unit.
6 Due to its design and construction, this is the limit of dismantling possible on the electronic ignition distributor. It is possible to renew the rotor, vacuum unit, ignition module and distributor cap separately, but if inspection shows any of the components remaining on the distributor to be in need of attention the complete distributor assembly must be renewed.
7 Check the distributor cap for corrosion of the segments and for signs of tracking, indicated by a thin black line between the segments. Make sure that the carbon brush in the centre of the cap moves freely and stands proud by at least 6.3 mm ($\frac{1}{4}$ in). Renew the cap if necessary.
8 If the metal portion of the rotor is badly burned or loose, renew the rotor. If slightly burnt it may be cleaned with a fine file.
9 Suck on the end of the vacuum unit outlet and check that the operating rod moves in as the suction is applied. Release the suction and check that the rod returns to its original position. If this is not the case, renew the vacuum unit.
10 Inspect the distributor body and shaft assembly for excessive side movement of the shaft in its bushes. Check that the advance weights are free to move on their pivot posts and that they return under the action of the springs. Check the security of all the components on the distributor shaft and finally check for wear of the lug on the drive coupling.

7.5 The condenser can be removed after undoing the retaining screw

7.6A Undo the vacuum unit retaining screws (arrowed) ...

7.6B ... then withdraw the vacuum unit while at the same time disengaging the operating arm (arrowed)

7.7A Undo the baseplate securing screws (arrowed) ...

7.7B ... and withdraw the baseplate

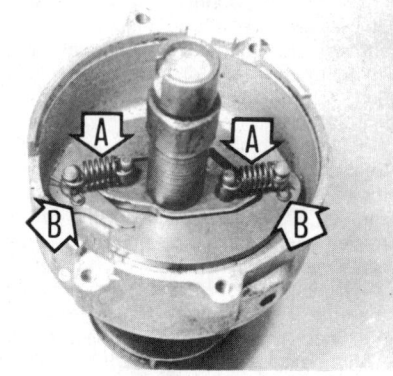

7.8 Centrifugal advance springs (A) and advance weight retaining clips (B)

8.3 Timing marks on 1.2 and 1.3 models

8.4 Disconnecting the distributor wiring connector at the coil

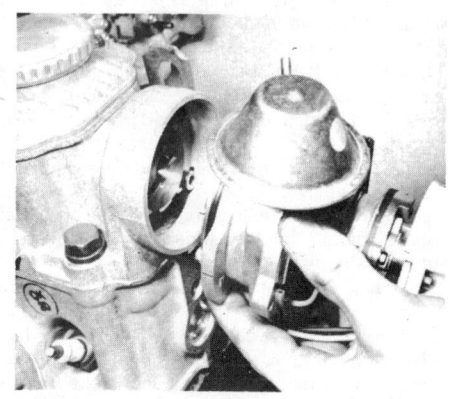

8.5 Removing the distributor from the camshaft housing

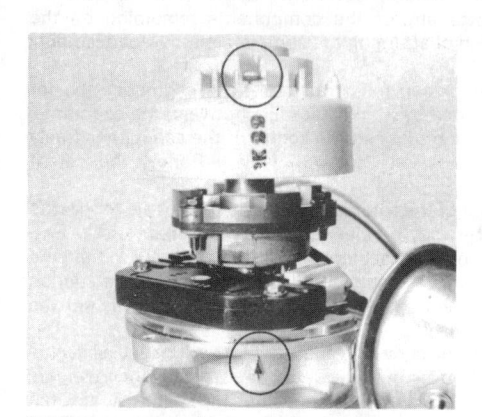

8.7A Correct positioning of distributor rotor contact in line with arrow on distributor body ...

8.7B ... which will enable the offset lug on the drive coupling ...

8.7C ... to engage the similarly offset slot in the camshaft

Chapter 4 Ignition system

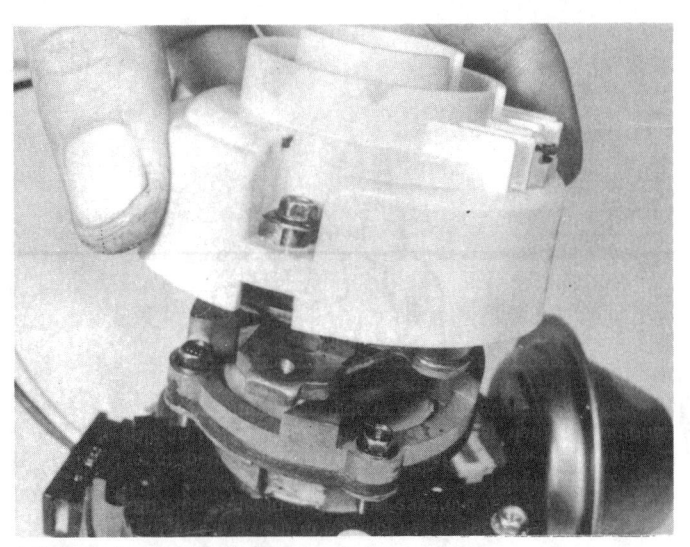

9.2 Removing the distributor rotor

9.3 Disconnect the ignition module electrical plugs ...

9.4 ... then undo the two screws (arrowed) to release the module

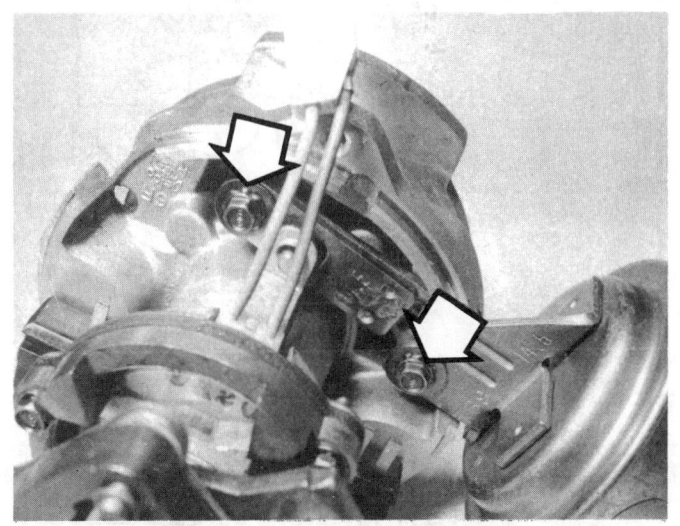

9.5 Distributor vacuum unit retaining screws (arrowed)

11 Reassembly of the distributor is the reverse sequence to dismantling, but apply a few drops of engine oil to the advance weight pivot posts before refitting the rotor. If a new ignition module is being refitted the new module will be supplied with a small quantity of silicone grease. This should be applied between the module and its housing to improve heat dissipation.

12 Refit the distributor, as described in Section 8, after reassembly.

10 Ignition timing – adjustment

1 For prolonged engine life, efficient running performance and economy it is essential for the fuel/air mixture in the combustion chambers to be ignited by the spark plugs at precisely the right moment in relation to engine speed and load. For this to occur the ignition timing must be set accurately and should be checked at the intervals given in Section 2 or whenever the position of the distributor has been altered. To make an accurate check of the ignition timing it is necessary to use a stroboscopic timing light, whereby the timing is checked with the engine running at idling speed.

2 If the distributor has been removed, or if for any reason its position on the engine has been altered, obtain an initial setting to enable the engine to be run, as described in Section 6 for 1.0 models or Section 8 for 1.2 and 1.3 models. Also, on 1.0 models, make sure that the contact breaker points gap or dwell angle are correctly set, as described in Section 3.

3 To check the timing, first highlight the timing marks using white chalk or quick-drying paint. On 1.0 models the marks are a notch on the crankshaft pulley and a raised mark on the timing cover (photo 6.3). On 1.2 and 1.3 models the marks consist of a notch on the crankshaft pulley and a pointer on the oil pump housing (photo 8.3). In both cases the engine is at the specified number of degrees BTDC when the marks are aligned, **not** TDC.

4 Run the engine up to its normal operating temperature and then switch off. Disconnect the vacuum pipe from the distributor vacuum unit.

5 Connect a timing light to the spark plug lead of No 1 cylinder – following the manufacturer's instructions.

6 With the engine idling, point the timing light at the timing marks. The marks will appear stationary, and if the timing is correct, they will be aligned.

7 If the marks are not aligned, slacken the distributor clamp retaining bolt and move the distributor body slowly, one way or the other until the marks line up. Tighten the clamp bolt and check that the setting has not altered.

8 Open the throttle slightly and note the movement of the timing marks. If the centrifugal advance in the distributor is working correctly the marks should appear to move away from each other as the engine speed increases. The same should happen if suction is applied to the vacuum advance pipe after disconnecting it from the carburettor, indicating that the distributor vacuum unit is satisfactory.

9 After checking the timing, switch off the engine and disconnect the timing light. If removed, refit the vacuum pipe to the carburettor.

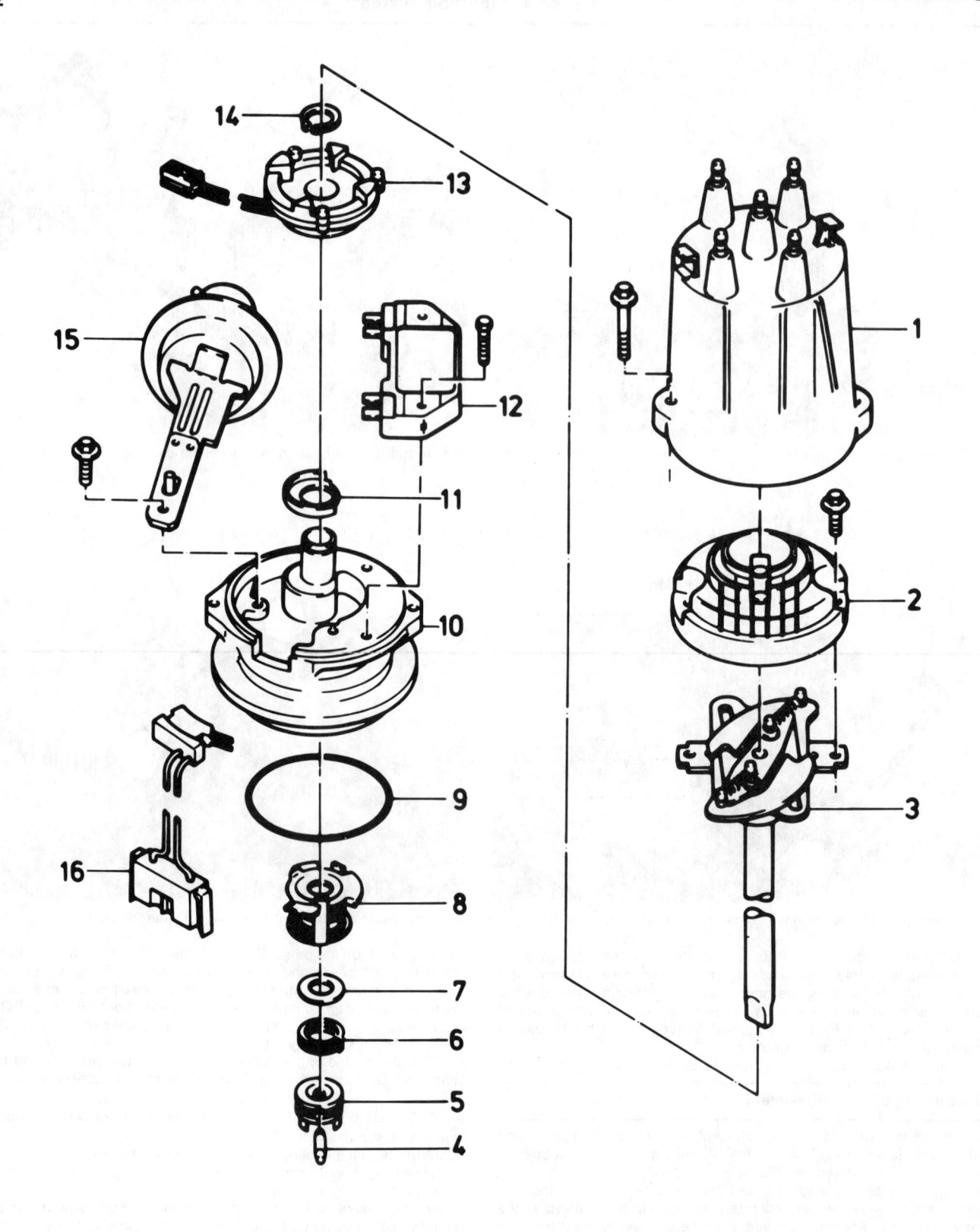

Fig. 4.1 Exploded view of the electronic breakerless distributor (Sec 9)

1 Distributor cap
2 Distributor rotor
3 Distributor shaft
4 Coupling pin
5 Drive coupling
6 Spring
7 Washer
8 Tension spring
9 O-ring seal
10 Distributor body
11 Seal
12 Ignition module
13 Induction sensor
14 Retaining ring
15 Vacuum unit
16 Electrical plug

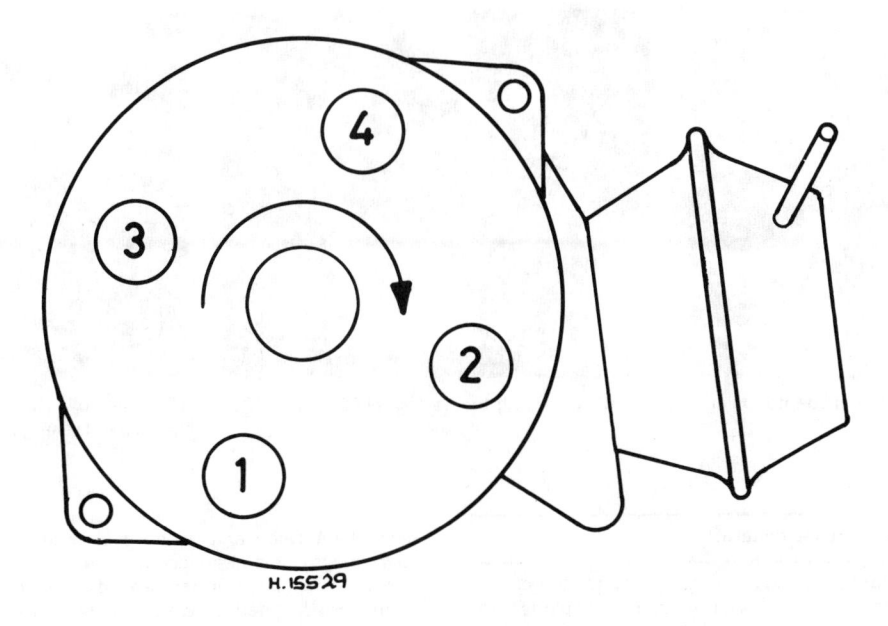

Fig. 4.2 Spark plug HT lead connecting diagram – 1.0 model (Sec 11)

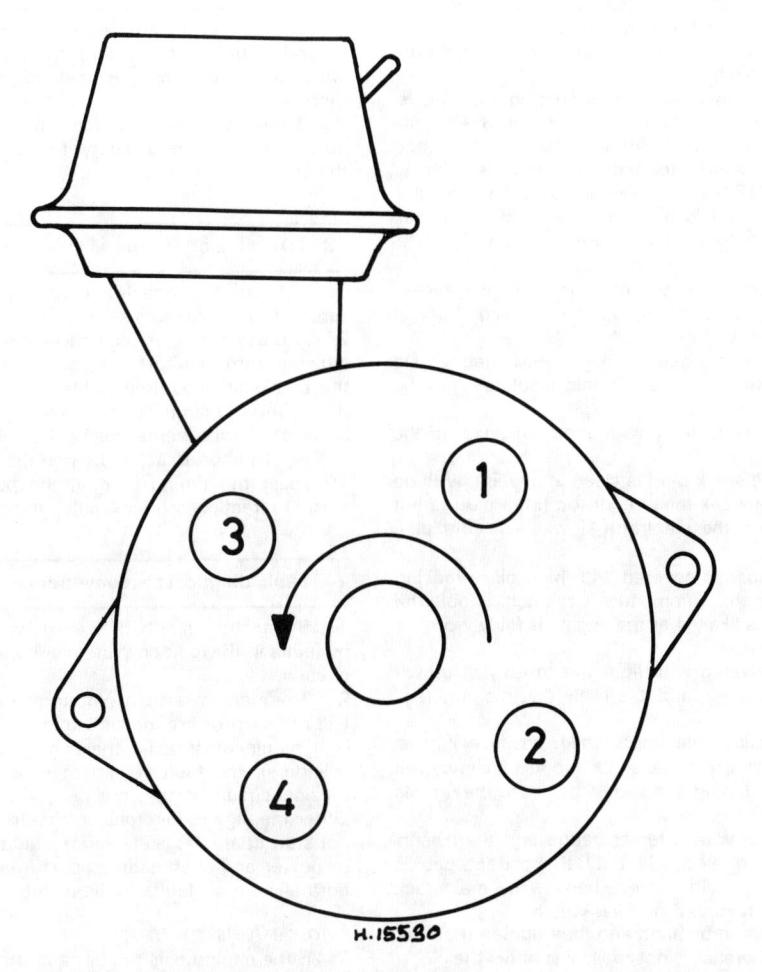

Fig. 4.3 Spark plug HT lead connecting diagram – 1.2 and 1.3 models (Sec 11)

Chapter 4 Ignition system

11.3A Prising up a heat shield adaptor using an open-ended spanner

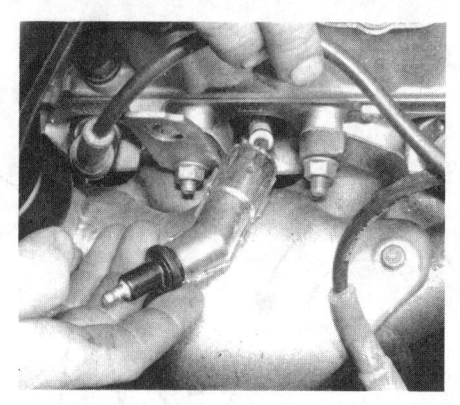

11.3B Removing the heat shield adaptor

12.1 The coil is located on the left-hand valance in the engine compartment

11 Spark plugs and HT leads – general

1 The correct functioning of the spark plugs is vital for the currect running and efficiency of the engine. It is essential that the plugs fitted are appropriate for the engine, and the suitable type is specified at the beginning of this chapter. If this type is used and the engine is in good condition, the spark plugs should not need attention between scheduled service replacement intervals. Spark plug cleaning is rarely necessary and should not be attempted unless specialised equipment is available, as damage can easily be caused to the firing ends.
2 At the intervals specified in Section 2, the old spark plugs should be removed and new plugs fitted.
3 To remove the spark plugs, open the bonnet and, on 1.0 models, remove the air cleaner assembly, as described in Chapter 3. Grip the rubber end fittings and pull the HT leads from the plugs. On 1.2 and 1.3 models, heat shield adaptors are fitted between the ends of No 1, 2 and 3 spark plugs and the HT leads. To remove these, pull off the HT lead and then prise off the shield using an open-ended spanner (photos). The shield will be quite tight and require a good pull to remove it, and be careful not to damage the plug.
4 Brush out any accumulated dirt or grit from the spark plug recess in the cylinder head, otherwise it may drop into the combustion chamber when the plug is removed.
5 Unscrew the spark plugs with a deep socket or box spanner. Do not allow the tool to tilt, otherwise the ceramic insulator may be cracked or broken.
6 Examination of the spark plugs will give a good indication of the condition of the engine.
7 If the insulator nose of the spark plug is clean and white, with no deposits, this is indicative of a weak mixture, or too hot a plug (a hot plug transfers heat away from the electrode slowly – a cold plug transfers heat away quickly).
8 If the top and insulator nose is covered with hard black-looking deposits, then this is indicative that the mixture is too rich. Should the plug be black and oily, then it is likely that the engine is fairly worn, as well as the mixture being too rich.
9 If the insulator nose is covered with light tan to greyish brown deposits, then the mixture is correct, and it is likely that the engine is in good condition.
10 The spark plug is of considerable importance, because if it is either too large or too small the size of the spark and its efficiency will be seriously impaired. The spark plug gap should be set to the figures given in the Specifications.
11 To set it, measure the gap with a feeler gauge and then bend open, or close, the outer plug electrode until the correct gap is achieved. The centre electrode should never be bent as this may crack the insulation and cause plug failure, if nothing worse.
12 To fit new plugs screw them in by hand and then tighten them to the specified torque. If a torque wrench is not available tighten the plugs by hand onto their seating and then tighten a further $\frac{1}{8}$ of a turn.
13 When reconnecting the HT leads, make sure that they are refitted in their correct order, 1-3-4-2, No 1 cylinder being nearest the crankshaft pulley end of the engine.
14 The plug leads themselves require no routine attention other than being kept clean and wiped over regularly. When attending to the spark plugs it is a good idea to remove each plug lead in turn from the distributor cap. Water can seep down into the joints giving rise to a white corrosive deposit which must be carefully removed from the end of each cable.

12 Coil – general

1 The coil is an auto-transformer and has two sets of windings wound around a core of soft iron wires. The resistance of the primary winding is given in Specifications at the beginning of this Chapter (photo).
2 If the coil is suspect then the resistance may be checked by an auto-electrician and if faulty it may readily be renewed after undoing the mounting bolts.

13 TDC sensor – general

1 1.2 and 1.3 models incorporate a facility for connecting a TDC sensor to the crankcase.
2 A brass sleeve is located in the crankcase to accept the sensor which in turn monitors the position of the two pins which are fitted to the crankshaft counterweights.
3 Without a suitable ignition tester it is unlikely that this device will be of use to the home mechanic, only to garages or service stations.
4 No 1 piston is at TDC when the notch in the crankshaft pulley is 10° past the timing pointer, the pulley having been turned in the normal direction of crankshaft rotation.

14 Fault diagnosis – conventional ignition system

1 By far the majority of breakdown and running troubles are caused by faults in the ignition system either in the low tension or high tension circuits.
2 There are two main symptoms indicating faults. Either the engine will not start or fire, or the engine is difficult to start and misfires. If it is a regular misfire (ie the engine is running on only two or three cylinders), the fault is almost sure to be in the secondary or high tension circuit. If the misfiring is intermittent the fault could be in either the high or low tension circuits. If the car stops suddenly, or will not start at all, it is likely that the fault is in the low tension circuit. Loss of power and overheating, apart from faulty carburation settings, are normally due to faults in the distributor or to incorrect ignition timing.

Engine fails to start

3 If the engine fails to start and the car was running normally when it was last used, first check that there is fuel in the petrol tank. If the engine turns over normally on the starter motor and the battery is evidently well charged, then the fault may be in either the high or low tension circuits. First check the HT circuit.
4 If the engine fails to start due to either damp HT leads or distributor

cap, a moisture dispersant such as Holts Wet Start can be very effective. To prevent the problem from recurring, Holts Damp Start can be used to provide a sealing coat, so excluding any further moisture from the ignition system. In extreme difficulty, Holts Cold Start will help to start a car when only a very poor spark occurs.

5 If the engine still fails to start, check that voltage is reaching the plugs by disconnecting each plug lead in turn at the spark plug end and holding the end of the cable with rubber or an insulated tool about $\frac{1}{4}$ mm (6 mm) away from the cylinder block. Spin the engine on the starter motor.

6 Sparking between the end of the cable and the block should be fairly strong with a regular blue spark. If voltage is reaching the plugs then remove, examine and regap if necessary. The engine should now start.

7 If there is no spark at the plug leads, take off the HT lead from the centre of the distributor cap and hold it to the block as before. Spin the engine on the starter once more. A rapid succession of blue sparks between the end of the lead and the block indicates that the coil is in order and that the distributor cap is cracked, the rotor arm is faulty or the carbon brush in the top of the distributor cap is not making good contact with the rotor arm.

8 If there are no sparks from the end of the lead from the coil, check the connections at the coil end of the lead. If it is in order start checking the low tension circuit. Possibly, the points are in bad condition. Clean and reset them as described in this Chapter, Section 3.

9 Use a 12V voltmeter or a 12V bulb and two lengths of wire. With the ignition switch on and the points open, test between the low tension wire to the coil and earth. No reading indicates a break in the supply from the ignition switch. Check the connections at the switch to see if any are loose. Refit them and the engine should run. A reading shows a faulty coil or condenser, or broken lead between the coil and the distributor.

10 Take the condenser wire off the points assembly and with the points open test between the moving point and earth. If there is now a reading then the fault is in the condenser. Fit a new one and the fault is cleared.

11 With no reading from the moving point to earth, take a reading between earth and the distributor terminal of the coil. A reading here shows a broken wire which will need to be replaced between the coil and the distributor. No reading confirms that the coil has failed and must be renewed, after which the engine will run once more. Remember to refit the condenser wire to the points assembly. For these tests it is sufficient to separate the points with a piece of dry paper while testing with the points open.

Engine misfires

12 If the engine misfires regularly run it at a fast idling speed. Pull off each of the plug caps in turn and listen to the note of the engine. Hold the plug cap in a dry cloth or with a rubber glove as additional protection against a shock from the HT supply.

13 No difference in engine running will be noticed when the lead from the defective circuit is removed. Removing the lead from one of the good cylinders will accentuate the misfire

14 Remove the plug lead from the plug which is not firing and hold it about $\frac{1}{4}$ in (6 mm) away from the block. Restart the engine. If the sparking is fairly strong and regular, the fault must lie in the spark plug.

15 The plug may be loose, the insulation may be cracked, or the points may have burnt away giving too wide a gap for the spark to jump. Worse still, one of the points may have broken off. Either renew the plug or clean it, reset the gap, and then test it.

16 If there is no spark at the end of the plug lead, or if it is weak and intermittent, check the ignition lead from the distributor to the plug. If the insulation is cracked or perished, renew the lead. Check the connections at the distributor cap.

17 If there is still no spark, examine the distributor cap carefully for tracking. This can be recognised by a very thin black line running between two or more electrodes, or between an electrode and some other part of the distributor. These lines are paths which now conduct electricity across the cap thus letting it run to earth. The only answer is a new distributor cap.

18 Apart from the ignition timing being incorrect, other causes of misfiring have already been dealt with under the Section dealing with the failure of the engine to start. To recap, these are that

(a) The coil may be faulty giving an intermittent misfire
(b) There may be a damaged wire or loose connection in the low tension circuit
(c) The condenser may be faulty
(d) There may be a mechanical fault in the distributor (broken driving spindle or contact breaker spring)

19 If the ignition timing is too far retarded, it should be noted that the engine will tend to overheat, and there will be a quite noticeable drop in power. If the engine is overheating and the power is down, and the ignition timing is correct, then the carburettor should be checked, as it is likely that this is where the fault lies.

20 If the resistor wire is broken or disconnected, the engine will fire when the starter motor is operating, but will refuse to run. Renewal of the resistor wire will cure the problem. Do not bypass the resistor wire with ordinary wire, or overheating of the coil will occur.

15 Fault diagnosis – electronic breakerless ignition system

Fault diagnosis in the HT circuit of the electronic ignition system is the same as described in the previous Section for conventional systems. Remember that the voltages in this system are much higher than in a mechanical breaker system, so insulate the HT leads and other components adequately before handling them when the engine is running. If a fault is suspected in the LT circuit, special test equipment is needed to accurately diagnose the problem and this work should be left to a GM dealer or suitably equipped auto-electrician.

Chapter 5 Clutch

For modifications, and information applicable to later models, see Supplement at end of manual

Contents

Clutch – adjustment 2	Clutch cable – removal and refitting 3
Clutch assembly – inspection 7	Clutch pedal – removal and refitting 4
Clutch assembly – removal and refitting (engine and transmission in car) 5	Clutch release bearing – removal and refitting 8
	Fault diagnosis – clutch 9
Clutch assembly – removal and refitting (engine or transmission removed from car) 6	General description 1

Specifications

General
Type	Cable-operated single dry plate with diaphragm spring
Adjustment	By adjuster nut on the cable at the release lever – for pedal stroke setting
Driven plate diameter	190 mm (7.5 in)
Release bearing	Sealed ball-bearing

Torque wrench settings
	Nm	lbf ft
Access plate to transmission casing	7	5
Clutch cover assembly to flywheel	15	11
Release fork-to-release lever pivot shaft clamp bolt	35	26
End cover screw plug:		
Four-speed transmission	50	37
Five-speed transmission	30	22
Input shaft socket headed screw	15	11

1 General description

All Nova models are equipped with a single plate diaphragm spring clutch which is enclosed in a pressed-steel cover bolted to the flywheel. The gearbox input shaft projects through the clutch assembly and is supported at its forward end by a spigot bearing within the centre of the crankshaft.

The clutch driven plate is located between the flywheel and the clutch pressure plate and slides on splines on the gearbox input shaft. When the clutch is engaged, the diaphragm spring forces the pressure plate to grip the driven plate against the flywheel and drive is transmitted from the crankshaft, through the driven plate, to the gearbox input shaft. On disengaging the clutch the pressure plate is lifted to release the driven plate with the result that the drive to the gearbox is disconnected.

The clutch is operated by a foot pedal suspended under the facia and a cable connected to the clutch release lever mounted on the transmission casing. Depressing the pedal causes the release lever to move the release bearing against the fingers of the diaphragm spring in the pressure plate assembly. The spring is sandwiched between two rings which act as fulcrums. As the centre of the spring is moved in, the periphery moves out to lift the pressure plate and disengage the clutch. The reverse takes place when the pedal is released.

As wear occurs on the driven plate with usage, the foot pedal will rise progressively relative to its original position. Periodic adjustment is not required.

An unusual feature of the design of this particular clutch is that the driven plate, pressure plate and release bearing assembly can be renewed without having to remove either the engine or transmission from the car.

Chapter 5 Clutch

2 Clutch – adjustment

1 The clutch is normally self-adjusting, but if the cable or driven plate are renewed, the following initial adjustment will be required.
2 Undo the three retaining screws and withdraw the parcel shelf from its location under the facia on the driver's side.
3 With the clutch pedal released, take a measurement from the outer edge of the steering wheel to the centre of the clutch pedal pad and record the dimension. Now take a second measurement with the clutch pedal fully depressed. These measurements can be taken using a suitable strip of wood or metal as the important figure is the difference between the two measurements, that is the movement (stroke) of the pedal itself. This should be between 124 and 131 mm (4.9 and 5.2 in) and if this is not the case adjustment is required. This is done at the threaded end of the cable where it enters the release lever on the transmission casing.
4 To carry out the adjustment first remove the spring clip that secures the cable to the release lever (photo). Pull the cable forward slightly and then turn the adjuster as necessary until the stroke dimension given in paragraph 3 is obtained. Refit the spring clip to the cable after completing the adjustment.
5 Note that there should be no play in the clutch pedal of these vehicles.

2.4 Spring clip securing clutch cable to release lever (A) and cable adjuster (B)

3 Clutch cable – removal and refitting

1 Before disturbing the clutch cable installation, take a measurement of the length of the threaded end of the cable protruding

Fig. 5.1 Clutch pedal released measurement – A (Sec 2)

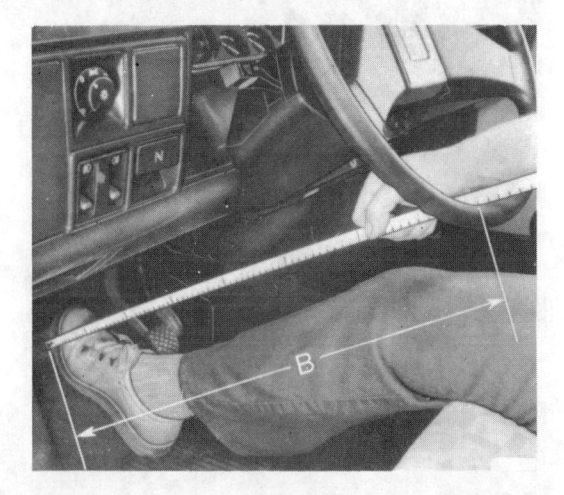

Fig. 5.2 Clutch pedal depressed measurement – B (Sec 2)

Fig. 5.3 Measure the exposed length of clutch cable before removal (Sec 3)

through the cable adjuster at the clutch release lever. This will enable an initial adjustment setting to be made when refitting the cable.
2 Remove the spring clip that secures the cable to the release lever, slacken the cable adjuster and slip the cable out of the release lever slot. Move the cable assembly rearwards and withdraw it from the lug on the transmission casing (photo).
3 Remove the cable from the support bracket behind the engine and, where applicable, release any clips or ties securing the engine wiring harness to the clutch cable.
4 Working inside the car undo the three retaining screws and withdraw the parcel shelf from its location under the facia on the driver's side.
5 Unhook the return spring from the clutch pedal and disconnect the cable from the pedal lever.
6 The cable can now be withdrawn into the engine compartment by pulling it through the bulkhead.
7 Refitting the clutch cable is the reverse sequence to removal. Position the cable adjuster initially so that the same amount of cable protrudes through the adjuster as noted during removal, then adjust the clutch, as described in Section 2.

4 Clutch pedal – removal and refitting

1 Refer to Chapter 12 and remove the heater air distribution housing.

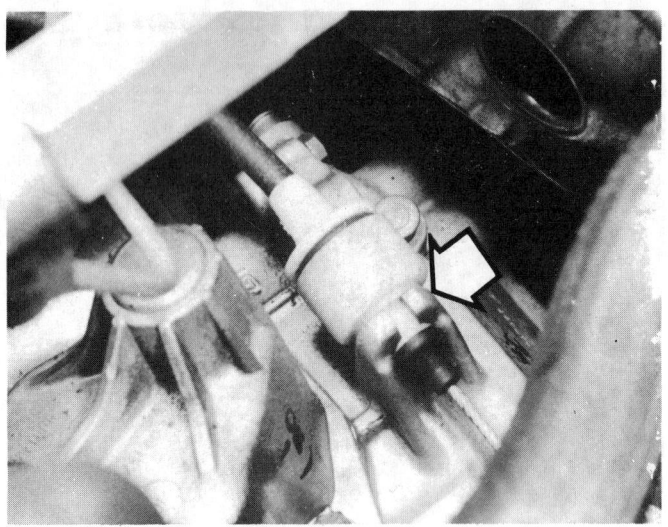

Chapter 5 Clutch

3.2 Clutch cable attachment at lug on transmission casing (arrowed)

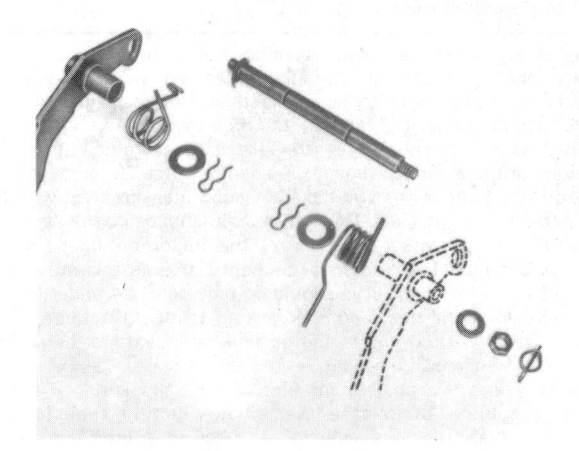

Fig. 5.5 Arrangement of the clutch pedal, pivot and springs – left-hand drive version illustrated (Sec 4)

Fig. 5.4 Clutch pedal return spring fitting (Sec 3 and 4)

2 Refer to the previous Section and disconnect the clutch cable from the release lever.
3 From inside the car unhook the return spring from the clutch pedal and disconnect the cable from the pedal lever.
4 Remove the wire locking clip from the pedal pivot retaining nut, then unscrew the nut and collect the washer.
5 Extract the two wire clips located between the clutch and brake pedals on the pedal pivot.
6 Push the pivot out of the support bracket, towards the centre of the car, and recover the two return springs, two spacing washers and the clutch pedal. Note that the brake pedal will remain in position as it is still secured by its pushrod to the brake relay lever.
7 Refitting is the reverse sequence to removal. Before inserting the pedal pivot, lightly smear the bearing surface with molybdenum disulphide grease. Refit the cable, as described in the previous Section, and adjust if necessary. Refer to Chapter 12 and refit the heater air distribution housing.

5 Clutch assembly – removal and refitting (engine and transmission in car)

Note: *The design of the engine/transmission unit on Nova models is such that the components of the clutch can be renewed without having to remove either the engine or the transmission. However, in order to keep the pressure plate in compression, which is necessary to facilitate removal, three small clamps are necessary. These can easily be made up, as described in the text and shown in the accompanying illustrations, if the manufacturer's special tool is not available.*

1 Prise off the wheel trim from the left-hand front roadwheel and slacken the wheel bolts. Jack up the car and support it securely on axle stands. Remove the roadwheel.
2 Working under the left-hand wheel arch unscrew the large plug from the transmission end cover using a 36 mm socket and long bar on four-speed transmissions (photo), or a suitable hexagon bar on five-speed transmissions. The socket-headed plug on the latter is very large, and if a suitable bar is not available, it should be possible to use a bolt with a head the correct size to fit the socket of the plug. The shank of the bolt is then bent at 90° to provide the necessary leverage to undo the plug.
3 Extract the circlip, now exposed, from the centre of the input shaft using circlip pliers (photo).
4 Underneath the circlip is a socket-headed screw which will require an 8.0 mm AF twelve-splined key or socket bit to extract it. Motor accessory shops can usually supply this type of key (photo). Undo and remove the screw from the input shaft (photo).
5 The gearbox input shaft can now be withdrawn sufficiently to allow the clutch driven plate to be removed. To do this screw a suitable

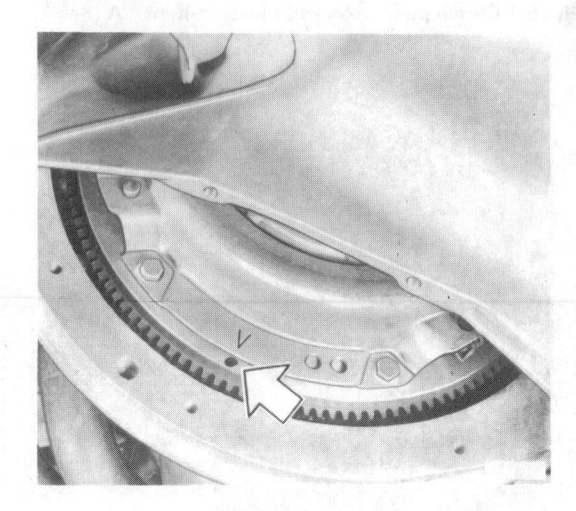

Fig. 5.6 Clutch cover and flywheel alignment marks (Sec 5)

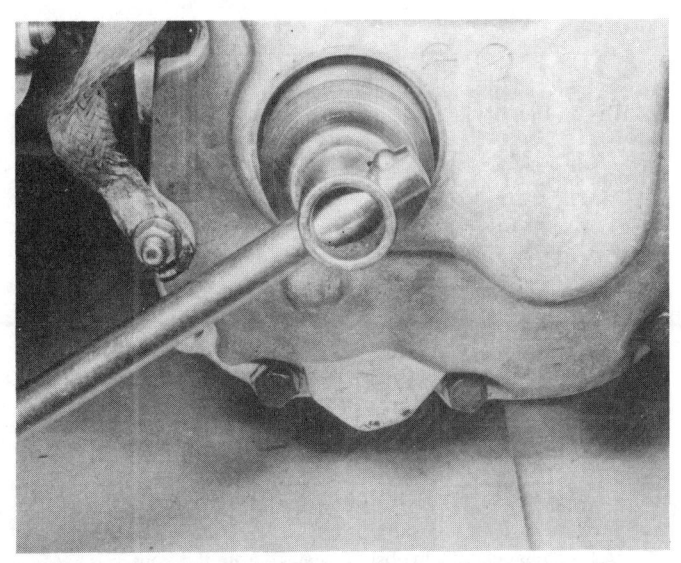

5.2 Removing the screw plug from the transmission end cover (four-speed transmission shown)

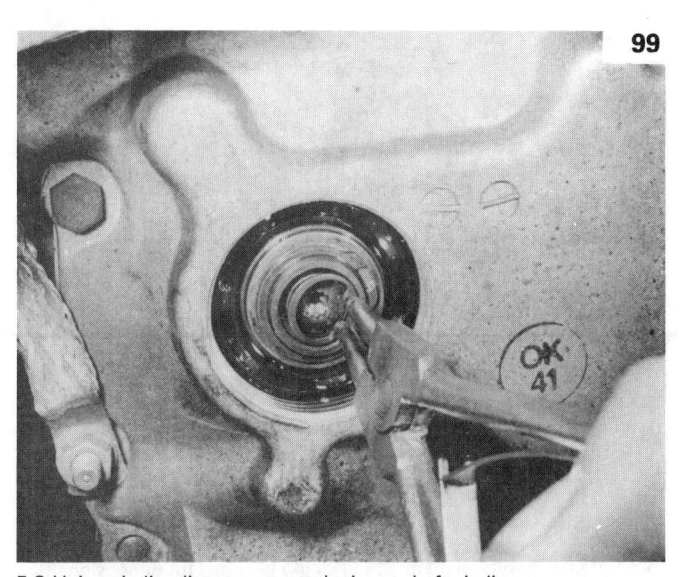

5.3 Using circlip pliers to extract the input shaft circlip

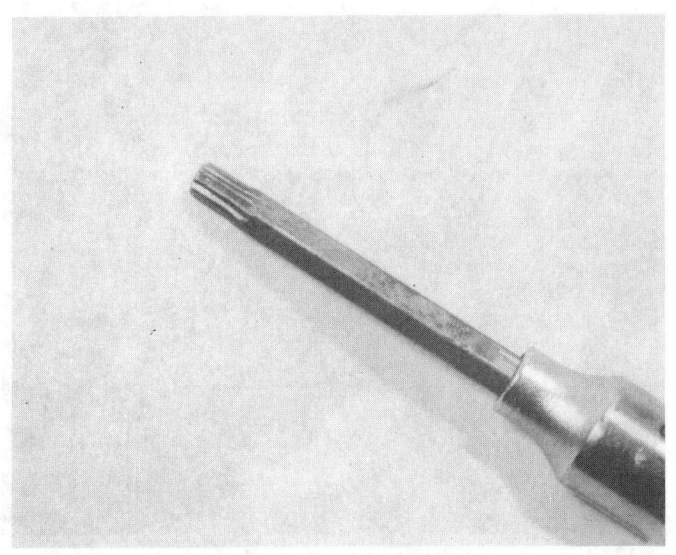

5.4A A typical twelve-splined drive socket bit ...

5.4B ... which is required for removal of the input shaft socket-headed screw

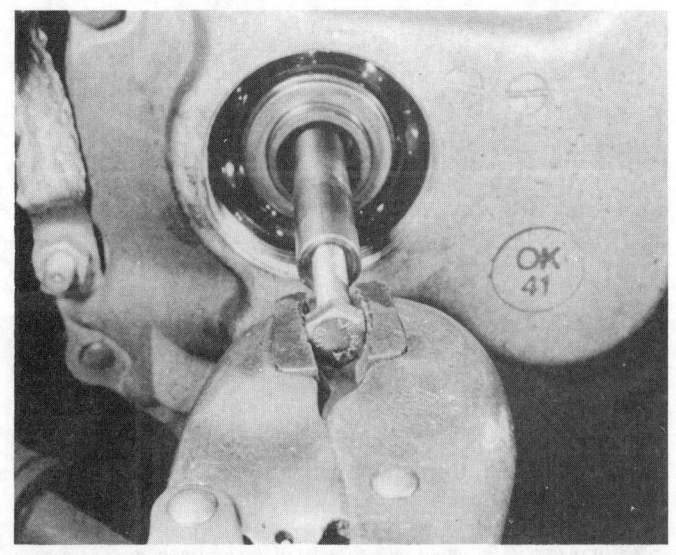

5.5 Using a self-locking wrench and bolt to withdraw the input shaft

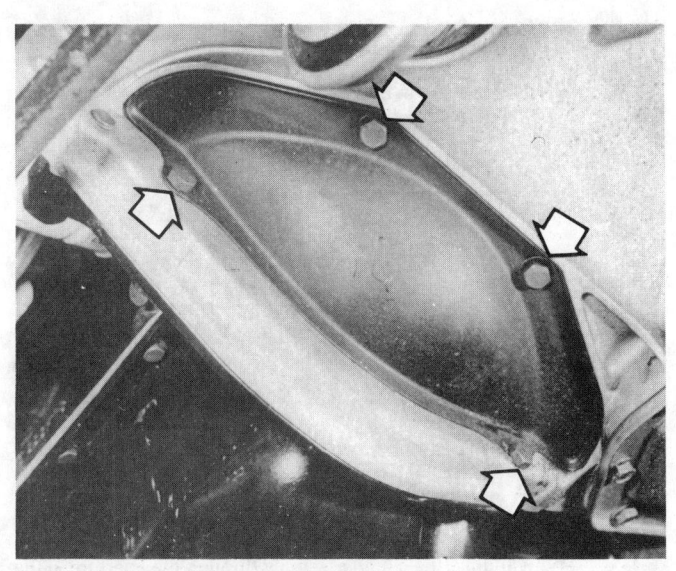

5.6 Clutch access plate retaining bolts (arrowed)

5.8 Clutch pressure plate retaining clamp (see text for dimensions)

5.10A Retaining clamps in position viewed from the front ...

5.10B ... and rear of the cover assembly

5.12A Remove the clutch cover retaining bolts ...

5.12B ... and withdraw the clutch assembly through the aperture in the transmission casing

5.23 Applying PTFE tape to the threads of the four-speed transmission end cover screw plug

bolt into the end of the input shaft and withdraw it using a self-locking wrench (photo). One of the bolts securing the gearchange mechanism cover to the top of the transmission casing can be used for this purpose if no other suitable bolt can be found. *Don't forget to refit it to the cover after withdrawing the input shaft.* If the shaft is tight, use a long screwdriver or suitable bar and lever between the wrench jaws and the end cover. Note that the input shaft cannot be removed completely and it is only necessary to withdraw it about halfway, until it reaches its stop, to allow removal of the clutch.

6 Working underneath the car unscrew the four bolts securing the clutch access plate to the transmission casing (photo) and lift off the plate.

7 Before the clutch can be removed the pressure plate must be compressed against the tension of the diaphragm spring, otherwise the clutch assembly will be too thick to pass through the space between the flywheel and the edge of the casing.

8 Three special clamps are available from the manufacturer for this operation, but suitable alternatives can be made up from strips of metal. The clamps should be U-shaped and conform to the following dimensions (photo):

Thickness of metal strip = 3.0 mm (0.12 in)
Distance between U legs = 15.0 mm (0.6 in)

9 Bevel the edges of the clamps to make them easier to fit and file or cut a notch in one side to clear the pressure plate rivets.

10 To fit the clamps have an assistant depress the clutch pedal fully and then fit each clamp securely over the edge of the cover/pressure plate assembly (photos). Ensure that they engage in the apertures spaced around the rim of the clutch cover. Turn the crankshaft by means of a socket on the pulley bolt to bring each aperture into an accessible position.

11 When all three clamps are in position and secure, the clutch pedal can be released.

12 Progressively slacken and remove each of the six bolts securing the clutch cover to the flywheel (photo). Withdraw the clutch assembly downwards and out through the aperture in the transmission casing (photo).

13 If necessary the three clamps can be removed from the cover/pressure plate by compressing the assembly in a vice between blocks of wood, one across the flywheel side of the cover and one against the diaphragm spring.

14 It is important that no grease or oil is allowed to come into contact with the driven plate friction linings, or the pressure plate and flywheel faces. It is advisable to handle the parts with clean hands and to wipe off the pressure plate and flywheel faces with a clean, dry rag before inspection or refitting commences.

15 If a new clutch cover assembly is being fitted it will be necessary to transfer the retaining clamps to the new unit.

16 To refit the clutch assembly first apply a smear of molybdenum disulphide grease to the splines of the driven plate hub and then position the plate against the flywheel so that the greater projecting side of the hub is facing away from the flywheel. Hold the driven plate against the flywheel and place the cover assembly over it.

17 Push the input shaft through the driven plate hub and engage its end with the spigot bearing in the end of the crankshaft.

18 Turn the flywheel and cover assembly until the paint mark on the flywheel and the notch on the edge of the cover are located. With these marks aligned insert the cover assembly retaining bolts and progressively tighten them, in a diagonal sequence, to the specified torque.

19 The clutch pedal can now be depressed and the retaining clamps removed.

20 Refit the access plate to the transmission casing and secure with the four bolts.

21 From under the wheel arch, make sure that the input shaft is fully inserted and then refit the socket-headed screw.

22 Refit the circlip, ensuring that it locates properly in the groove in the input shaft.

23 Apply a sealant or PTFE tape to the threads of the end cover plug (photo) and screw it into place. Tighten the plug to the specified torque. The plug should not protrude by more than 4.0 mm (0.16 in) from the face of the end cover when correctly fitted. If it does, the input shaft has not been pushed fully home.

24 Refit the roadwheel, lower the car to the ground and check the clutch adjustment, as described in Section 2.

6 Clutch assembly – removal and refitting (engine or transmission removed from car)

1 Having removed the transmission, as described in Chapter 6, the engine or the engine/transmission assembly, as described in Chapter 1, undo and remove the bolts securing the clutch cover assembly to the flywheel. Lift off the cover assembly and collect the driven plate (photo).

2 It is important that no grease or oil is allowed to come into contact with the driven plate friction linings, or the pressure plate and flywheel faces. It is advisable to handle the parts with clean hands and to wipe off the pressure plate and flywheel faces with a clean, dry rag before inspection or refitting commences.

3 To refit the clutch, place the driven plate against the flywheel so that the greater projecting side of the hub is facing away from the flywheel.

4 Locate the clutch cover so that the notch on the cover edge is aligned with the paint mark on the flywheel. Refit the cover retaining bolts and tighten them finger tight so that the driven plate is gripped, but can still be moved.

5 The driven plate must now be centralised so that when the engine and transmission are mated, the input shaft splines will pass through the splines in the driven plate hub.

6 Centralisation can be carried out quite easily by inserting a round

6.1 Removing the clutch cover and driven plate with the engine removed from the car

6.7 Universal aligning tool used to centralise the driven plate while the cover bolts are tightened

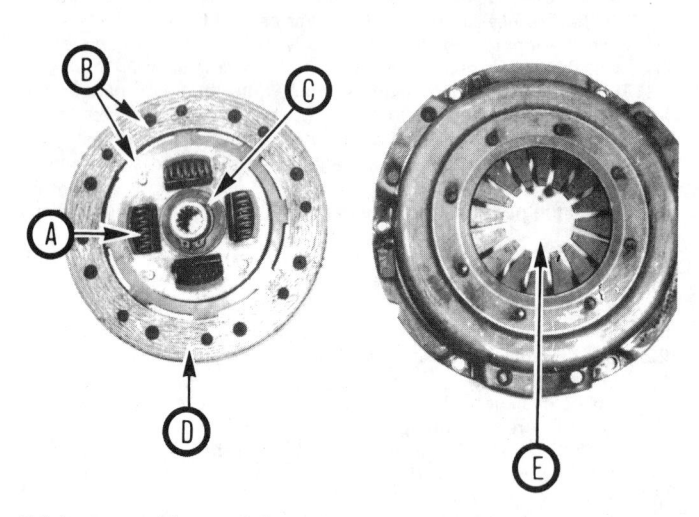

7.1 Items requiring careful inspection when checking the driven plate and cover assembly for wear

- A Cushioning springs
- B Rivets
- C Hub splines
- D Lining material
- E Diaphragm spring

8.1 Removing the release fork clamp bolt

8.3 Removing the release fork and release bearing

8.6 The groove in the pivot shaft (arrowed) must align with the clamp bolt hole in the release fork when refitting

bar or long screwdriver through the centre of the hub so that it rests in the hole in the end of the crankshaft containing the input shaft spigot bearing. Moving the bar sideways or up-and-down will move the driven plate in whichever direction is necessary to achieve centralisation.

7 Centralisation can be checked by removing the bar and viewing the driven plate splined hub boss in relation to the crankshaft spigot bearing. When the hub and spigot bearing are aligned, all is correct. Alternatively, if a universal clutch aligning tool can be obtained, this will eliminate all the guesswork and obviate the need for visual alignment (photo).

8 With the driven plate centralised, tighten the clutch cover retaining bolts progressively, in a diagonal sequence, to the specified torque.

9 Refit the engine or transmission, as described in Chapter 1 or Chapter 6.

7 Clutch assembly – inspection

1 With the clutch assembly removed, examine the driven plate friction linings for wear and loose rivets, and the hub for distortion, cracks, broken or weak cushioning springs and worn splines (photo). The surface of the friction linings may be highly glazed, but as long as the friction material pattern can be clearly seen, this is acceptable. If the friction material is worn down to the level of the rivet heads, the driven plate must be renewed. The plate must also be renewed if there is any sign of oil contamination of the friction material caused by a leaking transmission input shaft oil seal or crankshaft rear oil seal. If oil contamination is evident the cause of the trouble must be rectified immediately, as described in Chapter 1 or Chapter 6.

2 Check the machined faces of the flywheel and pressure plate. If either is grooved or heavily scored, renewal is necessary. If the pressure plate is cracked or split, or if the diaphragm spring is damaged or its pressure suspect, a new unit must be fitted.

3 Check the release bearing for smoothness of operation. There should be no harshness or slackness in it and it should spin freely without tight spots.

4 When considering renewing clutch components individually, bear in mind that new parts (or parts from different manufacturers) do not always bed into old ones satisfactorily. A clutch cover assembly or driven plate renewed separately may sometimes cause judder or snatch. Although expensive, the clutch cover assembly, driven plate

Chapter 5 Clutch

and release bearing should be renewed together wherever possible.

8 Clutch release bearing – removal and refitting

Note: *If this operation is being carried out with the engine/transmission in the car, it will first be necessary to remove the clutch assembly, as described in Section 5.*

1 With the clutch assembly removed, undo the clamp bolt securing the release fork to the release lever pivot shaft (photo).
2 Referring to Section 3 if necessary, disconnect the clutch cable from the release lever.
3 Pull the release lever upwards out of the transmission casing, lift out the release fork and slide the release bearing off the guide tube (photo).
4 Before refitting, smear the guide tube and the fingers of the release fork with molybdenum disulphide grease.
5 Slide the release bearing onto the guide tube, engage the release fork and push the pivot shaft down into engagement with the fork.
6 Align the clamp bolt hole in the fork with the machined groove on the pivot shaft (photo) and refit the clamp bolt. Tighten the clamp bolt to the specified torque.
7 Refit the clutch assembly, as described in Section 5.

9 Fault diagnosis – clutch

Symptom	Reason(s)
Judder when taking up drive	Loose or worn engine or transmission mountings Oil contamination of driven plate friction linings Excessive wear of driven plate friction linings Broken or weak driven plate cushioning springs Distorted or damaged pressure plate or diaphragm spring Clutch cable binding
Clutch spin (failure to disengage so that gears are difficult to engage)*	Incorrect clutch adjustment Driven plate sticking to flywheel face due to corrosion or oil contamination Input shaft seized in crankshaft spigot bearing
Clutch slip (increase in engine speed does not result in increase in road speed – particularly on gradients)	Incorrect clutch adjustment Oil contamination of driven plate friction linings Excessive wear of driven plate friction linings Weak or broken diaphragm spring
Noise evident on depressing clutch pedal	Worn or damaged release bearing Damaged or broken diaphragm spring Incorrect clutch adjustment
Noise evident as clutch pedal is released	Broken or weak driven plate cushioning springs Damaged or broken diaphragm Worn gearbox bearings or internal components (see Chapter 6)

* This condition may also be due to the driven plate being rusted to the flywheel or pressure plate. It is possible to free it by applying the handbrake, depressing the clutch, engaging top gear and operating the starter motor. If really badly corroded, then the engine will not turn over, but in the majority of cases the driven plate will free. Once the engine starts, rev it up and slip the clutch several times to clear the rust deposits.

Chapter 6 Transmission

For modifications, and information applicable to later models, see Supplement at end of manual

Contents

Fault diagnosis – transmission	18
Gearchange linkage – adjustment	3
Gearchange linkage – removal and refitting	4
Gearchange linkage remote control control housing – removal and refitting	5
Gear lever – removal and refitting	6
Gear lever rubber boot – removal and refitting	7
General description	1
Input shaft – overhaul	13
Mainshaft – overhaul	14
Maintenance and inspection	2
Selector housing cover – overhaul	11
Synchroniser units – overhaul	15
Transmission – removal and refitting	8
Transmission casing – overhaul	12
Transmission (five-speed) – dismantling and reassembly	17
Transmission (four-speed) – dismantling into major assemblies	10
Transmission (four-speed) – reassembly	16
Transmission overhaul – general	9

Specifications

General
Type .... Four or five forward speeds (all synchromesh) and reverse. Final drive integral with main gearbox

Designation:
- Four-speed transmission F10/4
- Five-speed transmission F10/5

Gear ratios (all models)
1st	3.55:1
2nd	1.96:1
3rd	1.30:1
4th	0.89:1
5th (F10/5)	0.71:1
Reverse	3.18:1

Final drive ratios

	F10/4	F10/5
1.0 models:		
Up to February 1983	3.74:1	4.18:1
From February 1983	3.94:1	4.18:1
1.2 models	3.74:1	3.94:1
1.3 models	—	3.94:1 or 4.18:1 (SR models only)

Lubricant capacity
- Four-speed transmission .... 1.7 litres (3.0 pts)
- Five-speed transmission .... 1.8 litres (3.1 pts)

Lubricant type/specification
Gear oil, viscosity SAE 80, to API-GL3 or GL4 or GM special oil 90 188 629 (Duckhams Hypoid 80, or Hypoid 75W/90S)

Torque wrench settings

	Nm	lbf ft
Transmission bellhousing to engine	75	55
End cover to transmission casing	22	16
Differential cover plate to transmission casing	30	22
Release bearing guide tube to transmission casing	5	4
Selector cover to transmission casing	15	11
Crownwheel to differential	85	63
Gearchange remote control housing to underbody	16	12
Reversing lamp switch to transmission casing	20	15
Left-hand engine/transmission mounting retaining bolts	65	48
Rear engine/transmission mounting bracket bolt	65	48
End cover shield retaining bolts (F10/5)	15	11
Interlock pin bridge piece bolts (F10/5)	5	4
Selector interlock pawl bolts (F10/5)	9	6
Selector fork bearing support (F10/5)	22	16

Chapter 6 Transmission

1 General description

A four-speed manual transmission is fitted to 1.0 and 1.2 litre models and a five-speed unit is fitted to 1.3 litre models. The five-speed transmission is also available as an option on the smaller engined models. Synchromesh gear engagement is used on all forward gears on both transmission types.

Drive is transmitted through the transmission and integral final drive/differential assembly, and then to the front wheels via the driveshafts. Gear selection is by means of a floor-mounted lever and remote control linkage.

If transmission overhaul is necessary, due consideration should be given to the costs involved, since it is often more economical to obtain a factory exchange or good second-hand unit rather than fit new parts to the existing transmission.

2.3 The differential cover plate must be removed to drain the transmission oil

3.3 Remote control rod clamp pinch-bolt (arrowed)

2 Maintenance and inspection

1 Every 6000 miles (15 000 km) or 6 months, whichever comes first, inspect the transmission joint faces and oil seals for any signs of damage, deterioration or oil leakage.
2 At the same service interval check and if necessary top up the transmission oil. The level checking plug is located on the side of the transmission casing just below the left-hand driveshaft inner constant velocity joint. The oil level should be maintained up to the level of the plug orifice. If topping up is necessary this is done at the filler/breather plug located on top of the gear selector cover. On no account should oil be added to the transmission through the level checking plug. Note that the transmission and final drive/differential both share the same lubricating oil.

Fig. 6.1 Transmission oil level checking plug location – arrowed (Sec 2)

Fig. 6.2 Transmission oil filler/breather plug location – arrowed (Sec 2)

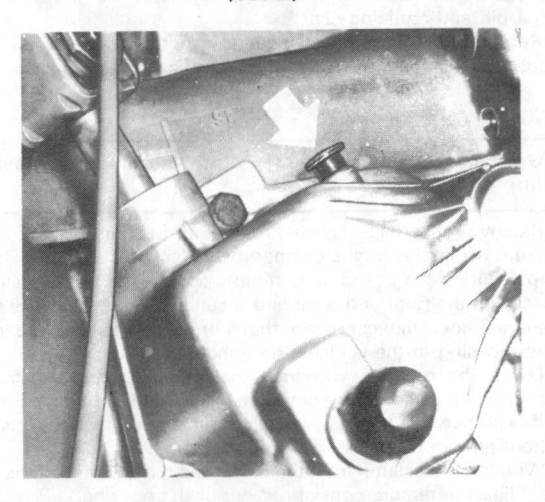

Fig. 6.3 Gearchange linkage adjustment blanking plug (arrowed) on selector cover (Sec 3)

3 Renewal of the transmission oil is not considered necessary by the manufacturer; however the diligent home mechanic may wish to do this in the interests of extended transmission life. If so, it will be necessary to remove the cover plate at the base of the differential to drain the oil (photo), as a drain plug is not provided. Use a new gasket on the cover plate when refitting and observe the specified torque for the retaining bolts. Ensure adequate cleanliness is maintained throughout the operation.

4 Every 18 000 miles (30 000 km) or 12 months, whichever comes first, check the gearchange linkage for excess freeplay or wear in the joints and adjust the linkage if necessary, as described in Section 3.

3 Gearchange linkage – adjustment

For adjustment of 5-speed linkage, refer to Chapter 13.

1 Remove the centre console, if fitted, as described in Chapter 12.
2 Move the gear lever to the neutral position.
3 From within the engine compartment slacken the pinch-bolt at the clamp securing the gearchange remote control rod to the linkage (photo).
4 Prise out the small blanking plug on the side of the transmission selector cover.
5 Looking towards the front of the car, turn the selector rod protruding from the selector cover in an anti-clockwise direction until a 4.5 mm (0.18 in) diameter drill bit can be inserted into the hole left by removal of the blanking plug. Push the drill fully into the hole until it enters a corresponding hole in the internal selector lever.
6 Pull the boot up the gear lever and move the lever across the neutral gate to the left until it is against the contact strip, with the lever pad inside the marked notch of the transmission tunnel cutout.
7 Have an assistant hold the gear lever in this position and then tighten the pinch-bolt at the gearchange rod clamp.
8 Remove the drill bit and refit the blanking plug. Use a new plug if the old one does not seal properly.
9 Reposition the gear lever boot and refit the centre console.

4 Gearchange linkage – removal and refitting

1 Using a screwdriver, prise off the ball socket ends of the linkage operating rod from the ball-studs on the levers, and remove the rod.
2 Slacken the pinch-bolt at the clamp securing the gearchange remote control rod to the linkage.
3 Withdraw the linkage pivot retaining pin from the pivots after releasing the spring retaining cap.
4 Extract the retaining clip and withdraw the connecting pin at the linkage universal joint.
5 Slide the linkage out of the remote control rod clamp and remove it from the car.
6 If necessary the pivot retaining pin bushes can be renewed after prising out the old ones, and the universal joint can also be renewed after grinding off the rivet heads. The new joint is then refitted using a pivot pin and retaining clip.
7 Refitting is the reverse sequence to removal, but adjust the linkage, as described in Section 3, before tightening the clamp pinch-bolt.

5 Gearchange linkage remote control housing – removal and refitting

1 Remove the gear lever, as described in Section 6.
2 From within the engine compartment slacken the pinch-bolt at the clamp securing the gearchange remote control rod to the linkage.
3 Jack up the front of the car and securely support it on axle stands.
4 From under the car undo the four bolts securing the remote control housing to the vehicle floor (photo).
5 Lower the housing, withdraw the remote control rod from the transmission linkage and remove the assembly from under the car.
6 If necessary the remote control rod rubber boot and the bushing in the housing can be renewed as follows:
7 Withdraw the clamp from the end of the rod and pull off the rubber boot. Slide the remote control rod out of the housing.
8 Push the bushing assembly out of the housing from inside to outside and then detach the bushing from the bearing ring.
9 Renew the bushing and rubber boot if they are worn, damaged or show signs of deterioration.
10 Refit the bushing into the bearing ring and refit this assembly to the housing from the inside.
11 Lubricate the bushing with silicone grease and slide in the control rod. Refit the rubber boot and the clamp.
12 Refit the remote control housing to the car using the reverse sequence to removal. With the gear lever in position adjust the gearchange linkage, as described in Section 3.

Fig. 6.4 Gearchange linkage clamp pinch-bolt and pivot retaining pin – arrowed (Sec 4)

Fig. 6.5 Universal joint connecting pin (arrowed) on the gearchange linkage (Sec 4)

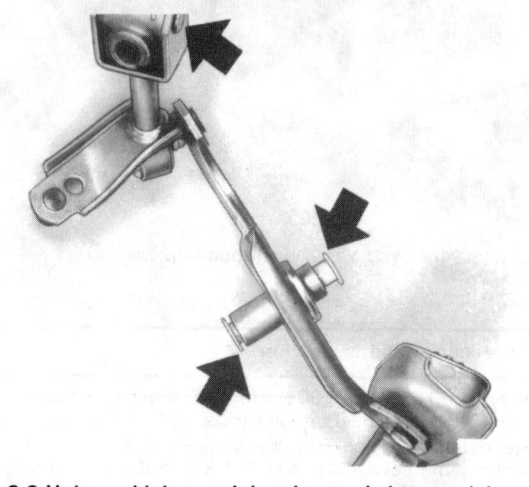

Fig. 6.6 Universal joint retaining rivet and pivot retaining pin bushes (arrowed) on the gearchange linkage (Sec 4)

Chapter 6 Transmission

5.4 Gearchange remote control housing retaining bolts (arrowed)

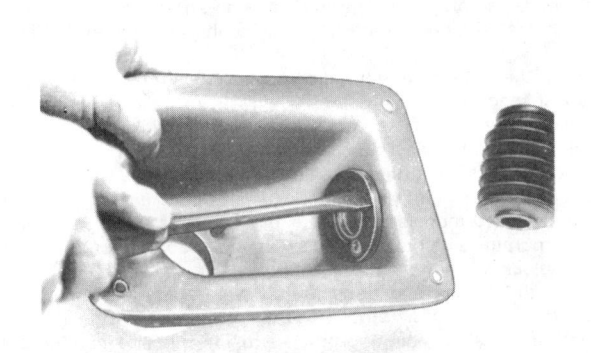

Fig. 6.7 Removing the gearchange rod bushing from the remote control housing (Sec 5)

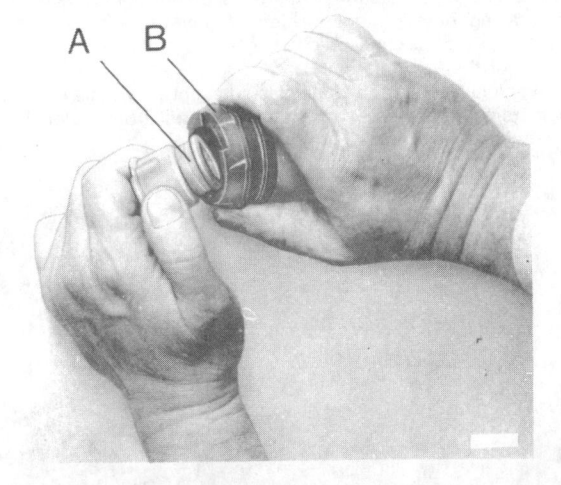

Fig. 6.8 Separating the gearchange rod bushing (A) from the retaining ring (B) (Sec 5)

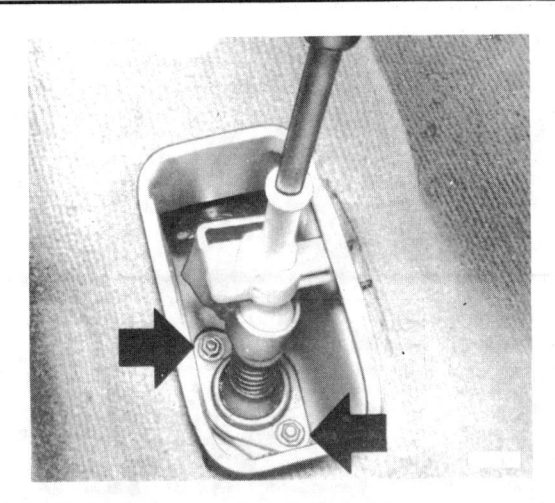

Fig. 6.9 Gear lever mounting plate retaining nuts – arrowed (Sec 6)

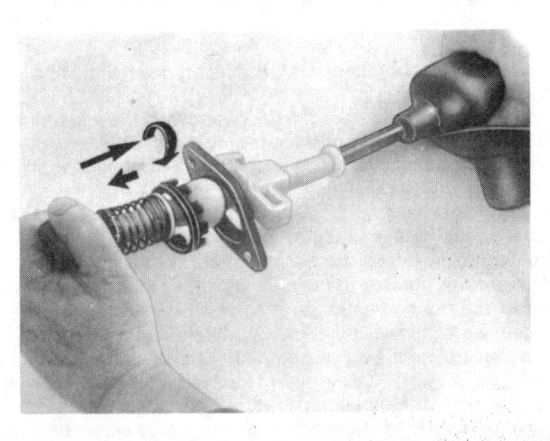

Fig. 6.10 Remove the gear lever from the pivot ball by pushing the ball up and turning clockwise – gear lever shown removed for clarity (Sec 6)

3 Undo the two nuts securing the gear lever mounting plate to the remote control housing.
4 Pull the gear lever up as far as possible and then reach down to the base of the lever and grasp the pivot ball. Push the pivot ball upwards against spring pressure and at the same time turn it clockwise. When the ball is released, lift out the gear lever and mounting plate assembly.
5 Refitting is the reverse sequence to removal, but lubricate the contact points with silicone grease prior to fitting.

7 Gear lever rubber boot – removal and refitting

1 Remove the gear lever, as described in Section 6.
2 Immerse the gear lever knob in water at approximately 80°C (176°F) for a few minutes and then remove the knob by twisting and tapping it off the lever.
3 Slide the rubber boot off the lever and then fit the new boot.
4 Heat the knob as previously described and refit it to the lever using a soft-faced mallet or hammer and block of wood to drive it on. Make sure that the shift gate pattern on the knob is correctly positioned in relation to the lever.
5 Refit the gear lever, as described in Section 6.

6 Gear lever – removal and refitting

1 Remove the centre console, if fitted, as described in Chapter 12.
2 Place the gear lever in neutral and slide the rubber boot up the lever.

8 Transmission – removal and refitting

1 Disconnect the battery negative terminal.
2 Refer to Chapter 5 and disconnect the clutch cable from the clutch release lever and transmission.

Chapter 6 Transmission

Fig. 6.11 Gearchange linkage and clutch cable disconnection points (arrowed) for transmission removal (Sec 8)

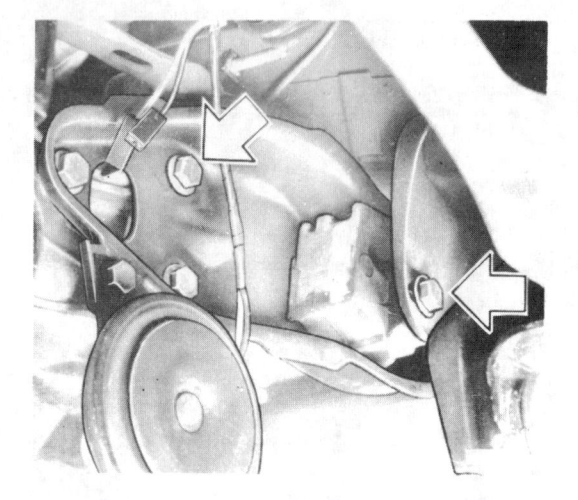

Fig. 6.12 Left-hand engine/transmission mounting retaining bolt locations – arrowed (Sec 8)

3 Disconnect the two electrical leads at the reversing lamp switch on the transmission casing.
4 Remove the speedometer cable by unscrewing the end fitting from the driven gear on the transmission casing.
5 Extract the retaining clip and withdraw the connecting pin from the gearchange linkage universal joint.
6 Withdraw the gearchange linkage pivot retaining pin from the pivots after releasing the spring retaining cap.
7 Remove the trim from both front wheels and slacken the wheel bolts. Jack up the front of the car, support it securely on axle stands and remove the front wheels.
8 At the base of the steering knuckle on the right-hand side of the car, undo the nut and bolt securing the shank of the control arm balljoint to the knuckle. Lever the control arm down to release the balljoint shank from the steering knuckle.
9 Working on the left-hand side of the car release the clamps securing the anti-roll bar (where fitted) to the tie-bar and remove the tie-bar front mounting from the mounting bracket. Remove the control arm inner mounting/pivot bolt securing the arm to its mounting bracket. Further details of these operations will be found in Chapter 11.
10 Release both driveshaft inner constant velocity joints from the transmission using a suitable tool, as described in Chapter 7. Be prepared for some oil spillage as the inner joints are removed. Suitably tie up or support the driveshafts to prevent undue strain being placed on the outer joints.

11 Remove the plug from the transmission end cover and withdraw the input shaft – using the procedure described in Chapter 5, Section 5.
12 Undo the nut and remove the earth strap from the transmission end cover (photo).
13 Take the weight of the engine on a hoist or jack. Place a second jack, preferably of trolley type, under the transmission.
14 Undo the retaining bolts and lift off the flywheel cover plate. This is the cover plate that faces the crankshaft pulley end of the engine. Note that on 1.0 models the plate is of more substantial construction and incorporates side braces which must be unbolted from the engine cylinder block.
15 With the engine and transmission well supported on the jacks undo the nut and remove the bolt securing the rear engine/transmission mounting bracket to the rubber mounting block (photo).
16 Disconnect the two electrical leads at the horn and then undo and remove the bolts securing the left-hand front engine/transmission mounting to the transmission casing and body frame side-member. Remove the mounting and bracket complete with horn.
17 Undo the bolts securing the transmission bellhousing to the engine, taking note of the position of any cable clips or support brackets. Move all disconnected wires, cables and controls to one side well clear of the transmission.
18 Lower the engine and transmission slightly, withdraw the transmission from the engine and withdraw the unit from under the car.

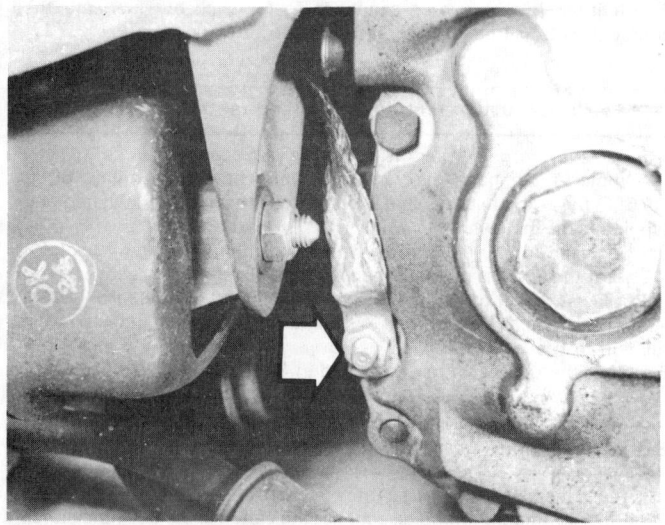

8.12 Earth strap attachment (arrowed) at transmission casing

8.15 Rear engine/transmission mounting retaining bolt (arrowed)

Chapter 6 Transmission

19 Refitting the transmission is the reverse sequence to removal, bearing in mind the following points:

(a) Ensure that all retaining nuts and bolts are tightened to the specified torque
(b) Refit the driveshaft inner constant velocity joints to the transmission, as described in Chapter 7
(c) Reconnect the front suspension control arm, tie-bar and anti-roll bar, using the procedures described in Chapter 11
(d) Refit the transmission input shaft and end cover plug then refit and adjust the clutch cable, using the procedures described in Chapter 5.
(e) Adjust the gearchange linkage and refill or top up the transmission oil, as described in this Chapter, Sections 3 and 2 respectively

9 Transmission overhaul – general

The overhaul of the transmission requires the use of certain special tools and some critical adjustments, if the job is to be done successfully. For this reason it is not recommended that a complete overhaul be attempted by the home mechanic unless he has access to the tools required and feels reasonably confident after studying the procedure. However the transmission can at least be dismantled into its major assemblies without too much difficulty, and the following Sections describe this and the overhaul procedures.

Before starting any repair work on the transmission, drain the oil by removing the differential cover plate, and also thoroughly clean the exterior of the casings using paraffin or a suitable solvent. Dry the unit with a lint-free rag. Make sure that an uncluttered working area is

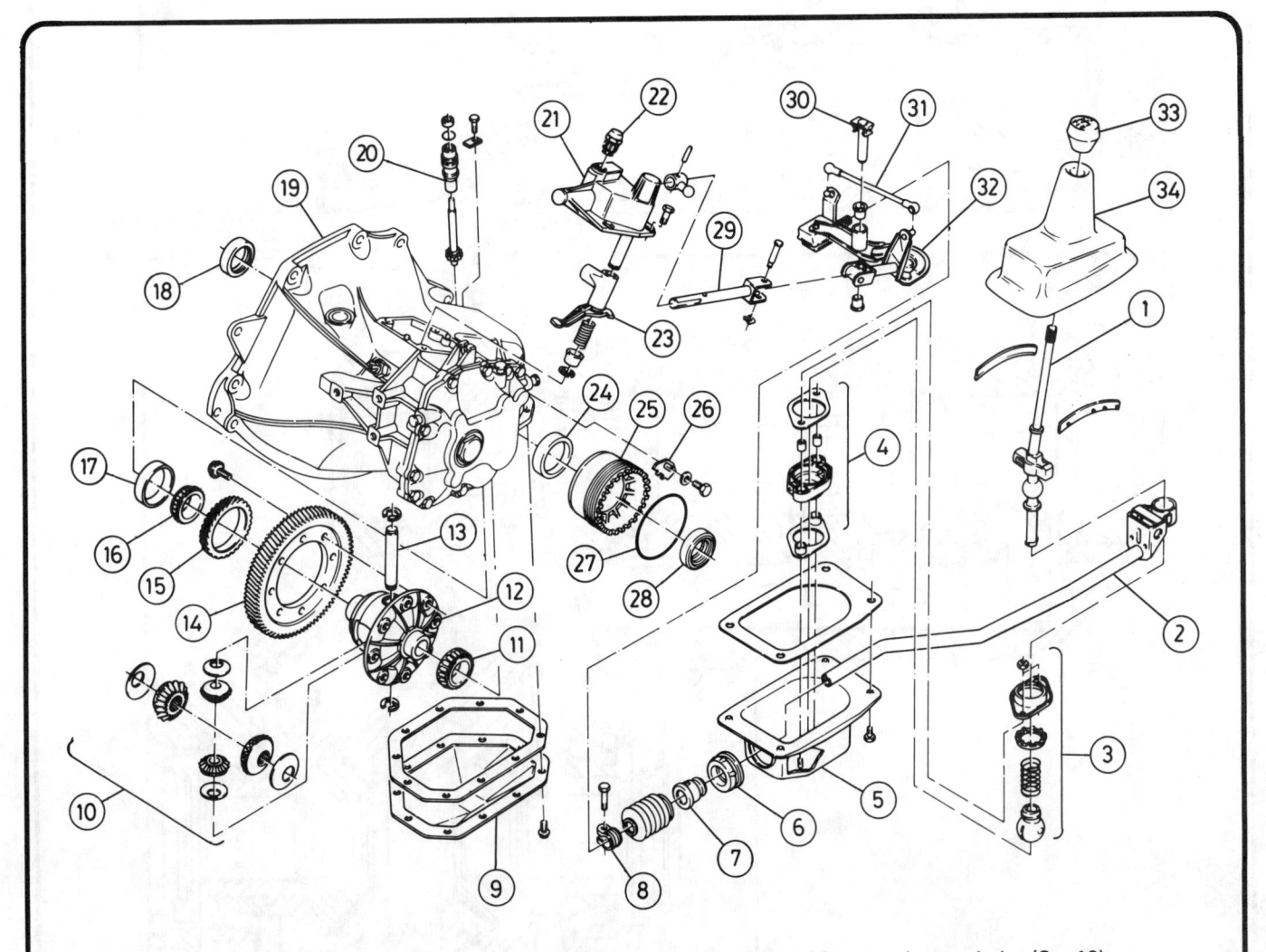

Fig. 6.13 Differential and gearchange mechanism components – four and five-speed transmission (Sec 10)

1 Gear lever
2 Remote control rod
3 Gear lever pivot ball assembly
4 Gear lever mounting plate assembly
5 Remote control housing
6 Bearing ring
7 Bushing
8 Remote control rod clamp
9 Differential cover plate
10 Differential pinion and side gears
11 Differential bearing
12 Differential case
13 Differential pinion shaft
14 Crownwheel
15 Speedometer drivegear
16 Differential bearing
17 Bearing outer track
18 Oil seal
19 Transmission casing
20 Speedometer driven gear assembly
21 Selector cover
22 Oil filler/breather plug
23 Selector lever
24 Bearing outer track
25 Bearing adjuster ring
26 Ring lock
27 O-ring seal
28 Oil seal
29 Gearchange rod
30 Pivot retaining pin
31 Linkage operating rod
32 Gearchange linkage assembly
33 Gear lever knob
34 Rubber boot

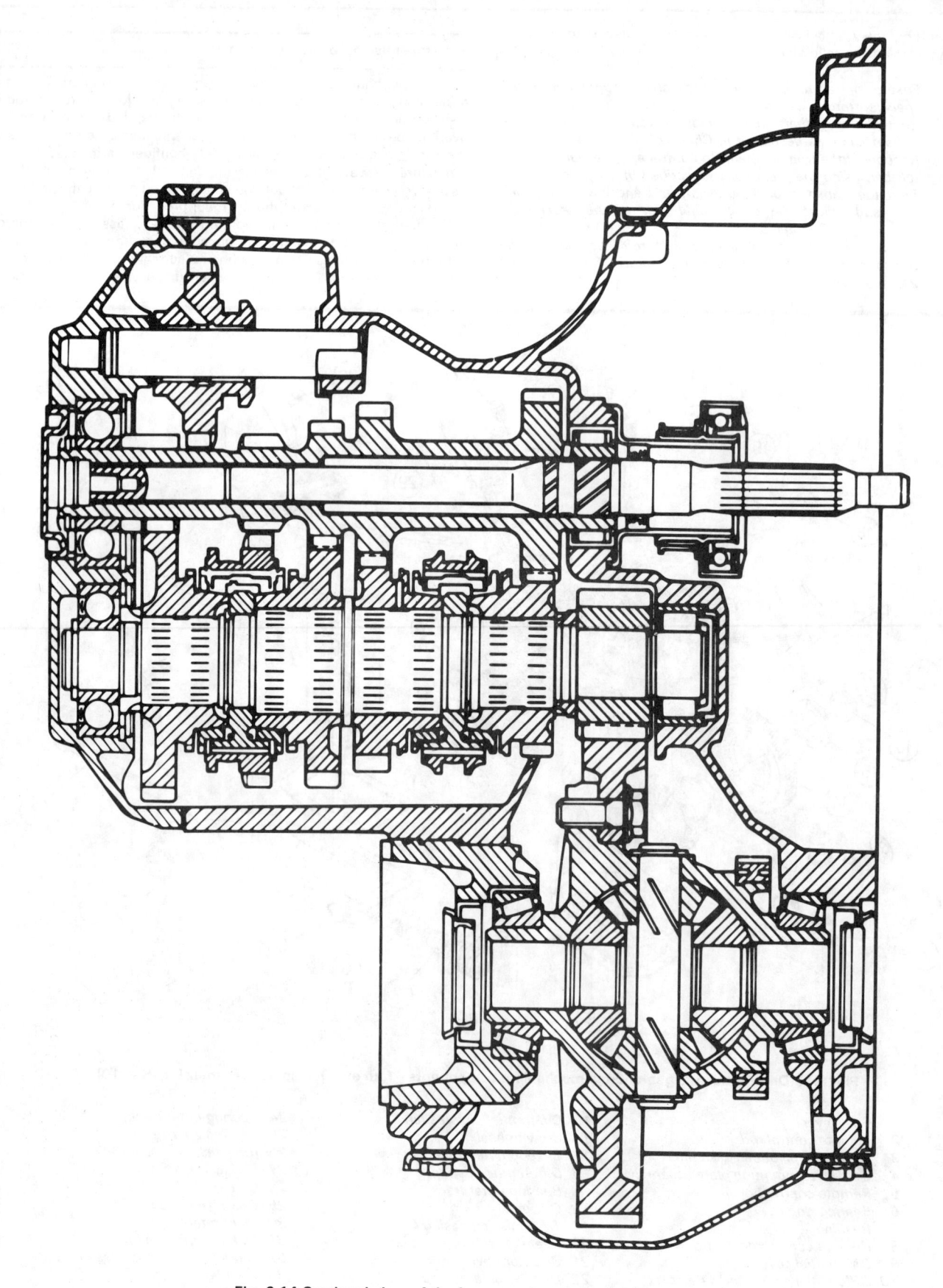

Fig. 6.14 Sectional view of the four-speed transmission (Sec 10)

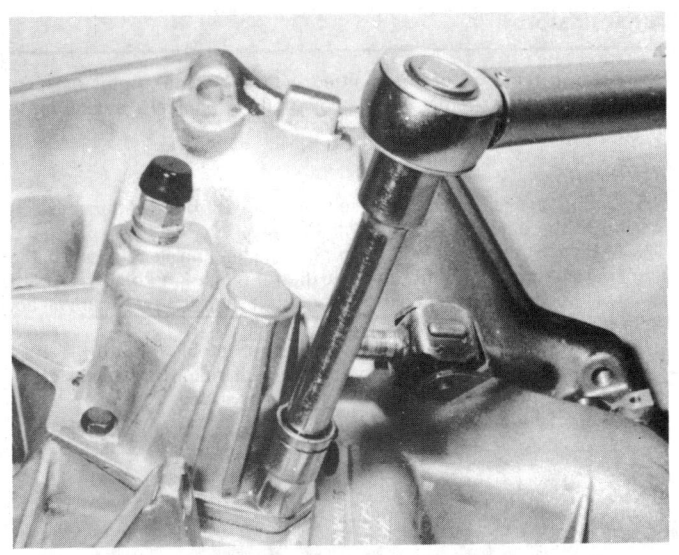

10.2A Undo the retaining bolts ...

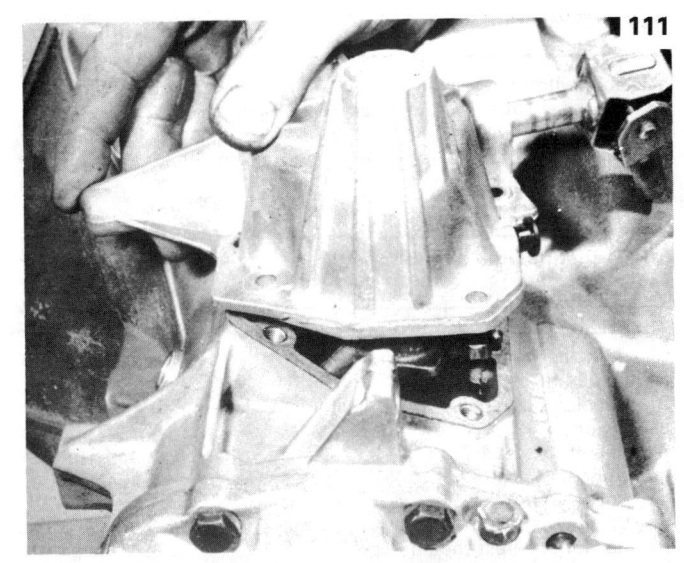

10.2B ... and remove the selector cover

10.3A Unbolt the retaining plate ...

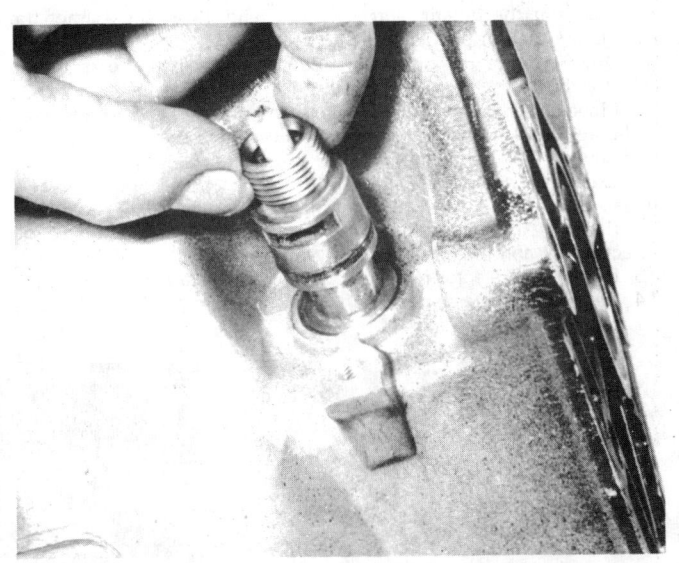

10.3B ... and withdraw the speedometer driven gear assembly

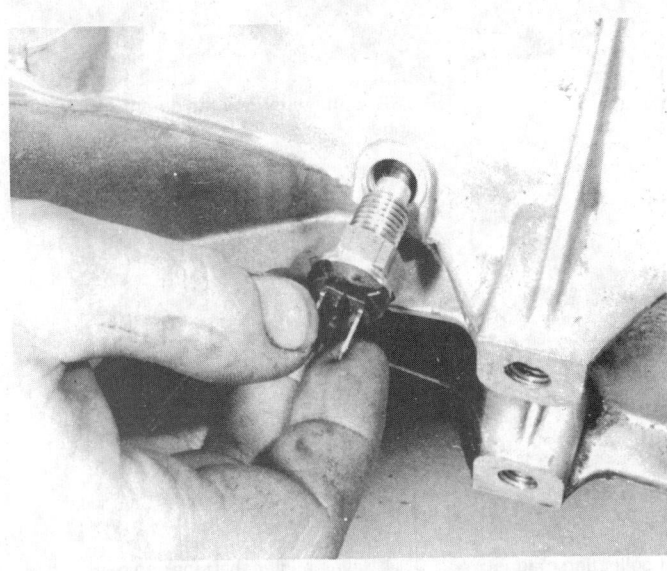

10.4 Unscrew the reversing lamp switch

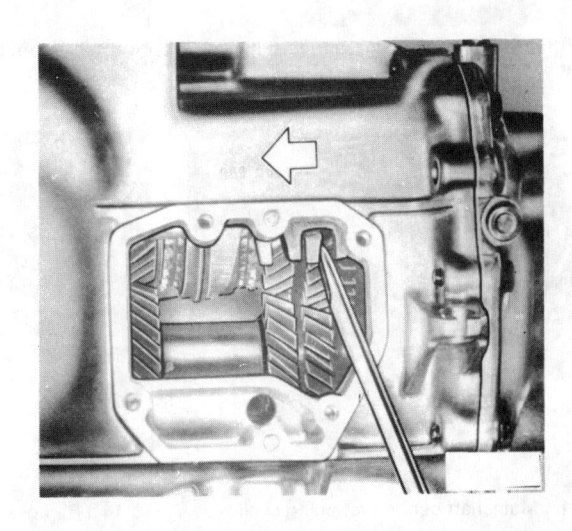

Fig. 6.15 Using a screwdriver to engage 2nd gear (Sec 10)

Chapter 6 Transmission

available with some small containers and trays handy to store the various parts. Label everything as it is removed.

Before starting reassembly, all the components must be spotlessly clean and should be lubricated with the recommended grade of gear oil during assembly.

10 Transmission (four-speed) – dismantling into major assemblies

Gearbox

1 With the transmission removed from the car and on the bench proceed as follows:
2 Unbolt and remove the selector cover from the transmission casing (photos).
3 Unbolt the retaining plate and withdraw the speedometer driven gear (photos).
4 Unscrew and remove the reversing lamp switch (photo).
5 Using a screwdriver as a lever, engage 2nd gear by moving the selector fork nearest the end cover.
6 Unscrew and remove the end cover bolts and nuts.
7 Withdraw the main casing from the end cover and geartrains (photo).
8 Prise out the detent plugs from the end cover and extract the springs and detent plungers (photos).
9 Drive out the roll pins which secure the selector forks to the selector rods.
10 Move the synchro sleeve back to the neutral position and then withdraw 3rd/4th and reverse selector forks and their rods from the end cover.
11 Extract the circlips which retain the mainshaft and input shaft gear trains. Extract the swarf collecting magnet (photos).
12 Remove the geartrain assemblies together with the 1st/2nd selector fork and rod simultaneously.
13 Extract the selector rod interlock pins from the end cover.
14 Remove the reverse idler shaft from the end cover. To do this, grip the shaft in the jaws of a vice fitted with soft metal protectors and, using a brass drift, gently tap the cover off the shaft. Take care not to lose the locking ball (photo).

Differential

15 Unbolt and remove the pressed-steel cover from the transmission casing at the base of the differential (photo).
16 Mark the position of the bearing adjuster ring in relation to the transmission casing (photo). Unbolt the ring lock (photo).
17 Unscrew the bearing adjuster ring. A piece of flat steel bar will act as a suitable wrench (photos).
18 Withdraw the differential/crownwheel assembly (photo).
19 Depending upon the need for further dismantling due to leaking oil seals or worn bearings, proceed in the following way.

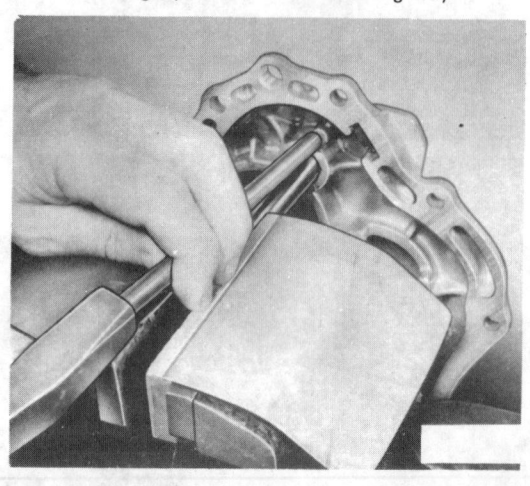

Fig. 6.16 Tapping the end cover off the reverse idler shaft (Sec 10)

10.7 Removing the main casing from the end cover

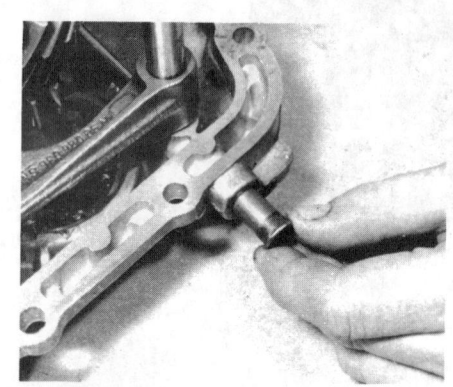

10.8A Prise out the detent plugs ...

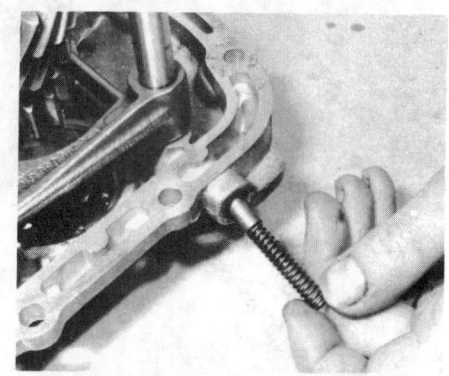

10.8B ... and withdraw the detent springs and plungers

10.11A Mainshaft bearing retaining circlip

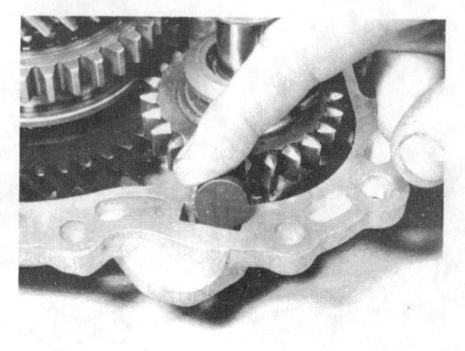

10.11B Remove the swarf collecting magnet

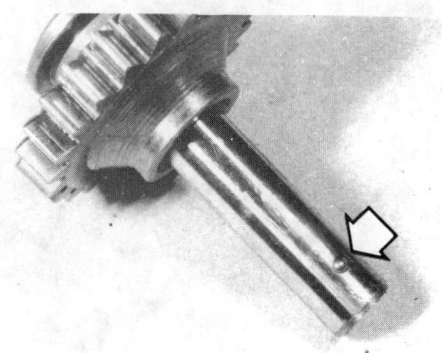

10.14 Reverse idler shaft locking ball (arrowed)

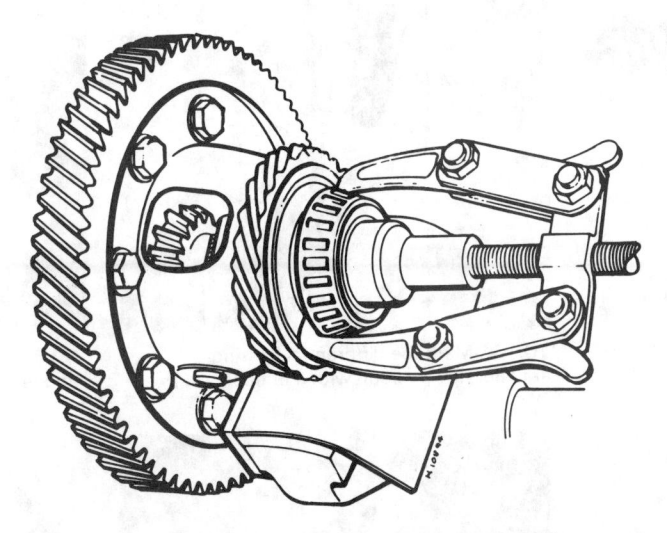

Fig. 6.17 Differential bearing removal (Sec 10)

Fig. 6.18 Removing the crownwheel retaining bolts (Sec 10)

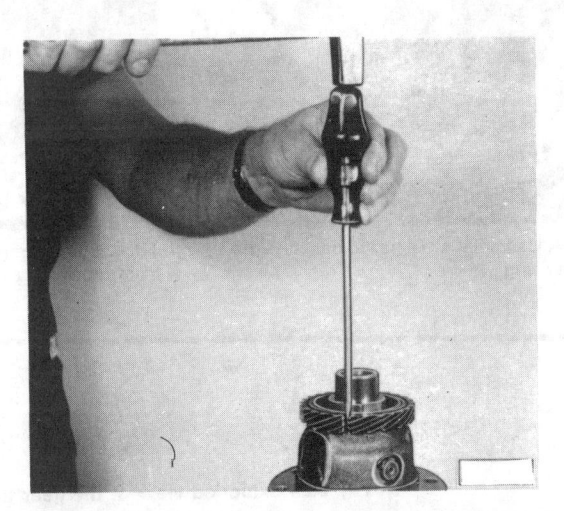

Fig 6.19 Splitting the speedometer drivegear (Sec 10)

Fig. 6.20 Removing the pinion shaft circlips (Sec 10)

Fig. 6.21 Differential pinion and side gear removal (Sec 10)

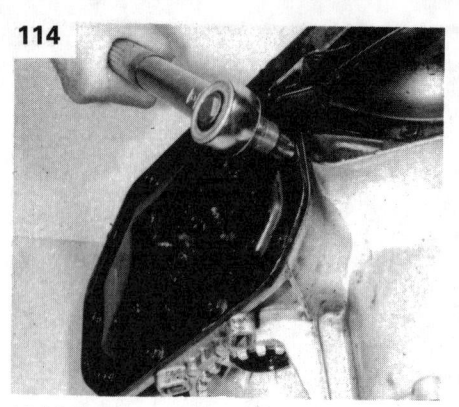

10.15 Remove the differential cover plate

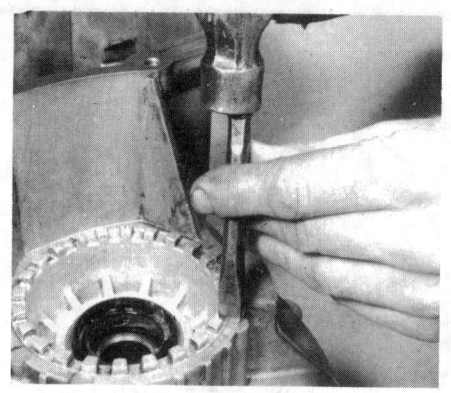

10.16A Mark the differential bearing adjuster ring position with a punch ...

10.16B ... then unscrew the bolt and remove the ring lock

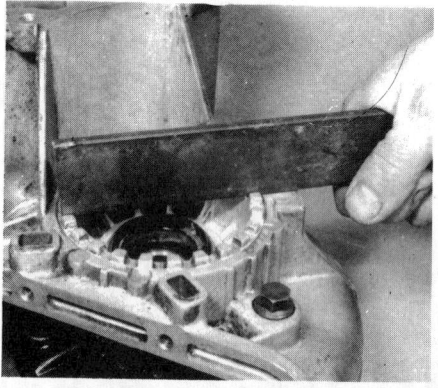

10.17A Use a flat steel bar to unscrew the adjuster ring ...

10.17B ... then remove the adjuster ring from the casing

10.18 Withdraw the differential crownwheel assembly

Fig. 6.22 Exploded view of the gear train assemblies – four-speed transmission (Sec 10)

1. Mainshaft roller bearing (in casing)
2. Circlip
3. Pinion drivegear
4. Spacer
5. 4th gear
6. Baulk ring
7. Synchro sleeve (3rd/4th)
8. Circlip
9. Synchro spring
10. Sliding key
11. Synchro hub (3rd/4th)
12. Synchro spring
13. Baulk ring
14. 3rd gear
15. 2nd gear
16. Baulk ring
17. 1st/2nd synchro sleeve with reverse gear
18. Synchro spring
19. Synchro hub (1st/2nd)
20. Synchro spring
21. Circlip
22. Baulk ring
23. 1st gear
24. Spacer
25. Mainshaft securing circlip
26. Bearing
27. Circlip
28. 3rd/4th selector fork
29. Roll pin
30. Selector rod
31. 1st/2nd selector fork
32. Roll pin
33. Selector rod
34. Input shaft roller bearing (in casing)
35. Plug
36. Transmission casing
37. Gasket
38. Plug
39. Input shaft
40. Input geartrain
41. Input geartrain retaining circlip
42. Bearing
43. Thrust washer
44. Circlip
45. Detent plug
46. Detent spring
47. Detent plunger
48. Long interlock plunger
49. Short interlock plunger
50. Detent plunger
51. Detent spring
52. Detent plug
53. End cover
54. Input shaft screw
55. Circlip
56. End cover screw plug
57. Detent plug
58. Detent spring
59. Detent plunger
60. Reverse selector rod
61. Reverse idler shaft
62. Reverse idler gear
63. Thrust washer
64. Reverse selector fork
65. Roll pin

115

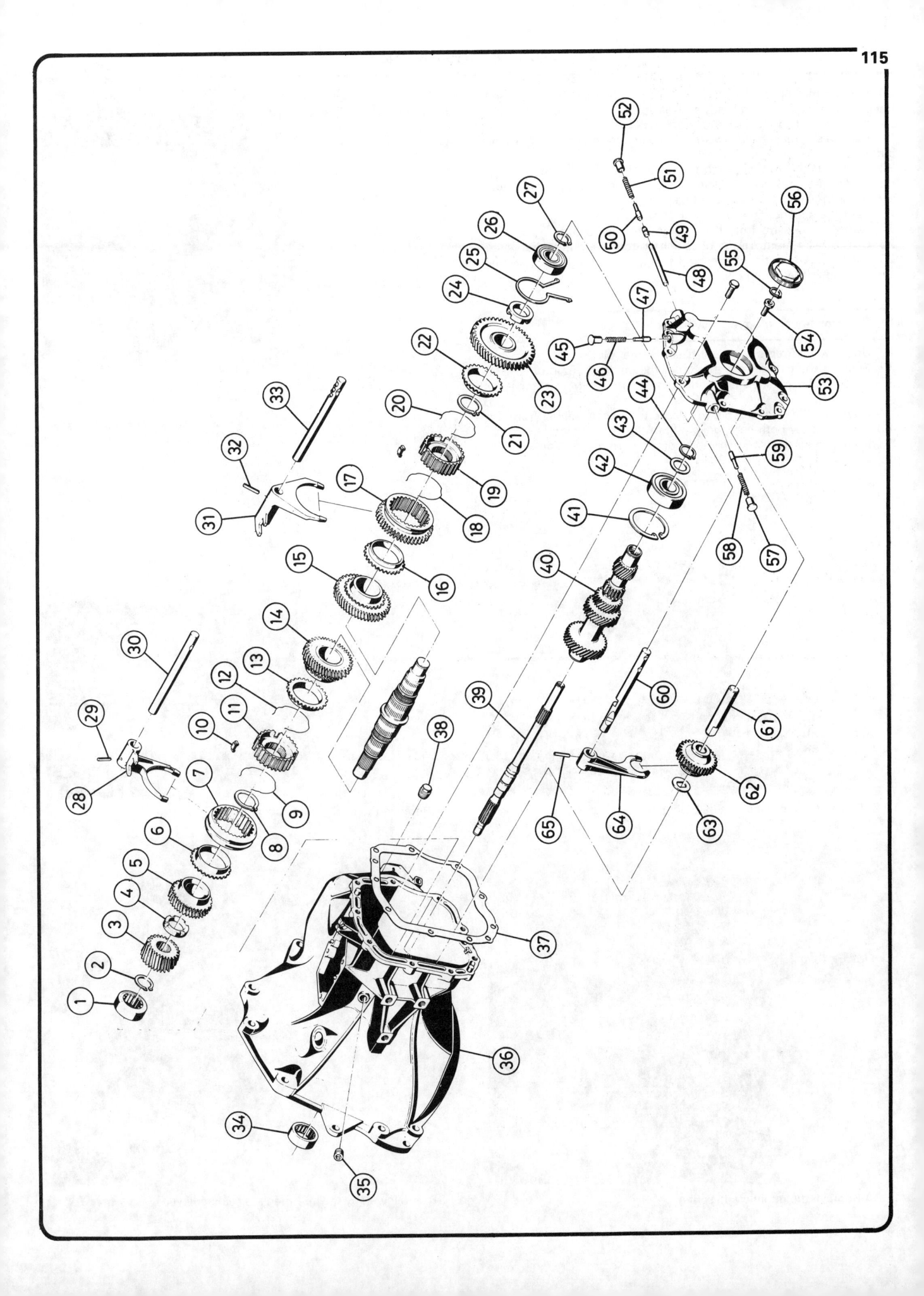

Chapter 6 Transmission

20 Renew the oil seals and bearings in the adjuster ring and transmission casing using a piece of tubing to remove the old components and to install the new (photos).
21 Using a suitable puller remove the tapered roller bearings from the differential.
22 Unbolt the crownwheel and tap it from its register using a brass drift. If the crownwheel or pinion gear are to be renewed, they must always be renewed as a matched pair.
23 Split the speedometer drivegear and discard it.
24 Extract the circlips from the differential pinion shaft.
25 Use a drift to remove the pinion shaft from the differential case.
26 Slide the differential pinions and side gears out of the differential case. Remove the spring discs.

11 Selector housing cover – overhaul

1 Unscrew and remove the oil filler/breather plug.
2 Remove the circlip from the top of the guide pin (photo).
3 Take off the retainer, coil spring and intermediate selector lever (photos).
4 Drive out the retaining pin to release the selector finger from the rod and withdraw both components from the cover (photo).
5 If the universal joint on the selector rod is worn, grind off the rivet

10.20A Using a tube, block of wood and hammer to renew the oil seal and bearing track in the adjuster ring ...

10.20B ... and in the transmission casing

11.2 Remove the circlip from the guide pin

11.3A Remove the retainer and spring ...

11.3B ... followed by the intermediate selector lever

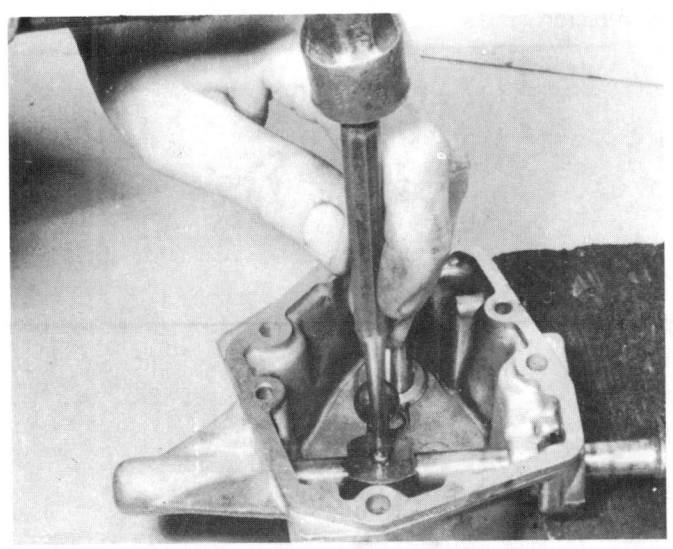

11.4 Drive out the pin to release the selector finger from the gearchange rod

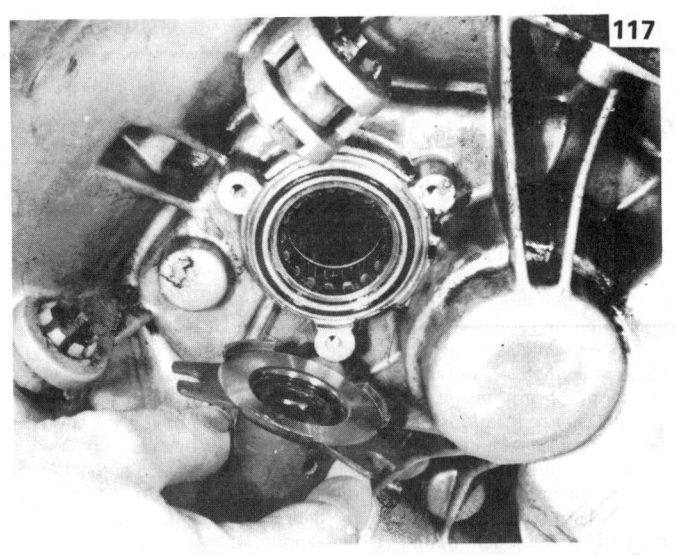

12.2 Remove the release bearing guide tube to gain access to the oil seal and roller bearing

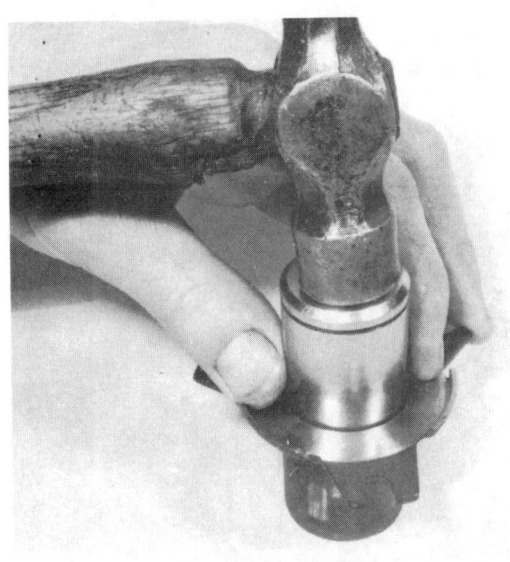

12.3 Fitting a new input shaft oil seal

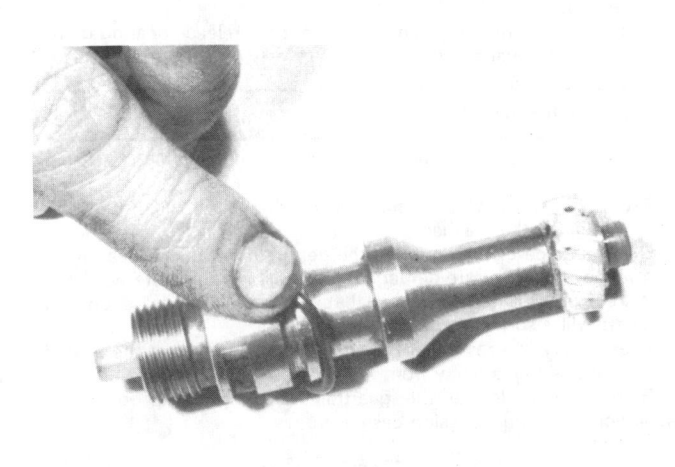

12.5A Speedometer driven gear O-ring ...

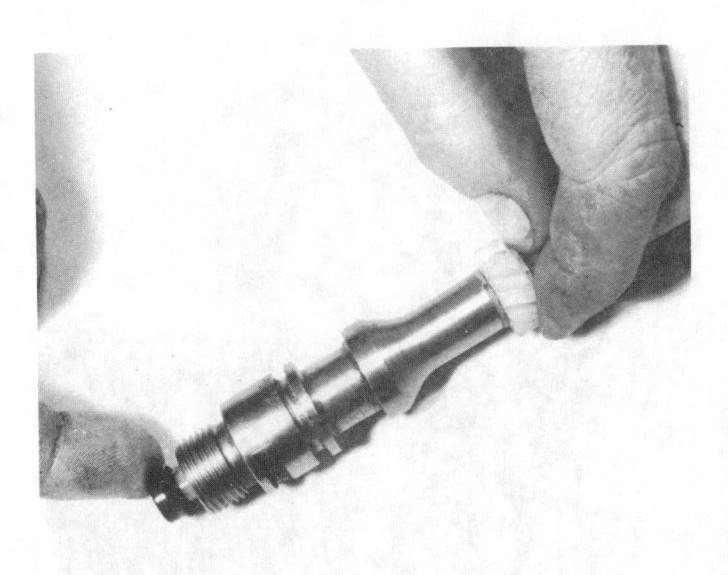

12.5B ... and small oil seal

12.7 Place a new O-ring in position before refitting the release bearing guide tube

to dismantle it and to fit new components. A pin and circlip are supplied as replacements.
6 Renew the oil seal in the selector cover.

12 Transmission casing – overhaul

1 Undo the clamp bolt securing the clutch release fork to the release lever pivot shaft. Pull the release lever upwards out of the transmission casing and withdraw the release fork.
2 Undo the three retaining bolts and remove the release bearing guide tube and input shaft oil seal (photo). Recover the O-ring seal.
3 Prise the old seal out of the guide tube and tap in a new seal using a tube of suitable diameter (photo). Fill the space between the lips with general purpose grease.
4 The release lever bushes in the casing can be removed by tapping them out with a suitable drift. Install the new bushes with their locating tongues engaged in the slots in the casing. Coat the bush inner surfaces with molybdenum disulphide grease.
5 Prise out the speedometer driven gear and its guide. Renew the O-ring and small oil seal before refitting (photos).
6 Inspect the roller bearing in the casing and renew if necessary by drifting it out with a tube of suitable diameter. Refit a new bearing in the same way.
7 With a new O-ring in place (photo) refit the release bearing guide tube and release lever and fork using the reverse procedure to removal.

13 Input shaft – overhaul

1 Support the end of the geartrain and tap or press the shaft from it.
2 Extract the circlip which secures the bearing to the end of the shaft (photo). Take off the washer.
3 Using a piece of tubing, drive the geartrain out of the bearing.
4 If any of the gears are damaged, the geartrain complete will have to be renewed. This will mean that the matching gears on the mainshaft will also have to be renewed.
5 Reassembly is a reversal of dismantling, but note that the sealed side of the bearing is away from the gear.
6 Remember to locate the geartrain securing circlip ready for installation in its transmission casing groove.

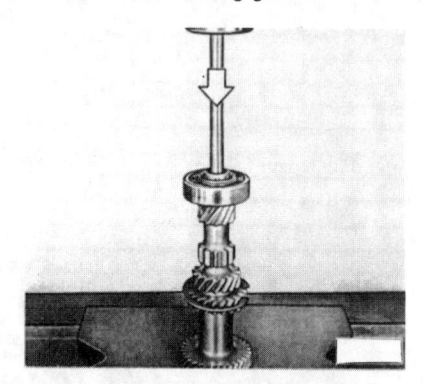

Fig. 6.23 Removing input shaft from input geartrain (Sec 13)

14 Mainshaft – overhaul

Note: *An hydraulic press, or suitable heavy-duty puller, may be required to carry out the following procedure. Before pressing the shaft out of a gear, synchro unit, etc, ensure that the lower face of the component is firmly supported. Similarly, if a puller is used, ensure that its claws are securely located on the main body of the component, **not** solely on the gear teeth. In extreme cases, removal of stubborn components may be aided by gentle heating, bearing in mind that it is not recommended to heat any component above 100°C (212°F).*

1 Extract the retaining circlip from the bearing at the end of the shaft (photo).
2 Support the 1st gear and drive the shaft out of the bearing and gear. Note the spacer washer between the bearing and gears (photos).
3 Support 2nd gear and then extract the circlip which secures the 1st/2nd synchro (photo).

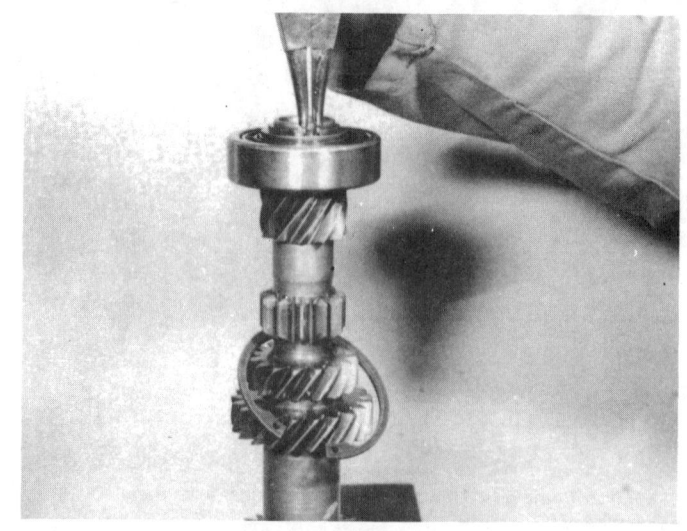

13.2 Removing the input shaft bearing circlip

14.1 Remove the mainshaft bearing circlip

14.2A Remove the mainshaft bearing ...

Chapter 6 Transmission

4 Take the synchro baulk ring from the shaft (photo).
5 Take the 1st/2nd synchro unit from the shaft; if a puller is used, apply pressure behind 2nd gear, and remove 2nd gear, the baulk ring, and the 1st/2nd synchro unit as an assembly. Note the reverse gear teeth on the sleeve (photo).
6 Remove the next baulk ring (photo).
7 Remove the 2nd gear (photo).
8 Now turn your attention to the opposite end of the mainshaft. Extract the circlip which secures the pinion drivegear to the shaft (photo).
9 Remove the pinion drivegear (photo).
10 Remove the spacer washer (photo).
11 Remove 4th gear (photo).
12 Remove the baulk ring (photo).
13 Extract the circlip which secures the 3rd/4th synchro unit to the mainshaft (photo).
14 Remove the 3rd/4th synchro unit (photo).
15 Remove the next baulk ring (photo).
16 Remove 3rd gear.
17 With the mainshaft completely dismantled, examine the gears for chipped or worn teeth and the shaft for deformation of splines. Renew all circlips.
18 If there has been a history of noisy gear changing or if the synchromesh could be easily beaten during changes, renew the synchro unit complete or overhaul, as described in the next Section.
19 With all parts clean and oiled, reassemble as follows. Where necessary, tap components onto the shaft using a suitably-sized tubular drift which bears on the component's main body, **not** solely on the gear teeth.
20 Fit 3rd gear to the shaft's pinion drivegear end, then place the first baulk ring on the gear's cone.
21 Heating it to 100°C (212°F) if necessary, refit the 3rd/4th synchro unit as noted in paragraph 19 above (photo).
22 Fit a new circlip to secure the synchro unit, then refit the remaining baulk ring, followed by 4th gear.
23 Heating them to 100°C (212°F) if necessary, refit the spacer washer and the pinion drivegear (photo). Fit a new circlip to secure the pinion drivegear.
24 Fit 2nd gear over the shaft's opposite end, then place the first baulk ring on the gear's cone.
25 Heating it to 100°C (212°F) if necessary and ensuring that the reverse gear teeth are next to 2nd gear, refit the 1st/2nd synchro unit as noted above (photo).
26 Fit a new circlip to secure the synchro unit, then refit the remaining baulk ring, followed by 1st gear.
27 Heating it to 100°C (212°F) if necessary, refit the spacer washer, with its grooves against 1st gear (photo).
28 Fit a new long-eared circlip, then the bearing; secure the bearing with a new circlip.

14.2C ... followed by 1st gear

14.3 Extract the circlip retaining 1st/2nd synchro unit

14.2B ... and the spacer washer ...

14.4 Lift off the baulk ring

14.5 Remove the 1st/2nd synchro unit

14.6 Lift off the next baulk ring

14.7 Remove 2nd gear

14.8 At the other end of the mainshaft, extract the pinion retaining circlip

14.9 Remove the pinion drivegear

14.10 Lift off the spacer washer

14.11 Remove 4th gear

14.12 Lift off the baulk ring

14.13 Extract the circlip retaining the 3rd/4th synchro unit

14.14 Remove the 3rd/4th synchro unit

14.15 Lift off the next baulk ring followed by 3rd gear

14.21 Use a suitable tube to refit 3rd/4th synchro unit

Chapter 6 Transmission

14.23 Fitting the spacer washer between 4th gear and the pinion drivegear

14.25 Use a suitable tube to refit the 1st/2nd synchro unit, ensuring that the reverse gear teeth are towards 2nd gear

14.27 Refit 1st gear and the spacer washer

15 Synchroniser units – overhaul

1 Components of 1st/2nd and 3rd/4th synchro units are interchangeable.
2 It is not good practice, however, to mix parts which have been in use for a high mileage and which have not run-in together.
3 If either the hub or sleeve show signs of wear in their teeth the individual part may be renewed, but general wear is best rectified by complete renewal of the unit.
4 To dismantle, push the sleeve off the hub, taking care not to allow the sliding keys to fly out.
5 Extract the circular springs and keys.
6 Reassembly is a reversal of dismantling. Make sure that the hooked ends of the springs engage in the same sliding key but run in opposite directions in relation to each other (photos).
7 To check the baulk rings for wear, twist them onto the gear cones. The ring should 'stick' to the cone and show a definite clearance between the ring and gear shoulder. If these conditions are not met, renew the baulk rings.

16 Transmission (four-speed) – reassembly

1 Liberally lubricate the differential components with the recommended grade of oil.
2 Install the side gears and pinions, the spring discs and the pinion shaft into the differential case.
3 Fit new retaining circlips.
4 If the speedometer drivegear was removed, warm the new gear in hot water at 80°C (176°F) and tap it onto the differential case with a piece of tubing until it snaps into position. Make sure that the lugs on the gear are aligned with the cut-outs in the differential case.
5 Warm the crownwheel to 80°C (176°F) and locate it on the differential case. Use new bolts and tighten them to the specified torque.
6 Fit the tapered roller bearings to the differential case (if removed at dismantling).
7 If not already done, fit the bearing outer tracks to the transmission casing.
8 Fit new driveshaft seals into the transmission casing (if not already done) and fill the lips with grease.
9 Lower the differential into the transmission casing.
10 Fit a new O-ring and oil seal to the bearing adjuster ring (photo). Apply grease to the seal lips and to the screw threads (photo).
11 Screw the adjuster ring into the transmission casing, hand tight at this stage.
12 Adjust the bearing in one of the following ways, depending upon whether the original bearings have been refitted or new ones installed.
13 **Original bearing**: Simply screw in the adjuster ring until the alignment marks made before dismantling are opposite to each other. Should any axial play exist, the ring may be further adjusted to give a turning torque of between 6.1 and 10.3 kgf cm (5.3 and 8.9 lbf in) as described for new bearings in the following paragraph.
14 **New bearings**: The bearing preload must be adjusted by means of the adjuster ring so that a torque of 15.3 and 18.3 kgf cm (13.3 to 15.9 lbf in) is required to keep the crownwheel and bearings turning slowly. Unless a special torsion or friction gauge is available push a tapered softwood rod into the splined side gear and then wrap a cord round it and attach it to a spring balance. Provided the cord leaves the rod at a point about 25 mm (1.0 in) from the centre point of its cross section, the torque will be fairly accurately measured on the 'lb' scale. (If using the metric scale a small calculation will be required.) Adjust the ring until the turning torque is within the specified range.
15 Fit the adjuster lock without moving the position of the ring. The ring lock is elongated to make this possible.
16 Use a new gasket and bolt the pressed-steel cover to the transmission casing (photo).
17 Fit the reverse idler shaft to the transmission end cover, making sure that the locking ball is in position.
18 Pin the 1st/2nd selector fork to its rod, but leave the pin projecting by approximately 2.0 mm (0.08 in).
19 Hold the mainshaft, input shaft and reverse geartrains meshed together, with the 1st/2nd selector fork and rod engaged in the groove of 1st/2nd synchro.
20 Locate the assembly into the end cover. The help of an assistant will facilitate the work. Fit the selector rod interlock plungers (photos).

15.6A Refit the synchro sleeve to the hub ...

15.6B ... locate the sliding keys in their grooves ...

15.6C ... and retain the keys with the circular springs

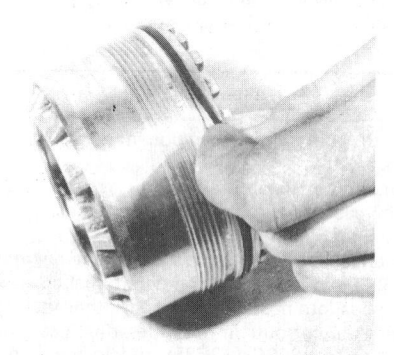

16.10A Fit a new O-ring to the bearing adjuster ring

16.10B Fill the space between the oil seal lips with grease

16.16 Refit the differential cover plate using a new gasket

16.20A Locate the short ...

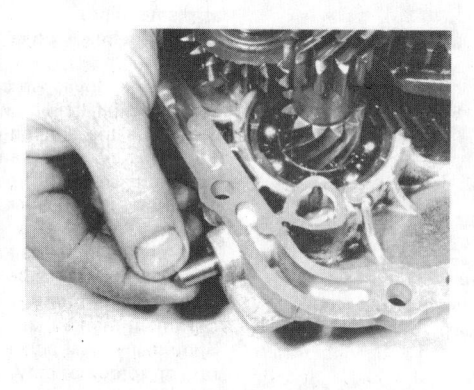

16.20B ... and long interlock plungers in the end cover

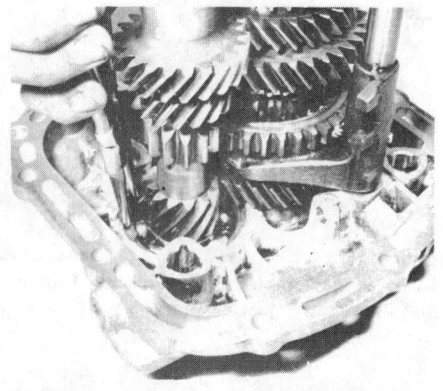

16.21A Engage the geartrain retaining circlips in their grooves ...

16.21B ... and refit the reverse idler thrust washer

16.22A Refit the 3rd/4th ...

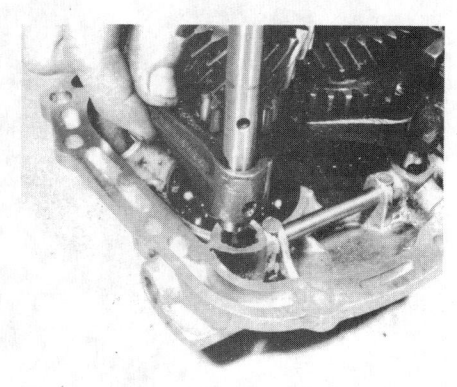

16.22B ... and reverse selector forks and rods

Chapter 6 Transmission

16.23 Secure the selector forks by tapping in the roll pins

16.26 Refit the casing to the end cover using a new gasket

21 Fit the circlips which retain the mainshaft and input shaft assemblies to the transmission casing (photo). Make sure that they engage positively in their grooves. Fit the thrust washer to the reverse idler gear (photo), also the swarf collecting magnet.
22 Check that the sleeve on 1st/2nd synchro is in neutral then fit the 3rd/4th and reverse selector forks and rods (photos).
23 Pin the forks to the rods (photo).
24 Refit the detent plungers and springs. If the sealing plugs are not a really tight fit, oversize ones should be obtained and driven in.
25 Using a screwdriver, move the sleeve of the appropriate synchro unit to engage 2nd gear.
26 Stick a new gasket, with grease, to the transmission casing (photo) and then insert the geartrains with end cover into the casing until the fixing bolts and nuts can be screwed in to the specified torque.
27 Fit the speedometer driven gear and bolt on its retainer plate.
28 Screw in the reversing lamp switch to the specified torque.
29 Set the transmission in neutral and stick a new selector cover gasket in position.
30 Bolt on the selector cover, tightening the bolts to the specified torque.
31 The transmission can be filled with oil now as described in Section 2 provided it is held in the in-vehicle attitude, otherwise wait until it has been refitted to the vehicle.

17 Transmission (five-speed) – dismantling and reassembly

1 The five-speed manual transmission fitted to Nova models is essentially the same as the four-speed limit with the exception of the additional fifth gear and synchroniser, and modifications to the selector mechanism. Dismantling and overhaul procedures for the four-speed unit should therefore be used in conjunction with the following supplementary information.
2 Undo the bolts securing the end cover shield to the end cover and then proceed with the dismantling, as described in Section 10, paragraphs 2 to 7 inclusive.
3 Using a suitable Allen key undo the two socket-headed bolts and remove the bearing support with selector fork from the end cover.
4 Engage 3rd and reverse gear by moving the selector forks.
5 Extract the 5th gear synchroniser retaining circlip and then remove the synchroniser using a two-legged puller. Recover the baulk ring.
6 Lift off the mainshaft 5th gear, needle roller bearing, the two thrust washer halves and the thrust washer retaining ring.
7 Extract the input shaft 5th gear retaining circlip and then draw off the gear using a large puller. Use a tubular distance piece for the puller centre screw so that it bears against the input geartrain and not directly on the input shaft itself.
8 Using an Allen key unscrew the socket-headed screws which secure the 5th gear selector interlock pawl to the end cover.
9 Using a forked tool as a lever, extract the four detent plugs from the edge of the end cover. Be prepared to catch the coil spring which will be ejected. Pull out the detent plungers. Renew the detent plugs if they were damaged during removal.
10 Move 5th gear selector rod to its engaged position and also move the 2nd gear selector fork to engage the gear.
11 Again using the Allen key unscrew the socket-headed bolts and remove the interlock pin bridge piece.
12 Return all the gears and selector rods to neutral.
13 Drive out the securing roll pin and remove the selector shaft and fork for 3rd/4th gears. Remove the reverse shaft and fork in the same way.
14 Pull the 5th gear selector driver from the end cover.
15 The remainder of the dismantling sequence now follows the procedure described in Section 10, paragraphs 11 to 26 inclusive.
16 Overhaul of the selector housing cover, transmission casing, input shaft, mainshaft and synchronisers also follow the procedures previously described.
17 To reassemble the transmission begin by following the procedure described in Section 16, paragraphs 1 to 21 inclusive.
18 Refit the reverse, and 3rd/4th gear selector shafts and forks, and the 5th gear selector driver to the end cover. Secure the forks with the roll pins.
19 Refit the interlock pin bridge piece and secure it in position using new microencapsulated socket headed bolts, tightened to the specified torque.
20 Refit the 5th gear selector interlock pawl to the end cover, noting that the slot in the 3rd/4th selector shaft must align with the pawl. Secure the assembly with new socket-headed bolts tightened to the specified torque.
21 Refit the detent plungers and springs and drive in the detent plugs, noting that the long plug is for the 3rd/4th selector shaft.
22 Press the input shaft 5th gear onto the geartrain with the longer hub toward the bearing.
23 Refit the retaining circlip.
24 Refit the two thrust washer halves, retaining ring, needle roller bearing and 5th gear to the mainshaft.
25 Place the baulk ring in position over 5th gear.
26 Heat the 5th gear synchroniser assembly in boiling water and then position it over the mainshaft. Press or drive it into place using a suitable tube, ensuring that the slots in the baulk ring engage with the sliding keys.
27 Refit the retaining circlip.
28 Refit the bearing support and selector fork to the end cover and secure with two new socket-headed bolts tightened to the specified torque.
29 The remainder of the reassembly procedure is now as described in Section 16, paragraphs 26 to 31 inclusive. Refit the end cover shield using a new gasket before refilling the transmission with oil.

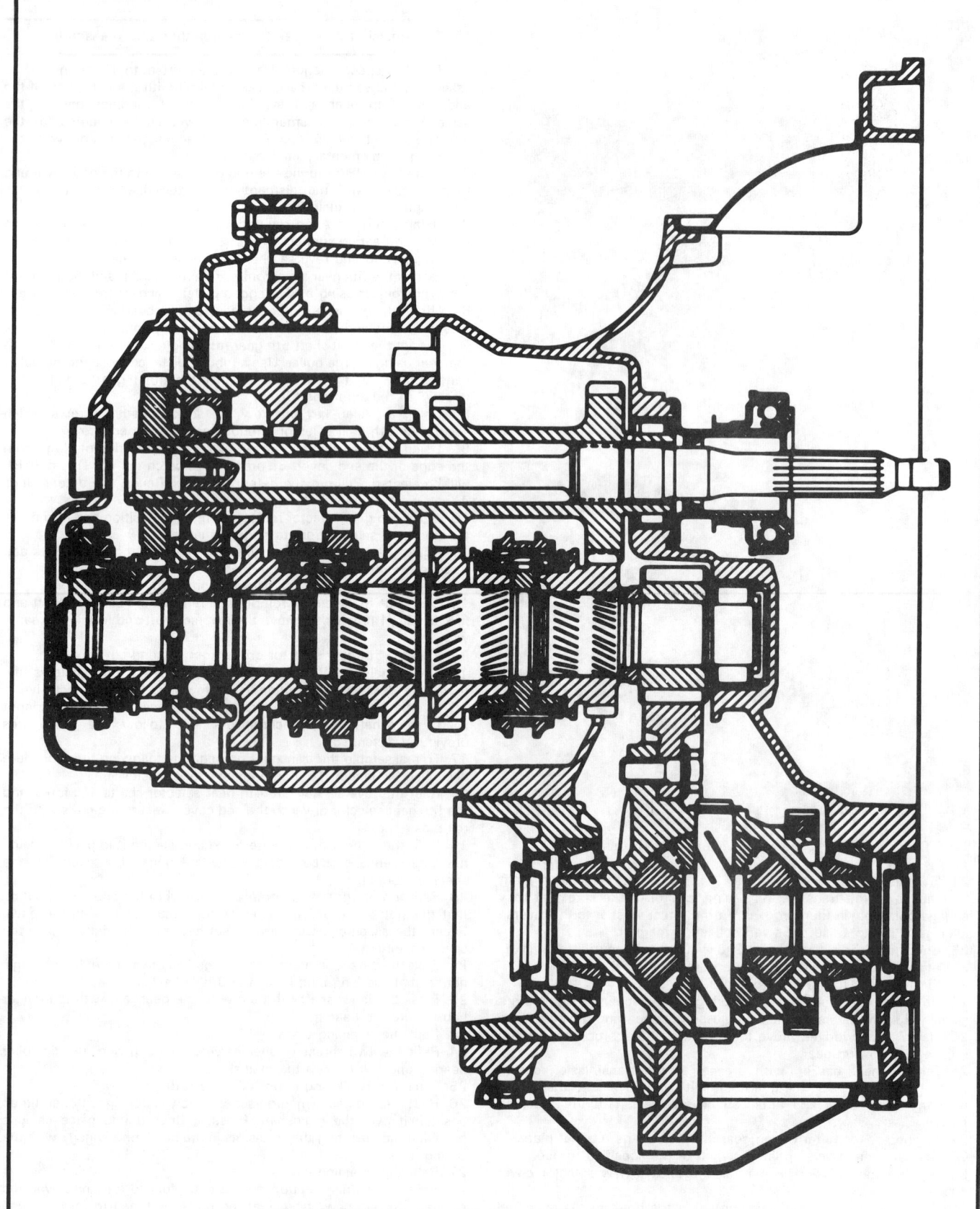

Fig. 6.24 Sectional view of the five-speed transmission (Sec 17)

125

Fig. 6.25 Selector fork and bearing support socket-headed retaining bolts (Sec 17)

Fig. 6.26 Mainshaft 5th gear, needle roller bearing and thrust washers (Sec 17)

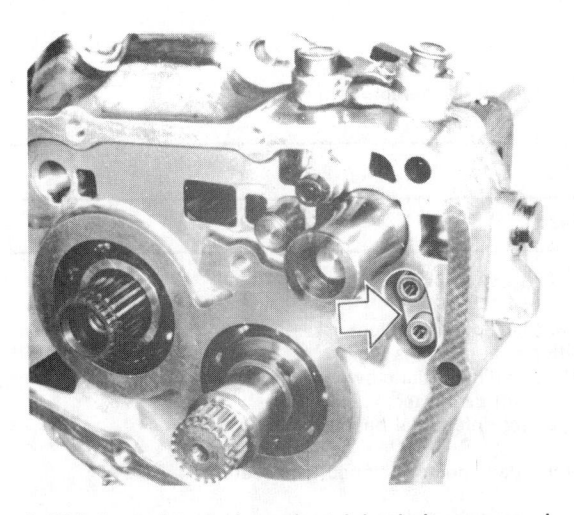

Fig. 6.27 Selector interlock pawl retaining bolts – arrowed (Sec 17)

Fig. 6.28 Interlock pin bridge piece retaining bolts – arrowed (Sec 17)

Fig. 6.29 Removing the 5th gear selector driver (Sec 17)

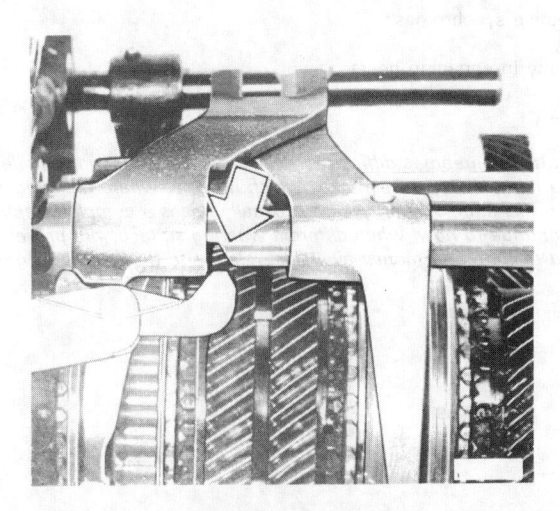

Fig. 6.30 Alignment of 3rd/4th selector shaft slot with interlock pawl – arrowed (Sec 17)

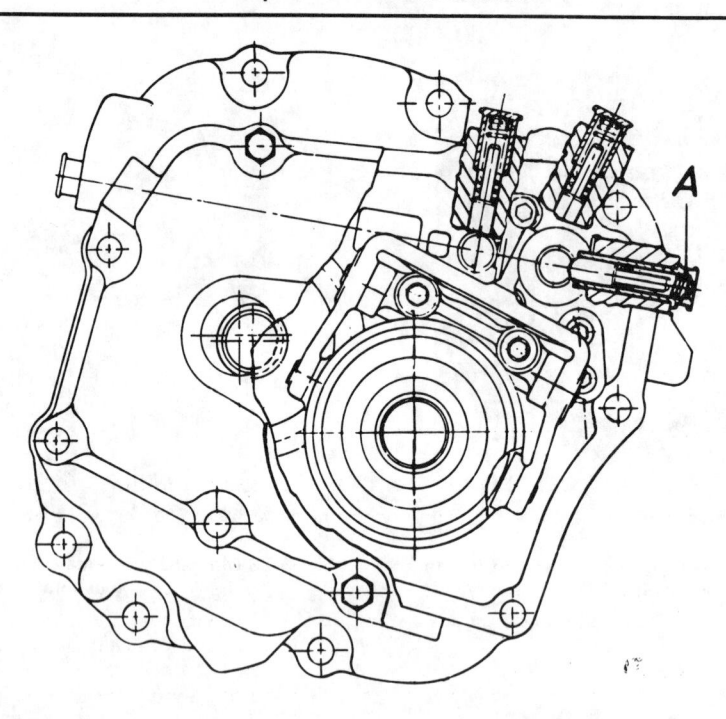

Fig. 6.31 The long detent plunger (A) is for the 3rd/4th selector shaft (Sec 17)

18 Fault diagnosis – transmission

Symptom	Reason(s)
Transmission noisy in neutral	Input shaft bearings worn
Transmission noisy only when moving (in all gears)	Mainshaft bearings worn Differential bearings worn Wear or differential crownwheel and mainshaft pinion teeth Incorrect differential bearing adjustment
Transmission noisy in only one gear	Worn, damaged or chipped gear teeth
Transmission jumps out of gear	Worn synchroniser units Worn selector shaft detent plungers or springs Worn selector forks
Ineffective synchromesh	Worn synchroniser units or baulk rings
Difficulty in engaging gears	Gear linkage adjustment incorrect Worn selector forks or selector mechanism Clutch fault (see Chapter 5)

Note: *It is sometimes difficult to decide whether it is worthwhile removing and dismantling the gearbox for a fault which may be nothing more than a minor irritant. Gearboxes which howl, or where the synchromesh can be beaten by a quick gearchange, may continue to perform a long time in this state. A worn gearbox usually needs a complete rebuild to eliminate noise because the various gears, if re-aligned on new bearings, will continue to howl when different wearing surfaces are presented to each other. The decision to overhaul therefore, must be considered with regard to time and money available, relative to the degree of noise or malfunction that the driver has to suffer.*

Chapter 7 Driveshafts

For modifications, and information applicable to later models, see Supplement at end of manual

Contents

Constant velocity joint – removal and refitting 5	Fault diagnosis – driveshafts 6
Constant velocity joint rubber boot – removal and refitting 4	General description 1
Driveshaft – removal and refitting 3	Maintenance and inspection 2

Specifications

Type .... Unequal length shafts with homokinetic inner and outer constant velocity joints

Torque wrench settings	**Nm**	**lbf ft**
Driveshaft retaining nut:	
Stage 1	100	74
Slacken, then Stage 2	20	15
Stage 3*	Tighten further through 90°
Control arm balljoint-to-steering knuckle retaining bolt	30	22
Torsional damper to driveshaft	10	7
Roadwheel bolts	90	66

After tightening, slacken nut if necessary to align split pin hole. Do not tighten to align

1 General description

Drive is transmitted from the differential to the front wheels by means of two unequal length driveshafts. A constant velocity joint is fitted to both ends of each shaft to cater for steering and suspension movement. The constant velocity joint comprises a driving member (splined to the driveshaft), six caged steel balls and a driven member (splined to the front hub flange). The driven member pivots freely on the steel balls, thus allowing the drive to be smoothly transmitted to the front wheels throughout the full range of steering and suspension travel. A torsional damper is fitted to the longer right-hand driveshaft to overcome torque reaction during acceleration.

2 Maintenance and inspection

1 Every 9000 miles (15 000 km) or 6 months, whichever comes first, carefully inspect the rubber boots protecting the inner and outer constant velocity joints. If they are torn, split, or show signs of deterioration they should be renewed as soon as possible, as described in Section 4. Very rapid wear of the joint internal components due to water and grit ingress will occur if a damaged rubber boot is not renewed quickly.
2 Wear in the outer constant velocity joints is detected by a regular knocking when accelerating from rest with the steering on full lock. In very severe cases it is only necessary to turn the steering slightly for the noise to begin. Wear in the inner joints is often felt as a vibration when accelerating hard in a straight line.
3 The constant velocity joints are lubricated and sealed during assembly and therefore, providing the rubber boots remain intact, require no further lubrication in service.

3 Driveshaft – removal and refitting

1 Remove the wheel trim, slacken the wheel retaining bolts and then jack up the front of the car and securely support it on axle stands. Remove the wheel bolts and the front roadwheel.
2 Extract the split pin from the driveshaft retaining nut then, using a socket and bar, unscrew the nut and recover the thrust washer. The driveshaft can be prevented from turning by having an assistant firmly depress the footbrake, or alternatively by attaching a suitably drilled bar to the hub using two wheel bolts, allowing the end of the bar to rest on an axle stand.
3 Undo and remove the nut and bolt securing the suspension control arm balljoint to the steering knuckle (photo). Lever the control arm down to release the balljoint shank from its location in the steering knuckle.
4 A tool will now be required for insertion between the transmission casing and the inner constant velocity joint. In the absence of the manufacturer's tool, a flat steel bar with a good chamfer on one end will serve as a substitute. Drive the tool into the gap between the joint and the casing to release the internal retaining circlip from the differential side gear. In some cases it may even be possible to release the joint using a large screwdriver or tyre lever (photo).
5 Once the circlip has been released move the driveshaft away from the centre of the car and withdraw the joint the rest of the way out of the transmission casing by hand. As the joint is removed there will be some oil spillage, so have a suitable container handy.
6 It should now be possible to push the outer constant velocity joint out of the hub flange by hand. If it is tight, carefully tap it out using a soft-faced mallet or in extreme cases use a hub puller. Once the outer joint is removed from the hub the driveshaft assembly can be withdrawn from under the car.

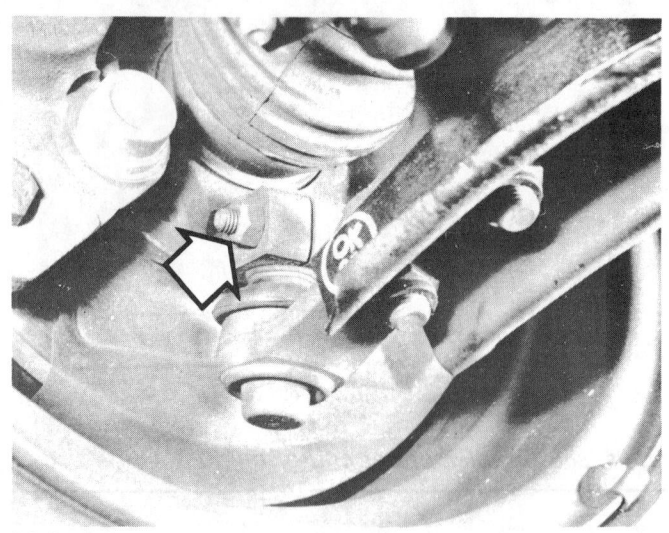

Chapter 7 Driveshafts

7 Before refitting the driveshaft lubricate the splines of both constant velocity joints and make sure that the contact faces of the outer joint and the hub bearing are absolutely clean. Check the condition of the circlip on the end of the inner joint and renew it if it is in any way deformed or damaged.

8 To refit the driveshaft first insert the outer constant velocity joint into the hub flange and firmly push it into place. Fit a new thrust washer and retaining nut finger tight at this stage.

9 Insert the inner constant velocity joint into the transmission casing as far as it will go. Drive the joint fully into place using a screwdriver in contact with the bead of the friction weld around the joint (not the metal cover). Pull on the joint to check that it is fully engaged.

10 Locate the shank of the control arm balljoint in the steering knuckle and insert the retaining bolt so that the bolt head is towards the rear of the car. Refit the nut and tighten to the specified torque.

11 Tighten the driveshaft retaining nut to the Stage 1 setting given in the Specifications to seat the components. Back off the nut and retighten it to the Stage 2 setting. Finally tighten the nut by a further 90° and secure with a new split pin. If the split pin hole in the constant velocity joint does not align with a slot in the nut, turn the nut back until the next slot is in alignment. *Do not tighten to align.*

12 Check the position of the torsional damper on the right-hand driveshaft (photo) and ensure that it is set as shown in Fig. 7.5.

13 Refit the roadwheel, wheel bolts and trim, top up the transmission oil, as described in Chapter 6, and then lower the car to the ground.

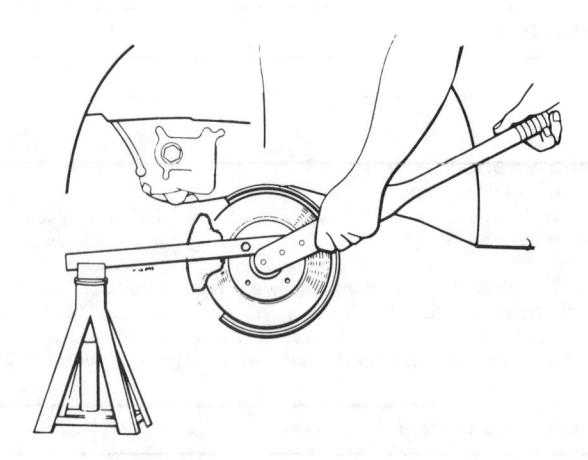

3.3 Control arm balljoint to steering knuckle retaining bolt and nut (arrowed)

3.4 Using a screwdriver to release the driveshaft inner constant velocity joint

Fig. 7.1 Driveshaft retaining nut removal (Sec 3)

3.12 Right-hand driveshaft torsional damper

Fig. 7.2 Manufacturer's special tool for releasing inner constant velocity joint (Sec 3)

Chapter 7 Driveshafts

and constant velocity joint. Fold the rubber boot back and slide it along the driveshaft to expose the internal components of the joint.

3 Secure the driveshaft in a vice, spread the retaining circlip with circlip pliers and tap the joint off the shaft using a soft-faced mallet.

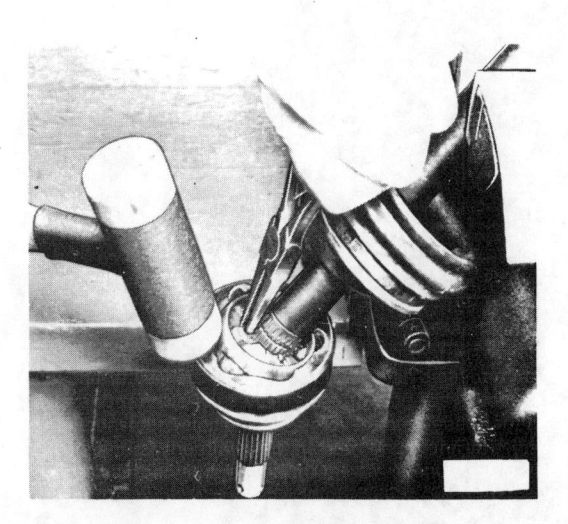

4 With the constant velocity joint removed the rubber boot can now be slid off the end of the driveshaft. If both boots are to be removed, the other boot can be withdrawn from the same end of the driveshaft without having to remove the other joint.

5 Thoroughly clean away all traces of the old grease from the constant velocity joint using paraffin or a suitable solvent. Obtain a new rubber boot, new retaining clips and a suitable quantity of the special lubricating grease from a GM dealer and refit the rubber boot as follows.

6 Slide the boot onto the shaft until the smaller diameter of the boot engages with the groove on the shaft.

7 Wrap the small metal clip around the boot and, pulling the clip as tight as possible, engage the lug on the end of the clip with one of the slots. Use a screwdriver if necessary to push the clip as tight as possible before engaging the lug and slot. Now finally tighten the clip by compressing the raised square portion of the clip with pincers or pliers.

8 Fold the rubber boot back and engage the constant velocity joint with the driveshaft splines. Using a soft-faced mallet, tap the joint onto the shaft until the circlip engages with its groove.

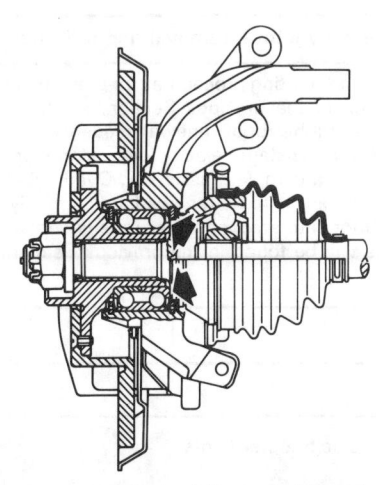

Fig. 7.3 Outer constant velocity joint-to-hub bearing contact faces (arrowed) must be clean before refitting (Sec 3)

Fig. 7.4 Driving the inner constant velocity joint fully into place to engage the retaining circlip (Sec 3)

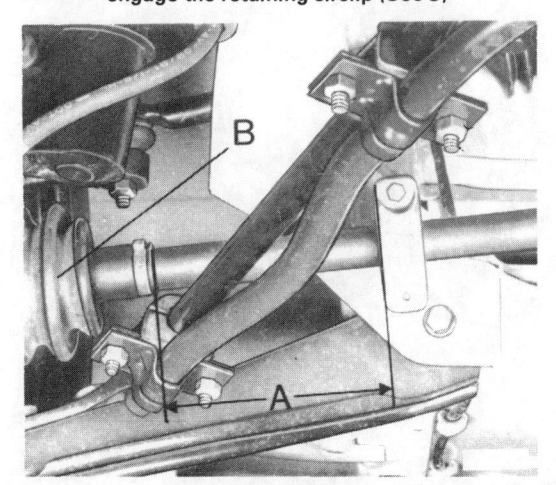

Fig. 7.5 Torsional damper setting position on right-hand driveshaft (Sec 3)

A = 130 mm (5.1 in) from edge of rubber boot (B)

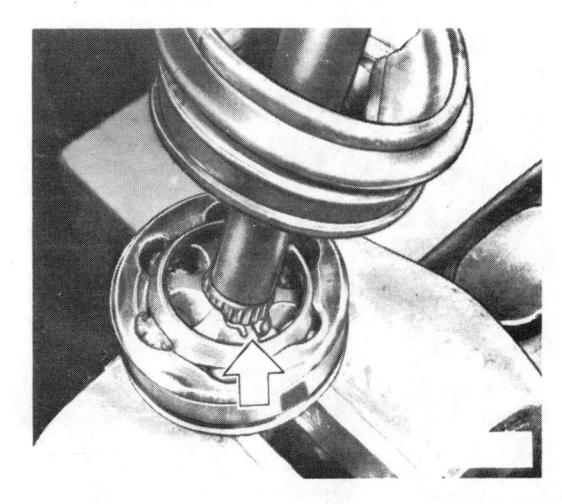

Fig. 7.6 Constant velocity joint retaining circlip – arrowed (Sec 4)

Fig. 7.7 Constant velocity joint removal (Sec 4)

4 Constant velocity joint rubber boot – removal and refitting

1 Remove the driveshaft from the car, as described in the previous Section.

2 Remove the metal clips securing the rubber boot to the driveshaft

9 Fill all the spaces in the joint with the special lubricating grease, moving the joint around at the same time. Apply a liberal quantity of grease and make sure it enters all the spaces and cavities between the joint members.
10 Fold the rubber boot back over the joint, making sure the larger diameter fits over the retaining clip groove in the joint outer diameter.
11 Expel all air from the rubber boot and then fit the large retaining clip using the same procedure as for the small one.
12 The driveshaft can now be refitted to the car, as described in Section 3.

5 Constant velocity joint – removal and refitting

1 The removal and refitting procedures for the constant velocity joint are covered in full in the previous Section, as this operation is an integral part of the rubber boot renewal sequence.
2 In the event of a constant velocity joint being worn or defective it must be renewed as a complete assembly. Overhaul of these joints is not possible. If the joint is to be renewed, it is strongly recommended that the rubber boot and retaining clips are renewed at the same time, again full details will be found in the previous Section.

6 Fault diagnosis – driveshafts

Symptom	Reason(s)
Knocking noise when accelerating with steering on lock	Worn constant velocity outer joints
Vibration	Worn constant velocity inner joints Damaged or distorted driveshaft
Knock or clunk when taking up drive	Incorrectly tightened driveshaft retaining nut Worn splines on constant velocity joint, hub flange or differential side gear Wear in suspension components, joints or attachments (see Chapter 11)

Chapter 8 Steering gear

Contents

Fault diagnosis – steering gear	13
Front wheel alignment – adjustment	12
General description	1
Maintenance and inspection	2
Rack and pinion steering gear – dismantling and reassembly	11
Rack and pinion steering gear – removal and refitting	10
Steering column – dismantling and reassembly	5
Steering column – removal and refitting	4
Steering column flexible rubber coupling – removal and refitting	6
Steering column lock – removal and refitting	7
Steering gear rubber gaiter – removal and refitting	9
Steering wheel – removal and refitting	3
Tie-rod outer balljoint – removal and refitting	8

Specifications

General
Type	Rack and pinion with collapsible, energy absorbing column
Steering ratio	21.4 : 1
Steering wheel turns – lock to lock	3.8
Steering gear centering dimension	419 to 422 mm (16.50 to 16.62 in)

Front wheel alignment (see text)
Reference value (unladen)	0.5 mm (0.02 in) toe-in to 1.5 mm (0.06 in) toe-out
Adjustment value (laden*)	Zero (parallel) to 2.0 mm (0.08 in) toe-out

*Laden indicates a vehicle containing two front seat occupants and a half-filled fuel tank

Torque wrench settings
	Nm	lbf ft
Steering wheel retaining nut	15	11
Flexible rubber coupling clamp bolts	25	18
Steering column lower mounting bolt	25	18
Steering column upper mounting nut	15	11
Steering gear clamp retaining nuts	15	11
Steering gear rack damper locking ring	60	44
Tie-rod inner balljoint housings to rack	60	44
Tie-rod-to-outer balljoint locking nut	50	37
Tie-rod outer balljoint-to-steering knuckle locknut	35	26

1 General description

The steering gear is of the conventional rack and pinion type and is secured to the engine compartment bulkhead by a U-shaped mounting bracket at each end of the rack housing. Steering wheel movement is transmitted to the steering gear via a collapsible, energy absorbing steering column. At the base of the column a flexible rubber coupling connects the column shaft to the pinion of the steering gear.

A tie-rod with rubber-gaiter-enclosed inner balljoint, and exposed outer balljoint, connects each end of the rack to the front wheels via the front suspension steering knuckles. The position of the outer balljoints on the tie-rods can be altered to cater for toe setting adjustment.

2 Maintenance and inspection

1 Every 9000 miles (15 000 km) or 6 months, whichever comes first, the following service operations should be carried out to the components of the steering gear.
2 Carry out a careful visual inspection of the condition and security of the rubber gaiters at each end of the steering gear. If there are any visible signs of cuts, splits or damage due to chafing of the gaiters, they should be renewed, as described in Section 9.
3 At the same service interval the steering balljoints should be checked for wear in the following way.
4 Observe the tie-rod outer balljoints while an assistant turns the steering wheel back and forth through an arc of about 20°. If there is

132   Chapter 8   Steering gear

any side to side movement of the balljoints as the steering is turned they should be renewed, as described in Section 8. Renewal is also necessary if the rubber dust covers around the balljoints are split, or damaged, or show any signs of deterioration.

5   The tie-rod inner balljoints can be checked in the same way by placing a hand over the rubber gaiter and feeling for any free play as the steering is turned. If excessive play is evident, renewal of the tie-rod is necessary, as described in Section 11.

6   Also inspect the condition of the flexible rubber coupling at the base of the steering column and renew this component if the rubber shows signs of deterioration or swelling, or if any cracks or splits are apparent. Full details will be found in Section 6.

7   Every 18 000 miles (30 000 km) or 12 months, whichever comes first, the front wheel toe setting should be checked and if necessary adjusted, as described in Section 12.

3   Steering wheel – removal and refitting

1   Disconnect the battery negative terminal.
2   Carefully prise the horn button contact pad from the centre of the steering wheel and disconnect the two electrical leads (photo).
3   Bend back the locktabs and using a socket and bar, undo the steering wheel retaining nut (photo).
4   Using a dab of paint or a small file, make an alignment mark between the steering wheel boss and the column shaft.
5   Using a small two-legged puller with outward facing legs inserted through the two holes in the wheel boss, withdraw the steering wheel from the shaft. Do not attempt to remove the steering wheel by striking it, otherwise the collapsible steering column may be damaged.
6   With the wheel removed, the horn contact ring may be renewed if necessary by unclipping it from the underside of the steering wheel boss. When refitting the new ring, make sure that the direction indicator switch cancelling segment is towards the left-hand side.
7   To refit the steering wheel first ensure that the shaft preload spring and washer are in position (photo), then place the wheel on the shaft with the previously made marks aligned.
8   Refit the lockwasher and retaining nut, tighten the nut to the specified torque and bend over the lockwasher tabs.
9   Reconnect the electrical leads to the horn button contact pad and push the assembly into its location in the centre of the steering wheel.
10   Reconnect the battery earth terminal.

3.2 Remove the two electrical leads after carefully prising out the horn button contact pad

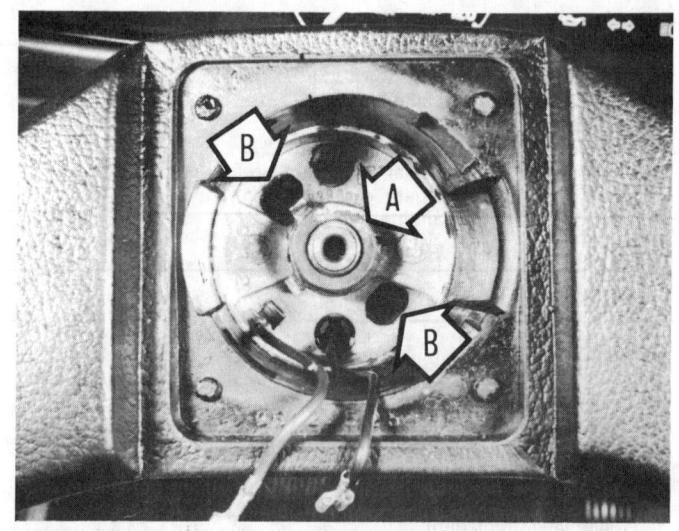

3.3 Steering wheel retaining nut (A) and holes for puller legs (B)

Fig. 8.1 Refitting the horn contact ring with the direction indicator switch cancelling segment to the left (Sec 3)

3.7 Ensure that the washer and preload spring are in place before refitting the steering wheel

4   Steering column – removal and refitting

1   Disconnect the battery earth terminal.
2   Undo the four retaining screws and lift off the steering column shroud lower half (photo).
3   Insert the ignition key into the ignition/steering lock cylinder and turn it to position 'II'.
4   Using a short length of small diameter wire inserted into the hole

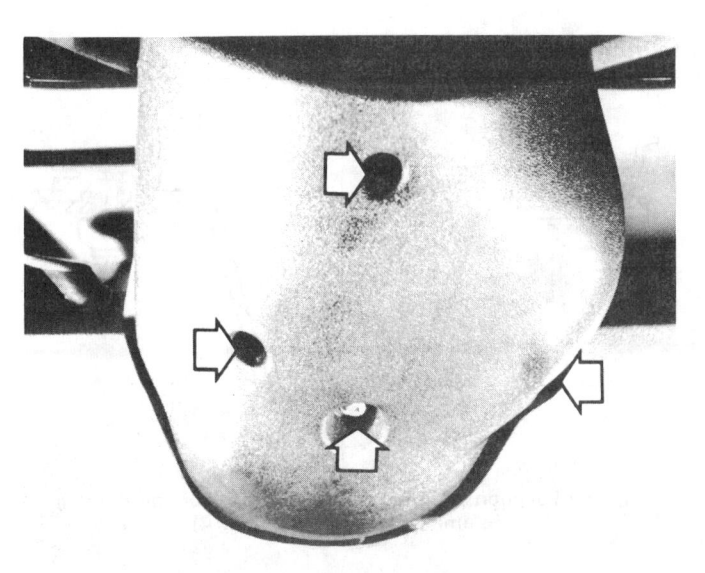

4.2 Steering column shroud lower half retaining screw locations (arrowed)

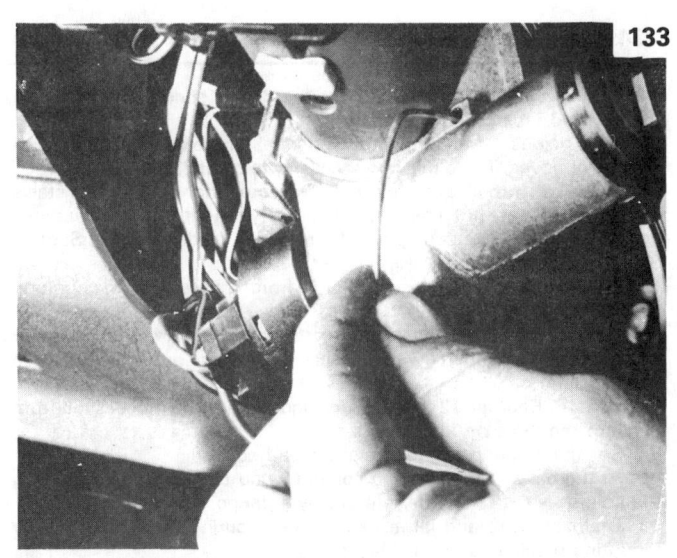

4.4A Depress the ignition/steering lock detent plunger using a short length of wire ...

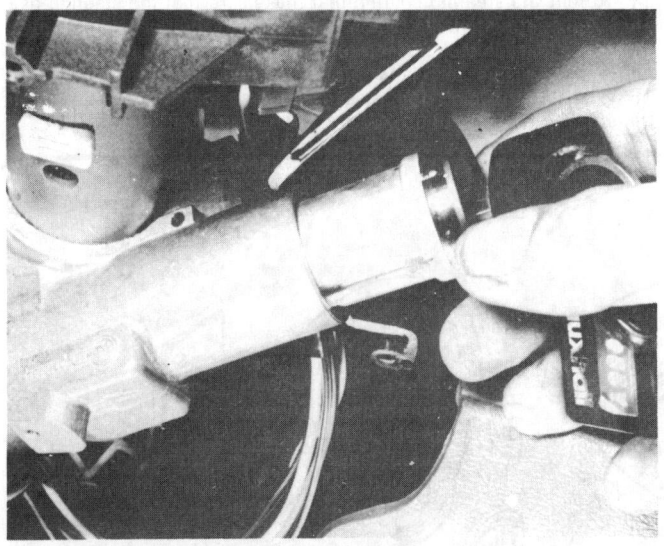

4.4B ... and remove the lock cylinder with the key at position 'II'

4.5 Remove the column shroud upper half by judicious manipulation. The steering wheel may be left in place

4.7 Depress the upper and lower catches to remove the column switches

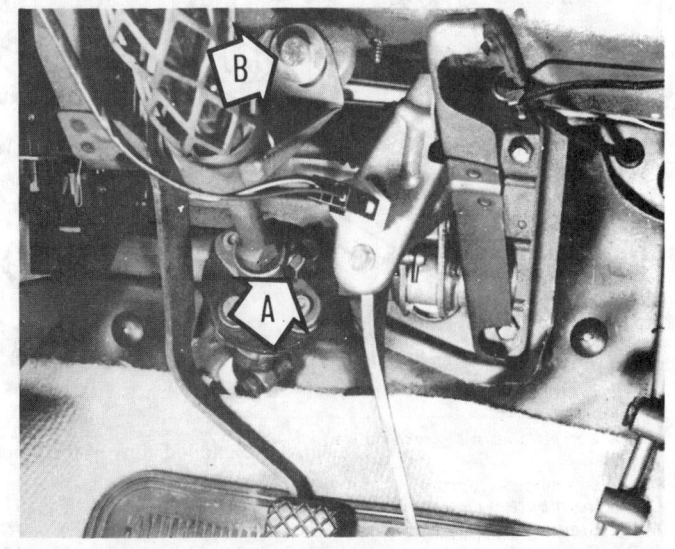

4.10 Column shaft-to-flexible coupling flange clamp bolt (A) and column lower mounting bolt (B)

134 Chapter 8 Steering gear

in the lock housing (photo), depress the detent spring and withdraw the key and lock cylinder (photo).

5 By judicious manipulation the upper half of the column shroud can now be removed. To do this it is necessary to carefully bend the shroud outward slightly to clear the direction indicator and wiper switch stalks (photo). Although not absolutely necessary, slightly more clearance can be gained if the steering wheel is removed, as described in Section 3.

6 Disconnect the wiring multi-plug from the rear of the ignition switch.

7 Depress the upper and lower catches on the direction indicator and windscreen wiper switches and slide them out of the switch housing (photo).

8 Refer to Chapter 12 if necessary, and remove the parcel shelf from beneath the facia on the driver's side.

9 Set the steering to the straight-ahead position.

10 At the base of the steering column, undo the clamp bolt securing the column shaft to the flexible coupling flange (photo).

11 Undo the bolt and large flat washer securing the column lower mounting bracket to the bulkhead.

12 The bolts securing the upper mounting bracket to the bulkhead can now be removed. On the left-hand side a shear-bolt is used and to remove this it will be necessary to centre punch the shear-head, drill a pilot hole and unscrew the bolt using a suitable stud extractor. On the right-hand side a conventional locknut and stud are used for retention and are removed in the normal way.

13 With the upper mounting released, withdraw the column slightly to release the column shaft from the flexible coupling flange and then remove the assembly from inside the car. Take care not to bump or knock the column after removal as this could damage the collapsible element.

14 To refit the steering column first make sure that the steering gear is still in the straight-ahead position – the clamp bolt hole on the flexible coupling upper flange must be uppermost and horizontal. If removed, refit the steering wheel, as described in Section 3.

15 With the steering wheel also in the straight-ahead position, enter the steering column shaft into the coupling flange.

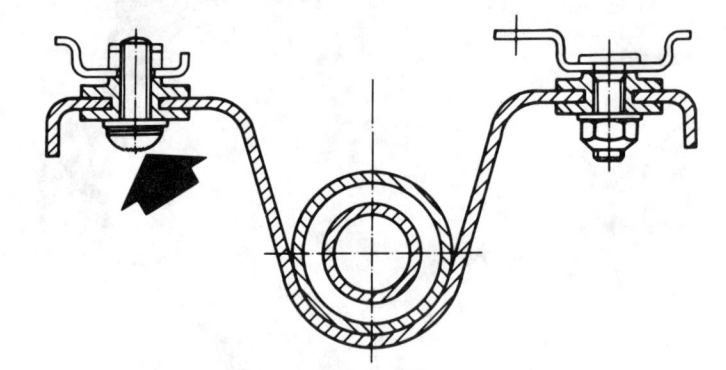

Fig. 8.2 Location of the shear-bolt (arrowed) on the steering column upper mounting (Sec 4)

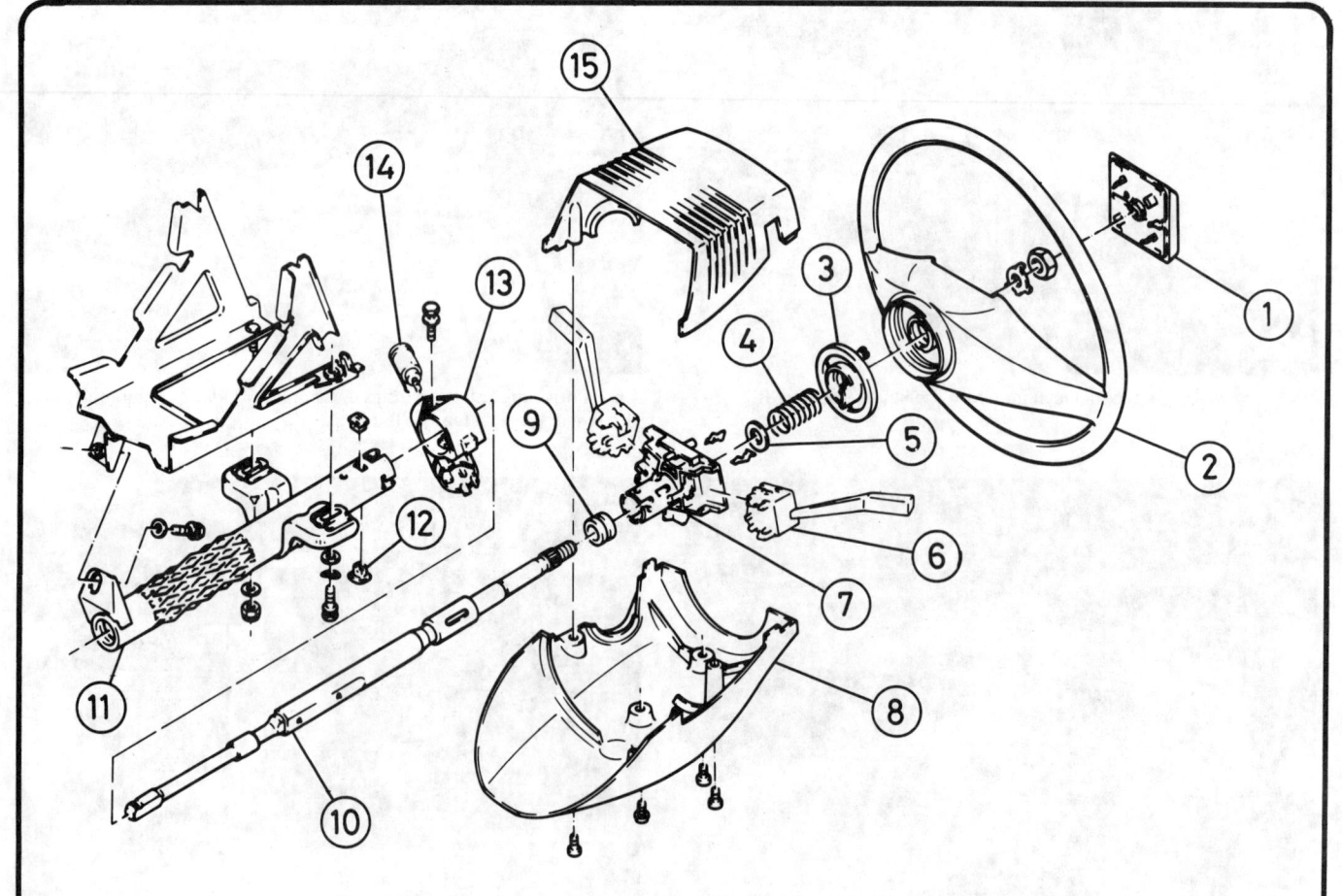

Fig. 8.3 Exploded view of the steering column assembly (Secs 4 and 5)

1 Horn button contact pad
2 Steering wheel
3 Contact ring
4 Preload spring
5 Washer
6 Direction indicator switch
7 Switch housing
8 Column shroud lower half
9 Shaft upper support ball-bearing
10 Steering column shaft
11 Steering column
12 Switch housing locating plug
13 Ignition/steering lock housing
14 Lock cylinder
15 Column shroud upper half

Chapter 8 Steering gear

16 Refit the bolt and washer securing the column lower mounting to the bulkhead. Tighten the bolt finger tight only at this stage.
17 Refit the upper mounting locknut and shear-bolt, but only tighten these finger tight also.
18 With the column assembly free from any stress, tighten the mountings to the specified torque or, in the case of the shear-bolt, until the head breaks off.
19 Refit the clamp bolt securing the flexible coupling flange to the steering column shaft. Tighten the bolt to the specified torque. Slacken the lower clamp bolt securing the coupling flange to the steering gear pinion shaft. Pull upwards on the steering wheel until the column shaft contacts its stop in the upper support bearing. Hold the shaft in this position and tighten the lower clamp bolt to the specified torque.
20 Slide the direction indicator and windscreen wiper switches into their locations in the switch housing.
21 Reconnect the wiring multi-plug to the rear of the ignition switch.
22 Position the upper half of the steering column shroud over the steering column.
23 Insert the ignition/steering lock cylinder, with key set at position 'II', into the lock housing. Depress the detent spring and push the cylinder fully home. Return the key to position 'B'.
24 Refit the lower half of the column shroud and secure it with the four retaining screws.
25 Refit the parcel shelf, as described in Chapter 12, and reconnect the battery negative terminal.

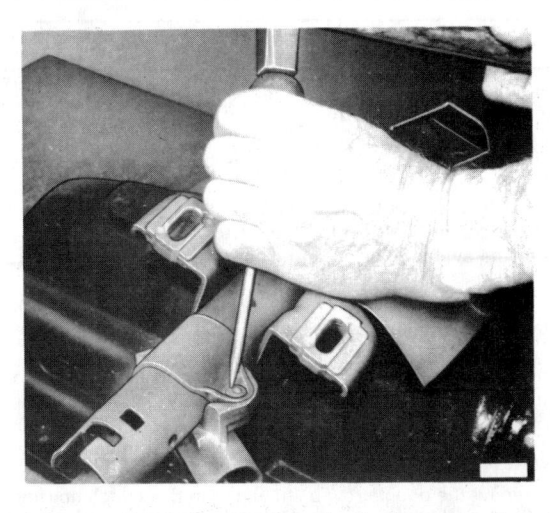

Fig. 8.6 Removing the ignition/steering lock housing shear-bolt (Sec 5)

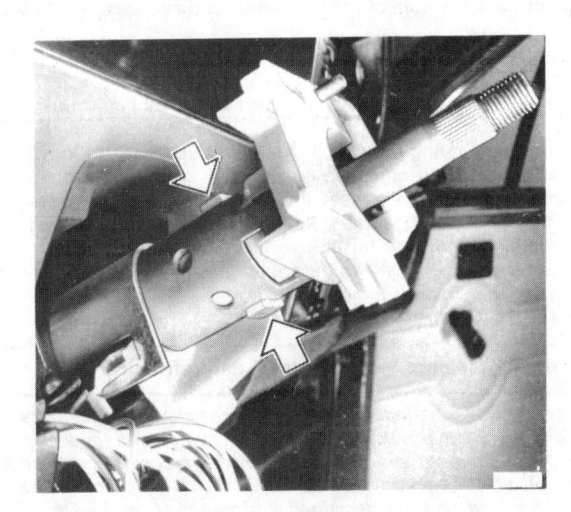

Fig. 8.4 Location of the switch housing locating plugs (Sec 5)

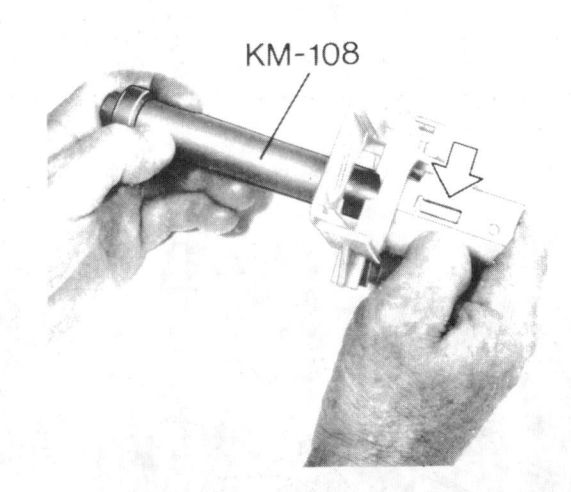

Fig. 8.7 The upper support bearing in the switch housing can be driven out whilst at the same time spreading the two spring catches – arrowed (Sec 5)

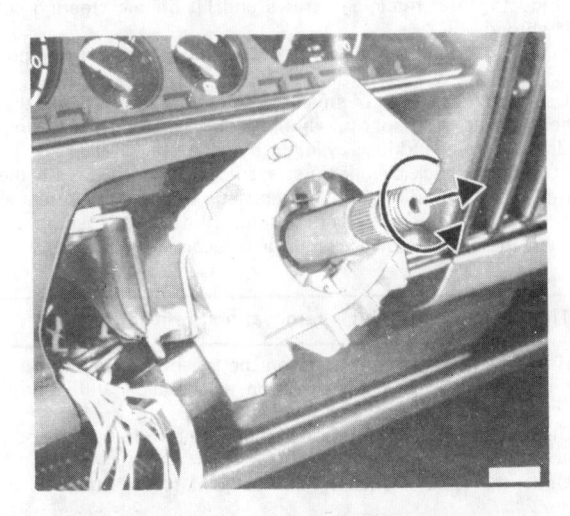

Fig. 8.5 With the locating plugs removed the switch housing can be turned anti-clockwise and withdrawn (Sec 5)

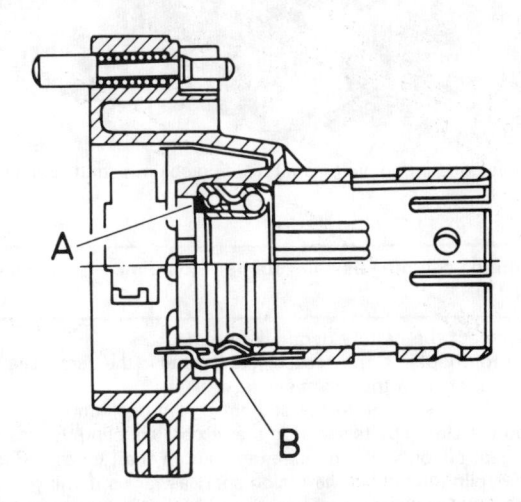

Fig. 8.8 Sectional view of the switch housing assembly (Sec 5)

A Thrust washer   B Bearing retaining spring catches

5 Steering column – dismantling and reassembly

1 Remove the steering column, as described in the previous Section.
2 Remove the steering wheel, if still in position, using the procedure described in Section 3.
3 Withdraw the preload spring and washer from the steering column shaft.
4 Carefully prise out the locating plugs from their locations on each side of the column unit behind the switch housing.
5 Turn the switch housing anti-clockwise and withdraw it from the column.
6 Carefully slide the steering column shaft out of the column, from bottom to top, and lay it aside where it is not likely to be knocked or dropped.
7 The ignition/steering lock housing can be removed from the column if required after removal of the shear-bolt. To do this either unscrew the bolt using a centre punch, or drill a pilot hole and unscrew it with a stud extractor. With the shear-bolt removed, lift the lock housing retaining plate and withdraw the housing.
8 To renew the double row ball-bearing in the switch housing, tap it out using a tube of suitable diameter whilst at the same time spreading the two spring catches on each side of the housing. Refit a new bearing in the same way, but make sure that the thrust washer is in position in the housing before fitting the bearing.
9 To reassemble the steering column refit the ignition/steering lock housing and secure with a new shear-bolt, tightened until the head breaks off.
10 Slide the steering column shaft into place and refit the switch housing by locating it on the column and turning clockwise. Secure the switch housing using two new locating plugs.
11 Position the washer and preload spring over the shaft and then refit the steering wheel, as described in Section 3.
12 The steering column can now be refitted to the car, as described in the previous Section.

6.4 Flexible coupling to pinion shaft (A) and column shaft (B) clamp bolts

6 Steering column flexible rubber coupling – removal and refitting

1 Disconnect the battery negative terminal.
2 Refer to Chapter 12 if necessary, and remove the parcel shelf from beneath the facia on the driver's side.
3 Position the steering in the straight-ahead position.
4 Undo the clamp bolts securing the flexible coupling flanges to the steering gear pinion shaft and steering column shaft (photo). Push the flexible coupling down on the pinion shaft as far as it will go.
5 Undo the four retaining screws and lift off the lower half of the steering column shroud.
6 Insert the ignition key into the ignition/steering lock and turn it to position 'II'.
7 Using a short length of small diameter wire inserted into the hole in the lock housing (photo 4.4a), depress the detent spring and withdraw the key and lock cylinder (photo 4.4b).
8 By judicious manipulation the upper half of the column shroud can now be removed. To do this it is necessary to carefully bend the shroud outward slightly to clear the direction indicator and wiper switch stalks. Although not absolutely necessary, slightly more clearance can be gained if the steering wheel is removed, as described in Section 3.
9 Carefully prise out the locating plugs from their locations on each side of the column just behind the switch housing.
10 Turn the switch housing anti-clockwise to release it from the column and then carefully pull the steering wheel, steering column shaft and switch housing out of the column, until the shaft is clear of the flexible rubber coupling.
11 Slide the coupling up and off the steering gear pinion shaft.
12 To refit the coupling, slide the lower flange over the pinion shaft and then engage the steering column shaft into the upper flange.
13 Locate the switch housing on the column and turn it clockwise into its correct position. Secure the housing using two new locating plugs.
14 Place the steering column shroud upper half in position over the column.
15 Insert the ignition/steering lock cylinder, with key set at position 'II', into the lock housing. Depress the detent spring and push the cylinder fully home. Return the key to position 'B'.
16 Refit the lower half of the column shroud and secure with the four retaining screws.
17 Refit the steering wheel, if previously removed, as described in Section 3.
18 Refit the clamp bolt securing the flexible rubber coupling upper flange to the column shaft. Tighten the bolt to the specified torque.
19 Turn the steering through its full travel from lock to lock to settle the flexible coupling and then refit the lower flange clamp bolt. Pull upwards on the steering wheel until the column shaft contacts its stop in the upper support bearing. Hold the shaft in this position and tighten the coupling clamp bolt to the specified torque.
20 Refit the parcel shelf, as described in Chatper 12, and reconnect the battery negative terminal.

7 Steering column lock – removal and refitting

Note: *The steering column lock consists of three parts – the ignition/steering lock cylinder, the ignition switch and the lock housing around the steering column into which these components are fitted. Removal of the ignition switch and the lock cylinder are quite straightforward and are described in Chapter 10 and in this Section respectively. Removal of the lock housing necessitates removal of the steering column and reference should be made to Section 4 of this Chapter.*

1 To remove the lock cylinder first disconnect the battery negative terminal.
2 Undo the four retaining screws and lift off the steering column lower shroud.
3 Insert the ignition key into the lock cylinder and turn it to position 'II'.
4 Using a short length of small diameter wire inserted into the hole in the lock housing (photo 4.4a), depress the detent spring and withdraw the key and lock cylinder (photo 4.4b).
5 To refit the lock cylinder first set it at position 'II' with the key inserted and position it in the lock housing. Depress the detent spring and push the cylinder fully home. Return the key to position 'B'.
6 Refit the column lower shroud and reconnect the battery.

8 Tie-rod outer balljoint – removal and refitting

1 Remove the wheel trim, slacken the wheel bolts and jack up the front of the car. Support the car on axle stands and remove the roadwheel.
2 Slacken the locking nut on the steering tie-rod a quarter of a turn (photo).
3 Undo the locknut securing the balljoint to the steering knuckle arm.
4 Using a universal balljoint separator, release the taper of the balljoint shank and then lift the joint out of the arm (photos).

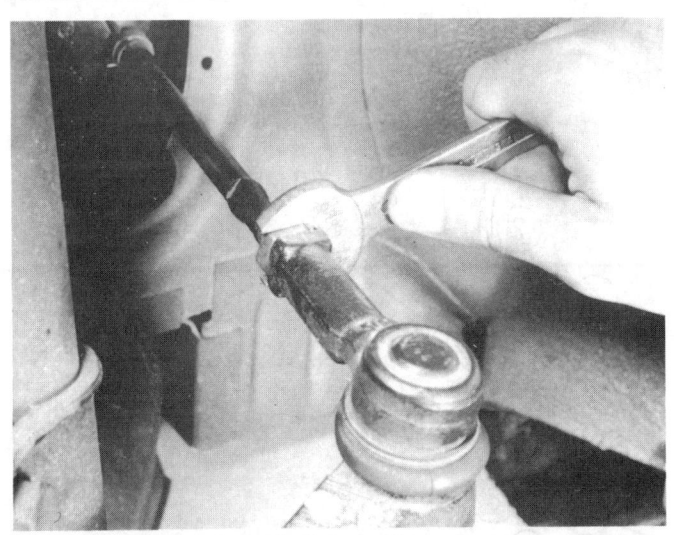

Chapter 8 Steering gear

5 Unscrew the balljoint from the tie-rod using a spanner on the flat of the tie-rod to stop it turning, if necessary.
6 Refit the balljoint by screwing it on up to the locking nut on the tie-rod.
7 Engage the shank of the balljoint into the steering knuckle arm and secure with the locknut tightened to the specified torque.
8 Secure the balljoint to the tie-rod by tightening the locking nut.
9 Refit the roadwheel and lower the car to the ground.
10 Check and if necessary reset the toe setting, as described in Section 12.

9 Steering gear rubber gaiter – removal and refitting

1 Remove the appropriate tie-rod outer balljoint, as described in the previous Section.
2 Mark the position of the locking nut on the tie-rod so that it can be refitted in the same place and then unscrew it from the tie-rod.
3 Release the two rubber gaiter retaining clips and then slide the gaiter off the rack housing and tie-rod.
4 Slide a new gaiter into position, ensuring that it locates in the recesses in the rack housing and tie-rod. Secure the gaiter in position using new clips or a few turns of soft iron wire.
5 Refit the locking nut to the position marked during removal.
6 Refit the tie-rod outer balljoint, as described in the previous Section. Check the front wheel alignment (Section 12).

10 Rack and pinion steering gear – removal and refitting

1 Disconnect the battery negative terminal.
2 Remove the wheel trim, slacken the wheel bolts and jack up the front of the car. Support the car on axle stands and remove the front roadwheels.
3 From under the left-hand wheel arch, undo the locknut securing the tie-rod outer balljoint to the steering knuckle arm. Using a universal balljoint separator, release the balljoint from the arm. Repeat this procedure on the right-hand tie-rod balljoint.
4 From inside the car, undo the clamp bolts securing the flexible rubber coupling flanges to the pinion and steering column shafts. Slide the coupling up the steering column shaft.
5 Using a pencil accurately mark the outline of the two clamps securing the steering gear to the bulkhead.
6 Undo the two retaining nuts and washers each side and lift off the steering gear clamps (photo). If access proves difficult on 1.3 models, refer to Chapter 3 and remove the air cleaner assembly.
7 Withdraw the steering gear into the engine compartment and then remove it from the car through the right-hand wheel arch.
8 Before refitting the steering gear centralise the rack by turning the pinion shaft until the distance between the left-hand shoulder of the

8.2 Slacken the tie-rod outer balljoint locking nut a quarter of a turn

8.4A Release the balljoint shank using a universal balljoint separator ...

8.4B ... then lift out the balljoint and unscrew it from the tie-rod

10.6 Right-hand side steering gear clamp and lower retaining locknut (arrowed)

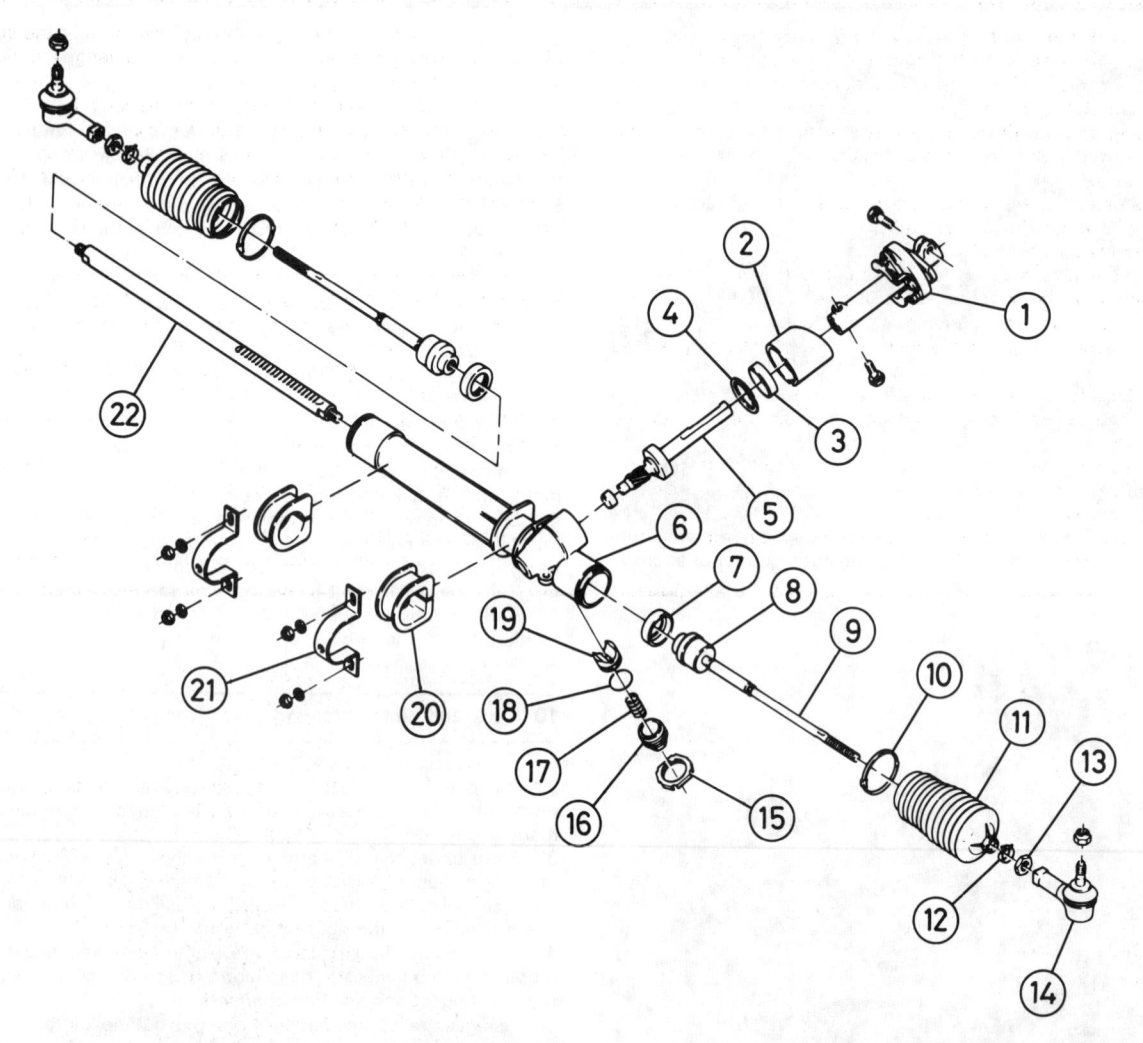

Fig. 8.9 Exploded view of the rack and pinion steering gear – left-hand drive unit shown (Secs 9 and 10)

1. Flexible rubber coupling
2. Bulkhead rubber seal
3. Pinion shaft seal
4. Pinion bearing retaining circlip
5. Pinion
6. Rack housing
7. Tie-rod stop ring
8. Tie-rod inner balljoint housing
9. Tie-rod
10. Rubber gaiter retaining clip (large)
11. Rubber gaiter
12. Rubber gaiter retaining clip (small)
13. Tie-rod balljoint locking nut
14. Tie-rod outer balljoint
15. Rack damper locking ring
16. Rack damper screw plug
17. Coil spring
18. Damper slipper seal ring
19. Rack damper slipper
20. Mounting clamp rubber block
21. Mounting clamp
22. Rack

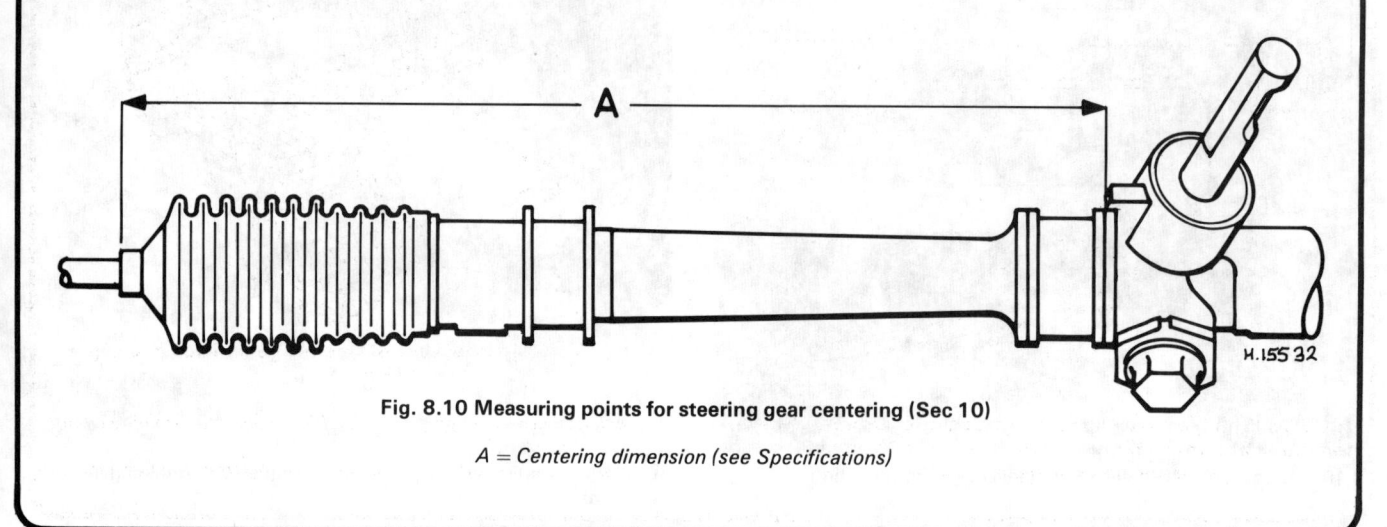

Fig. 8.10 Measuring points for steering gear centering (Sec 10)

A = Centering dimension (see Specifications)

rubber gaiter and the right-hand mounting shoulder of the rack housing (Fig. 8.10) is equal to the centering dimension given in the Specifications. In this position the longitudinal recess on the pinion shaft should be pointing to the right, ie away from the longer part of the rack housing. In both these cases it is assumed that the steering gear is being viewed in its normal fitted position, ie from the rear to the front of the vehicle.

9 Turn the steering wheel so that it is in the straight-ahead position and the clamp bolt on the flexible rubber coupling upper flange is uppermost and lying horizontally.

10 Locate the steering gear in the engine compartment and engage the pinion shaft with the coupling lower flange.

11 Refit the steering gear retaining clamps and secure with the locknuts and washers. Tighten the locknuts finger tight only at this stage.

12 Push the flexible rubber coupling down onto the pinion shaft and refit the coupling upper flange clamp bolt. Tighten the bolt to the specified torque.

13 Turn the steering wheel through the full extent of its travel from lock to lock to align the components. Move the steering gear retaining clamps as necessary until they are aligned with the outline marks made during removal. In this position tighten the clamp retaining nuts to the specified torque.

14 Pull the steering wheel up as far as it will go, refit the clamp bolt securing the steering gear pinion to the coupling flange and tighten the bolt to the specified torque.

15 Refit the tie-rod outer balljoints to the steering knuckle arms and secure with the locknuts, tightened to the specified torque.

16 Refit the roadwheels and lower the car to the ground.

17 Reconnect the battery and, if removed, the air cleaner, as described in Chapter 3.

18 Check the front wheel alignment (Section 12).

11 Rack and pinion steering gear – dismantling and reassembly

1 Remove the steering gear from the car, as described in the previous Section. Wipe away any external dirt and grease and then place the steering gear on a clean uncluttered work bench.

2 Remove the two mounting clamp locating rubber blocks and the pinion shaft-to-bulkhead rubber seal.

3 Slacken the tie-rod outer balljoint locking nuts by a quarter of a turn and unscrew the balljoints from the tie-rods.

4 Mark the position of the locking nuts on the tie-rods as an aid to reassembly and then unscrew the two locking nuts.

5 Release the retaining clips and withdraw the left-hand and right-

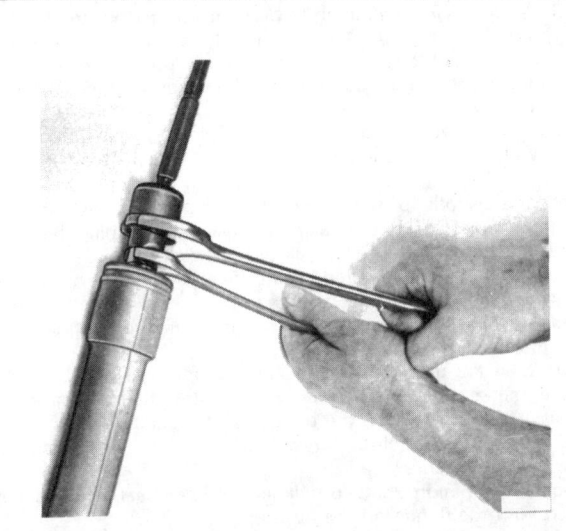

Fig. 8.11 Removing the tie-rod inner balljoint housing (Sec 11)

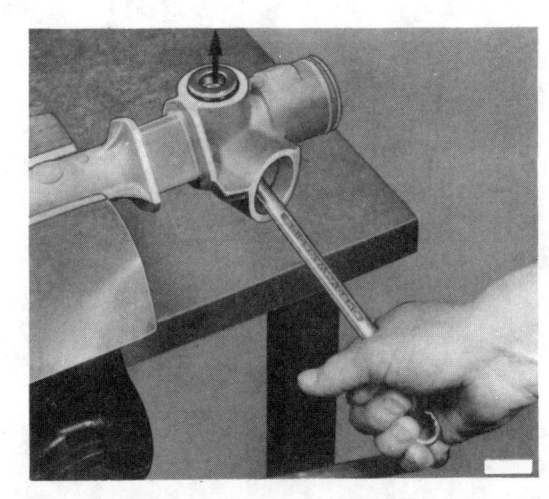

Fig. 8.12 Removing the rack damper slipper (Sec 11)

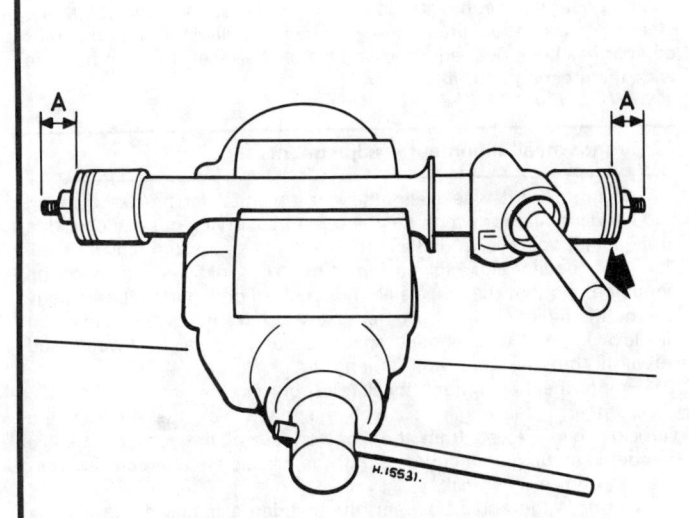

Fig. 8.13 With the pinion correctly fitted the longitudinal recess on the shaft (arrowed) must be to the right, with dimension A equal on both sides (Sec 11)

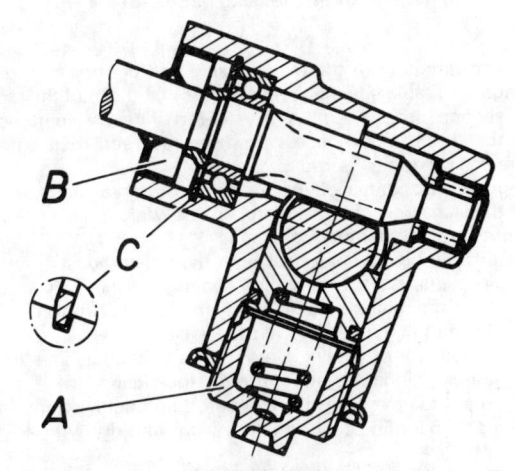

Fig. 8.14 Sectional view of steering gear pinion housing (Sec 11)

A Rack damper screw plug
B Pinion shaft seal
C Pinion bearing retaining circlip
Inset shows chamfer of circlip facing away from the bearing

Fig. 8.15 Lock the tie-rod balljoint housings onto the rack by flattening their edges onto the rack machined flats (Sec 11)

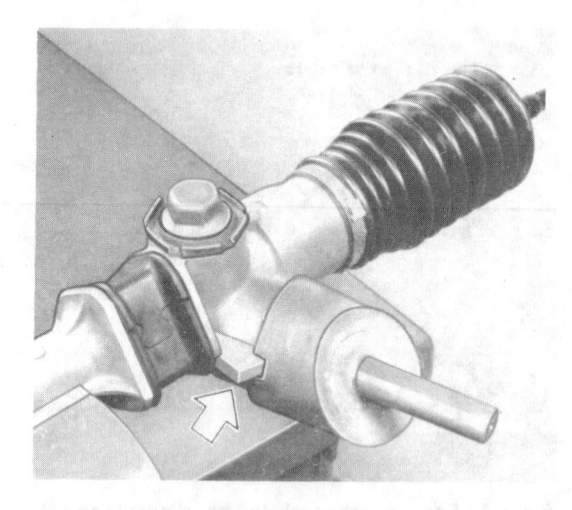

Fig. 8.16 The notch on the bulkhead rubber seal (arrowed) must engage with the housing flange (Sec 11)

hand rubber gaiters from the rack housing and tie-rods.
6 Engage a suitable spanner over the flats in the end of the rack and, using a second spanner, undo the tie-rod inner balljoint housing. Remove the tie-rod and housing from the rack and then repeat the procedure for the other tie-rod.
7 Using a large spanner or socket, slacken the rack damper locking ring and then unscrew the rack damper screw plug.
8 Lift out the rack damper coil spring.
9 Prise out the pinion shaft seal from the pinion housing and then, using circlip pliers, extract the pinion bearing retaining circlip.
10 Withdraw the pinion, complete with bearing from the pinion housing and then carefully slide out the rack.
11 Insert a spanner or screwdriver into the pinion housing and push the rack damper slipper up and out of its location.
12 With the steering gear dismantled, thoroughly clean all the components in paraffin or a suitable solvent and dry with a lint-free rag.
13 Inspect the parts for obvious signs of wear, damage or distortion. Check the pinion upper support bearing for smooth operation and also check the needle roller bearing in the pinion housing for wear or pitting of the rollers and for smooth operation. If the needle roller bearing is in need of renewal, press it out of the housing from outside to inside. Press in a new bearing from inside to outside. If the upper support bearing is in need of renewal it will be necessary to obtain a new pinion assembly with the bearing already in place. Similarly if the rack bearing bush in the rack housing is worn, a new housing with the bush in place must be obtained.
14 Obtain the new parts necessary for reassembly, including a new pinion shaft seal, rock damper slipper seal ring and a suitable quantity of steering gear grease available from GM dealers.
15 Begin reassembly by thoroughly lubricating the teeth of the rack and the pinion with steering gear grease. Also uniformly fill the inside of the rack housing with approximately 50g (1.8 oz) of the grease.
16 Position the rack centrally in the housing so that its ends extend uniformly from each end of the housing. With the rack in this position insert the pinion so that when it is fully engaged with the rack teeth, the longitudinal recess in the pinion shaft points to the right, ie away from the longer side of the rack housing (Fig. 8.13).
17 Refit the pinion bearing retaining circlip with its chamfered side away from the bearing. Ensure that the circlip seats fully into its groove.
18 Fill the space above the bearing with steering gear grease and then drive in a new pinion shaft seal using a tube of suitable diameter.
19 Position a new seal ring on the rack damper slipper and then insert the slipper into the housing.
20 Place the coil spring over the slipper and refit the screw plug. Tighten the screw plug until firm resistance is felt or, if a small torque wrench is available, to 5 Nm (4 lbf ft). Now unscrew the plug by 30° and check that the pinion will turn smoothly over the full length of rack travel. It is normal for the pinion to be tight to turn, but the tightness should be uniform over the full length of travel without any areas of binding. If the pinion does bind, unscrew the plug until the resistance is uniform. Note that the plug must only be unscrewed by a maximum of 60° from the starting point.
21 When the movement of the rack and pinion is satisfactory, hold the screw plug with a spanner and then refit and tighten the locking ring.
22 Refit the tie-rods to each end of the rack and securely tighten the balljoint housings. Secure the balljoint housings by tapping their edges onto the machined flats of the rack using a drift.
23 Refit the rubber gaiters and secure them to the rack housing and tie-rods using new clips or a few turns of soft iron wire. Make sure that the ends of the gaiters seat squarely in their grooves in the housing and on the tie-rods.
24 Screw on the tie-rod outer balljoint locking nuts and align them with the marks made during removal. If the tie-rods have been renewed, use the marks on the old parts as a guide.
25 Refit the outer balljoints, screw them up to the locking nuts and tighten the nuts securely.
26 Locate the pinion shaft-to-bulkhead rubber seal with its notch engaged with the flange on the housing.
27 Refit the mounting clamp locating rubber blocks and then refit the steering gear to the car, as described in the previous Section. After the steering gear has been installed, test drive the car and check that the steering self-centres after lock has been applied. If not, the rack damper has been over adjusted and should be reset. This can be done with the steering gear in the car.

12 Front wheel alignment – adjustment

1 Accurate front wheel alignment is essential for precise steering and handling and for even tyre wear. Before carrying out any checking or adjusting operations, make sure that the tyres are correctly inflated, that all steering and suspension joints and linkages are in sound condition and that the wheels are not buckled or distorted, particularly around the rims. It will also be necessary to have the car positioned on flat level ground with enough space to push the car backwards and forwards through about half its length.
2 Front wheel alignment consists of four factors:
Camber is the angle at which the roadwheels are set from the vertical when viewed from the front or rear of the vehicle. Positive camber is the angle (in degrees) that the wheels are tilted outwards at the top from the vertical.
Castor is the angle between the steering axis and a vertical line when viewed from each side of the vehicle. Positive castor is indicated when the steering axis is inclined towards the rear of the vehicle at its upper end.
Steering axis inclination is the angle, when viewed from the front or rear of the vehicle, between the vertical and an imaginary line drawn between the upper and lower front suspension strut mountings.

Chapter 8 Steering gear

Toe setting is the amount by which the distance between the front inside edges of the roadwheel rims differs from that between the rear inside edges, when measured at hub height. If the distance between the front edges is less than that at the rear, the wheels are said to toe-in. If it is greater than at the rear, the wheels toe-out.

3 On Nova models, camber, castor and steering axis inclination are set during manufacture and are not adjustable. Unless the vehicle has suffered accident damage, or there is gross wear in the suspension mountings or joints, it can be assumed that these settings are correct. If for any reason it is believed that they are not correct, the task of checking them should be left to a GM dealer who will have the necessary special equipment needed to measure the small angles involved.

4 It is, however, within the scope of the home mechanic to check and adjust the front wheel toe setting. To do this a tracking gauge must first be obtained. Two types of gauges are available and can be obtained from motor accessory shops. The first type measures the distance between the front and rear inside edges of the roadwheels, as previously described, with the car stationary. The second type, known as a scuff plate, measures the actual position of the contact surface of the tyre, in relation to the road surface, with the vehicle in motion. This is done by pushing or driving the front tyre over a plate which then moves slightly according to the 'scuff' of the tyre and shows this movement on a scale. Both types have their advantages and disadvantages, but either can give satisfactory results if used correctly and carefully.

5 Having obtained a suitable gauge, measure and record the toe setting of the car with the wheels in the straight-ahead position, according to the maker's instructions. If the setting is not in accordance with the reference value given in the Specifications, the toe should be set to the adjustment value with the car in a laden condition.

6 First clean the area aound the tie-rod balljoint locking nut using a wire brush and then count the number of exposed threads on the tie-rod. Now repeat this for the other tie-rod. If the numbers of exposed threads on the tie-rods are not identical then they will have to be equalised during the adjusting operation by turning one tie-rod slightly more than the other, as necessary. If the numbers of threads are the same then during adjustment both tie-rods must be turned by the same amount. With this design of rack and pinion steering, not only must the toe-setting be correct, but the length of the tie-rods must be equal on both sides of the car and this is indicated by an equal number of exposed threads on both sides.

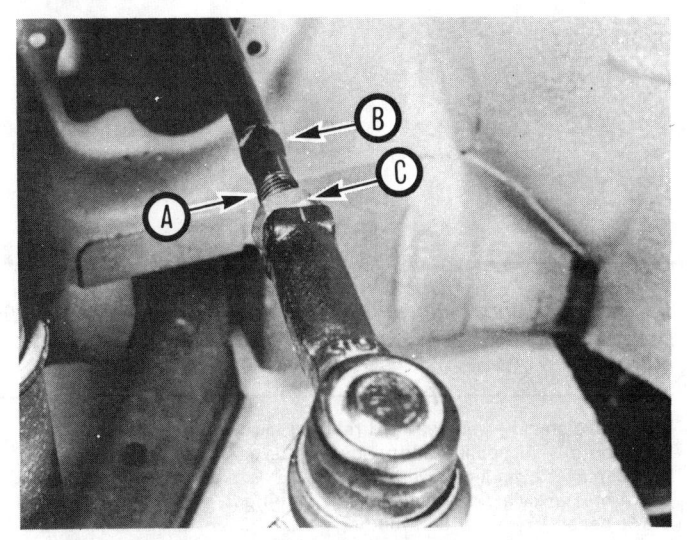

12.7 Tie-rod exposed threads (A), flat on tie-rod for spanner engagement (B) and balljoint locking nut (C)

7 To adjust the setting slacken the tie-rod outer balljoint locking nut and then turn the tie-rod clockwise (when viewed end on from under the wheel arch) to increase the toe-in or anti-clockwise to increase the toe-out (photo). Only turn the tie-rods by $\frac{1}{8}$ turn at a time and then recheck the setting. When the toe setting is correct and the tie-rod threads are equal, tighten the locking nuts. If the rubber gaiters on the steering gear have become twisted due to turning the tie-rods, straighten them out, otherwise they will deteriorate rapidly due to chafing.

13 Fault diagnosis – steering gear

Symptom	Reason(s)
Steering stiff or heavy	Incorrect tyre pressures Incorrect front wheel alignment Lack of lubricant in steering gear Distorted or damaged rack, rack housing or steering column Seized steering or suspension balljoint
Excessive play in steering	Wear in tie-rod balljoints Wear in suspension balljoints or mountings (see Chapter 11) Worn or perished steering column flexible rubber coupling
Steering wander	Incorrect tyre pressures Incorrect front wheel alignment Wear in tie-rod balljoints Wear in suspension balljoints, mountings or hub bearings (see Chapter 11) Damaged steering or suspension components
Wheel wobble and vibration	Damaged or out of balance roadwheels (see Chapter 11) Worn or damaged driveshaft or constant velocity joint (see Chapter 7)

Chapter 9 Braking system

For modifications, and information applicable to later models, see Supplement at end of manual

Contents

Brake disc shield – removal and refitting 8	Handbrake cable – removal and refitting 24
Brake drum – inspection and renovation 13	Handbrake lever – removal, overhaul and refitting 25
Brake master cylinder – overhaul 15	Hydraulic pipes and hoses – inspection, removal and refitting 17
Brake master cylinder – removal and refitting 14	Hydraulic system – bleeding 4
Brake pedal – removal and refitting 21	Maintenance and inspection 2
Brake pedal relay lever – removal and refitting 22	Pressure regulating valves – general 16
Braking system – adjustment 3	Rear brake backplate – removal and refitting 12
Fault diagnosis – braking system 26	Rear brake shoes – inspection and renewal 9
Front brake caliper – removal, overhaul and refitting 6	Rear wheel cylinder – overhaul 11
Front brake disc – removal and refitting 7	Rear wheel cylinder – removal and refitting 10
Front brake pads – inspection and renewal 5	Vacuum servo unit – removal and refitting 20
General description 1	Vacuum servo unit – testing 19
Handbrake – adjustment 23	Vacuum servo unit hose – renewal 18

Specifications

System type
Diagonally split, dual-circuit hydraulic with discs at the front and drums at the rear. Pressure regulating valves in rear hydraulic circuits. Cable operated handbrake on rear wheels. Servo-assistance on all models.

Front brakes
Type	Disc with single piston sliding calipers
Disc diameter	236.0 mm (9.29 in)
Disc thickness	10.0 mm (0.39 in)
Minimum thickness after refinishing	9.0 mm (0.35 in)
Maximum variation in thickness	0.01 mm (0.0004 in)
Maximum run-out	0.1 mm (0.004 in)
Minimum brake pad thickness (including backing plate)	7.0 mm (0.28 in)

Rear brakes
Type	Single leading shoe drum
Drum internal diameter	200.0 mm (7.88 in)
Maximum internal diameter after refinishing	201.0 mm (7.91 in)
Maximum allowable drum ovality	0.1 mm (0.004 in)
Minimum brake lining thickness	0.5 mm (0.02 in) above rivet heads

General
Master cylinder bore diameter	19.05 mm (0.75 in)
Servo unit type	178.0 mm (7.0 in) GMF single-diaphragm type
Pressure regulating valve identification code	3/25

Hydraulic fluid type/specification
Hydraulic fluid to SAE J1703F or DOT 4 (Duckhams Universal Brake and Clutch Fluid)

Torque wrench settings
	Nm	lbf ft
Brake servo to support bracket	18	13
Brake servo support bracket to bulkhead	18	13
Master cylinder to servo unit	18	13
Pressure regulating valves to master cylinder	40	30
Brake caliper mounting bolts	95	70
Bleed screws to caliper or wheel cylinder	9	7
Flexible brake hose to caliper banjo union	25	18
Wheel cylinder retaining bolt	9	7
Vacuum hose to inlet manifold	15	11
Brake pipe union nuts	11	8
Holding frame to brake caliper bolts	95	70

Chapter 9 Braking system

1 General description

The braking system is of the servo-assisted, dual-circuit hydraulic type with disc brakes at the front and drum brakes at the rear. A diagonally split dual-circuit hydraulic system is employed in which each circuit operates one front and one diagonally opposite rear brake from a tandem master cylinder. Under normal conditions both circuits operate in unison; however, in the event of hydraulic failure in one circuit, full braking force will still be available at two wheels. Each circuit is provided with a pressure regulating valve which is screwed into the rear brake outlets of the master cylinder. These valves regulate the hydraulic pressure applied to each rear brake and reduce the possibility of the rear wheels locking under heavy braking.

The front disc brakes are operated by single piston sliding type calipers. At the rear, leading and trailing brake shoes are operated by twin piston wheel cylinders and are manually adjustable by eccentric adjusters.

The cable-operated handbrake provides an independent mechanical means of rear brake application.

Driver warning lights are provided for low brake hydraulic fluid level and handbrake application.

2 Maintenance and inspection

1 The brake hydraulic fluid level should be checked weekly and, if necessary, topped up with the specified fluid to between the MIN and MAX marks on the reservoir. Any need for frequent topping-up indicates a fluid leak somewhere in the system which must be investigated and rectified immediately.

2 Every 9000 miles (15 000 km) or 6 months, whichever comes first, check the front brake pads and rear brake linings for wear, as described in Sections 5 and 9. At the same time check the condition of the handbrake cables, lubricate the exposed cables and linkages and if necessary adjust them, as described in Section 23. The condition and security of the hydraulic pipes and hoses should also be inspected, further details will be found in Section 17.

3 It is advisable to check the operation of the braking system warning lights at periodic intervals. The handbrake applied warning light can be simply checked by pulling up the handbrake lever with the ignition switched on. The light should illuminate as the lever is applied and go out as it is released. With the handbrake released, the ignition on and the wheels chocked, unscrew the master cylinder reservoir filler cap and withdraw the cap and sensor. The warning light should

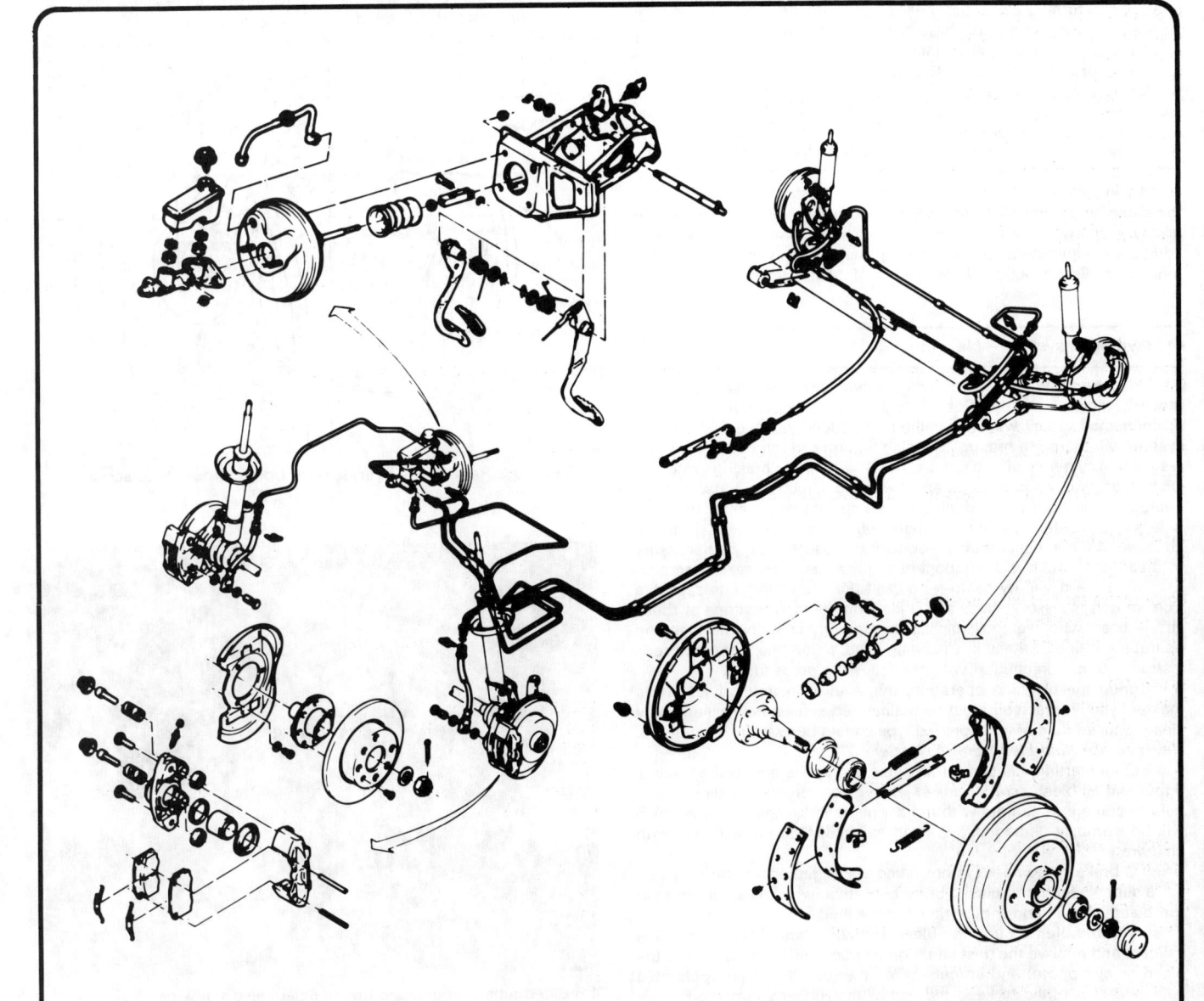

Fig. 9.1 Layout of the braking system hydraulic and mechanical components – left-hand drive shown (Sec 1)

144 Chapter 9 Braking system

illuminate as the sensor is lifted clear of the fluid. If the lights fail to work during these tests then either the bulb is blown, a fuse is at fault, or there is a fault in the circuit.

4 Annually, regardless of mileage, renew the brake hydraulic fluid by draining the system and refilling with fresh fluid, as described in Section 4.

5 Although not specified by the manufacturers, it is advisable to renew the flexible brake hoses and rubber seals in the brake calipers, wheel cylinders and master cylinder at 54 000 miles (87 000 km) or 3 years whichever comes first. Details of these operations will be found in the relevant Section of this Chapter.

3 Braking system – adjustment

1 As wear takes places on the friction material of the front brake pads and rear brake shoes, adjustment will be necessary to reduce the extra brake pedal travel that will result.

2 The front brake pads adjust themselves automatically, by moving closer to the disc as wear takes place on the pad material.

3 The rear brake shoes are individually adjusted, by two eccentric cam adjusters located in each brake backplate. The adjusters are operated by turning a hexagon-headed bolt attached to each adjuster, accessible from the rear of the backplates (photo).

4 To adjust the rear brakes, jack up the rear of the car and support it on axle stands. Ensure that the handbrake is released.

5 Using a suitable spanner or socket, turn one of the adjusters in whichever direction is necessary to lock the wheel. Now back off the adjuster until the wheel is just free to turn without binding.

6 Carry out this operation on the second adjuster then repeat the procedure for the other wheel.

7 An independent means of handbrake adjustment is provided on the cable, and this procedure is described in Section 23. Adjustment of the rear brakes will also reduce the handbrake lever travel and additional adjustment of the cable is usually only necessary after renewal or disconnection of the cable, or if it has stretched excessively.

4 Hydraulic system – bleeding

1 Whenever the brake hydraulic system has been dismantled to renew a pipe or hose (or if any other component has been disconnected in any way), air will be introduced into the system. The system will therefore require 'bleeding' in order to remove this air and restore the system's effectiveness. The design of the braking system is such that two entirely separate hydraulic circuits are used, each operating one front and one diagonally opposite rear brake. Therefore unless the master cylinder has been removed or the fluid is being changed at the annual renewal period, it will usually only be necessary to bleed one circuit if a component or pipe has been disconnected.

2 There are a variety of do-it-yourself brake bleeding kits available from motor accessory shops and it is recommended that one of these kits is used wherever possible as they greatly simplify the bleeding operation. If one of these kits is being used, follow the manufacturer's instructions in conjunction with the following procedure.

3 During the bleeding operation, the level of hydraulic fluid in the master cylinder reservoir must be maintained at least half-full and only clean, unused fluid of an approved type should be used for topping up. *Never re-use fluid bled from the system.*

4 Before starting, check that all pipes and hoses are secure, unions tights and all bleed screws closed. Take great care not to allow brake fluid to come into contact with the car paintwork, otherwise the finish will be seriously damaged. Wash off any spilt fluid immediately with cold water.

5 If a brake bleeding kit is not being used, gather together a clean jar, a suitable length of plastic or rubber tubing which is a tight fit over the bleed screws and a new tin of brake fluid.

6 Clean the area around the bleed screw on one of the front brake calipers and remove the dust cap. Connect one end of the tubing to the bleed screw (photo) and immerse the other end in the jar containing sufficient brake fluid to keep the end of the rubber submerged.

7 Open the bleed screw half a turn and have an assistant depress the brake pedal to the floor and then slowly release it. Tighten the bleed screw at the end of each downstroke to prevent the expelled air and fluid from being drawn back into the system. Continue this

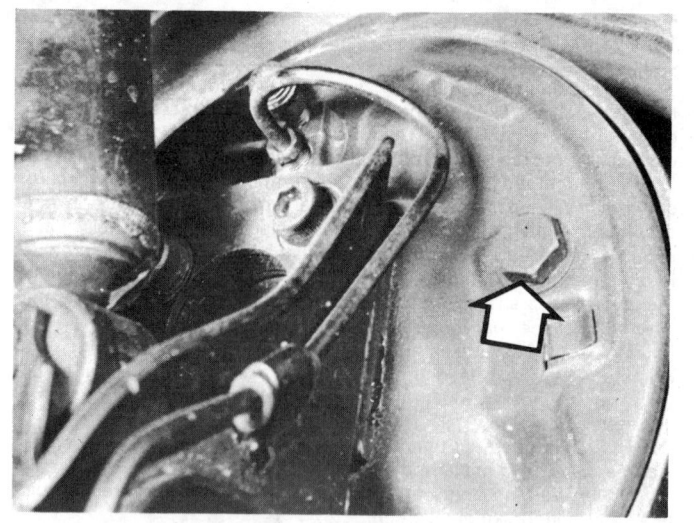

3.3 The rear brake adjusters (one adjuster arrowed) are accessible from the rear of the backplates

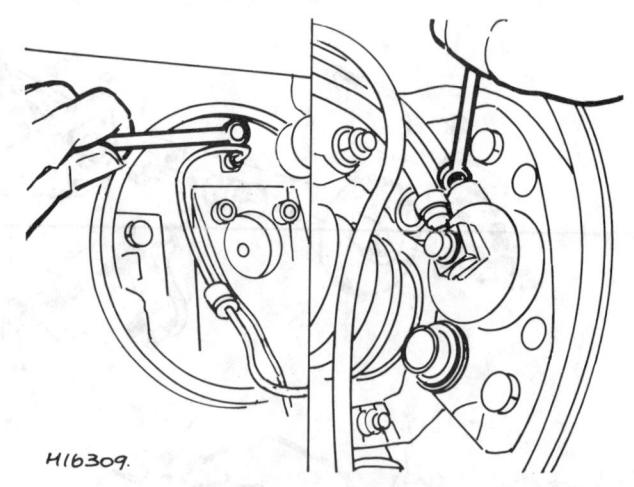

Fig. 9.2 Rear wheel cylinder and front caliper bleed screw locations (Sec 4)

4.6 Bleed tube connected to front caliper bleed screw

procedure until clean brake fluid, free from air bubbles, can be seen flowing into the jar, and then finally tighten the bleed screw.

8 Remove the tube, refit the dust cap and repeat this procedure on

Chapter 9 Braking system

the diagonally opposite rear wheel and then if necessary on the front and rear wheels of the other circuit.

9 When bleeding is complete, top up the fluid level in the master cylinder reservoir and refit the cap.

5 Front brake pads – inspection and renewal

1 Remove the wheel trim, slacken the wheel bolts and jack up the front of the car. Support the car on axle stands and remove the front roadwheels.

2 Inspect the thickness of the friction material on each pad. If any one is at or below the specified minimum, renew the pads as an axle set (four pads) in the following way.

3 Drive out the pad retaining pins by applying a suitable punch to their inboard ends. Be prepared for the pad tension springs to be released as the pins are driven out.

4 With the tension springs removed, withdraw the pads, one at a time from the caliper, using pliers if necessary (photos).

5 With the pads removed inspect the friction material for signs of oil or hydraulic fluid contamination, heavy scoring or cracking and for the security of the friction material on the metal backing. Renew all four pads if any one exhibits these conditions.

6 Brush away the dust and dirt from the caliper, piston and disc, but **do not** inhale as it is injurious to health.

7 Rotate the brake disc by hand and scrape away any rust and scale. Carefully inspect the entire surface of the disc and if there are any signs of cracks, deep scoring or abrasions, the disc must be renewed. Also inspect the caliper for signs of fluid leaks around the piston, corrosion or other damage. Renew the piston seals or the caliper body if necessary.

8 If new brake pads are being fitted it will be necessary to push the caliper piston fully into its bore to accommodate the new thicker pads. Using a flat bar such as a tyre lever or a large screwdriver, carefully lever the piston squarely into its bore as far as it will go. The action of depressing the piston will cause a quantity of hydraulic fluid to be returned to the master cylinder reservoir. Remove the reservoir filler cap and place absorbent rags around the master cylinder to collect any fluid that may overflow, or preferably syphon some fluid from the reservoir using a syringe.

9 Place the new pads in position in the caliper with the friction material against the disc.

10 Position the tension springs over the pads and drive in the retaining pins from the outside to the inside (photos).

11 Refit the roadwheels and lower the car to the ground.

12 Depress the footbrake several times to bring the piston into contact with the pads and centralise the caliper.

13 Top up the master cylinder reservoir if necessary and refit the filler cap. **Note:** *If new brake pads have been fitted, avoid unnecessary heavy braking for the first 120 miles (200 km).*

5.4A Removing the inboard pad

5.4B Removing the outboard pad

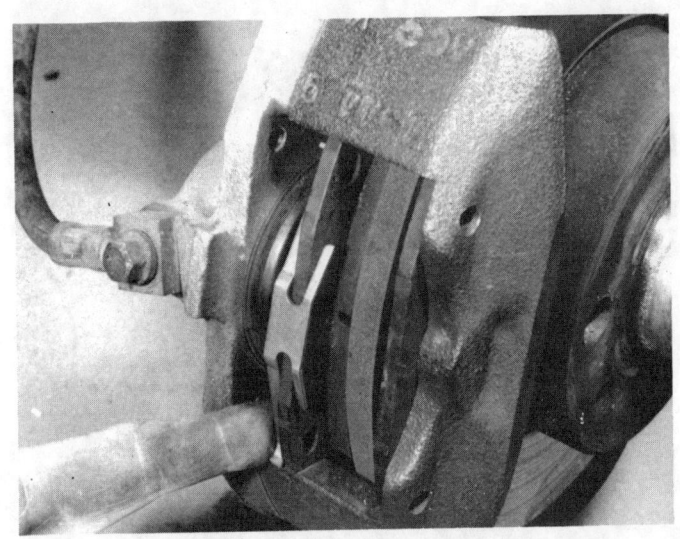

5.10A Position the tension springs against the pads ...

5.10B ... and drive in the retaining pins

Chapter 9 Braking system

6 Front brake caliper – removal, overhaul and refitting

1 Remove the wheel trim, slacken the wheel bolts and jack up the front of the car. Support the car on axle stands and remove the front roadwheel.
2 Using a brake hose clamp or self-grip wrench with its jaws suitably protected, clamp the flexible brake hydraulic hose. This will minimise brake fluid loss during subsequent operations.
3 Undo the union bolt securing the brake hose to the caliper (photo) and recover the two sealing washers. Plug or tape over the hose and caliper orifices to prevent dirt entry.
4 Using a chisel release the protective caps over the two caliper mounting bolts and then prise the caps off with a screwdriver (photo).
5 Using an Allen key or socket bit, undo the two caliper mounting bolts and remove the caliper, complete with brake pads from the steering knuckle and disc (photo).
6 If the caliper has been removed for overhaul, remove the brake pads, as described in the previous Section.
7 Using a chisel, release the sliding sleeve inner dust caps from the caliper housing flanges.
8 Again using a chisel, release the piston dust cover from the caliper housing.
9 Push the sliding sleeves inward and remove the dust caps from the sleeve grooves.
10 Remove the piston dust cover from the housing and piston, and remove the two sliding sleeves from their locations.
11 Place a piece of wood approximately 15 to 20 mm (0.6 to 0.8 in) thick over the end of the piston and apply air pressure to the brake hose union on the side of the caliper. Only low pressure air is needed to eject the piston, such as is generated by a tyre foot pump. Remove the wood and withdraw the piston the rest of the way out of its bore by hand.
12 Undo the two bolts securing the holding frame to the caliper housing and separate the two parts.
13 Carefully remove the fluid seal from the caliper bore using a blunt instrument such as a knitting needle.
14 Thoroughly clean all the parts in clean brake fluid or methylated spirit and dry with a clean lint-free rag.
15 Inspect the surfaces of the piston and caliper bore for corrosion, scoring or scuff marks and, if evident, renew the complete caliper. If these parts are in a satisfactory condition, obtain a caliper repair kit which will contain all the necessary seals needed for reassembly.
16 Begin reassembly by lubricating the fluid seal, piston and caliper bore with clean brake fluid.
17 Fit the new fluid seal to the caliper bore and locate it squarely in its groove.
18 Place the piston in position and push it squarely into its bore approximately half-way.
19 Position a new dust cover over the piston and in engagement with the piston groove. Push the piston fully into its bore and tap the larger diameter of the dust cover into position on the housing using a tube of suitable diameter.
20 Refit the holding frame to the caliper housing and secure with the two bolts tightened to the specified torque.
21 Remove the old seals from the sliding sleeves using a blunt instrument and fit the new seals after lubricating them with the grease supplied in the repair kit. Ensure that the seals are not twisted in their grooves.
22 Place the sliding sleeves in their locations in the caliper housing so that the inner dust cap groove is toward the piston side of the housing. Before pushing the sleeves fully home, place the dust caps over them so that they locate in the sleeve grooves and then tap the dust caps onto the housing flange using a small tube. Now centre the sliding sleeves in the housing.
23 Refit the assembled caliper to the steering knuckle and secure with the two mounting bolts, having first cleaned their threads and applied thread locking compound. Tighten the bolts to the specified torque and refit the protective caps. New caps must be used if the old ones were in any way damaged or deformed during removal. Tap the caps fully into place using a tube of suitable diameter.
24 Refit the flexible brake hose and secure with the union bolt. The copper sealing washers must be positioned on each side of the hose union. Tighten the bolt to the specified torque.
25 Refit the brake pads, as described in Section 5, and bleed the hydraulic system, as described in Section 4.
26 Refit the roadwheel and lower the car to the ground.

6.3 Brake hose-to-caliper banjo union bolt (arrowed)

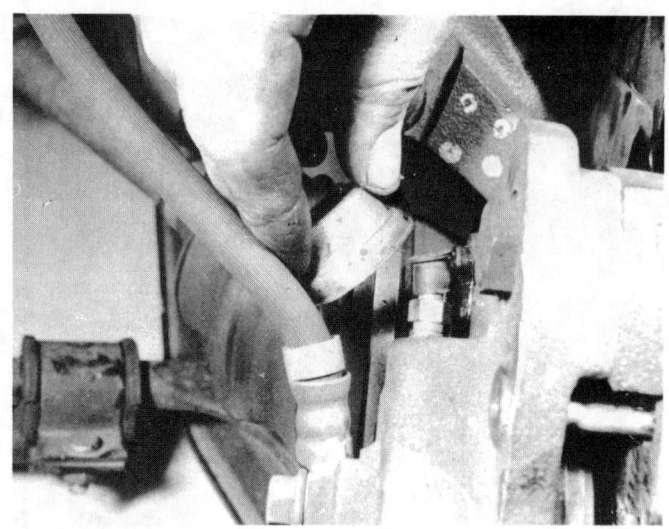

6.4 The protective caps must be removed to gain access to the caliper mounting bolts

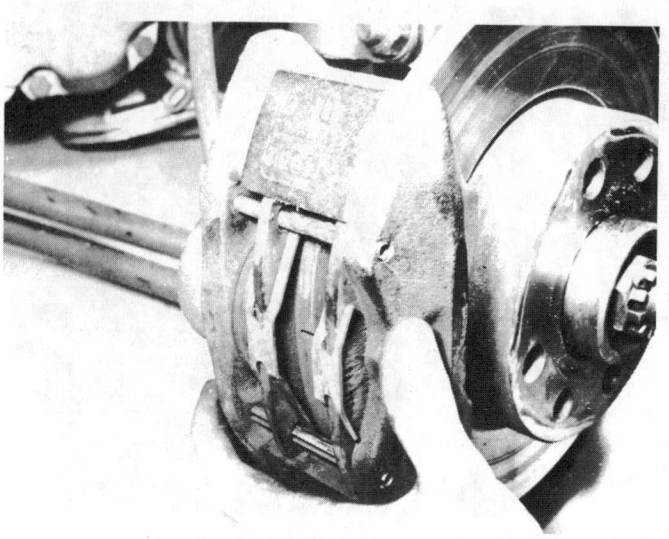

6.5 Removing the front brake caliper

147

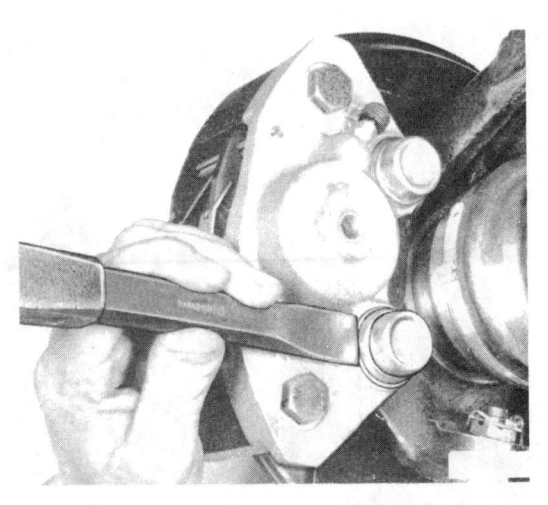

Fig. 9.3 Removing of the caliper mounting bolt protective caps using a chisel (Sec 6)

Fig. 9.4 An Allen key or Allen key socket bit is needed to undo the caliper mounting bolts (Sec 6)

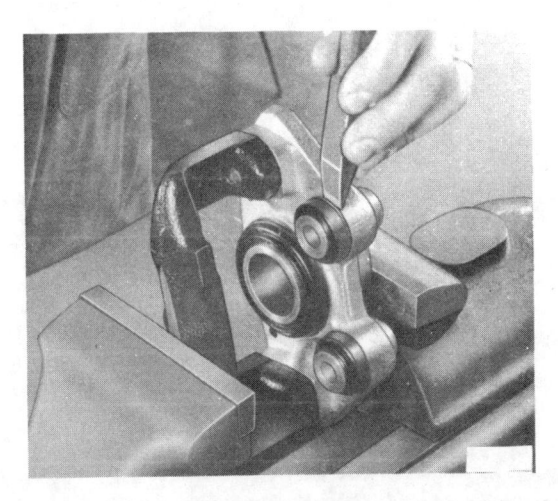

Fig. 9.5 Release the caliper sliding sleeve inner dust caps with a chisel (Sec 6)

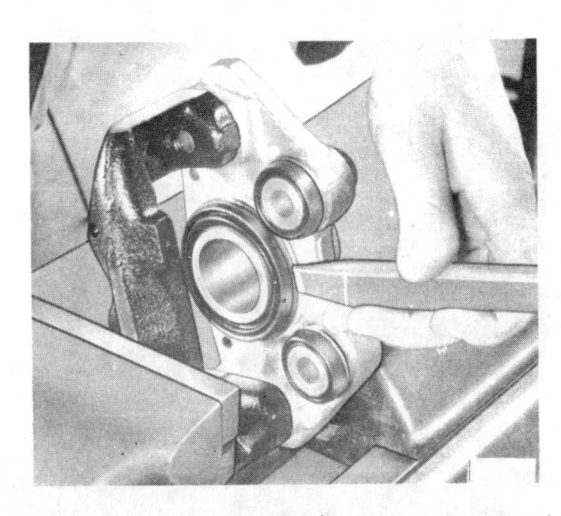

Fig. 9.6 A chisel will also be needed to release the piston dust cover from the caliper housing (Sec 6)

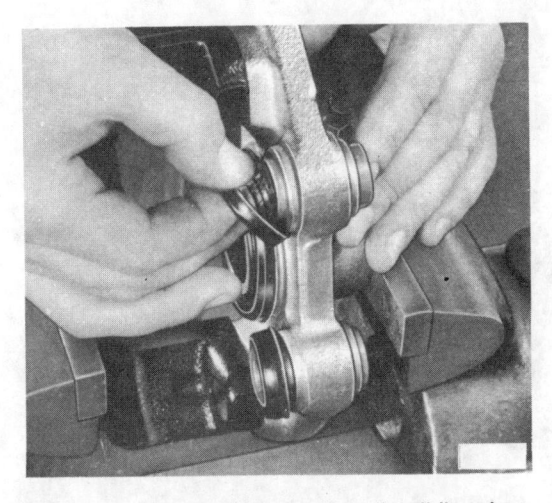

Fig. 9.7 Removing the dust caps from the sliding sleeve grooves (Sec 6)

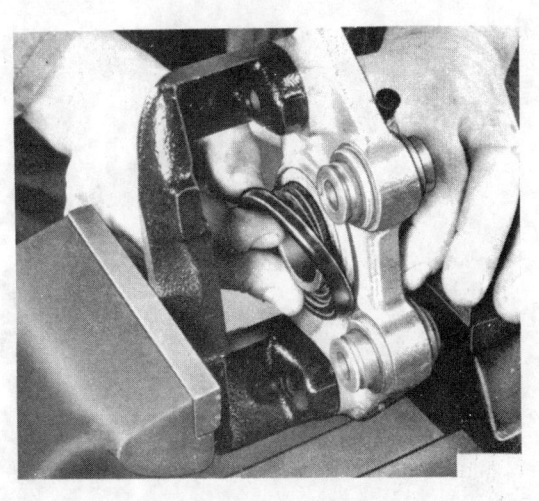

Fig. 9.8 Removing the piston dust cover from the housing and piston (Sec 6)

Fig. 9.9 Use a low pressure air line to eject the piston from the housing (Sec 6)

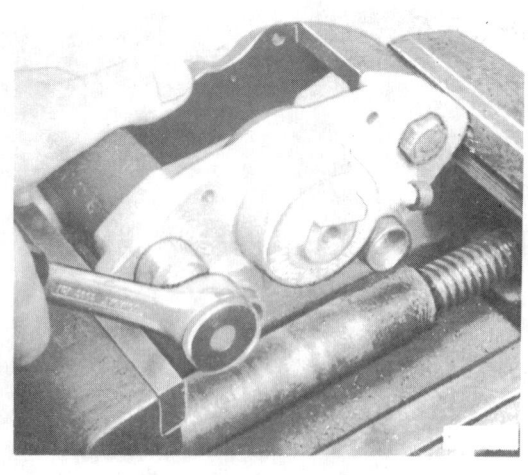

Fig. 9.10 Undo the two bolts to separate the caliper housing from the holding frame (Sec 6)

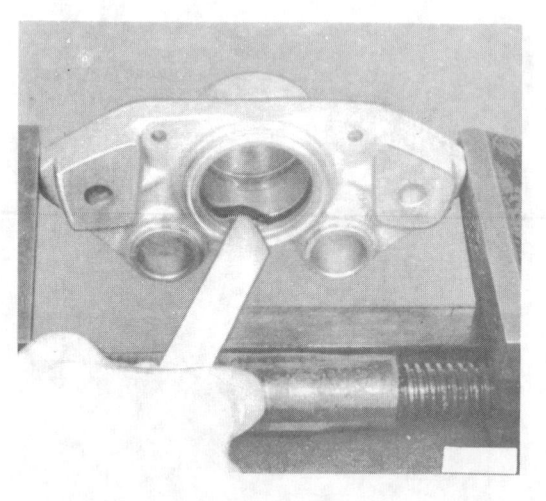

Fig. 9.11 Remove the fluid seal from the caliper bore using a blunt instrument (Sec 6)

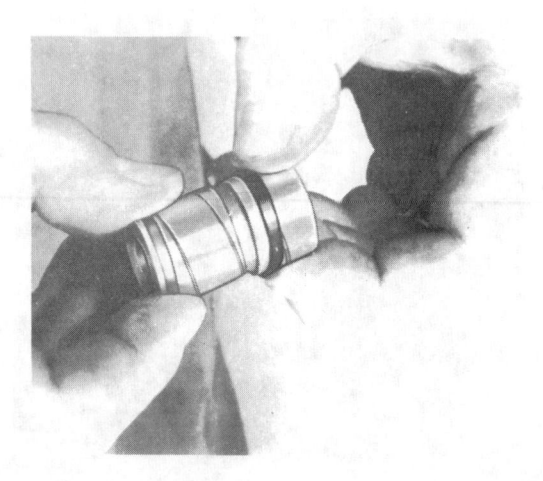

Fig. 9.12 Refitting new seals to the sliding sleeves (Sec 6)

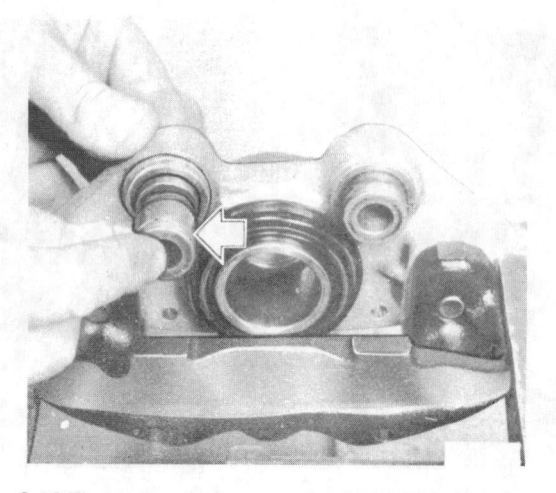

Fig. 9.13 The sliding sleeves must be fitted with their dust cap grooves (arrowed) toward the piston side of the housing (Sec 6)

Chapter 9 Braking system

7 Front brake disc – removal and refitting

1 Remove the wheel trim, slacken the wheel bolts and jack up the front of the car. Support the car on axle stands and remove the roadwheel.
2 Rotate the disc by hand and examine it for deep scoring, grooving or cracks. Light scoring is normal, but if excessive the disc must be refinished or renewed.
3 If it is thought that the disc is distorted, check it for run-out using a dial gauge or using feeler blades between the disc and a fixed point as the disc is rotated. If the run-out exceeds the tolerance shown in the Specifications, refinishing or renewal is necessary. Bolt the disc to the hub using wheel bolts for this operation.
4 To remove a brake disc, first remove the brake pads, as described in Section 5.
5 On models fitted with alloy roadwheels, remove the protective shield from around the disc centre hub.
6 Undo the small screw securing the disc to the hub, tilt the disc slightly and remove it from the hub and caliper. If necessary pull the caliper outwards slightly to provide extra clearance.
7 If the disc has been removed for renewal or machining, it is recommended that both discs are renewed or refinished at the same time otherwise uneven braking may occur.
8 Refitting is the reverse sequence to removal.

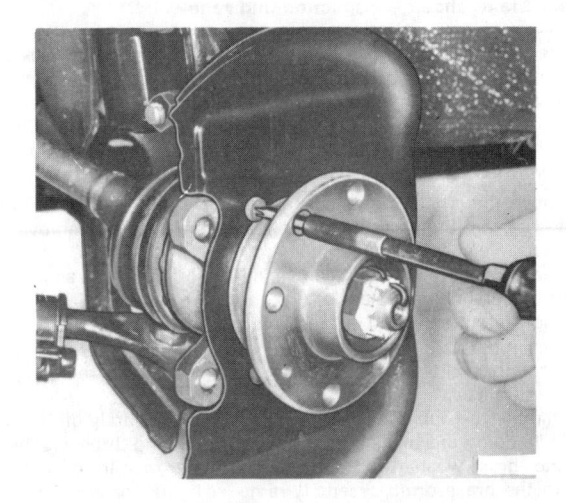

Fig. 9.14 Insert a screwdriver through the holes in the hub to remove the disc shield retaining screws (Sec 8)

Fig. 9.15 The disc shield connecting web must be cut off to allow removal of the shield (Sec 8)

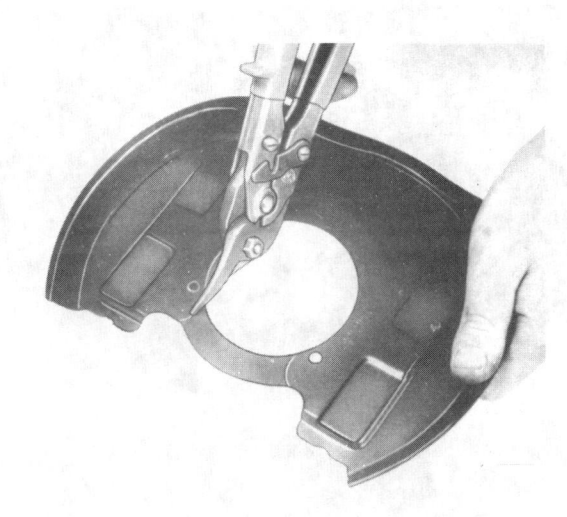

Fig. 9.16 Removal of the connecting web prior to fitting a new disc shield (Sec 8)

8 Brake disc shield – removal and refitting

1 Remove the wheel trim, slacken the wheel bolts and jack up the front of the car. Support the car on axle stands and remove the front roadwheel.
2 Using a chisel, release the protective caps over the brake caliper mounting bolts and then prise the caps off using a screwdriver.
3 Using an Allen key or socket bit, undo the two caliper mounting bolts and withdraw the caliper, complete with brake pads off the steering knuckle and disc. Suspend the caliper from a convenient place under the wheel arch using string or wire. Avoid straining the flexible brake hose.
4 Undo the small retaining screw and withdraw the disc from the hub.
5 Using a screwdriver inserted through the holes in the hub flange, undo the disc shield retaining screws.
6 Using tin snips, cut off the shield connecting web, as shown in Fig. 9.15, and then slide the shield off the hub.
7 Before fitting a new shield, the connecting web must be cut off in the same way to allow the shield to be slid over the hub. Paint the cut edges of the shield after removing the web to prevent corrosion.
8 Refitting is the reverse sequence to removal, but clean the threads of the caliper mounting bolts and apply thread locking compound before refitting. Tighten the bolts to the specified torque and refit the protective caps using a small diameter tube to drive them into position. Use new caps if the old ones were in any way damaged or deformed during removal.

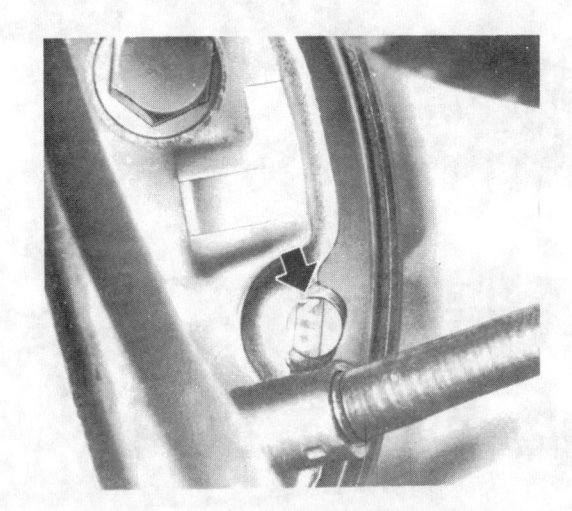

Fig. 9.17 Location of the brake shoe inspection hole in the rear brake backplate (Sec 9)

9 Rear brake shoes – inspection and renewal

Note: *An inspection hole is provided on each rear brake backplate and after removal of the plug the lining thickness can be observed through the hole. This must be regarded purely as a quick check because only one brake shoe can be viewed. It is also not possible to inspect the lining condition, only its thickness. To carry out a thorough inspection, the hub/drum assembly must be removed, as follows:*

1 Remove the wheel trim, slacken the wheel bolts and jack up the rear of the car. Support the car on axle stands and remove the roadwheels. Release the handbrake.
2 Refer to Section 3, if necessary, and back off the brake adjuster so that the brake shoes are clear of the drum.
3 Using a stout screwdriver, prise off the dust cap from the centre of the hub (photo).
4 Extract the split pin and unscrew the hub bearing retaining nut from the stub axle.
5 Withdraw the hub and brake drum assembly squarely off the stub axle. While doing this place a thumb over the bearing thrust washer to retain the thrust washer and outer bearing inner race in the hub.
6 With the brake drum assembly removed, brush or wipe the dust from the drum, brake shoes, wheel cylinder and backplate. **Take great care** not to inhale the dust as it is injurious to health.
7 Measure the brake shoe lining thickness. If it is worn down to the specified minimum amount, or if it is nearly worn down to the rivets, renew all four brake shoes. The shoes must also be renewed if any are contaminated with brake fluid or grease, or show signs of cracking or glazing. If contamination is evident, the cause must be traced and cured before fitting new brake shoes.
8 If the brake shoes are in a satisfactory condition, proceed to paragraph 21; if removal is necessary, proceed as follows:
9 First make a careful note of the location and position of the various springs and linkages as an aid to refitting (photo).
10 Using pliers, release the upper brake shoe return spring from the leading shoe and then disengage the spring from the trailing shoe (photo).
11 Ease the leading shoe from its lower pivot location and remove the lower return spring from both shoes (photo).
12 Withdraw the leading shoe from its upper and lower locations and from the steady clip and pushrod (photo).
13 Withdraw the trailing shoe from its upper and lower locations, from the steady clip and pushrod, and then detach the handbrake cable end from the shoe lever (photo). Remove the pushrod.
14 Before fitting the new brake shoes, it will be necessary to transfer the trailing shoe handbrake lever to the new shoe. This can be done after removal of the retaining circlip and pivot pin.
15 Prior to installation, clean off the brake backplate with a rag and apply a trace of silicone grease to the shoe contact areas and pivots (photo).

9.3 Lever off the dust cap to gain access to the hub bearing retaining nut

9.9 Note the location and fitted position of the rear brake components before removal

9.10 Using pliers to release the upper brake shoe return spring

9.11 Ease the leading shoe out of its lower pivot location to facilitate removal of the lower return spring

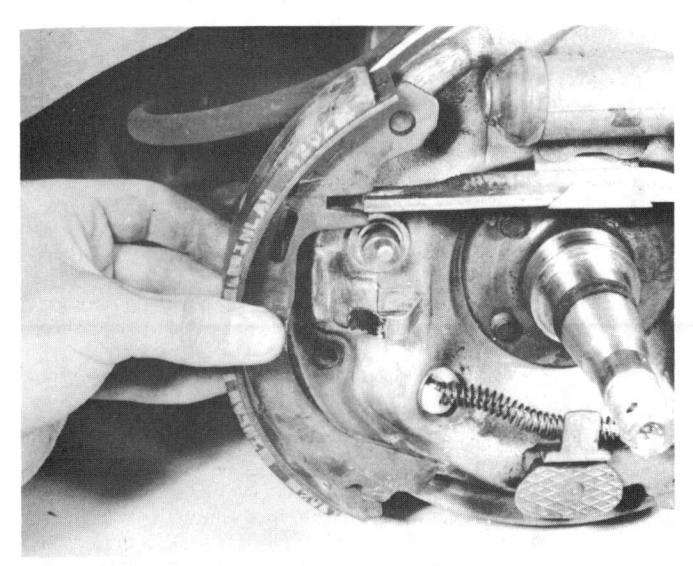

9.12 Remove the leading shoe ...

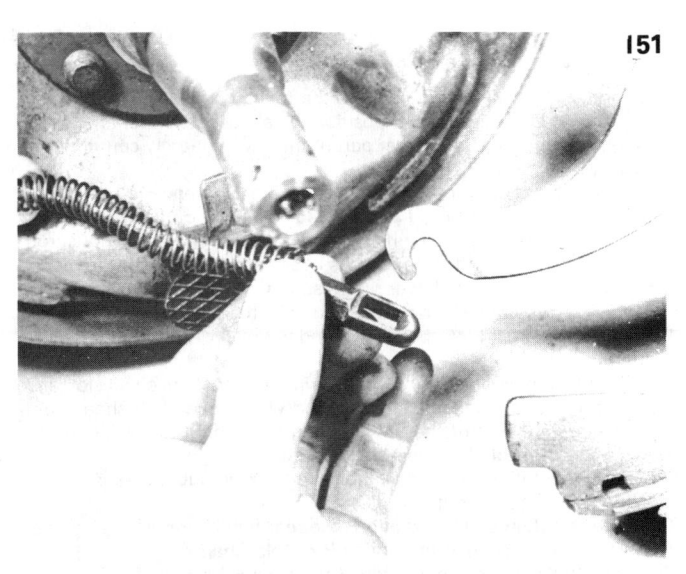

9.13 ... detach the handbrake cable and remove the trailing shoe

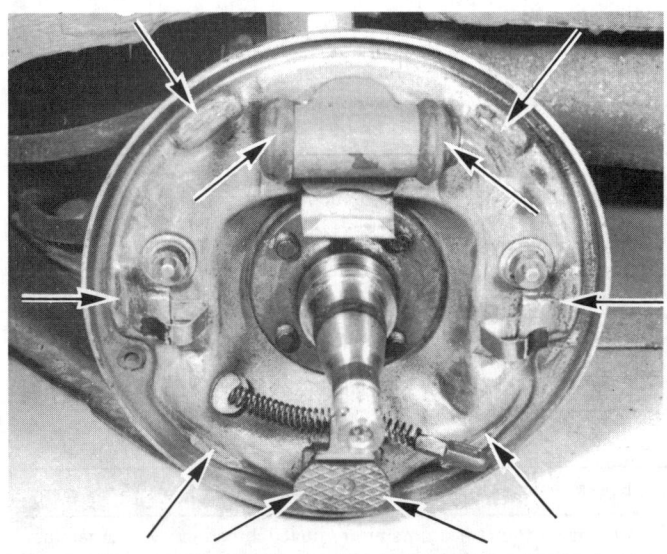

9.15 Apply a trace of silicone grease to the brake shoe pivot and contact areas (arrowed) prior to refitting the shoes

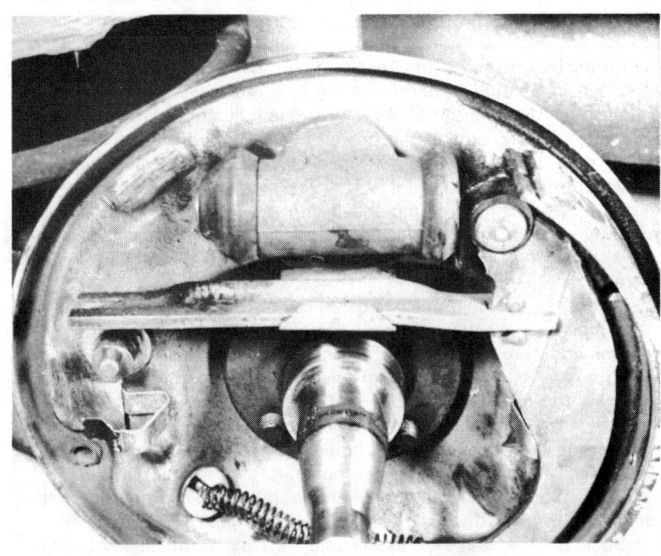

9.17 With the trailing shoe in position, refit the pushrod

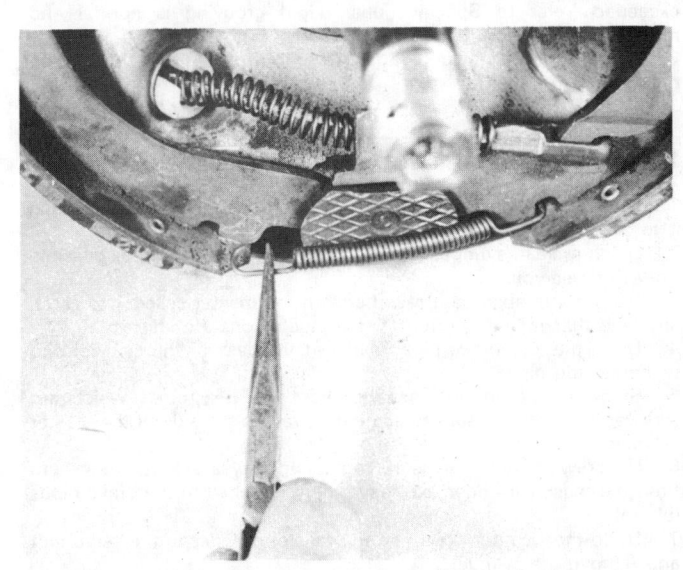

9.19A Connect the lower return spring to both brake shoes ...

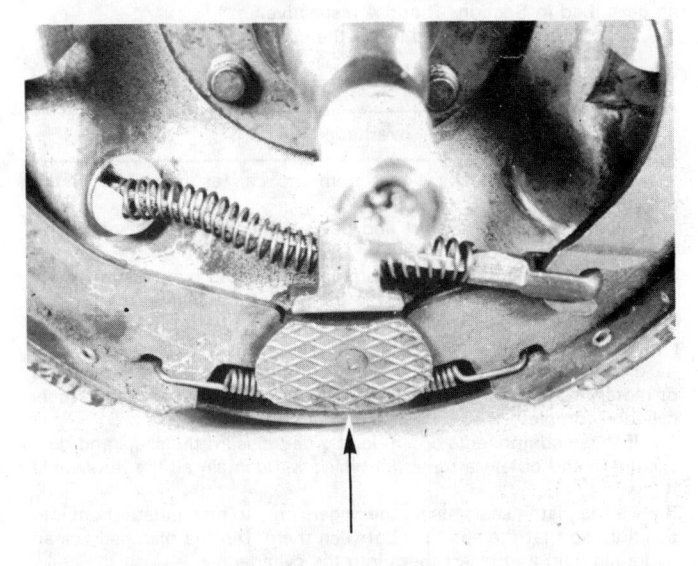

9.19B ... and then ease the coiled part of the spring under the lip of the lower pivot mounting (arrowed)

Chapter 9 Braking system

16 Connect the handbrake cable to the trailing shoe lever, locate the shoe under the steady clip and position the shoe in its upper and lower locations. Ensure that the upper part of the shoe squarely contacts the wheel cylinder piston.
17 Slide the pushrod into its support bracket and engage the forked end between the two raised pips on the trailing shoe lever (photo).
18 Locate the leading shoe under the steady clip, engage the pushrod fork and then position the shoe in its upper and lower pivot locations.
19 Hook one end of the lower return spring into the trailing shoe slot, and using pliers connect the other end with the leading shoe (photo). Ease the coiled part of the spring under the lip of the lower pivot mounting (photo).
20 Engage the upper return spring with the trailing shoe slot and again using pliers, connect the other end with the leading shoe. Note that the straight part of the upper spring fits behind the trailing shoe web and in front of the leading shoe web.
21 Refit the brake drum assembly with inner hub bearing, thrust washer and retaining nut.
22 Refer to Chapter 11 and adjust the rear hub bearings, then adjust the brakes, as described in Section 3 of this Chapter.
23 Refit the roadwheels and lower the car to the ground.

10 Rear wheel cylinder – removal and refitting

1 Begin by removing the rear hub and brake drum assembly, as described in Section 9, paragraphs 1 to 5 inclusive.
2 With the drum removed, turn the brake adjusters at the rear of the backplate to fully expand the brake shoes. This will move the upper part of each brake shoe away from the wheel cylinder sufficiently to allow its removal.
3 Using a brake hose clamp or self-grip wrench with its jaws suitably protected, clamp the flexible brake hose located adjacent to the rear coil spring. This will minimise brake fluid loss during subsequent operations.
4 At the rear of the brake backplate undo the union nut securing the brake pipe to the wheel cylinder. Carefully ease the pipe out of the cylinder and plug or tape over its end to prevent dirt entry.
5 Undo the bolt securing the wheel cylinder to the backplate and withdraw the cylinder from between the brake shoes.
6 To refit the wheel cylinder, place it in position on the backplate and engage the brake pipe and union. Screw in the union nut two or three times to ensure the thread has started.
7 Refit the wheel cylinder retaining bolt and tighten it to the specified torque. Now fully tighten the brake pipe union nut.
8 Back off the brake adjusters and refit the drum asembly, inner bearing and thrust washer. Refit the hub retaining nut and adjust the hub bearings, as described in Chapter 11.
9 Adjust the rear brakes and bleed the brake hydraulic system, as described in Sections 3 and 4 respectively.
10 Refit the roadwheel and lower the car to the ground.

11 Rear wheel cylinder – overhaul

1 Remove the wheel cylinder from the car, as described in the previous Section.
2 Clean away external dirt and pull off the rubber dust excluders from the cylinder body.
3 The pistons and seals will normally be ejected by pressure of the coil spring, but if they are not, tap the end of the cylinder on a piece of hardwood or apply low air pressure from a tyre foot pump at the pipeline connection.
4 Inspect the surfaces of the piston and the cylinder bore for scoring or metal-to-metal rubbed areas. If these are evident, renew the wheel cylinder complete.
5 If these components are in good order, discard the seals and dust excluders and obtain a repair kit which will contain all the renewable items.
6 Fit the piston seals (using the fingers only to manipulate them into position) so that the spring is between them. Dip the pistons in clean hydraulic fluid and insert them into the cylinder.
7 Refit the dust excluders and then refit the wheel cylinder to the car, as described in the previous Section.

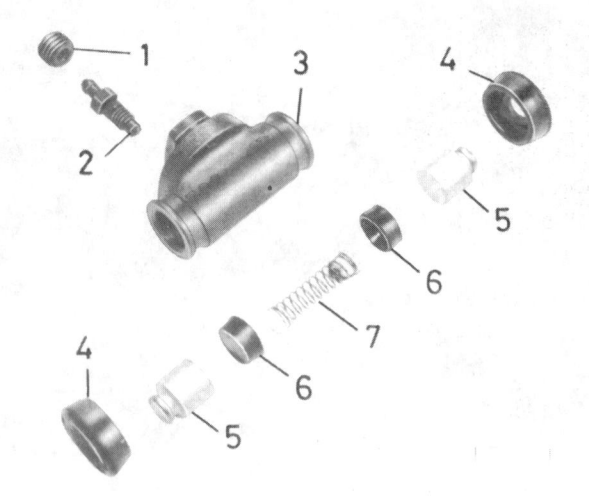

Fig. 9.18 Exploded view of the rear wheel cylinder (Sec 11)

1 Dust cap
2 Bleed screw
3 Cylinder body
4 Dust excluder
5 Piston
6 Piston seal
7 Spring

12 Rear brake backplate – removal and refitting

1 To remove the backplate it is also necessary to remove the rear stub axle as both are retained by common bolts. Full details of this procedure will be found in Chapter 11.
2 Prior to removal of the stub axle, remove the brake shoes, as described in Section 9, and the wheel cylinder, as described in Section 20. Also remove the handbrake cable from the backplate by pulling the plastic guide sleeve towards the end of the cable, against spring pressure, and slipping it off the cable. Now pull the cable out of the backplate.
3 Refitting the backplate is the reverse of the above mentioned removal procedures.

13 Brake drum – inspection and renovation

1 Whenever the rear drums are removed, brush away internal dust, **taking care** *not to inhale it,* and inspect the lining rubbing surface. If the shoes have worn grooves in the metal, then it may be possible to re-grind the inside provided the maximum internal diameter is not exceeded (refer to Specifications). Light grooving is normal and requires no attention.
2 If the drum is suspected of being out-of-round, it should be renewed.

14 Brake master cylinder – removal and refitting

1 Unscrew the filler cap from the master cylinder reservoir and place it to one side. Remove the windscreen washer reservoir.
2 Using a clean syringe, syphon off as much brake fluid as possible from the reservoir.
3 Place some absorbent rags beneath the master cylinder to catch any fluid that may drip out after the pipe unions are undone.
4 Undo the two brake pipe unions at the base of the cylinder and withdraw the pipes.
5 Slacken, but do not remove, the two brake pipes which are screwed into the pressure regulating valves on the side of the master cylinder.
6 Unscrew the two pressure regulating valves and remove them from the master cylinder, but leave them attached to the brake pipes (photo).
7 Undo the two nuts securing the master cylinder to the servo unit and remove the cylinder.
8 If the master cylinder has been removed for overhaul or renewal then it will be necessary to remove the reservoir. This is necessary

Chapter 9 Braking system

before overhaul can commence or for fitting to the new unit which will be supplied without a reservoir.

9 To do this, carefully support the cylinder in a vice and push back the two retaining clips with a screwdriver. Tip the reservoir to one side and lift it upwards and off the master cylinder.

10 Refitting is the reverse sequence to removal, but use new reservoir rubber seals and tighten the retaining nuts and unions to the specified torque.

11 Bleed the complete hydraulic system, as described in Section 4.

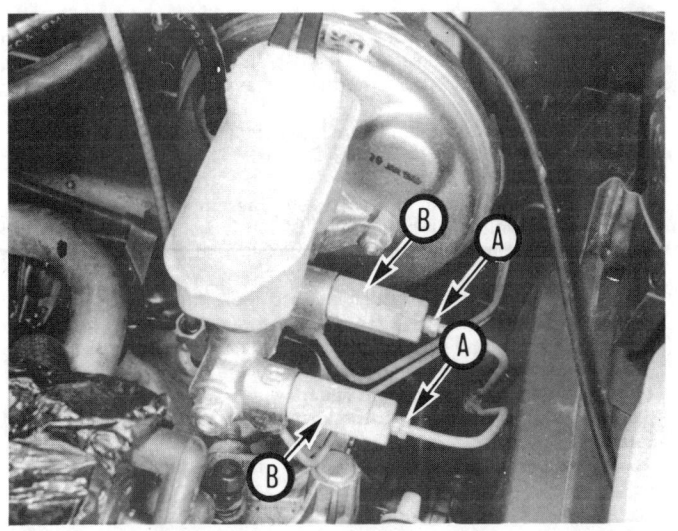

14.6 Brake pipe (A) and pressure regulating valve (B) connections at the master cylinder

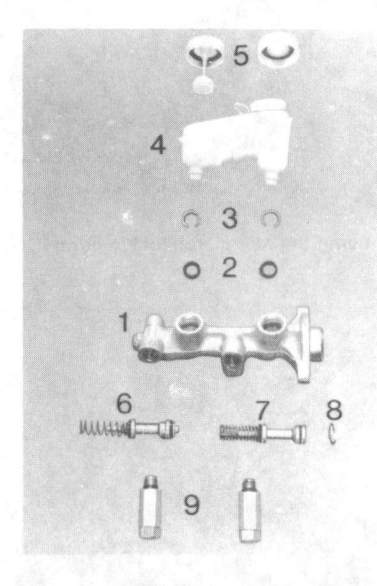

Fig. 9.21 Exploded view of the master cylinder (Sec 15)

1 Cylinder body
2 Reservoir seals
3 Reservoir retaining clips
4 Reservoir
5 Reservoir filler cap (shown with and without warning light sensor assembly)
6 Front (secondary) piston assembly
7 Rear (primary) piston assembly
8 Locking ring
9 Pressure regulating valves

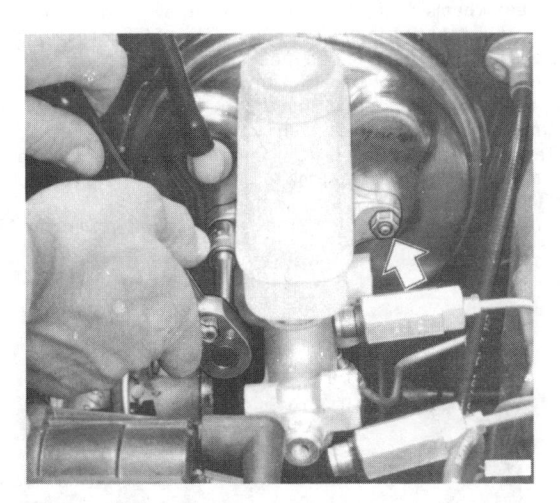

Fig. 9.19 Removing the master cylinder retaining nuts (Sec 14)

15 Brake master cylinder – overhaul

1 Remove the master cylinder from the car, as described in the previous Section, and prepare a clean uncluttered work surface.

2 Mount the cylinder in a soft-jawed vice, and using a well rounded bar or tube of suitable diameter, push the rear piston down the cylinder bore approximately 10 mm (0.4 in). Retain the piston in this position, by inserting a blunt-ended length of stiff wire into the reservoir inlet port nearest the mounting flange.

3 With the rear piston held down, use two screwdrivers to prise out the locking ring from the internal groove in the cylinder body.

4 Release the tension on the piston and remove the length of retaining wire. Lift out the rear piston and spring assembly from the cylinder bore.

5 Tap the cylinder on a block of wood, and retrieve the front piston as it emerges from the cylinder bore. Recover the seals, spring support and spring.

6 To dismantle the rear piston assembly, use a small socket or similar tool to compress the spring sufficiently to expose the retaining circlip. Using a small screwdriver, extract the circlip and withdraw the spring support sleeve, spring and seals from the rear piston.

7 Thoroughly clean all components in methylated spirit or clean brake fluid, and lay them out in the order in which they were removed, ready for inspection.

8 Carefully examine the internal cylinder bore and pistons for scoring or wear, and all components for damage or distortion. In order that the

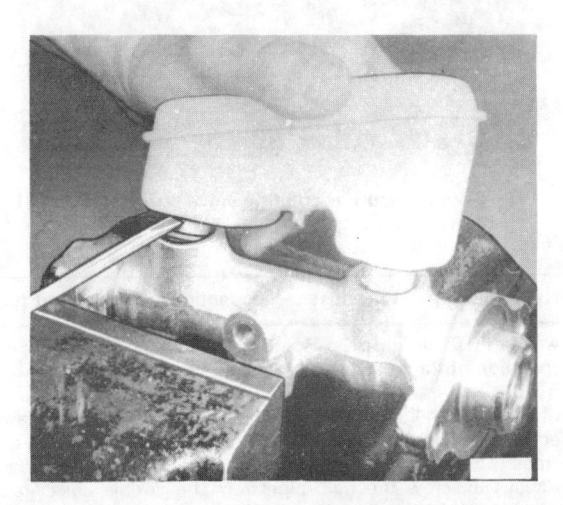

Fig. 9.20 Releasing the master cylinder reservoir retaining clips (Sec 14)

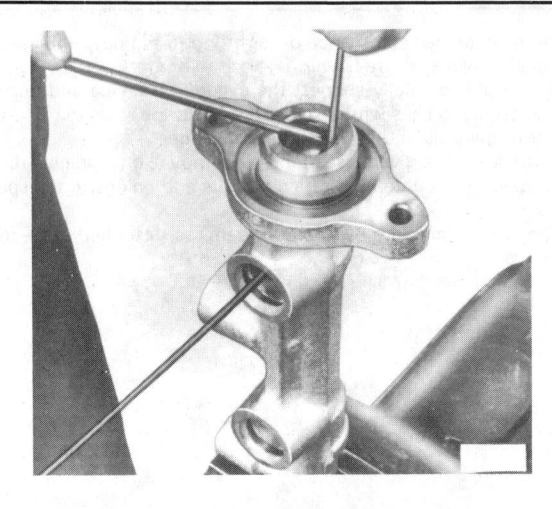

Fig. 9.22 Removing the rear piston locking ring (Sec 15)

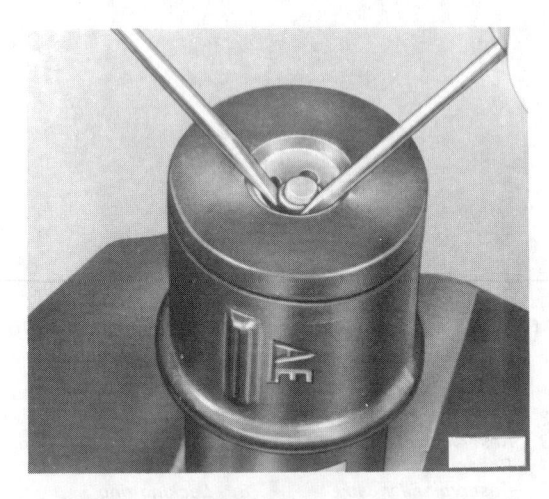

Fig. 9.23 Removing the rear piston retaining circlip (Sec 15)

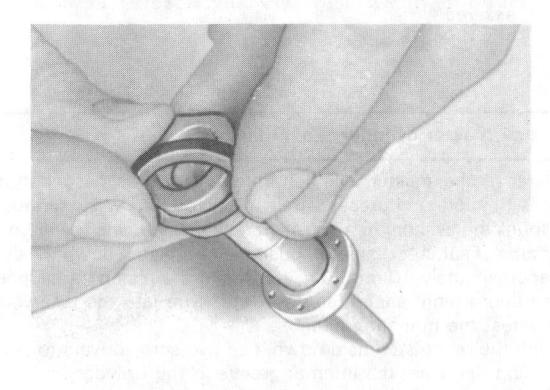

Fig. 9.24 Fit new seals to the pistons using the fingers only (Sec 15)

seals may adequately maintain hydraulic fluid pressure without leakage, the condition of the pistons and cylinder bore must be perfect. If in any doubt whatsoever about the condition of the components, renew the complete master cylinder.
9 If the master cylinder is in a satisfactory condition, a new set of seals must be obtained before reassembly. These are available in the form of a master cylinder repair kit, obtainable from your local dealer.
10 To reassemble the master cylinder, note which way round the old seals are fitted and slide them off the pistons using the fingers only. Refit the new seals in the same manner, after coating them in clean brake fluid.
11 Compress the spring on the rear piston, using a small socket, then secure the spring and support sleeve in position, using a new retaining circlip.
12 Thoroughly lubricate the two piston assemblies and the master cylinder bore with clean brake fluid. Slide the front piston spring and spring support into the cylinder bore, followed by the front piston assembly.
13 Insert the rear piston assembly into the cylinder bore. Push the piston down the bore and temporarily retain it in place, using the rounded bar and stiff wire as described in the dismantling sequence.
14 Fit a new piston retaining circlip into the groove in the cylinder bore, ensuring that it is seated squarely.
15 Release the tension on the piston and remove the length of retaining wire.
16 The master cylinder may now be refitted to the car.

16 Pressure regulating valves – general

1 The pressure regulating valves are screwed into the side of the master cylinder, one valve for each diagonally split rear circuit.
2 The purpose of these valves is to restrict hydraulic fluid pressure to the rear wheels during heavy braking to prevent rear wheel lock up.
3 No adjustment of the valves is possible and if it is suspected that a valve is faulty then it must be renewed. The valve operating pressure can be tested by a GM dealer using special equipment. If a valve is being renewed, it is essential that the numbers denoting gradient and pressure are identical on both valves.
4 The valves are removed from the master cylinder by unscrewing them after first disconnecting the relevant brake pipe. Syphon the fluid from the reservoir using a syringe before removal.
5 After refitting bleed the system, as described in Section 4.

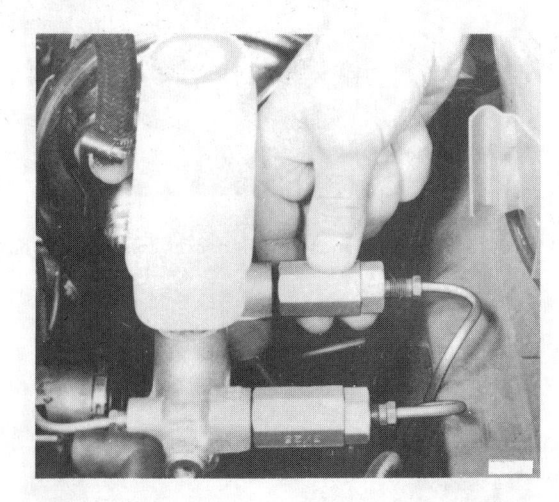

Fig. 9.25 Pressure regulating valve removal (Sec 16)

17 Hydraulic pipes and hoses – inspection, removal and refitting

1 At the intervals given in Routine Maintenance, carefully inspect the condition of the hydraulic pipes and hoses, hose connections and pipe unions.
2 Check the condition of the four hoses and if they appear swollen, chafed or when bent double with the fingers tiny cracks appear, then they must be renewed (photo).
3 Always unscrew the rigid pipe from the flexible hose first, then release the end of the flexible hose from the support bracket. To do this, pull out the retaining clips using pliers (photos).

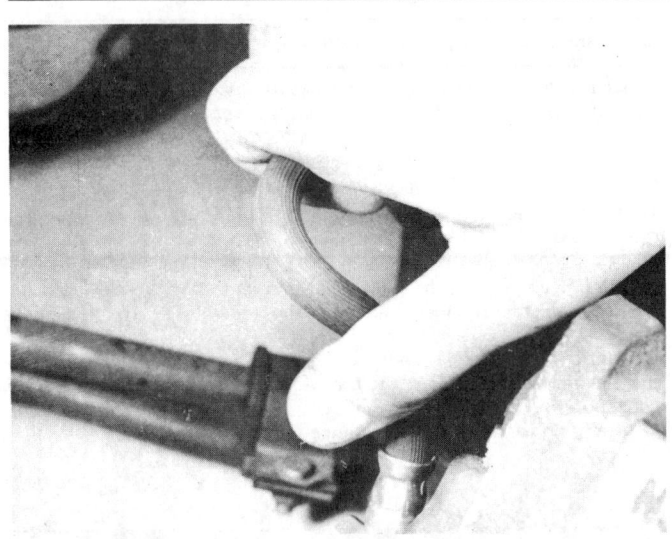

Chapter 9 Braking system

4 Now unscrew the flexible hose from the caliper or connector. On calipers a banjo type hose connector is used. When installing the hose, always use new sealing washers.
5 When installation is complete, check that the flexible hose does not rub against the tyre or other adjacent components. Its attitude may be altered to overcome this by pulling out the clip at the support bracket and twisting the hose in the required direction by not more than one quarter turn.
6 Wipe the steel brake pipes clean and examine them for signs of corrosion or other damage.
7 Examine the fit of the pipes in their insulated securing clips and bend the tongues of the clips if necessary to ensure a positive fit.
8 Check that the pipes are not touching any adjacent components or rubbing against any part of the vehicle. Where this is observed, bend the pipe gently away to clear.
9 Any section of pipe which is rusty or chafed should be renewed. Brake pipes are available to the correct length and fitted with end unions from most dealers and can be made to pattern by many accessory suppliers. When installing the new pipes use the old pipes as a guide to bending and do not make any bends sharper than is necessary.
10 Bleed the hydraulic system as described in Section 4, after renewal of a pipe or hose.

17.2 Check the brake hoses for cracks or deterioration of the rubber

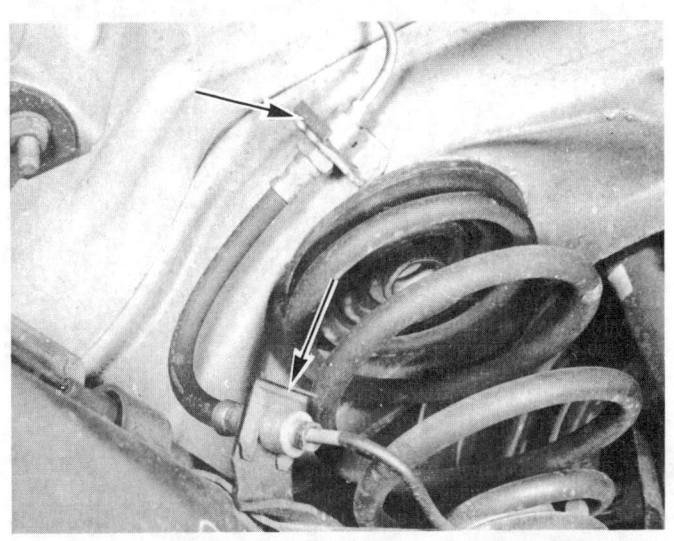

17.3A The rear brake hoses are secured by clips (arrowed) to the support brackets

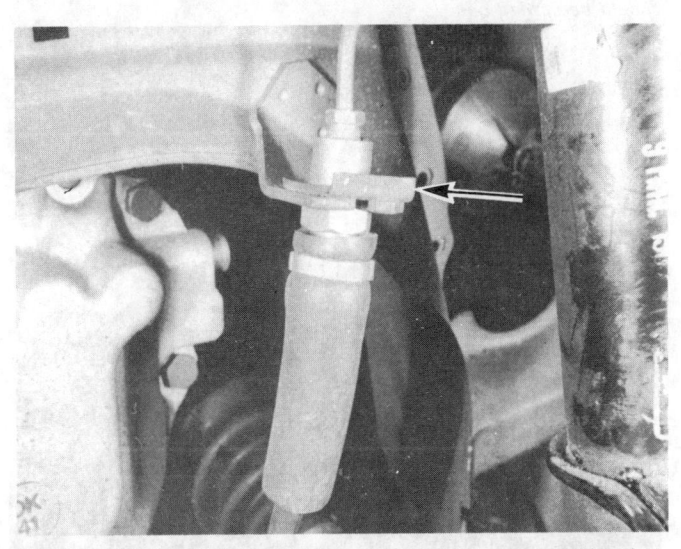

17.3B The front brake hoses are secured by a clip to a support bracket at their upper ends

Fig. 9.26 Renew the vacuum servo unit hose by cutting off the old hose (Sec 18)

Fig. 9.27 New vacuum hoses are of the fabric-covered type and must be secured by clips (Sec 18)

18 Vacuum servo unit hose – renewal

1 Unscrew the union nut securing the hose to its connection on the inlet manifold and lift off the hose.

2 Carefully release the elbow union at the servo by carefully prising it out using a screwdriver.
3 Remove the non-return valve and the two connectors from the hose by carefully cutting the hose lengthwise, but take care not to damage the connectors and valve.
4 The new hose supplied by GM dealers is of the fabric-covered type and must be secured to the connectors and non-return valve using suitable hose clips.
5 After cutting the hose to length and fitting the connectors and valve, refit the hose to the car using the reverse sequence to removal.

19 Vacuum servo unit – testing

1 With the engine switched off depress the foot brake several times and then hold it down. Start the engine and as the engine starts there should be a noticeable 'give' in the brake pedal.
2 Allow the engine to run for at least two minutes and then switch it off. If the brake pedal is now depressed again, it should be possible to hear a loud hiss from the unit when the pedal is depressed. After about four or five applications no further hissing will be heard and the pedal will feel considerably firmer.
3 If the servo unit does not operate as described, check for leaks at the vacuum hose unions and check the non-return valve. If a leak cannot be found an internal fault in the servo must be suspected. The unit cannot be dismantled for overhaul and, if faulty, renewal will be necessary.

20 Vacuum servo unit – removal and refitting

1 Disconnect the warning light wiring connection on the master cylinder reservoir filler cap.
2 Undo the two nuts securing the master cylinder to the servo and carefully ease the cylinder forwards off the mounting studs. Take care not to place excessive strain on the brake pipes.
3 Disconnect the vacuum pipe from the servo by prising the elbow connector out of its grommet using a screwdriver.
4 From inside the car remove the parcel shelf from under the facia on the passenger's side, referring to Chapter 12 if necessary.
5 Remove the retainer securing the servo pushrod clevis pin to the brake pedal relay lever. Withdraw the clevis pin.
6 Undo the nuts securing the servo to the support bracket in the engine compartment and remove the servo from the car.
7 Refitting is the reverse sequence to removal, but tighten all nuts to the specified torque.

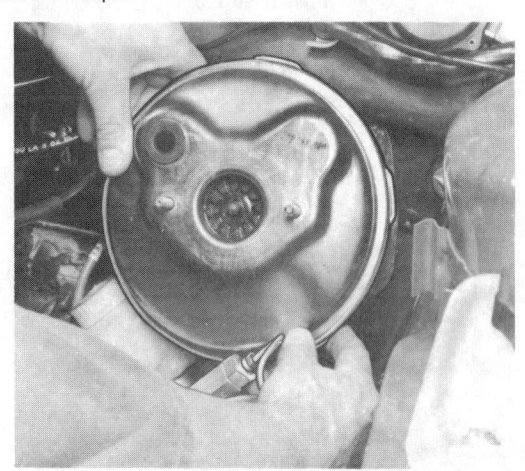

Fig. 9.28 Removing the vacuum servo unit (Sec 20)

21 Brake pedal – removal and refitting

1 Removal and refitting of the brake pedal follows the same procedure as described in Chapter 5 for the clutch pedal, as both pivot on the same axis. Additionally it will be necessary to remove the clevis pin securing the relay lever pushrod to the pedal.

22 Brake pedal relay lever – removal and refitting

Note: *On all Nova models, the brake master cylinder is mounted on the left-hand side of the engine compartment. On right-hand drive cars, brake pedal movement is transmitted to the master cylinder and servo by a relay lever running transversely across the car under the facia. Removal and refitting of this lever is as follows.*

1 Remove the heater air distribution housing, as described in Chapter 12.
2 Remove the retainers securing the brake pedal and servo unit pushrod clevis pins to the relay lever and withdraw the clevis pins.
3 Remove the retainers securing the relay lever to its support brackets.
4 Undo the nuts and bolts securing the left-hand support bracket to the bulkhead and withdraw the bracket and relay lever from inside the car (photo).
5 Refitting is the reverse sequence to removal.

22.4 Brake pedal relay lever left-hand support bracket

23 Handbrake – adjustment

Note: *The handbrake will normally be kept in correct adjustment by the routine adjustment of the rear brake shoes. If however the cable has been disconnected or renewed, or if the travel of the lever becomes excessive due to cable stretch, the following operations should be carried out.*

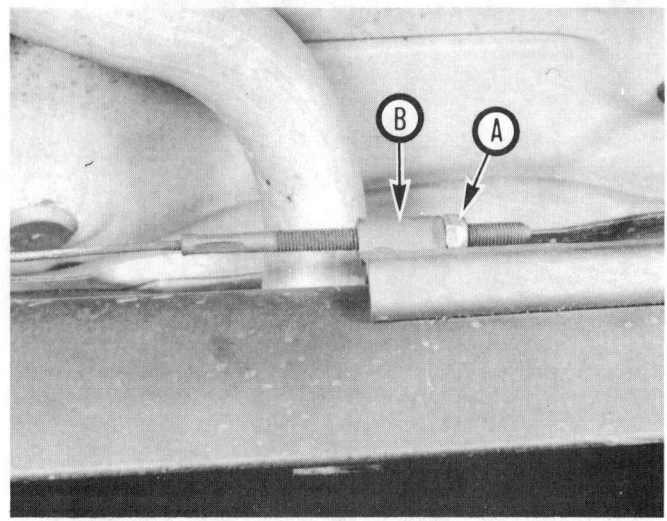

23.4 Handbrake cable adjusting locknut (A) and compensating yoke (B) on rear axle

1 Jack up the rear of the car and support it on axle stands. Release the handbrake.
2 Make sure that the rear brake shoes are correctly adjusted, as described in Section 3.
3 Pull the handbrake lever onto the second notch of the ratchet.
4 Tighten the adjusting locknut securing the handbrake cable to the compensating yoke on the rear axle (photo), until the brake linings can just be heard to rub as the wheels are turned. While tightening the adjusting nut, hold the cable with a second spanner or pliers to prevent it turning.
5 After adjustment, release the handbrake lever and make sure that the wheels are free to turn without binding, and then lower the car to the ground.

24 Handbrake cable – removal and refitting

1 Remove the relevant rear brake drum assembly, as described in Section 9, paragraphs 1 to 5 inclusive.

Left-hand (short) cable
2 Measure and record the length of exposed thread protruding through the handbrake cable adjusting locknut at the compensating yoke on the rear axle.
3 Hold the cable with pliers or a spanner, unscrew the adjusting nut and remove the cable end from the yoke.
4 Detach the cable from the cable guide on the rear axle.

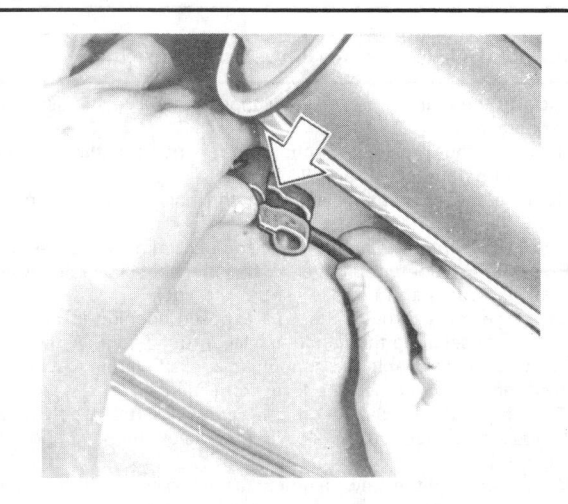

Fig. 9.29 Handbrake cable to handbrake lever connecting link – arrowed (Sec 24)

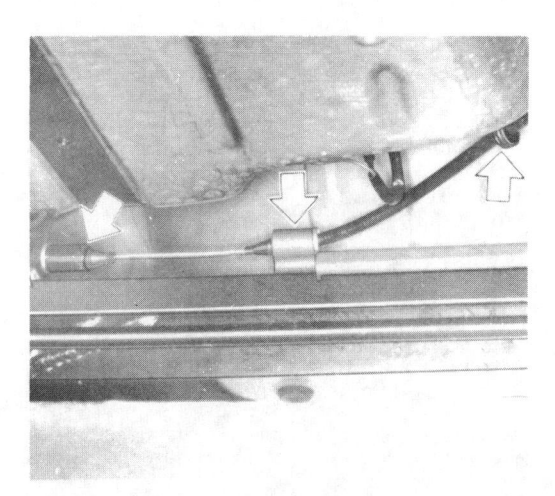

24.5 Removal of handbrake cable plastic guide sleeve from the brake backplate

5 Pull the plastic guide sleeve out of its location in the backplate and move it toward the cable end against spring pressure. Remove the guide sleeve from the cable (photo).
6 Unhook the handbrake cable from the brake shoe lever, pull the cable through the backplate and remove it from the car.
7 Refitting the cable is the reverse sequence to removal bearing in mind the following points:

 (a) Adjust the rear hub bearings, as described in Chapter 11
 (b) Set the cable adjusting locknut at the position recorded during removal and then if necessary adjust the handbrake, as described in Section 23, after completing the refitting operations.

Right-hand (long) cable
8 Carry out the operations described in paragraphs 2 and 3.
9 Remove the retainer and detach the cable from the connecting link located just to the rear of the handbrake lever rod.
10 Release the cable from the retaining clips on the underbody and from the compensating yoke and cable guide on the rear axle.
11 Carry out the operations described in paragraphs 5 to 7 inclusive.

Fig. 9.30 Handbrake cable compensating yoke and cable guide locations (arrowed) on rear axle (Sec 24)

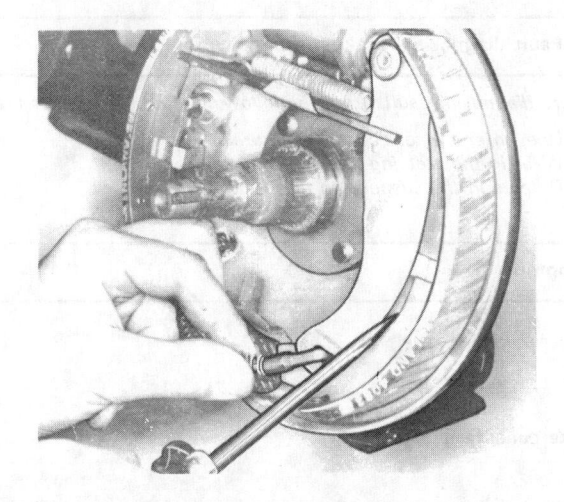

Fig. 9.31 Releasing the handbrake cable from the trailing brake shoe lever (Sec 24)

25 Handbrake lever – removal, overhaul and refitting

1 Jack up the rear of the car and support it on axle stands. Release the handbrake.
2 Refer to the previous Section and carry out the operations described in paragraphs 8 and 9.
3 Remove the connecting link from the handbrake lever rod and withdraw the grommet.
4 Refer to Chapter 12 and remove the passenger's front seat.
5 Press back the detent plunger of the inner seat slide rail, move the slide rail rearwards and remove it from the floor.
6 Using a socket through the now exposed opening in the carpet, undo the bolts securing the handbrake lever to the floor. Withdraw the lever sufficiently to enable the handbrake applied warning switch to be removed, and then remove the lever from the car.
7 To remove the ratchet segment, screw a suitable bolt into the pivot sleeve and drive out the sleeve. Push back the pawl and lift out the segment.
8 When refitting the new segment a new sleeve will be supplied which should be driven in the same way, just far enough to permit a little play between segment and lever.
9 To renew the pawl, drill out the pivot pin and rivet the new pin so that the pawl will move freely.
10 Refitting the handbrake lever is the reverse sequence to removal. On completion adjust the handbrake, as described in Section 23.

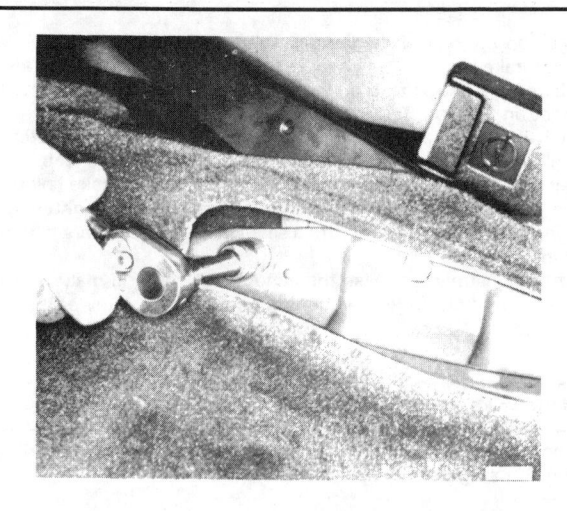

Fig. 9.33 Remove the handbrake lever retaining bolts through the seat slide rail opening (Sec 25)

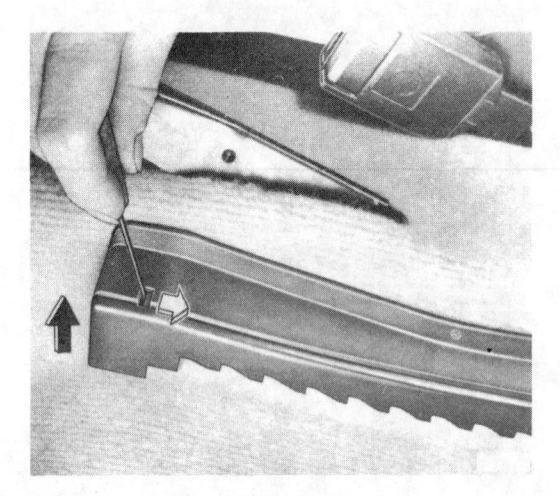

Fig. 9.32 Press back the detent plunger of the inner seat slide rail and move the rail to the rear to remove it (Sec 25)

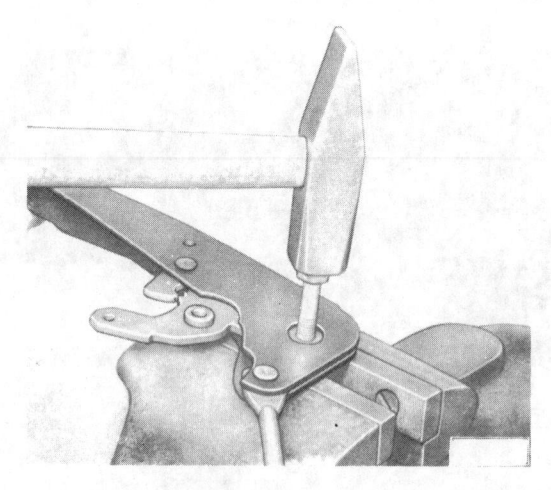

Fig. 9.34 Removing the handbrake lever pivot sleeve (Sec 25)

26 Fault diagnosis – braking system

Before diagnosing faults from the following chart, check that braking irregularities are not caused by:

   Uneven and incorrect tyre pressures
   Wear in the steering mechanism
   Defects in the suspension or shock absorbers

Symptom	Reason(s)
Excessive pedal travel	Incorrect rear brake adjustment Air in hydraulic system Faulty master cylinder
Brake pedal feels spongy	Air in hydraulic system Faulty master cylinder
Judder felt through brake pedal or steering wheel when braking	Excessive run-out or distortion of front discs or rear drums Brake pads or linings worn Brake backplate or disc caliper loose

Chapter 9 Braking system

Symptom	Reason(s)
Excessive pedal pressure required to stop car	Faulty servo unit, disconnected or damaged vacuum hose Brake pads or linings worn or contaminated Brake shoes incorrectly fitted Incorrect grade of pads or linings fitted Failure of one hydraulic circuit
Brakes pulls to one side	Brake pads or linings worn or contaminated Wheel cylinder or caliper piston seized Brake caliper seized on sliding sleeves Brake pads or linings renewed on one side only Brake disc or drum refinished or renewed on one side only
Brakes binding	Incorrect rear brake adjustment Incorrect handbrake adjustment Wheel cylinder or caliper piston seized Faulty master cylinder
Rear wheels locking under normal braking	Rear brake linings contaminated Faulty pressure regulating valve(s)

Chapter 10 Electrical system

For modifications, and information applicable to later models, see Supplement at end of manual

Contents

Alternator – fault diagnosis	8
Alternator – general description	6
Alternator – removal and refitting	7
Alternator brushes (Bosch) – inspection and renewal	10
Alternator brushes (Delco-Remy) – inspection and renewal	9
Battery – charging	5
Battery – removal and refitting	4
Bulbs – renewal	32
Cigarette lighter – removal and refitting	28
Clock – removal and refitting	27
Courtesy lamp switch – removal and refitting	25
Direction indicator and hazard flasher system – general	19
Electrical system – precaution	2
Facia switches – removal and refitting	20
Fault diagnosis – electrical system	43
Fuses and relays – general	18
General description	1
Handbrake applied warning switch – removal and refitting	24
Headlamp beams – adjustment	34
Headlamp lens assembly – removal and refitting	33
Headlamp washer system – general	39
Horn and switch – removal and refitting	26
Instrument panel – dismantling and reassembly	30
Instrument panel – removal and refitting	29
Maintenance and inspection	3
Radio or radio/cassette player (factory fitted unit) – removal and refitting	41
Reversing lamp switch – removal and refitting	22
Speedometer cable – renewal	31
Starter motor – dismantling and reassembly	17
Starter motor – general description	11
Starter motor – removal and refitting	13
Starter motor – testing in the car	12
Starter motor brushes – inspection and renewal	16
Starter motor renovation – general	14
Starter solenoid – removal and refitting	15
Steering column switches – removal and refitting	21
Stop-lamp switch – removal and refitting	23
Tailgate wiper motor – removal and refitting	37
Washer fluid – general	40
Windscreen and tailgate washer system – general	38
Windscreen wiper motor and linkage – removal and refitting	36
Wiper blades and arms – removal and refitting	35
Wiring diagram – explanatory notes	42

Specifications

System type
12 volt negative earth

Battery
Type .... Delco 'Freedom', maintenance-free
Capacity .... 36 Ah, 44 Ah or 55 Ah according to model

Alternator
Type .... Bosch or Delco-Remy
Output .... 45 to 55A, depending upon model
Minimum brush length:
    Bosch .... 5.0 mm (0.20 in)
    Delco-Remy .... 11.0 mm (0.43 in)

Wiper blades
Front and rear .... Champion X-4103

Starter motor
Type .... Bosch or Delco-Remy
Minimum brush length:
    Bosch .... 11.5 mm (0.45 in)
    Delco-Remy .... 5.0 mm (0.20 in)
Minimum commutator diameter:
    Bosch .... 31.2 mm (1.23 in)
    Delco-Remy .... 37.0 mm (1.46 in)

Chapter 10 Electrical system

Fuses

Fuse	Circuit protected	Rating (A)
2	Left-hand main beam	10
3	Right-hand main beam	10
4	Left-hand dipped beam	10
5	Right-hand dipped beam, rear fog guard lamp	10
7	Direction indicator lamps, stop-lamps	10
8	Heated rear window	20
9	Heater fan	20
10	Radiator fan	20
12	Front foglamps	15
13	Left-hand side and tail lamps	7.5
14	Right-hand side and tail lamp, number plate lamp, instrument and switch illumination, engine compartment lamp	7.5
16	Interior courtesy lamp, luggage compartment lamp, hazard flashers, clock, radio	15
17	Windscreen wipers, horn, tailgate window wiper	30
18	Reversing lamp, cigarette lighter, glovebox lamp, fuel and temperature gauges	20

Bulbs

Lamp	Wattage
Headlamp	60/55
Sidelamp	4
Direction indicators	21
Stop/tail lamp	21/5
Number plate lamp	10
Reversing lamp	21
Interior and courtesy lamps	10
Alternator warning lamp	3
Instrument panel warning lamps, ashtray lamp	1.2
Cigarette lighter lamp	1.2
Instrument illumination	1.2 or 3
Switch illumination	0.5
Hazard flasher switch and heater control illimination	1.2
Rear fog guard lamp	21

Torque wrench settings

	Nm	lbf ft
Alternator pulley nut:		
Bosch	40	29
Delco-Remy	70	52
Alternator mounting bracket bolts	40	29
Alternator pivot and adjustment bolts	34	25
Starter mounting bolts	25	18

1 General description

The electrical system is of the 12 volt negative earth type, and consists of a 12 volt battery, alternator, starter motor and related electrical components, accessories and wiring. The battery is of the maintenance-free 'sealed for life' type and is charged by an alternator which is belt-driven from the crankshaft pulley. The starter motor is of the pre-engaged type incorporating an integral solenoid. On starting, the solenoid moves the drive pinion into engagement with the flywheel ring gear before the starter motor is energised. Once the engine has started, a one-way clutch prevents the motor armature being driven by the engine until the pinion disengages from the flywheel.

Further details of the major electrical systems are given in the relevant Sections of this Chapter.

Caution: *Before carrying out any work on the vehicle electrical system, read through the precautions given in Safety First! at the beginning of this manual and in the following section of this Chapter.*

2 Electrical system – precautions

It is necessary to take extra care when working on the electrical system to avoid damage to semi-conductor devices (diodes and transistors) and to avoid the risk of personal injury. In addition to the precautions given in Safety First! at the beginning of this manual, observe the following when working on the system.

1 Always remove rings, watches, etc before working on the electrical system. Even with the battery disconnected, capacitive discharge could occur if a component live terminal is earthed through a metal object. This could cause a shock or nasty burn.

2 Do not reverse the battery connections. Components such as the alternator or any other having semi-conductor circuitry could be irreparably damaged.

3 If the engine is being started using jump leads and a slave battery, connect the batteries as described in 'Safety First!' at the beginning of this manual.

4 Never disconnect the battery terminals, or alternator wiring connections, when the engine is running.

5 The battery leads and alternator connections must be disconnected before carrying out any electric welding on the car.

6 Never use an ohmmeter of the type incorporating a hand-cranked generator for circuit or continuity testing.

3 Maintenance and inspection

1 Every 9000 miles (15 000 km) or 6 months, whichever comes first, carry out the following maintenance and inspection operations on the electrical system components.

2 Check the operation of all the electrical equipment, ie wipers, washers, lights, direction indicators horn etc. Refer to the appropriate Sections of this Chapter if any components are found to be inoperative.

3 Visually check all accessible wiring connectors, harnesses and retaining clips for security, or signs of chafing or damage. Rectify any problems encountered.

4 Check the alternator drivebelt for cracks, fraying or damage.

Renew the belt if worn or, if satisfactory, check and adjust the belt tension. These procedures are covered in Chapter 2.
5 Check the condition of the wiper blades and, if they are cracked or show signs of deterioration, renew them, as described in Section 35. Check the operation of the windscreen and tailgate washers and adjust the nozzles if necessary.
6 Check the battery terminals and, if there is any sign of corrosion, disconnect and clean them thoroughly. Smear the terminals and battery posts with petroleum jelly before refitting. If there is any sign of corrosion on the battery tray, remove the battery, clean the deposits away and treat the affected metal with an anti-rust preparation. Repaint the tray in the original colour after treatment.
7 Top up the windscreen and tailgate washer reservoirs and check the security of the pump wires and water pipes.
8 It is advisable to have the headlight aim adjusted using optical beam setting equipment.
9 While carrying out a road test, check the operation of all the instruments and warning lights, and the operation of the direction indicator self-cancelling mechanism.

4 Battery – removal and refitting

1 The battery is located on the rear right-hand side of the engine compartment.
2 Slacken the retaining nuts and first lift off the negative and then

4.2 Disconnect the battery negative terminal first

the positive terminals from the battery posts (photo).
3 Release the battery clamp plate and carefully lift the battery out of its location.
4 Refitting is a direct reversal of this procedure. Reconnect the positive terminal first and smear both terminals with petroleum jelly to prevent corrosion; *never use ordinary grease.* Do not overtighten the terminal retaining nuts or hammer the terminals on. The terminals and battery posts are made of lead and are easily damaged.

5 Battery – charging

1 In winter when a heavy demand is placed on the battery, such as when starting from cold and using more electrical equipment, it may be necessary to have the battery fully charged from an external source. Note that both battery leads must be disconnected before charging in order to prevent possible damage to any semi-conductor electrical components.
2 The terminals of the battery and the leads of the charger must be connected positive-to-positive and negative-to-negative.
3 Due to the design of certain maintenance-free batteries, rapid charging is not recommended. Additionally the significance of the

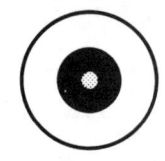

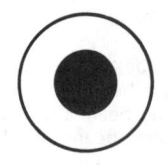

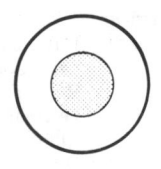

Darkened Indicator WITH GREEN DOT — Darkened Indicator NO GREEN DOT
MAY BE JUMP STARTED

Light or Bright Yellow Indicator NO GREEN DOT
DO NOT JUMP START

Fig. 10.1 Maintenance-free battery condition indicator (Sec 5)

battery condition indicator on top of the battery will be appreciated by reference to Fig. 10.1.
4 If a yellow indicator without a green dot is observed, do not use jump start leads or attempt to use a mains charger as the battery is almost certainly unfit for further use. If in any doubt about the condition of the battery or the suitability of certain types of charging equipment, consult your dealer.

6 Alternator – general description

Cars covered by this manual are fitted with either a Bosch or a Delco-Remy alternator; the two types are similar in construction and in output. The alternator generates alternating current (ac) which is rectified by diodes into direct current (dc) as this is the current needed for charging the battery.

The alternator is of the rotating field, ventilated design and comprises principally a laminated stator on which is wound the output winding, a rotor carrying the field winding, and a diode rectifier. A voltage regulator is incorporated in the Delco-Remy alternator but on the Bosch machine it is separately mounted at the rear. The alternator generates its current in the stator windings and the rotor carries the field. The field brushes therefore are only required to carry a light current and as they run on simple slip rings they have a relatively long life. This design makes the alternator a reliable machine requiring little servicing.

The rotor is belt-driven from the crankshaft pulley through a pulley keyed to the rotor shaft. A fan adjacent to the pulley draws cooling air through the unit. Rotation is clockwise when viewed from the drive end.

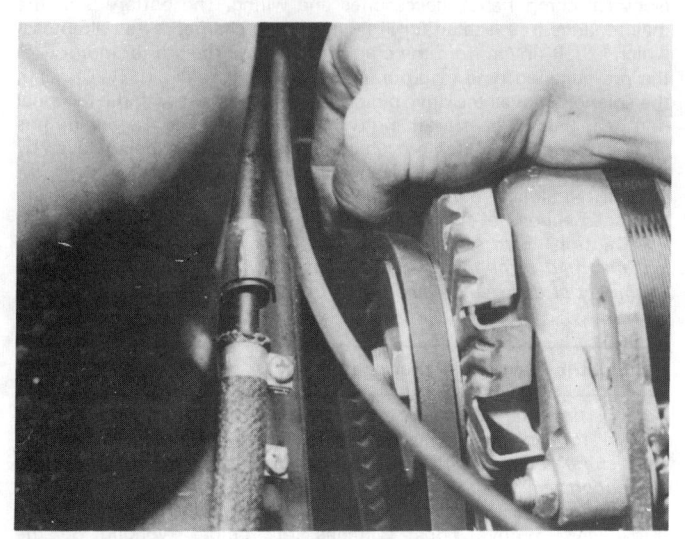

7.3 Disconnect the electrical leads and slip off the drivebelt

Chapter 10 Electrical system

7 Alternator – removal and refitting

1 Disconnect the battery negative terminal.
2 Make a note of the electrical connections at the rear of the alternator and disconnect the plug, multi-pin connector or terminals as appropriate.
3 Undo and remove the alternator adjustment arm bolt, note the short earth wire. Slacken the lower pivot bolt and swing the alternator in towards the engine. Lift the drivebelt off the alternator pulley (photo).
4 Remove the lower pivot bolt and lift the alternator away from the engine (photo). Take care not to drop or knock the alternator as this can cause irreparable damage.
5 Refitting the alternator is the reverse of the removal sequence. Tension the drivebelt as described in Chapter 2.

7.4 Undo the mounting bolts and remove the alternator

8 Alternator – fault diagnosis

Due to the specialist knowledge and equipment required to test or service an alternator it is recommended that, if the performance is suspect, the car be taken to an automobile electrician who will have the facilities for such work. Because of this recommendation, information is limited to the inspection and renewal of the brushes. Should the alternator not charge or the system be suspect the following points may be checked before seeking further assistance:

(a) Check the drivebelt tension as described in Chapter 2
(b) Check condition of battery and its connections (see Section 3)
(c) Inspect all electrical cables and connections for condition and security

9 Alternator brushes (Delco-Remy) – inspection and renewal

1 Remove the alternator from the engine, as described in Section 7.
2 Scribe a line across the drive end housing and slip ring end housing to ensure correct location when refitting.
3 Undo the three through-bolts and separate the two housings (photo). Check the condition of the two slip rings on the rotor (photo). If they are dirty, clean them with a petrol-moistened rag. In extreme cases polish with very fine glass paper.
4 Undo the three nuts and washers securing the stator leads to the rectifier and lift away the stator assembly (photo). Remove the terminal screw and lift out the diode bracket.
5 Undo the two screws retaining the brush holder and voltage regulator to the end housing and remove the brush holder assembly. Note insulation washers under the screw heads.
6 Check that the brushes move freely in the guides and that the length is within the limit given in the Specifications. If any doubt exists regarding the condition of the brushes the best policy is to renew them.
7 To fit new brushes, unsolder the old brush leads from the brushholder and solder on the new leads in exactly the same place.
8 Check that the new brushes move freely in the guides.
9 Before refitting the brushholder assembly retain the brushes in the retracted position using a piece of stiff wire or a small twist drill (photo).
10 Refit the brush holder so that the stiff wire or twist drill protrudes through the small slot in the end housing (photo).
11 Refit the diode bracket and stator to the end housing, making sure the stator leads are in their correct positions.
12 Assemble the two housings, ensuring that the scribe marks are aligned. Insert the three through-bolts and tighten fully.
13 Now carefully pull the stiff wire or twist drill out of the slot in the housing so that the brushes drop onto the rotor slip ring.
14 The alternator can now be refitted to the car and tested.
15 It should be remembered that if an alternator is being changed for a new or reconditioned unit then the pulley/fan assembly will be required from the original alternator. To release the pulley nut, hold the rotor shaft with an Allen key while the nut is unscrewed (photo).

10 Alternator brushes (Bosch) – inspection and renewal

1 Undo and remove the two screws, spring and plain washers that secure the brush box to the rear of the slip ring end housing. Lift away the brush box and voltage regulator.
2 Check that the carbon brushes are able to slide smoothly in their guides without any sign of binding.
3 Measure the length of the brushes. If they have worn below the specified limit, they must be renewed.
4 Hold the brush wire with a pair of engineer's pliers and unsolder it from the brush box. Lift away the two brushes.
5 Insert the new brushes and check to make sure that they are free to move in their guides. If they bind, lightly polish with a very fine file.
6 Solder the brush wire ends to the brush box taking care that solder is not allowed to pass to the stranded wire.
7 Whenever new brushes are fitted new springs should also be fitted.
8 Refitting the brush box is the reverse sequence to removal.

11 Starter motor – general description

The starter motor is mounted on the rear face of the crankcase and may be of either Delco-Remy or Bosch manufacture. Both makes are of the pre-engaged type, ie the drive pinion is brought into mesh with the starter ring gear on the flywheel before the main current is applied.
When the starter switch is operated, current flows from the battery to the solenoid which is mounted on the starter body. The plunger in the solenoid moves inwards, so causing a centrally pivoted lever to push the drive pinion into mesh with the starter ring gear. When the solenoid plunger reaches the end of its travel, it closes an internal contact and full starting current flows to the starter field coils. The armature is then able to rotate the crankshaft, so starting the engine.
A special freewheel clutch is fitted to the starter drive pinion so that as soon as the engine fires and starts to operate on its own it does not drive the starter motor.
When the starter switch is released, the solenoid is de-energised and a spring moves the plunger back to its rest position. This operates the pivoted lever to withdraw the drive pinion from engagement with the starter ring.
The construction of the two makes of starter motor is quite similar and the removal, refitting, dismantling, inspection and reassembly procedures detailed here will serve for both motors. Significant differences will be noted.

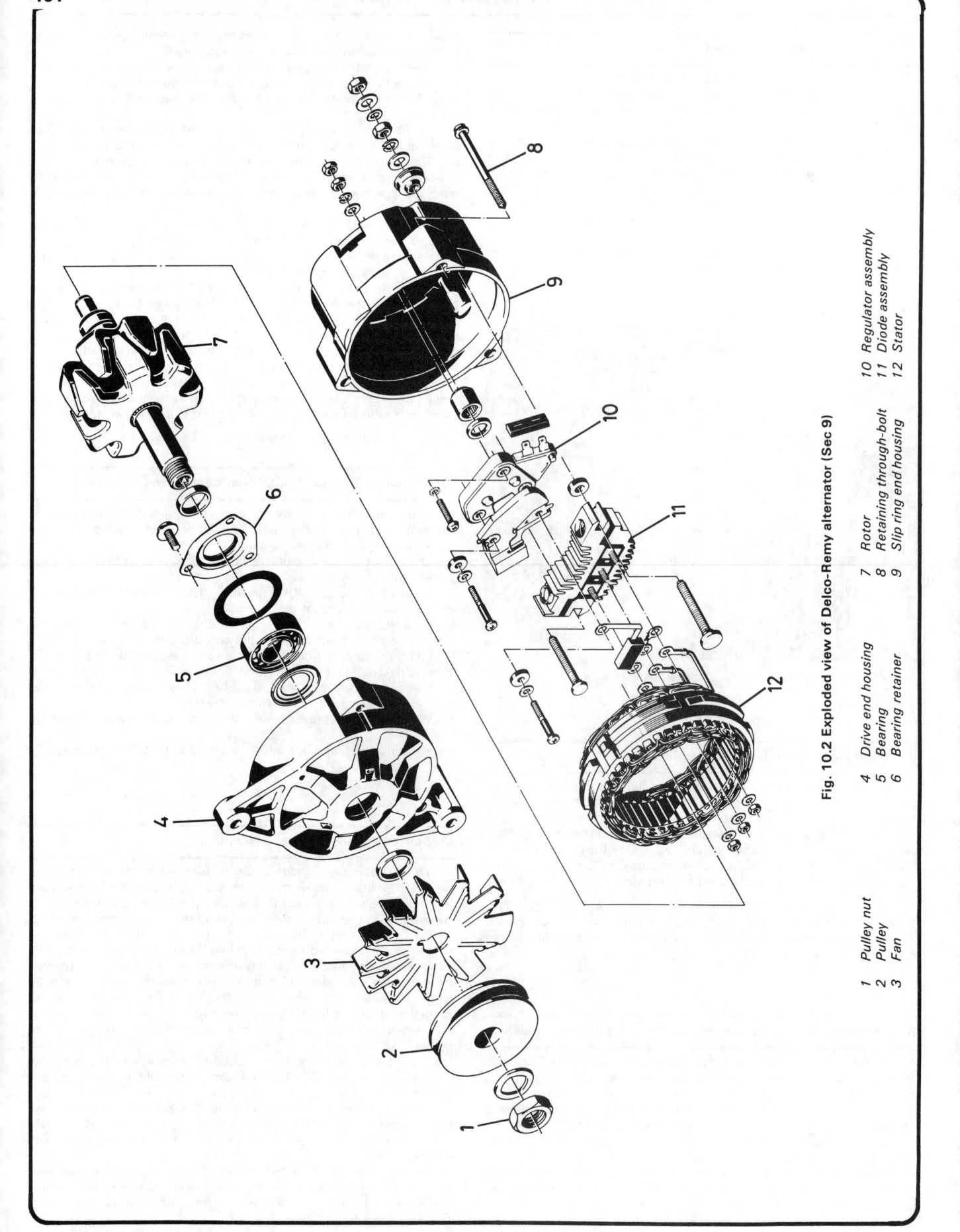

Fig. 10.2 Exploded view of Delco-Remy alternator (Sec 9)

1 Pulley nut
2 Pulley
3 Fan
4 Drive end housing
5 Bearing
6 Bearing retainer
7 Rotor
8 Retaining through-bolt
9 Slip ring end housing
10 Regulator assembly
11 Diode assembly
12 Stator

165

9.3A Undo the through-bolts and separate the alternator housings

9.3B Clean the alternator slip rings with a petrol-moistened rag

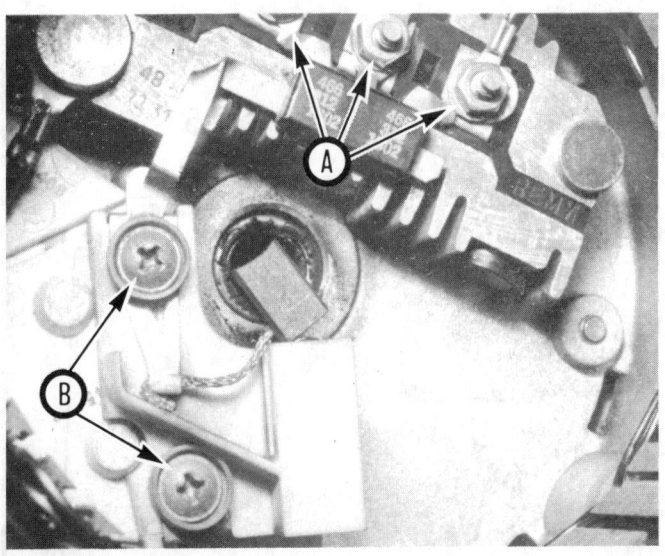

9.4 Alternator stator lead retaining nuts (A) and brush holder securing screws (B)

9.9 Alternator brushes held retracted with a small twist drill (arrowed)

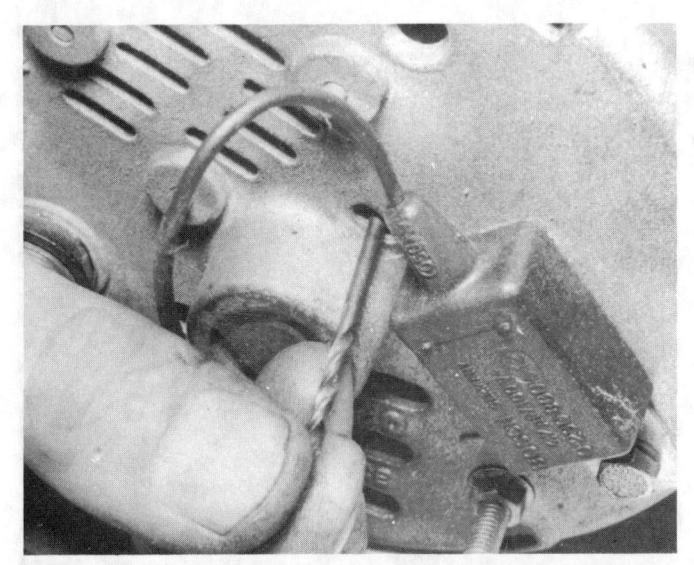

9.10 With the brush holder fitted the twist drill protrudes through the slot in the housing

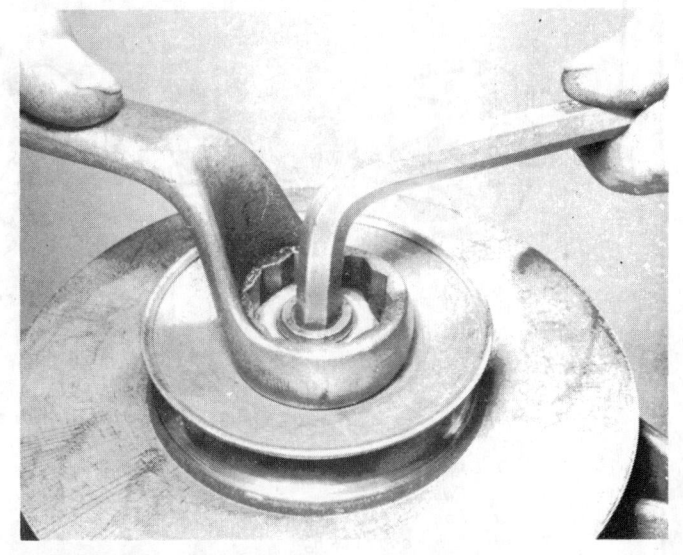

9.15 Alternator pulley nut removal

Fig. 10.3 Brush holder/regulator retaining screws – Bosch alternator (Sec 10)

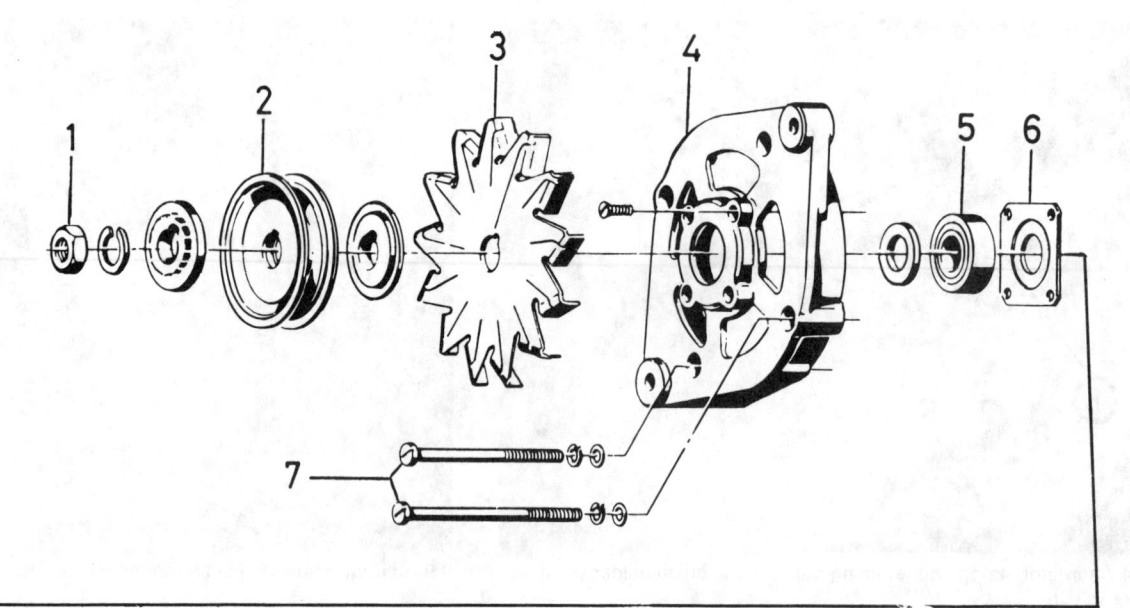

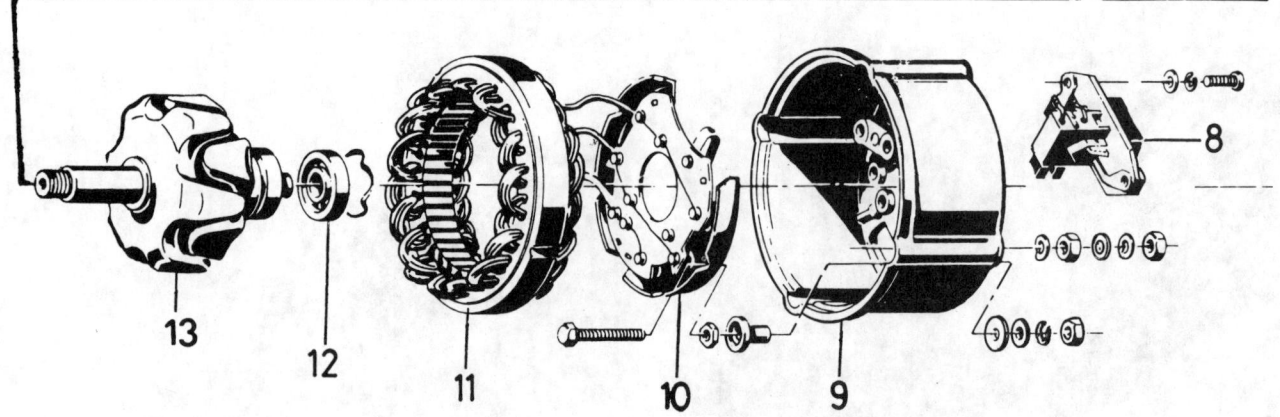

Fig. 10.4 Exploded view of Bosch alternator (Sec 10)

1. Pulley nut
2. Pulley
3. Fan
4. Drive end housing
5. Bearing
6. Bearing retainer
7. Through-bolts
8. Brush holder/regulator
9. Slip ring end housing
10. Collector ring endplate
11. Stator
12. Bearing
13. Rotor

Chapter 10 Electrical system

12 Starter motor – testing in the car

1 If the starter motor fails to turn the engine when the switch is operated there are five possible causes:

   (a) The battery is faulty
   (b) The electrical connections between the switch, solenoid, battery and starter motor are somewhere failing to pass the necessary current from the battery through the starter to earth
   (c) The solenoid switch is faulty
   (d) The starter motor is mechanically or electrically defective
   (e) The starter motor pinion and/or flywheel ring gear is badly worn and in need of replacement

2 To check the battery, switch on the headlights. If they dim after a few seconds the battery is in a discharged state. If the lights glow brightly, operate the starter switch and see what happens to the lights. If they dim then you know that power is reaching the starter motor but failing to turn it. If the starter turns slowly when switched on, proceed to the next check.

3 If, when the starter switch is operated the lights stay bright, then insufficient power is reaching the motor. Remove the battery connections, starter/solenoid power connections and the engine earth strap and thoroughly clean them and refit them. Smear petroleum jelly around the battery connections to prevent corrosion. Corroded connections are the most frequent cause of electric system malfunctions.

4 When the above checks and cleaning tasks have been carried out but without success, you will possibly have heard a clicking noise each time the starter switch was operated. This was the solenoid switch operating, but it does not necessarily follow that the main contacts were closing properly (if no clicking has been heard from the solenoid, it is certainly defective). The solenoid contact can be checked by putting a voltmeter or bulb across the main cable connection on the starter side of the solenoid and earth. When the switch is operated, there should be a reading or lighted bulb. If there is no reading or lighted bulb, the solenoid unit is faulty and should be renewed.

5 If the starter motor operates but doesn't turn the engine over then it is most probable that the starter pinion and/or flywheel ring gear are badly worn, in which case the starter motor will normally be noisy in operation.

6 Finally, if it is established that the solenoid is not faulty and 12 volts are getting to the starter, then the motor is faulty and should be removed for inspection.

13 Starter motor – removal and refitting

1 With the engine installed in the car it is easier to get to the starter motor from underneath, as it is located low on the rear side of the engine. If you prefer not to work under the car then it will be essential to remove the air cleaner to gain access to the starter.

13.3 Starter motor removal

2 Start by disconnecting the battery earth lead and then disconnect the solenoid and starter electrical leads (photo). Take note of their respective locations to ensure correct reassembly.
3 Unscrew and remove the starter motor unit retaining bolts and withdraw the unit from the clutch housing (photo).
4 Refitting the starter motor assembly is a direct reversal of the removal procedure.

14 Starter motor renovation – general

1 Such is the inherent reliability and strength of the starter motors fitted it is very unlikely that a motor will need dismantling until it is totally worn out and in need of replacement as a whole.
2 If, however, the motor is only a couple of years old or so and a pinion carriage, solenoid system or brush fault is suspected then remove the motor from the engine and dismantle as described in the following Sections.

15 Starter solenoid – removal and refitting

1 At the rear of the solenoid undo the retaining nut and washer and slip the starter electrical feed wire off the solenoid stud.
2 Undo the two screws securing the solenoid to the starter drive end housing.
3 If working on the Delco unit withdraw the solenoid and spring

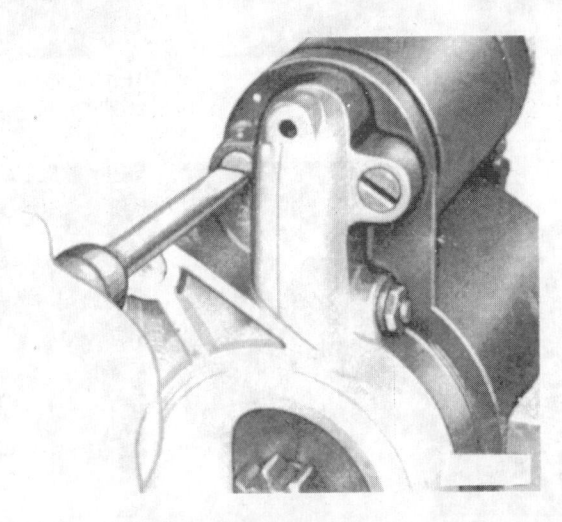

13.2 Electrical connections at the starter solenoid

Fig. 10.5 Removing the solenoid retaining screws – Bosch starter (Sec 15)

Chapter 10 Electrical system

Fig. 10.6 Releasing the solenoid from the actuating arm – Bosch starter (Sec 15)

from the housing. If working on the Bosch starter, disengage the solenoid shaft from the pinion carriage actuating arm and withdraw the solenoid

4   In both cases refitting is the reverse sequence to removal.

16 Starter motor brushes – inspection and renewal

Bosch
1   With the starter removed from the engine and on a clean bench, begin by removing the armature end cap which is secured by two small screws on the end of the motor (photo). Remove the armature retaining clip, washers and the rubber sealing ring which were exposed (photo). Undo and remove the two long bolts which hold the motor assembly together (photo). The end cover can now be removed to reveal the brushes and mounting plate (photos).
2   Take the brushes from the holder and slip the holder off the armature shaft. Retrieve the spacer washers between the brush plate and the armature block, where fitted.
3   Inspect the brushes; if they are worn down to less than the minimum length given in Specifications, they should be renewed. Replacement brushes to the latest standard have no shunt wire and to fit this type first crush the old brush in a vice, or with a hammer, to remove all the carbon from the shunt wire and scrape the wire to clean it ready for soldering. Insert the wire in the hole in the new brush and spread the end out to fill the countersunk hole in the brush. Hold the wire close under the brush with a pair of pliers to locate the wire properly for soldering and to prevent solder from penetrating the wire further than necessary as this would reduce its flexibility. A 12 to 15 watt pencil soldering iron is adequate for this job. After soldering the wire in place remove any excess solder with a file and check that the brush is an easy fit in the brush holder.
4   Wipe the starter motor armature and commutator with a non-fluffy rag wetted with petrol.
5   Reassemble the brushes into the holder (photo) and refit the holder over the armature shaft, remembering to fit the two washers between the holder and armature, where fitted.
6   Refit the motor end cover and secure with two long bolts.
7   Refit the armature shaft end cap after fitting the rubber sealing ring, washer and shaft clip.

Delco Remy
8   With the motor removed from the engine and on a clean bench, begin by undoing and removing the two long bolts which hold the motor assembly together. Punch mark the relative positions of the end cover and the yoke to ensure correct relocation on assembly.
9   Undo the two small screws which secure the end cover to the brush holder plate and lift off the end cover.
10 Lift the brush springs to remove the positive brushes and then remove the brush holder plate from the motor.

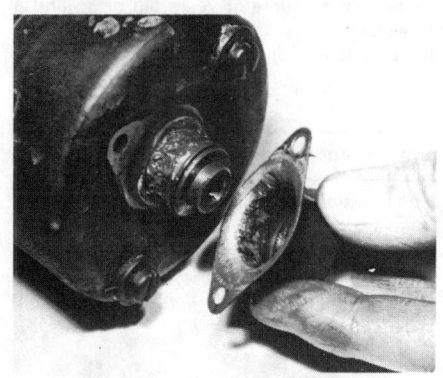

16.1A Remove the armature end cap ...

16.1B ... followed by the retaining clip, washers and sealing ring

16.1C Undo and remove the two long through-bolts ...

16.1D ... and lift off the end cover

16.1E Commutator and field coil brushes

16.5 Brush holder with brushes ready for refitting

169

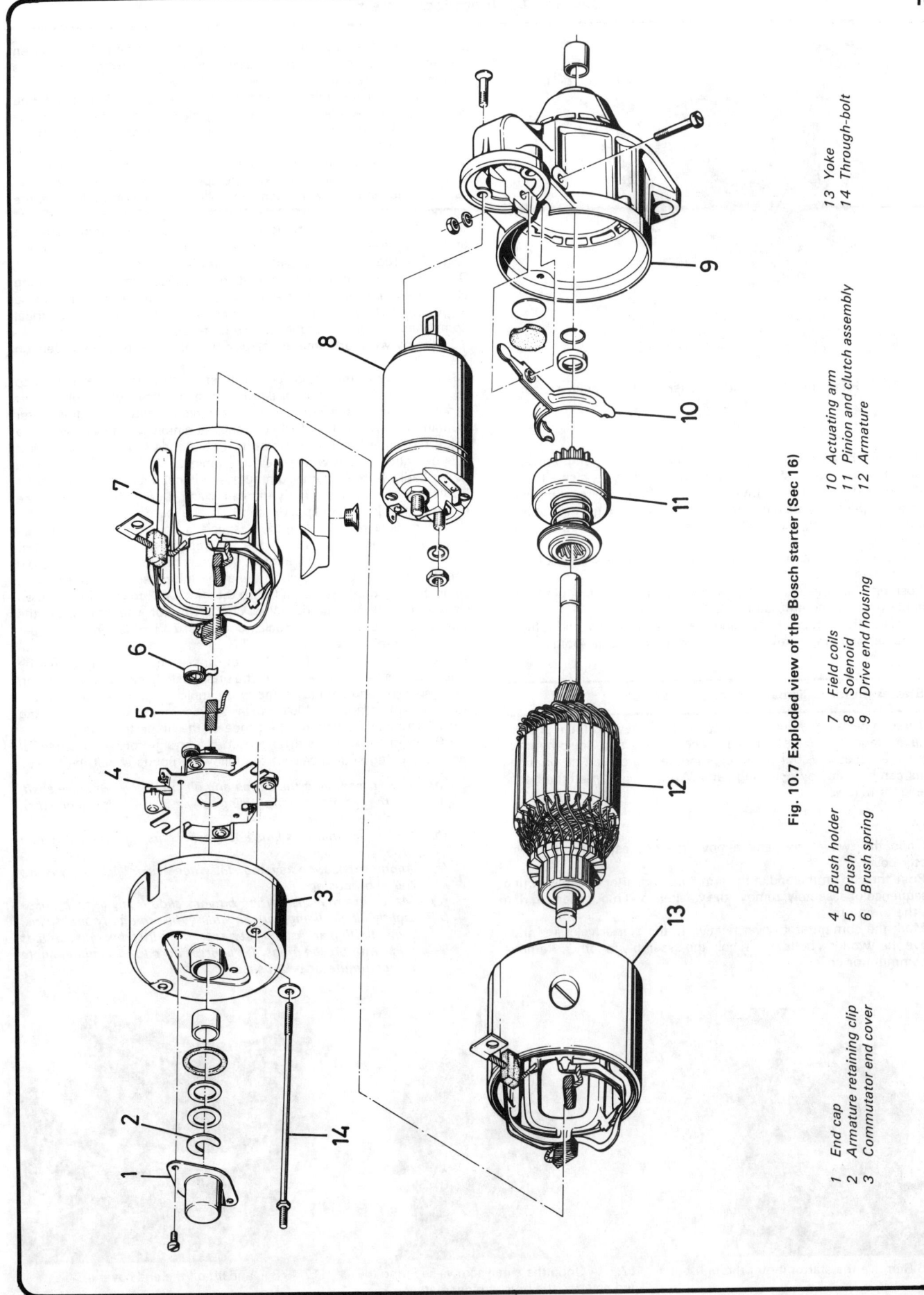

Fig. 10.7 Exploded view of the Bosch starter (Sec 16)

1 End cap
2 Armature retaining clip
3 Commutator end cover
4 Brush holder
5 Brush
6 Brush spring
7 Field coils
8 Solenoid
9 Drive end housing
10 Actuating arm
11 Pinion and clutch assembly
12 Armature
13 Yoke
14 Through-bolt

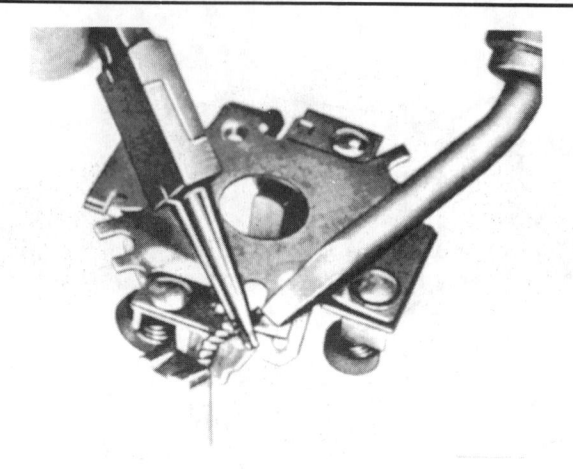

Chapter 10 Electrical system

5 Lift the brush springs to remove the positive brushes and then remove the brushplate from the assembly. Note and remove any shims that may be fitted.
6 Disconnect the field winding lead from the solenoid terminal and then undo the two retaining screws to release the solenoid from the assembly. As the solenoid is removed unhook the end fitting from the actuating arm.
7 Unscrew and remove the actuating arm pivot and then remove the drive end housing from the yoke assembly. As this is done remove the rubber plug and the actuating arm. Slide the armature out of the casing.
8 If it is required to remove the pinion or the clutch from the armature, press the retaining ring back on the shaft to enable the snap-ring to be removed. Then slide the components off the shaft.
9 With the starter motor dismantled the various components can be cleaned and inspected for general wear and/or signs of damage. Use a petrol-damped cloth for cleaning, but avoid wetting electrical components. Dry thoroughly with a fluff-free cloth.
10 Renew worn or damaged carbon brushes as explained in Section 16.
11 If the starter motor has shown a tendency to jam or a reluctance to disengage then the starter pinion is almost certainly the culprit. Dirt accumulation on the shaft or on the pinion could cause this. After cleaning off any such dirt, check that the pinion can move freely in a spiral movement along the shaft. If it still tends to bind or stick, or if it is defective in any way, renew the pinion.
12 A badly worn or burnt commutator will need skimming on a lathe but if it is only dirty or lightly marked, clean it up with a piece of fine grade glass paper wrapped round. If the commutator has to be skimmed have the job done by a specialist, but make sure that the minimum diameter, as listed in the Specifications, is maintained. After skimming, the separators should be undercut using a piece of old hacksaw blade ground down to the same thickness as the separators. Undercut to a depth of about 0.5 to 0.8 mm (0.02 to 0.03 in) and then clean up with fine grade glass cloth. Do not use emery on the commutator as abrasive particles could get embedded in the copper and cause rapid brush wear.
13 An armature with a bent shaft or other signs of damage must be renewed. Electrical checks should be undertaken by an auto-electrician with special equipment. Although simple continuity checks are possible with a lamp and low power source, more extensive checking is needed which is beyond the scope of the home mechanic.
14 Reassembly of the starter motor is a straightforward reversal of the dismantling sequence, but the following points should be noted.

(a) After assembling the clutch and pinion to the armature shaft, fit the retaining ring using a new snap-ring and then reposition the retainer
(b) Make sure that all shims and washers are fitted in the correct order
(c) Align the locating key and slot when assembling the yoke to the end housing
(d) Make sure that the carbon brushes slide freely in their boxes
(e) Lightly oil all sliding parts including the armature spiral spline, the actuating arm sliding surfaces, the clutch bearing surfaces and armature bearings. Of course, no oil must contaminate the commutator or brushes

Fig. 10.8 Soldering starter motor brush leads – Bosch starter (Sec 16)

11 If the brushes are worn to less than the minimum length given in the Specifications, they should be renewed; always replace all four.
12 When soldering new positive brushes, hold the connecting wires in a pair of pliers to prevent the solder from running into the wire strands and reducing its flexibility. Use a 12 to 15 watt pencil soldering iron.
13 Clean the motor armature and commutator with a non-fluffy rag moistened with petrol.
14 Position the brushes in the brush holder plate and place this assembly over the commutator.
15 Relocate the end cover, aligning the relevant marks, and refit the brush holder securing screws and the two long through-bolts.

17 Starter motor – dismantling and reassembly

1 The complete overhaul of a starter motor is beyond the resources of the average home mechanic as special tools and equipment for testing are necessary, but if the appropriate spares can be obtained repairs can be made by renewing parts. With the starter on the bench proceed as follows.

Bosch

2 Undo the two screws and remove the end cap from the commutator cover.
3 Prise the clip off the end of the armature and, after carefully noting the sequence of assembly, remove the washers and rubber sealing ring from the armature.
4 Mark the commutator cover relative to the starter yoke and then remove the two long bolts which bold the assembly together. Remove the commutator cover.

17.16 Remove the starter motor through-bolts

17.17A Undo the small screws securing the end cover to the brush holder plate ...

17.17B ... and lift off the end cover

Chapter 10 Electrical system

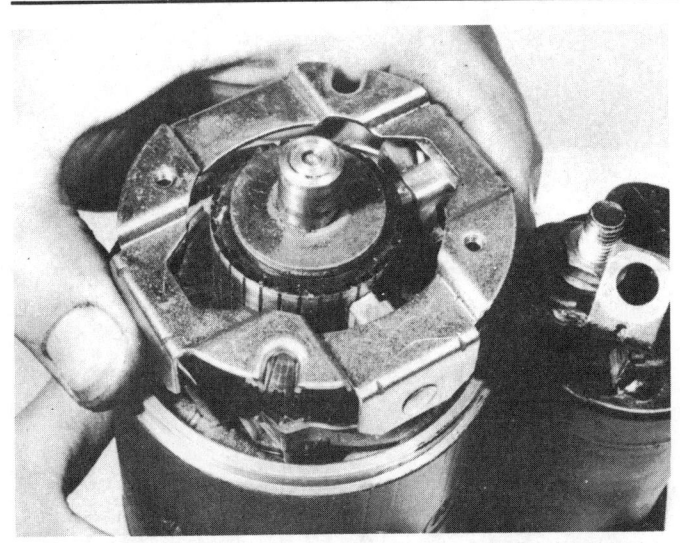

17.18 Remove the positive brushes and withdraw the brush holder plate

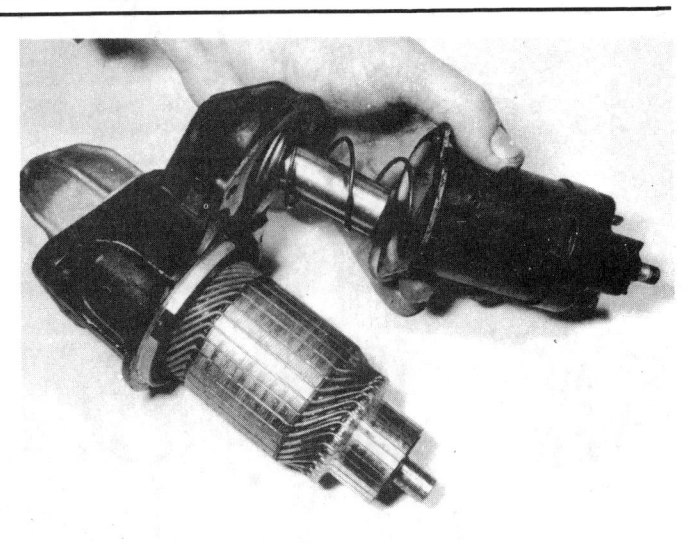

17.19A Remove the solenoid and spring ...

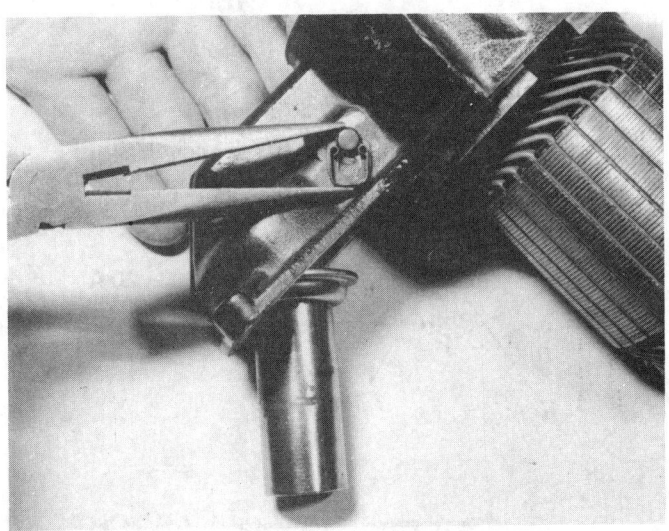

17.19B ... extract the actuating arm spindle retaining clip ...

17.19C ... withdraw the spindle ...

17.19D ... and remove the armature and actuating arm together from the end housing

Delco Remy

15 Mark the commutator end cover and the drive end housing relative to the yoke to ensure correct reassembly, and then disconnect the field winding connection from the lower stud on the solenoid.
16 Undo and remove the two long bolts which hold the motor assembly together (photo). Punch mark the relative positions of the end cover and the yoke to ensure correct relocation on assembly.
17 Undo the two small screws which secure the end cover to the brush holder plate and lift off the end cover (photos).
18 Lift the brush springs to remove the positive brushes and then remove the brush holder plate from the motor (photo).
19 Undo the two retaining screws and remove the solenoid and its spring from the drive end housing (photo). Extract the clip from the actuating arm spindle (photo) and tap the spindle out of the housing (photo). This will allow the armature and the actuating arm to be removed together (photo). The actuating arm can then be removed from the armature assembly (photo).
20 With the exception of the above, the remainder of the dismantling and reassembly procedures for this starter are the same as those described in paragraphs 8 to 14, to which reference should now be made. When the solenoid has been refitted to the drive end housing, use a little plastic sealing compound to seal the slot in the housing to prevent water entering. Then continue the reassembly as described.

Chapter 10 Electrical system

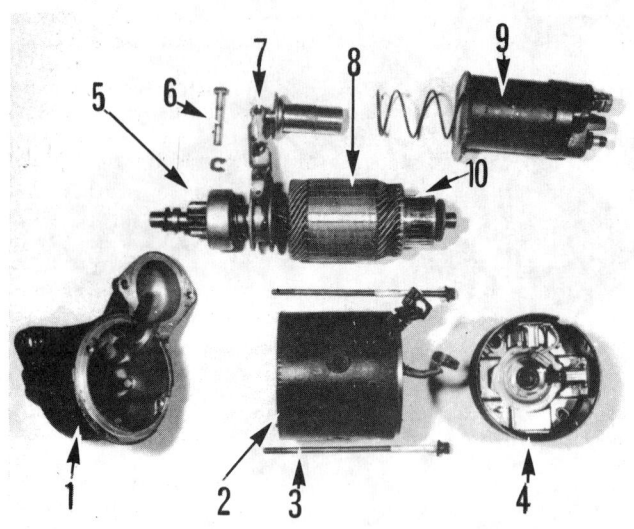

17.19E Exploded view of the Delco-Remy starter motor

1. Drive end housing
2. Yoke
3. Starter through-bolt
4. End cover and brush holder plate
5. Pinion and clutch assembly
6. Actuating arm spindle
7. Actuating arm
8. Armature
9. Solenoid
10. Commutator

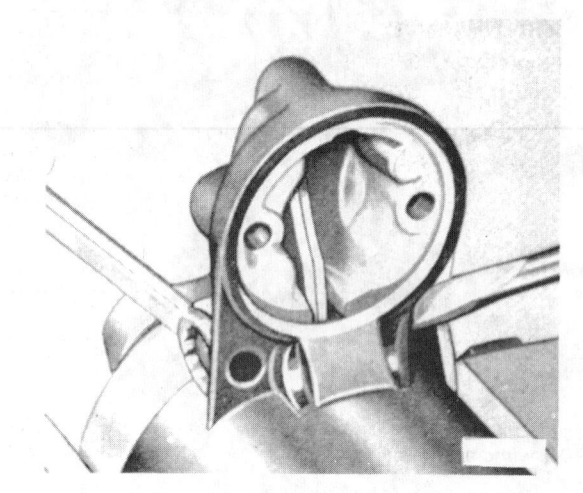

Fig. 10.9 Removing the starter actuating arm pivot – Bosch starter (Sec 17)

Fig. 10.10 Driving pinion retaining ring down armature shaft to gain access to the snap ring – Bosch starter (Sec 17)

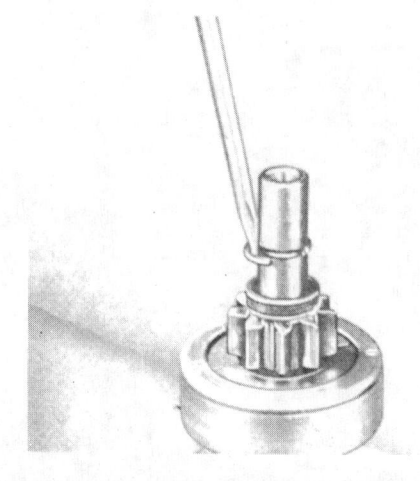

Fig. 10.11 Pinion retaining snap ring removal – Bosch starter (Sec 17)

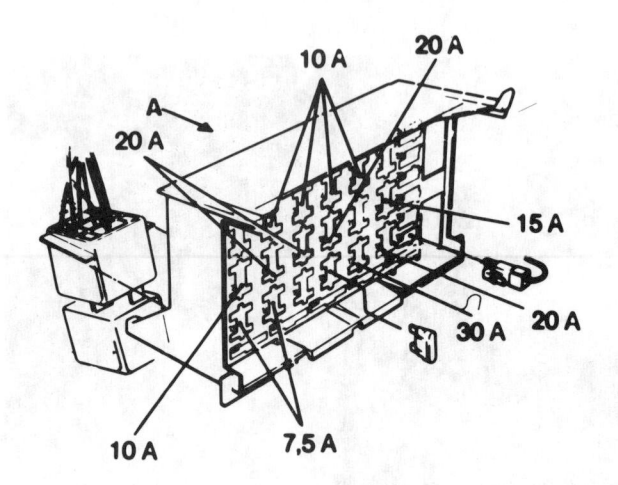

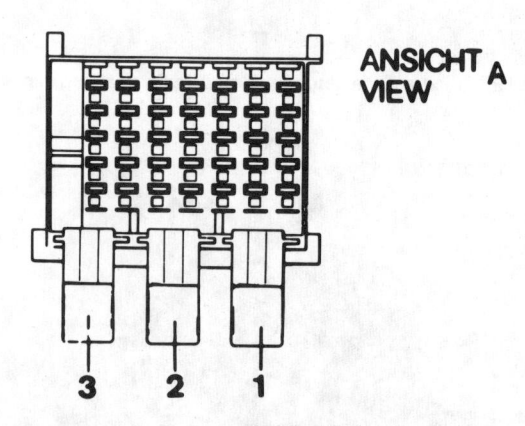

Fig. 10.12 Fuse and relay arrangement in the fusebox (Sec 18)

1. Windscreen wiper time delay relay
2. Direction indicator flasher unit
3. Heated rear window relay

18 Fuses and relays – general

Fuses

1 The fusebox is situated behind a cover located at the lower right-hand corner of the facia. To gain access to the fuses simply lift off the fusebox cover (photo). The fuse locations, current rating and circuits

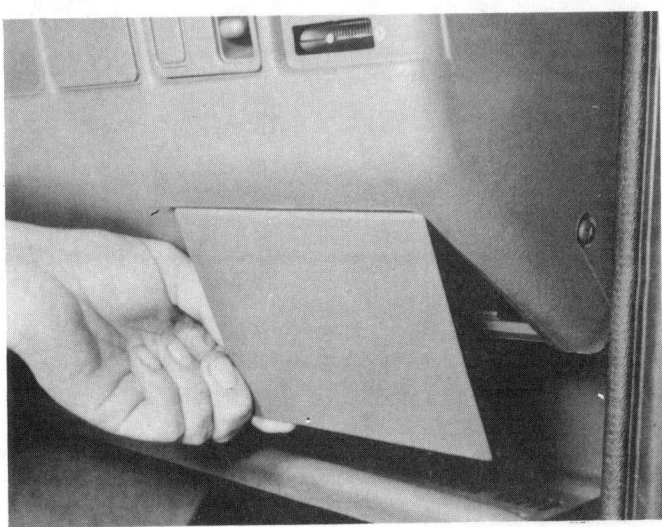

Chapter 10 Electrical system

protected are shown on the cover (photo). Each fuse is colour coded and has its rating stamped on it. To remove a fuse, use a coin if necessary to prise it free and then lift it out (photo).
2 Before renewing a blown fuse, trace and rectify the cause and always use a fuse of the correct value. Never substitute a fuse of a higher rating or use such things as a piece of wire, metal foil or a pin to act as a makeshift fuse as more serious damage or even fire may result.
3 Spare fuses are normally provided in the clip at the bottom of the fusebox, and can be obtained from a dealer.

Relays
4 The relays are of the plug-in type and are situated beneath the fusebox. Access is gained through the space between the lower edge of the facia and the parcel shelf.
5 Relays are provided to operate the windscreen wiper time delay, direction indicator/hazard flasher system and the heated rear window. Additional relays may be fitted according to vehicle equipment operations. The relays can be removed by simply pulling them from their respective locations.
6 If a system controlled by a relay becomes inoperative, and the relay is suspect, operate the system and if the relay is functioning it should be possible to hear it click as it is energized. If this is the case the fault lies with the components of the system. If the relay is not being energized then the relay is not receiving a main supply voltage or a switching voltage, or the relay itself is faulty.

18.1A The fusebox is located beneath a cover on the right-hand side of the facia

19 Direction indicator and hazard flasher system – general

1 Should the flashers become faulty in operation, check the bulbs for security and make sure that the contact surfaces are not corroded. If one bulb blows or is making a poor connection due to corrosion, the system will not flash on that side of the car.
2 If the flasher unit operates in one direction and not the other, the fault is likely to be in the bulbs or wiring to the bulbs. If the system will not flash in either direction, operate the hazard flashers. If these function, check for a blown fuse in position 7. If the fuse is satisfactory the fault is likely to lie with the flasher relay.

20 Facia switches – removal and refitting

Before removing a switch, disconnect the battery earth terminal.

Lighting switch
1 Using a small screwdriver, depress the upper and lower catches and withdraw the switch. Disconnect the electrical multi-plug and remove the switch (photo).
2 Refitting is the reverse sequence to removal.

18.1B The fuse rating, location, and circuits protected are shown on the inside of the cover

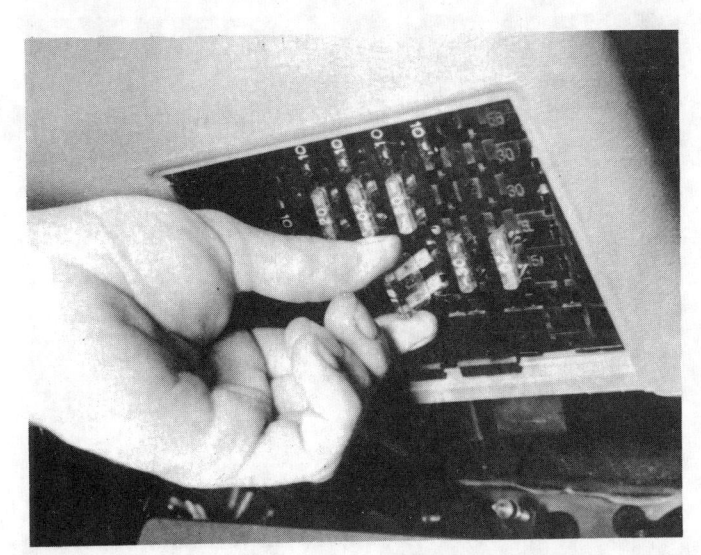

Foglamp switch
3 Depress the catches on each side of the switch panel using a screwdriver and withdraw the panel (photos).
4 Disconnect the electrical multiplug, release the detents and remove the switch from the panel (photo).
5 Refitting is the reverse sequence to removal.

Tailgate wiper/washer switch
6 The procedure for removal of the tailgate wiper/washer switch is identical to that for the foglamp switch.

Fan and heated rear window switch
7 Remove the ashtray
8 Undo the two screws securing the switch and vent panel to the facia (photo).
9 Depress the two catches at the bottom of the panel and withdraw the panel from the facia.
10 Disconnect the electrical wiring connectors at the rear of the switch, depress the two catches and withdraw the switch from the panel (photo).
11 Refitting is the reverse sequence to removal.

Hazard flasher switch
12 The procedure for removal of the hazard flasher switch is identical to that of the fan and heated rear window switch.

18.1C Remove a fuse by simply pulling it out of its location

20.1 Lighting switch removal. Upper retaining catch arrowed

20.3A Depress the catches on the foglamp switch panel ...

20.3B ... withdraw the panel from the facia ...

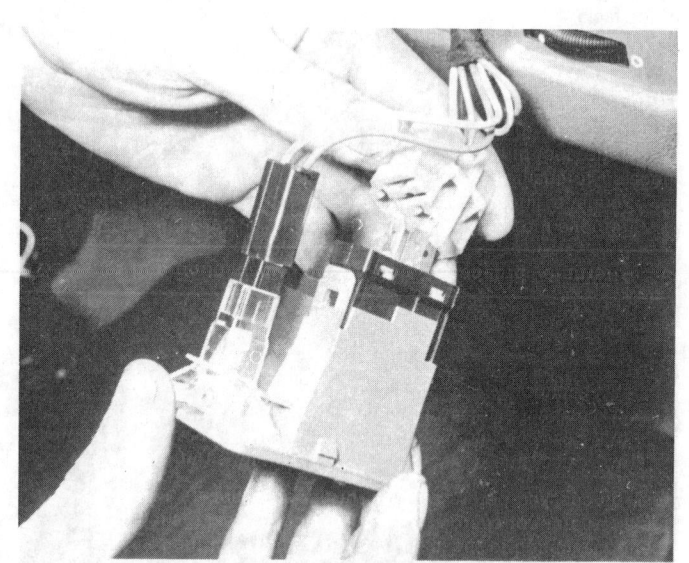

20.4 ... disconnect the multi-plug and remove the switch

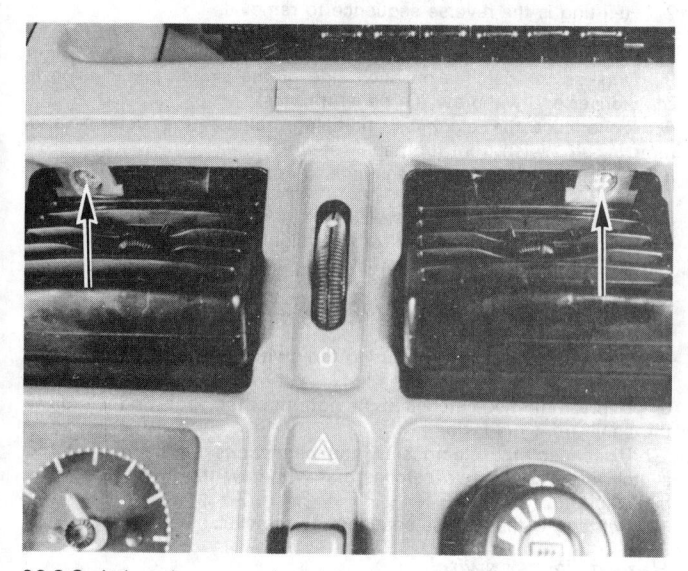

20.8 Switch and vent panel retaining screws (arrowed)

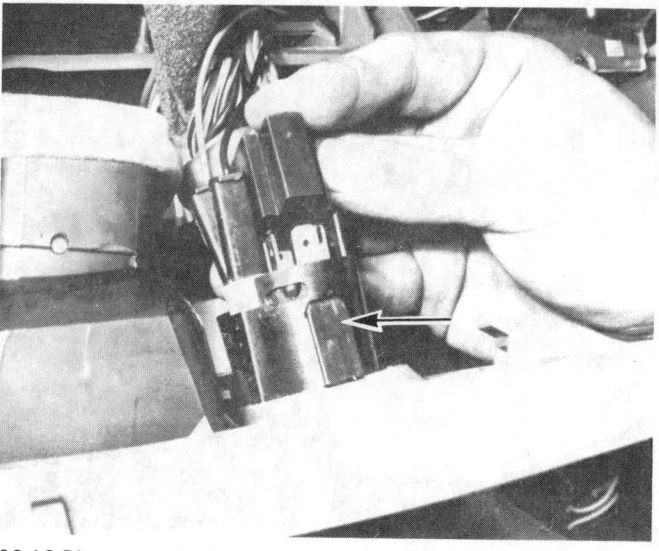

20.10 Disconnect the wiring connectors and remove the fan and heated rear window switch by depressing the catches (arrowed)

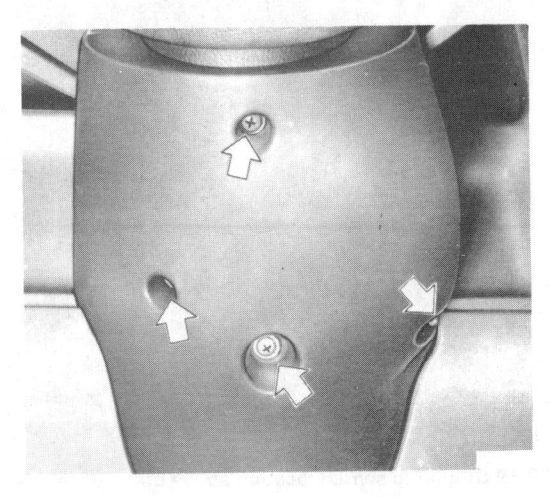

Fig. 10.13 Steering column shroud retaining screw locations (Sec 21)

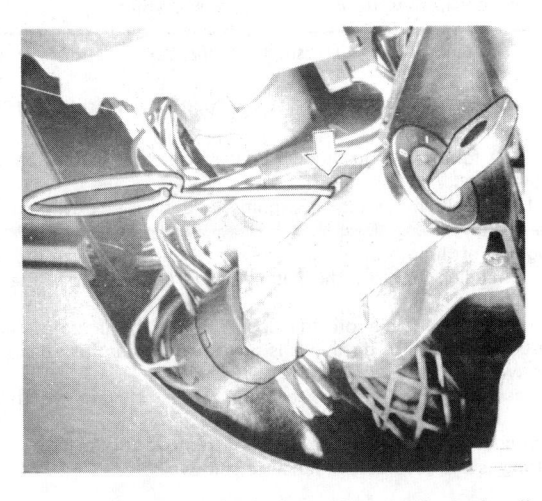

Fig. 10.14 Using a length of wire to depress the lock cylinder detent spring (Sec 21)

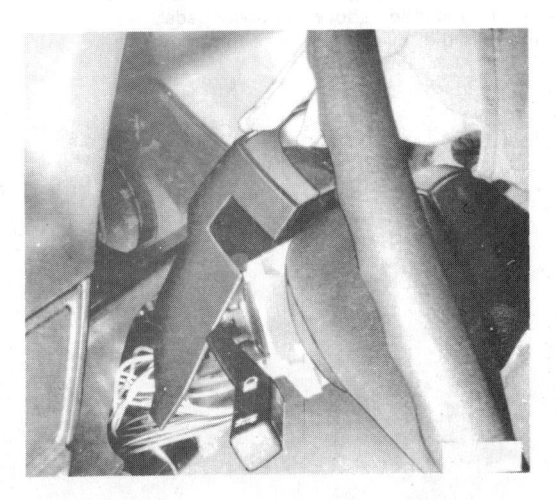

Fig. 10.15 Removing the steering column shroud upper half (Sec 21)

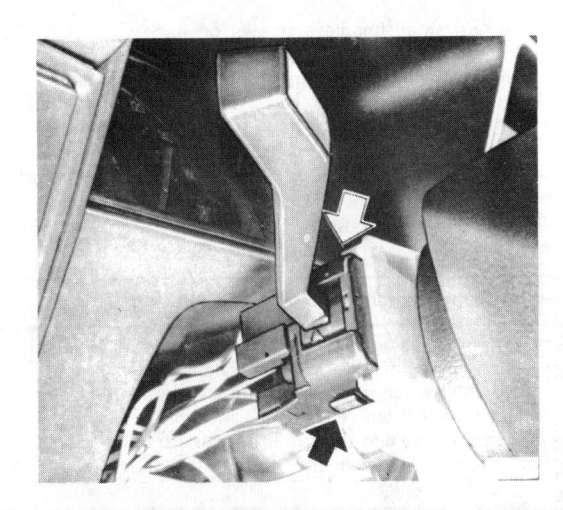

Fig. 10.16 Steering column switch upper and lower retaining catches – arrowed (Sec 21)

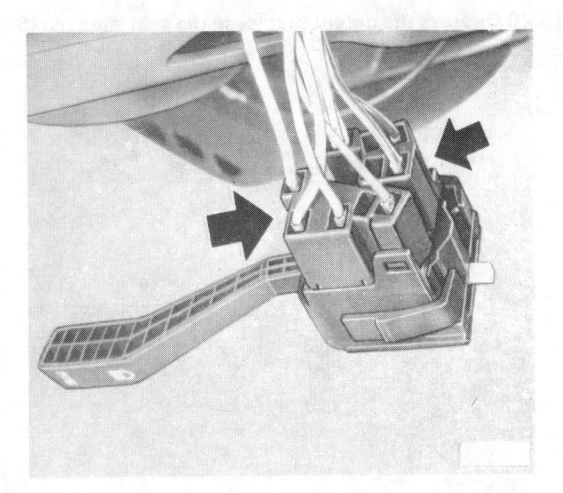

Fig. 10.17 Steering column switch wiring connectors – arrowed (Sec 21)

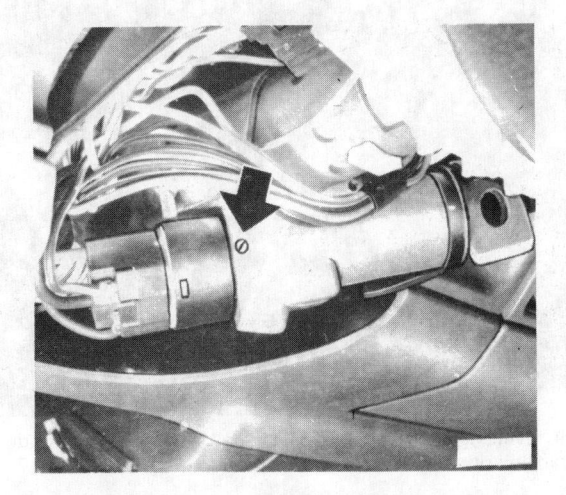

Fig. 10.18 Ignition switch retaining screw – arrowed (Sec 21)

21 Steering column switches – removal and refitting

Before removing a switch, disconnect the battery earth terminal.

Direction indicator and windscreen wiper/washer switches

1 Undo the four retaining screws and lift off the steering column shroud lower half.
2 Insert the ignition key into the ignition/steering lock cylinder and turn it to position 'II'.
3 Using a short length of small diameter wire inserted into the hole in the lock housing, depress the detent spring and withdraw the key and lock cylinder.
4 By judicious manipulation, the upper half of the column shroud can now be removed. To do this it is necessary to carefully bend the shroud outward slightly to clear the switch stalks.
5 Depress the upper and lower catches on the relevant switch and slide it out of the switch housing.
6 Disconnect the wiring connectors and remove the switch.
7 Refitting is the reverse sequence to removal.

Ignition switch

8 Undo the four screws and lift off the steering column shroud lower half.
9 Undo the two small headless screws on each side of the lock housing and withdraw the ignition switch.
10 Disconnect the wiring multi-plug and remove the ignition switch.
11 Removal of the ignition switch lock cylinder is described earlier in this Section and details of the lock housing removal procedure will be found in Chapter 8. *Note that under no circumstances must the ignition switch and lock cylinder be removed from the housing at the same time.*
12 Refitting is the reverse sequence to removal.

22 Reversing lamp switch – removal and refitting

1 The reversing lamp switch is located on the forward facing side of the transmission casing below the clutch release lever (photo).
2 To remove the switch, disconnect the battery negative terminal and disconnect the electrical leads at the switch.
3 Unscrew the switch and remove it from the transmission.
4 Refitting is the reverse sequence to removal.

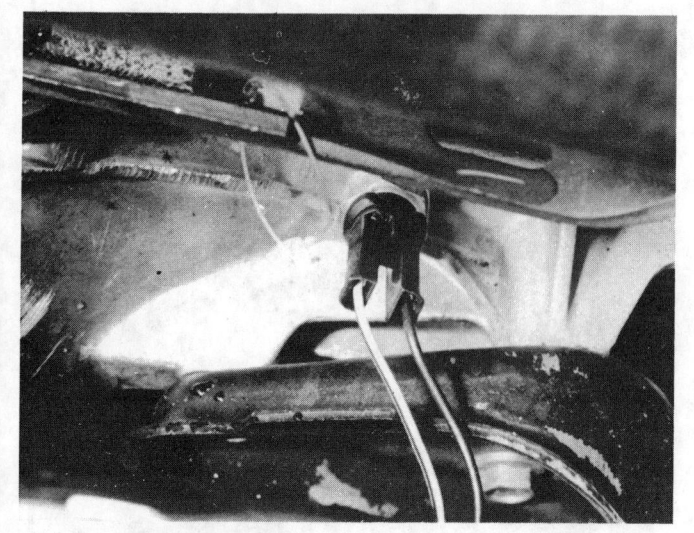

22.1 The reversing lamp switch is located on the forward facing side of the transmission

23 Stop-lamp switch – removal and refitting

1 Disconnect the battery negative terminal.
2 For greater access, remove the parcel shelf under the facia on the driver's side, referring to Chapter 12 if necessary.
3 Disconnect the wiring plug from the switch and unscrew the switch from its location above the brake pedal.
4 Refitting is the reverse sequence to removal.

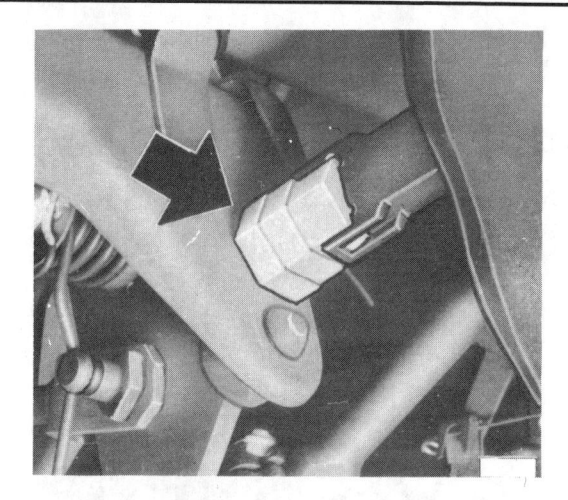

Fig. 10.19 Stoplamp switch location above brake pedal (Sec 23)

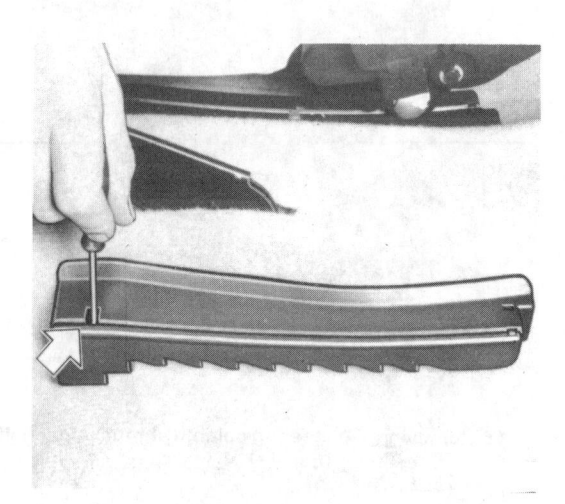

Fig. 10.20 Depress the detent to release the seat slide rail (Sec 24)

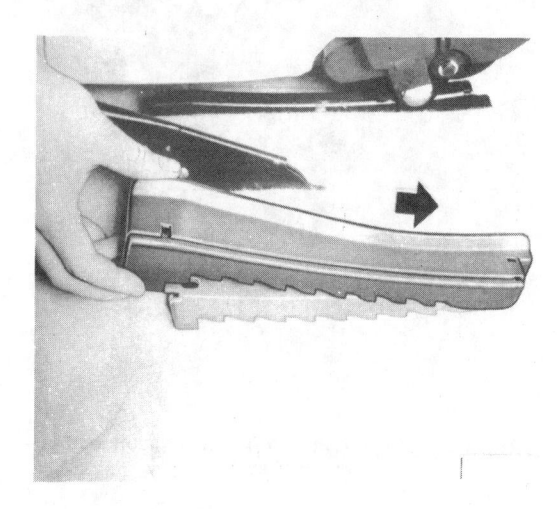

Fig. 10.21 Remove the seat slide rail by moving it upwards and to the rear (Sec 24)

Chapter 10 Electrical system

Fig. 10.22 Handbrake warning lamp switch retaining bolt – arrowed (Sec 24)

26.2 Horn and mounting bracket location at the front of the transmission

24 Handbrake applied warning switch – removal and refitting

1 Disconnect the battery negative terminal.
2 Refer to Chapter 12 and remove the passenger's front seat.
3 Remove the right-hand seat slide rail by depressing the detent with a screwdriver and moving the slide rail upwards and to the rear.
4 Using a socket through the now exposed slit in the carpet, undo the two handbrake lever retaining bolts.
5 Lift up the lever just sufficiently to gain access to the warning switch retaining bolt. Undo the bolt, disconnect the switch wires and remove the switch.
6 Refitting is the reverse sequence to removal.

26 Horn and switch – removal and refitting

1 The horn is located just ahead of the transmission casing. No maintenance or adjustment is required except to occasionally check the security of the mountings and electrical leads.
2 To remove the horn, disconnect the electrical leads, undo the retaining nut and remove the horn from its mounting bracket (photo).
3 To remove the horn button, disconnect the battery negative terminal and carefully prise the horn button contact pad from the centre of the steering wheel. Disconnect the electrical leads and remove the button.
4 Refitting, in both cases, is the reverse sequence to removal.
5 For access to the horn button contact ring it will be necessary to remove the steering wheel using the procedure described in Chapter 8.

25 Courtesy lamp switch – removal and refitting

1 The interior courtesy lamp switches are of the plunger type and are located in the door pillars (photo).
2 To remove a switch, disconnect the battery earth terminal, undo the switch retaining screw and withdraw the switch from the pillar. Disconnect the electrical lead and remove the switch. Be careful not to lose the lead inside the door pillar.
3 Refitting is the reverse sequence to removal.

27 Clock – removal and refitting

1 Disconnect the battery negative terminal.
2 Remove the ashtray.
3 Undo the two screws securing the switch and vent panel to the facia (photo 20.8).
4 Depress the two catches at the bottom of the panel and withdraw the panel from the facia.
5 Disconnect the electrical wiring at the rear of the clock, carefully

25.1 The courtesy lamp switches are located in the door pillars

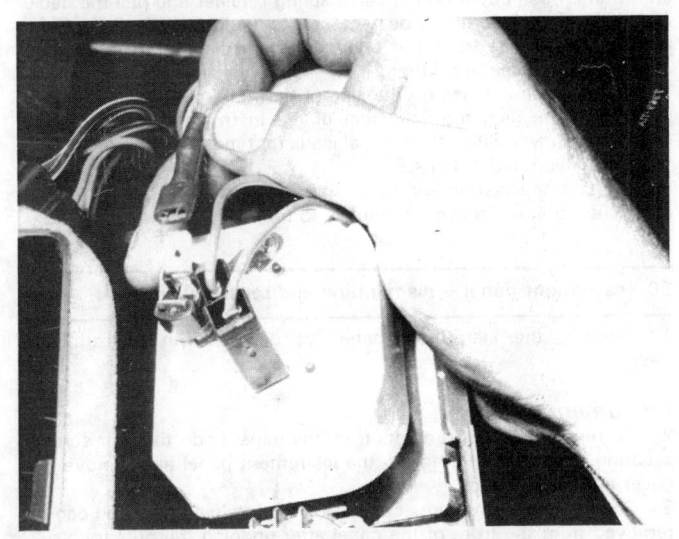

27.5 Electrical wiring connections at the rear of the clock

release the catches securing the clock to the panel and remove the clock (photo).
6 Refitting is the reverse sequence to removal.

28 Cigarette lighter – removal and refitting

1 Disconnect the battery negative terminal.
2 Refer to Chapter 12 and remove the glove compartment.
3 Remove the ashtray.
4 Pull out the cigarette lighter element and then, working through the glove compartment opening, disconnect the electrical wiring and illumination plug.
5 Remove the lighter socket from the facia.
6 Refitting is the reverse sequence to removal.

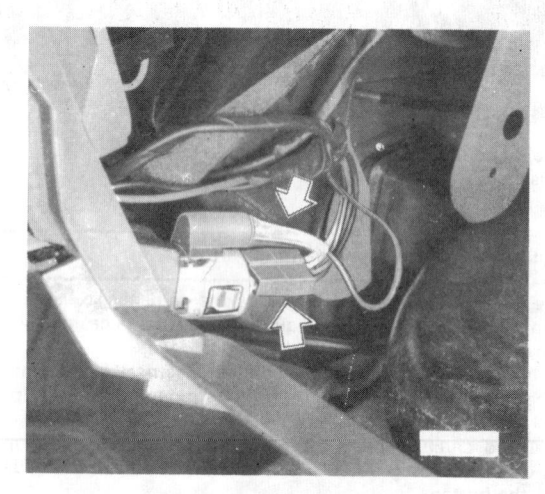

Fig. 10.23 Cigarette lighter illumination plug and wiring connectors (Sec 28)

29 Instrument panel – removal and refitting

1 Disconnect the battery negative terminal.
2 Very carefully prise up the underside of the instrument panel and pull the lower part of the panel away from the facia (photos). On SR models clearance is limited and it may be necessary to remove the upper and lower steering column shroud halves. Details of this procedure will be found in Section 21.
3 Release the panel upper locating catches and ease the panel away from the facia just sufficiently to gain access to the speedometer cable end fitting. Push down on the cable spring retainer and pull the cable off the speedometer. It may be necessary to feed the cable through the bulkhead grommet from the engine compartment side to provide access to the cable end fitting.
4 Depress the catches on the side of the multi-plug connector and disconnect the plug from the rear of the instrument panel (photo). Disconnect any additional electrical leads that may be fitted according to vehicle type and options.
5 Remove the instrument panel from the facia.
6 Refitting is the reverse sequence to removal.

30 Instrument panel – dismantling and reassembly

1 Remove the instrument panel, as described in the previous Section.

Instruments
2 To remove the instruments from the panel undo the four screws securing the cover and lens to the instrument panel and remove the cover and lens (photo).
3 The various instruments are secured by retaining nuts and can be removed from the front of the panel after undoing the nuts (photo).
4 Refitting is the reverse sequence to removal.

29.2A Carefully prise up the underside of the instrument panel ...

29.2B ... and pull the lower part of the panel away from the facia

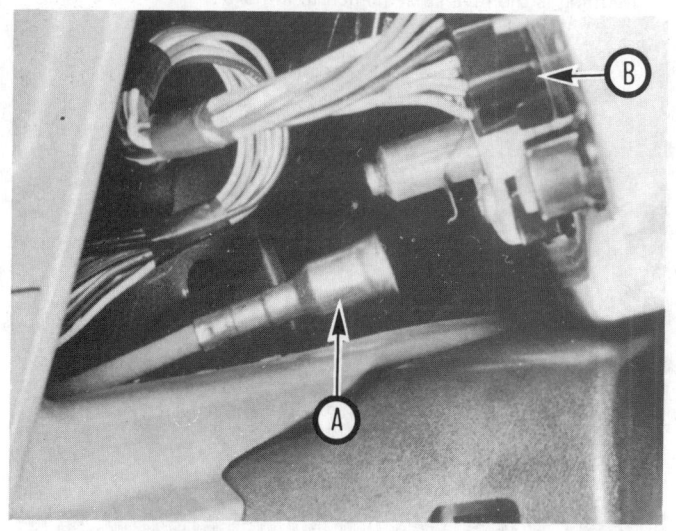

29.4 Disconnect the speedometer cable end (A) and the wiring multi-plug (B), and remove the instrument panel

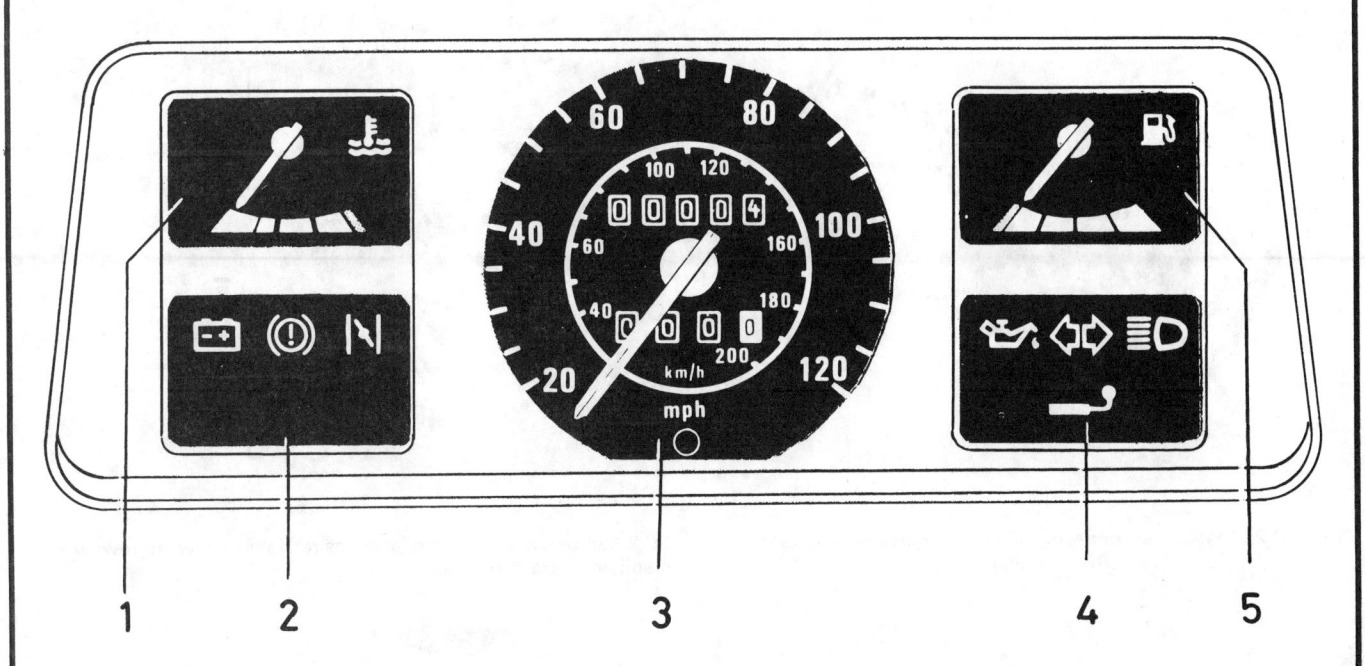

Fig. 10.24 Layout of the standard instrument panel fitted to Nova and Nova L models (Sec 30)

1 Temperature gauge
2 Warning lamps
3 Speedometer
4 Warning lamps
5 Fuel gauge

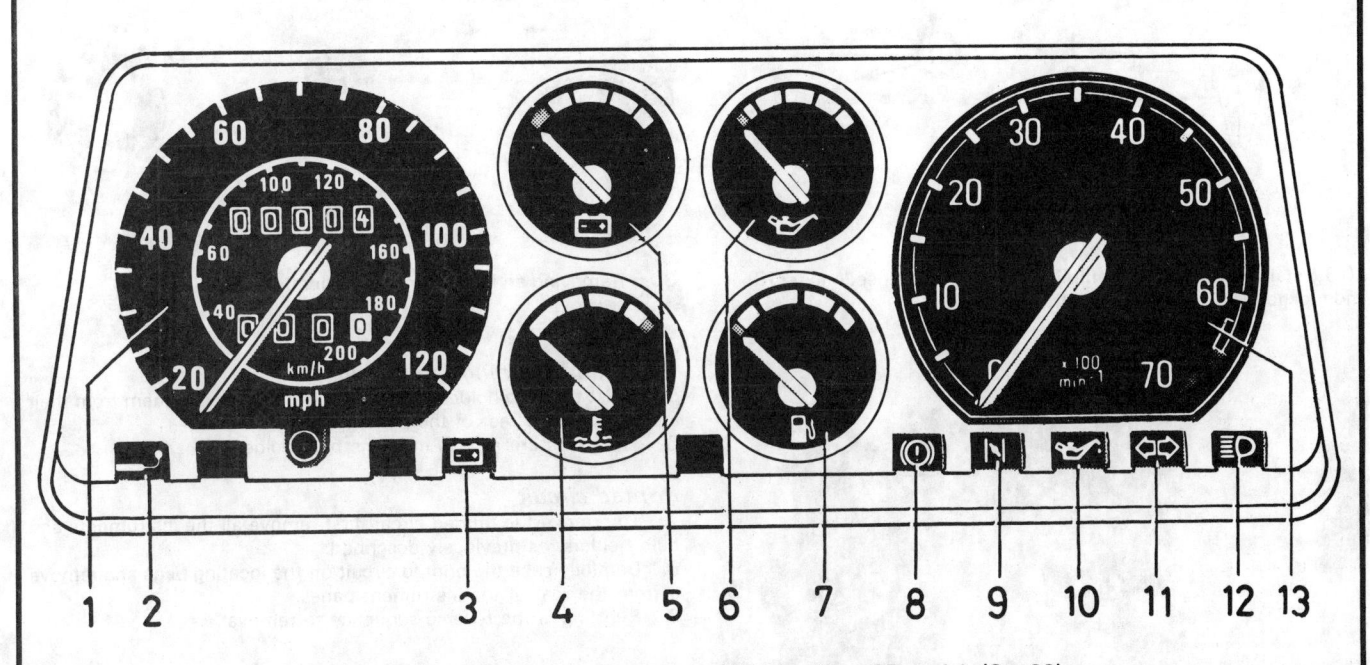

Fig. 10.25 Layout of the instrument panel fitted to Nova SR models (Sec 30)

1 Speedometer
2 Trailer direction indicator warning lamp
3 Alternator warning lamp
4 Temperature gauge
5 Voltmeter
6 Oil pressure gauge
7 Fuel gauge
8 Brake failure or handbrake applied warning lamp
9 Choke warning lamp
10 Oil pressure warning lamp
11 Direction indicator warning lamp
12 Main beam warning lamp
13 Tachometer

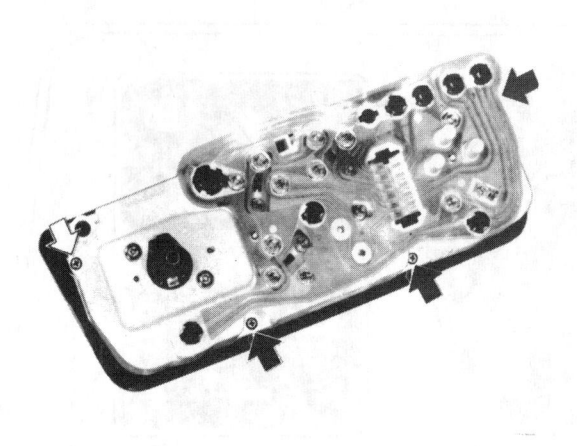

180 Chapter 10 Electrical system

Fig. 10.26 Instrument panel cover and lens retaining screws – Nova SR models (Sec 30)

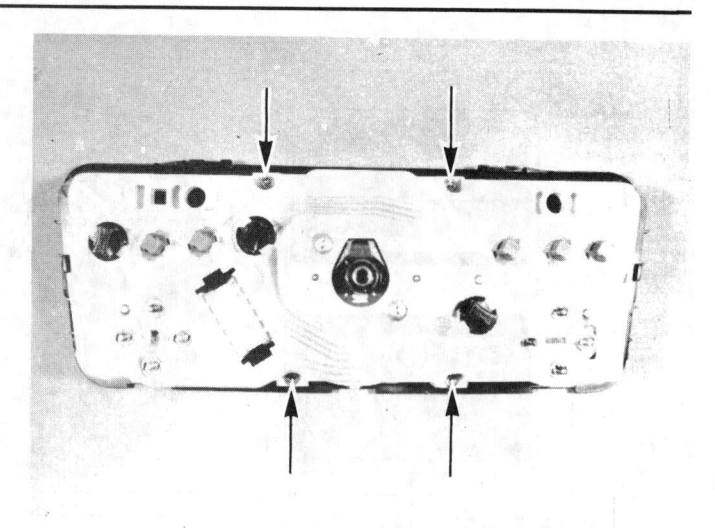

30.2 Instrument panel cover and lens retaining screws (arrowed) – standard instrument panel

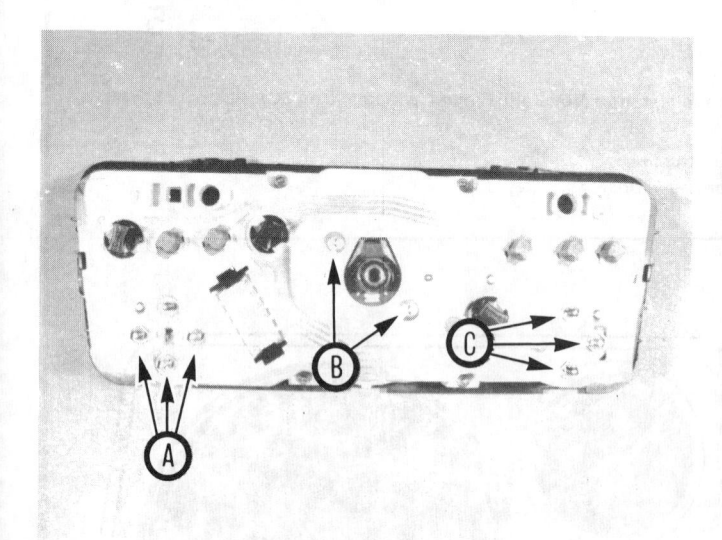

30.3 Retaining nuts for the temperature gauge (A), speedometer (B) and fuel gauge (C) – standard instrument panel

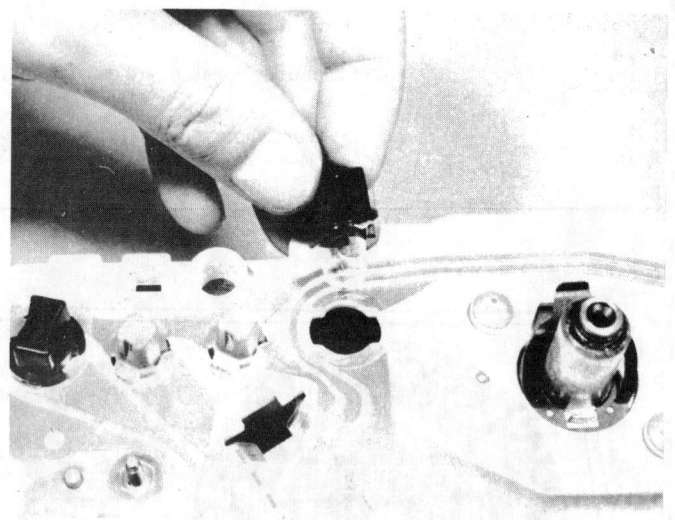

30.5 Removing an instrument panel illumination bulb holder

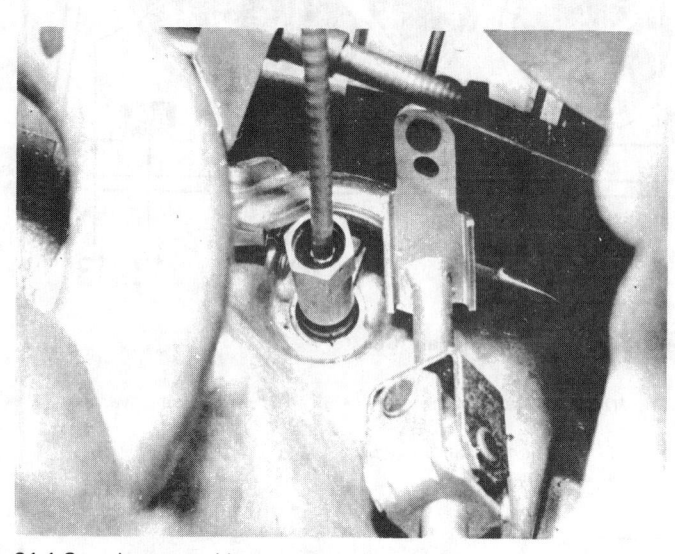

Panel and warning lamp bulbs
5 Twist the bulb holders anti-clockwise and remove them from their locations on the rear of the panel (photo).
6 The bulbs are a push fit in the bulb holders.

Printed circuit
7 To remove the printed circuit first remove all the instruments and bulb holders, as previously described.
8 Carefully ease the printed circuit off the locating pegs and remove it from the rear of the instrument panel.
9 Refitting is the reverse sequence to removal.

31 Speedometer cable – renewal

1 Working in the engine compartment, unscrew the speedometer cable end fitting from its attachment at the transmission (photo).
2 Release the other end of the cable from the instrument panel, as described in Section 29, paragraphs 1 to 3 inclusive.
3 Release the bulkhead grommet from its location and remove the cable, complete with grommet from the car.
4 Refitting is the reverse sequence to removal.

31.1 Speedometer cable to transmission end fitting

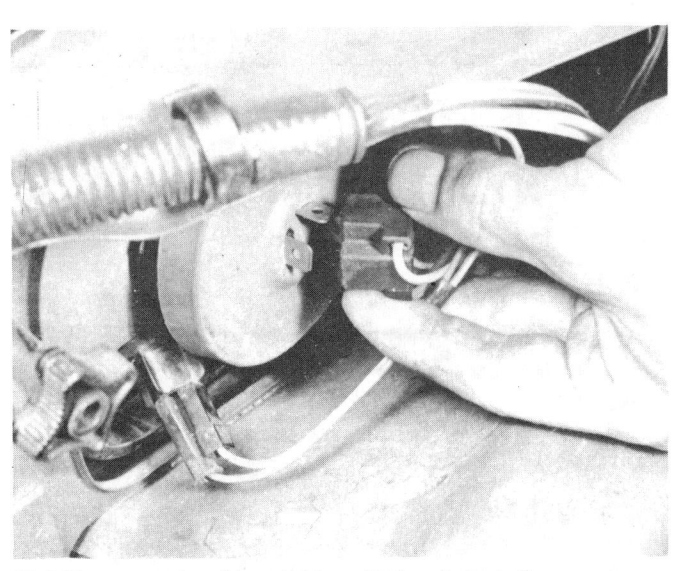

32.1 Disconnect the wiring plug from the headlamp bulb

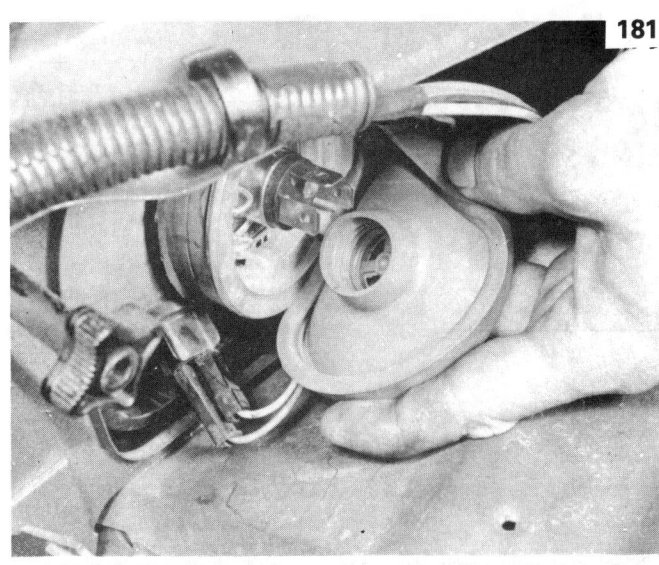

32.2 Pull off the rubber cover ...

32.3 ... and release the retaining clip (arrowed) to gain access to the bulb holder

32.8 Sidelamp bulb holder and bulb

32.15A Depress the rear lamp cluster retaining catches (arrowed) ...

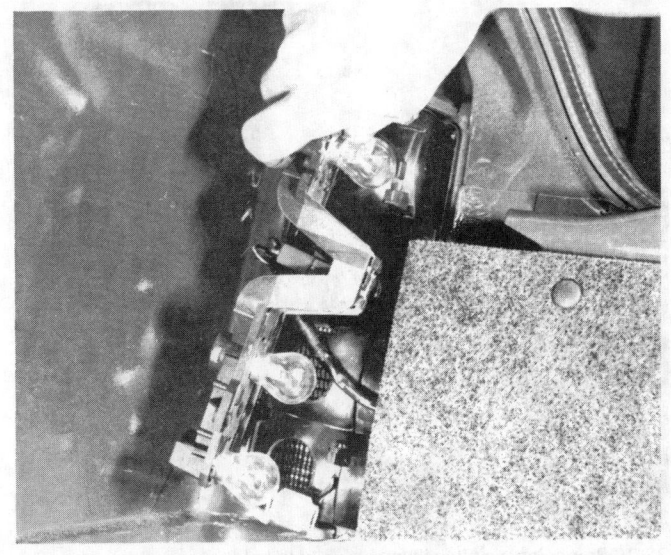

32.15B ... and withdraw the bulb holder to gain access to the bulbs

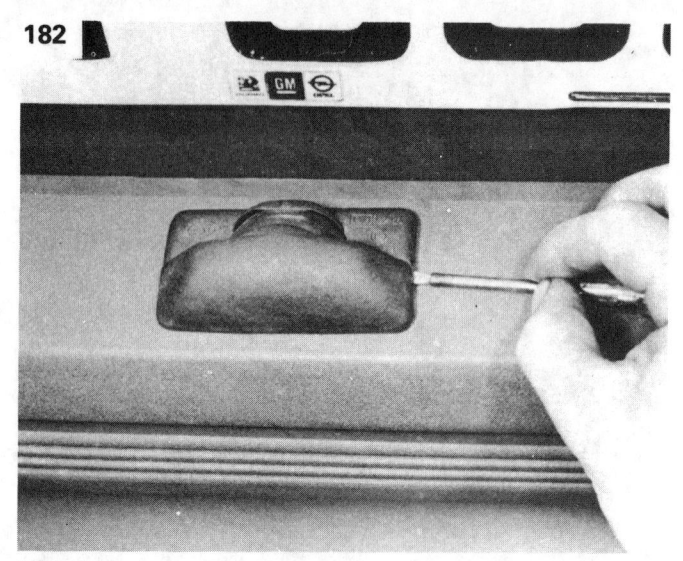

32.18 Insert a thin screwdriver in the notch to release the number plate lamp assembly

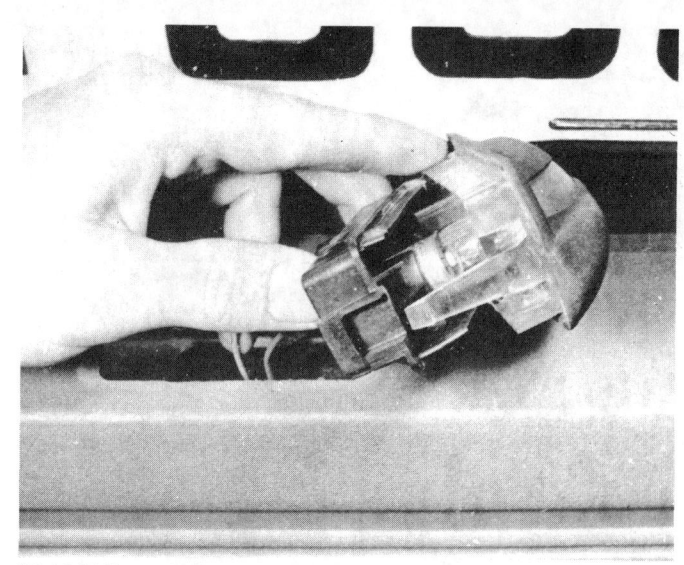

32.19A Lift out the lamp and remove the lens

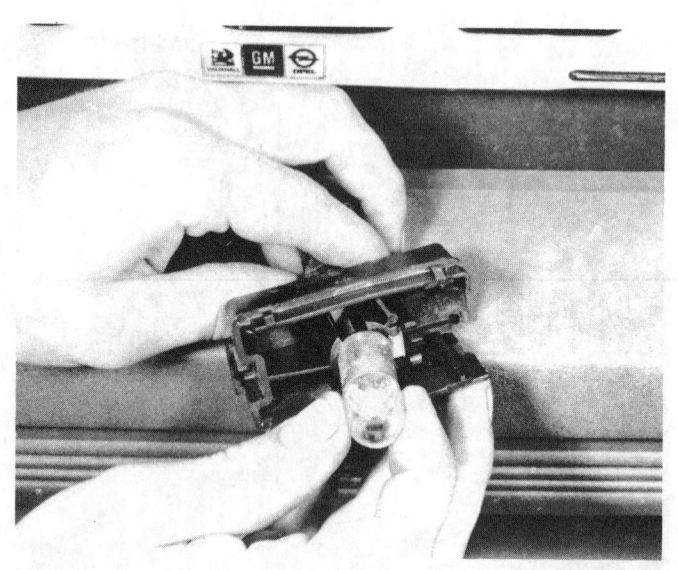

32.19B ... to gain access to the bulb

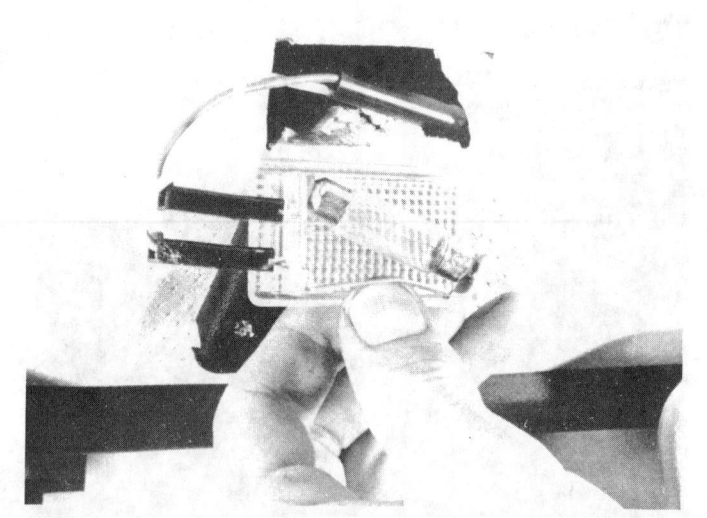

32.21 Carefully prise out the interior lamp bulb holder to renew the bulb

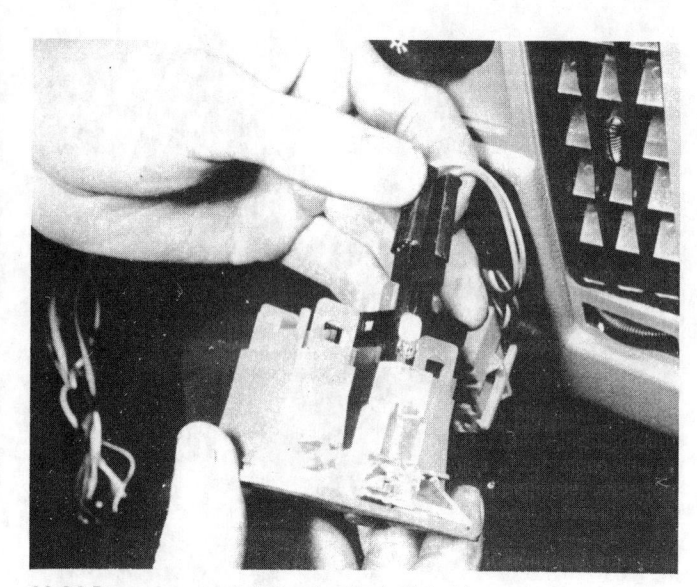

32.26 Removing a switch illumination bulb holder

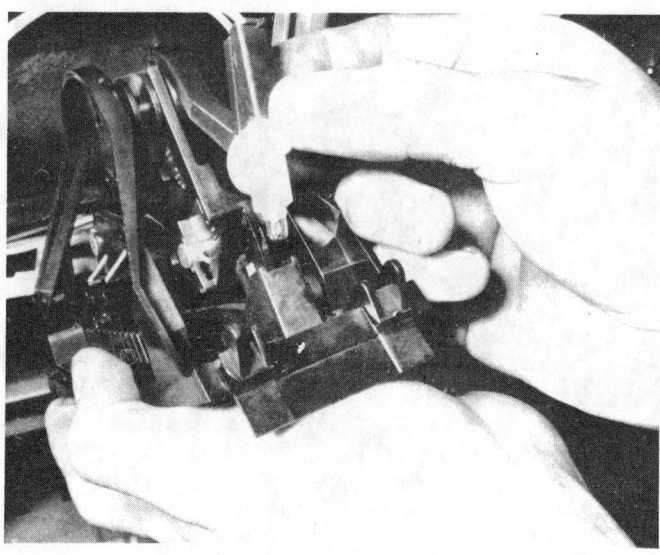

32.27 Heater control illumination bulb holder and bulb

Chapter 10 Electrical system

32 Bulbs – renewal

Headlamp
1 Open the bonnet and disconnect the wiring plug from the rear of the bulb (photo).
2 Pull off the rubber cover to gain access to the bulb (photo).
3 Compress the ends of the bulb retaining clip and swivel the clip to one side (photo).
4 Withdraw the bulb from the lens assembly.
5 The headlamp bulbs are of the halogen type and the glass should not be touched with the fingers as this could cause premature failure of the bulb. If the glass is touched, wipe it clean using a cloth moistened with methylated spirit.
6 Refitting the bulb is the reverse sequence to removal, but make sure that the lugs on the holder engage with the recesses in the lens assembly.

Sidelamp
7 Open the bonnet.
8 Reach down below the headlamp and release the sidelamp bulb holder from the headlamp lens assembly by turning it anti-clockwise (photo).
9 Remove the sidelamp bulb from the bulb holder.
10 Refitting is the reverse sequence to removal.

Front direction indicator lamp
11 Open the bonnet.
12 Release the direction indicator bulb holder from the lens by turning anti-clockwise.
13 Remove the bulb from the holder by releasing the bayonet fitting.
14 Refitting is the reverse sequence to removal.

Rear lamp cluster
15 Access to these bulbs is gained by unclipping the trim panel within the luggage compartment and releasing the bulb holder assembly by depressing the retaining lugs (photos).
16 With the bulb holder withdrawn the bulbs can be removed after releasing their bayonet fitting. The upper bulb is the stop/tail bulb, the centre is the direction indicator bulb and the lower bulb is the rear fog guard or reversing lamp bulb.
17 Refitting the rear lamp bulbs and cluster is the reverse sequence to removal.

Rear number plate lamp
18 Insert a thin screwdriver in the notch provided and prise the lamp from the bumper (photo).
19 Lift off the lamp lens and release the bayonet fitting to remove the bulb (photos).
20 Refitting is the reverse sequence to removal.

Interior lamp
21 The lamp lens complete with festoon type bulb can be removed from the headlining by carefully prising it out (photo).
22 Slip the bulb out of the contacts to renew.
23 Refitting is the reverse sequence to removal.
24 Additional interior lamps of a similar type may be fitted, depending on vehicle specification, and renewal of the bulbs in these is carried out in the same way.

Instrument panel and facia illumination
25 Renewal of the instrument panel warning and illumination lamp bulbs is described in Section 30.
26 The bulbs used for illumination of the facia switches can be renewed after removal of the relevant switch, as described in Section 20. The bulb holders are simply withdrawn from the rear of the switch and the bulb removed from the holder after releasing its bayonet fitting (photo).
27 To renew the heater control illumination bulb first remove the control unit, as described in Chapter 12, and then remove the bulb holder and bulb from the rear of the unit (photo).
28 Refitting is the reverse sequence to removal.

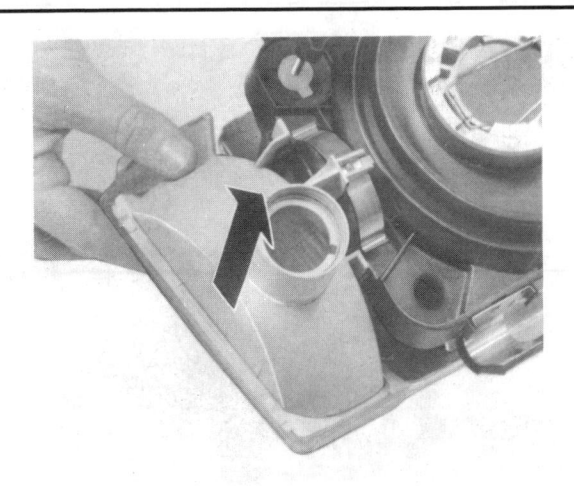

Fig. 10.27 Removing the direction indicator lens from the headlamp assembly (Sec 33)

33.2 Headlamp lens assembly retaining screws (arrowed)

33 Headlamp lens assembly – removal and refitting

1 From within the engine compartment, disconnect the headlamp bulb wiring plug and remove the sidelamp and direction indicator lamp bulb holders by turning them anti-clockwise.
2 Undo the two screws securing the lens assembly to the body panel (photo), tip the lens forward and remove it from the car.
3 With the lens assembly removed, lift off the rubber cover, compress the retaining clip and move the clip aside. Lift out the headlamp bulb. Take care not to touch the bulb glass with your fingers, but if you do, wipe the bulb with a cloth moistened with methylated spirit.
4 Depress the two retaining catches and slide the direction indicator lens off the headlamp assembly.
5 Refitting is the reverse sequence to removal.

34 Headlamp beams – adjustment

1 The headlamp beam adjustment is most important, not only for your own safety but for that of other road users as well. Accurate beam alignment can only be obtained using optical beam setting equipment and you should regard any adjustments made without such equipment as purely temporary.
2 To make a temporary adjustment, position the car on level ground about 3 metres (10 ft) in front of a vertical wall or a piece of board secured vertically. The wall or board should be square to the centre-

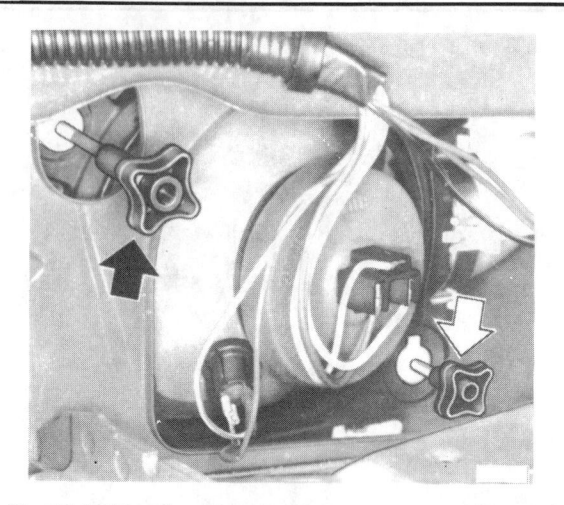

Chapter 10 Electrical system

Fig. 10.28 Headlamp beam adjusters – arrowed (Sec 34)

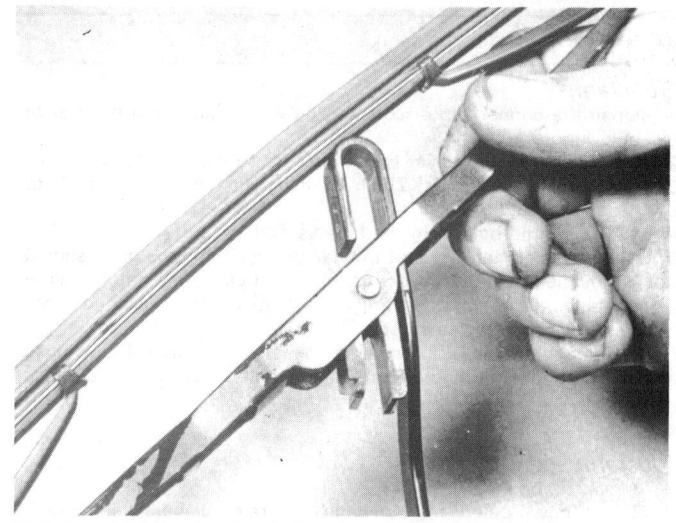

35.1 Removing a wiper blade from its arm

line of the car and the car should be normally laden. Check that the tyre pressures are correct.
3 Draw a vertical line on the board or card in line with the centre-line of the car.
4 Bounce the car on its suspension several times to ensure correct levelling and then accurately measure the height between the ground and the centre of the headlamps.
5 Draw a horizontal line across the wall or board at the same height as the headlamp centres and on this line mark a cross on either side of the centre line at the same distance apart as the headlamp centres.
6 Now locate the adjusters on each headlamp. There are two, diagonally opposite each other on each headlamp.
7 Switch the headlamps on to full beam and, using the adjusters, adjust each headlamp to align the beam to shine just below the corresponding cross on the wall or board.
8 Bounce the car on its suspension again to check that the beams return to the correct position. At the same time check the operation of the dipswitch to confirm that the beams dip to the nearside. Switch off the headlamps on completion.
9 Holts Amber Lamp is useful for temporarily changing the headlight colour to conform with the normal usage on Continental Europe.

35.4A To remove a wiper arm, lift up the plastic cover ...

Fig. 10.29 Wiper arm setting dimension (Sec 35)

A = 20.0 to 25.0 mm (0.8 to 1.0 in) between blades and windscreen rubber surround

35 Wiper blades and arms – removal and refitting

1 Pull the wiper arm away from the glass, swivel the blade on the arm and then depress the catch on the U-shaped retainer and slide the blade from the wiper arm (photo).
2 Before removing a wiper arm make sure that it is in its parked position having been switched off by the wiper switch and not the ignition key.

35.4B ... unscrew the nut and prise the arm off the spindle

Chapter 10 Electrical system

3 To facilitate re-alignment of the arms on the screen, stick a length of masking tape on the glass parallel to the blade before removing the arm.
4 Flip up the plastic cover and unscrew the arm retaining nut (photos).
5 Pull the arm from the splined driving spindle.
6 Refitting is a reversal of removal, do not overtighten the nut. If tape was not applied to the glass, observe the setting dimensions in the diagram (Fig, 10.29).

36 Windscreen wiper motor and linkage – removal and refitting

Wiper motor
1 Open the bonnet and disconnect the electrical multi-plug from the wiper motor (photo). Release the wiring harness cable tie from around the motor.
2 Undo the nut securing the linkage bellcrank arm to the motor drive spindle (photo). Remove the arm.
3 Undo the three screws securing the motor to the bulkhead and remove the motor.
4 Refitting is the reverse sequence to removal.

Wiper linkage
5 Remove the wiper arms, as described in the previous Section.
6 Undo the nuts securing the wiper arm spindles to the scuttle and lift off the shaped spacing block.
7 Undo the nut securing the linkage bellcrank arm to the motor spindle and remove the arm.
8 Withdraw the linkage assembly from the scuttle.
9 Refitting is the reverse sequence to removal.

37 Tailgate wiper motor – removal and refitting

1 Open the tailgate and carefully remove the trim panel by prising out the retaining buttons.
2 Disconnect the electrical wiring at the multi-plug connector and undo the earth lead retaining screw (photo).
3 Remove the wiper arm, as described in Section 35.
4 Remove the protective cap from the motor shaft and then unscrew the shaft retaining nut. Recover the washer.
5 Undo the two bolts securing the motor to the tailgate, recover the washers and remove the motor.
6 Refitting is the reverse sequence to removal.

38 Windscreen and tailgate washer system – general

1 On Saloon models the system consists of a fluid reservoir located on the left-hand side of the engine compartment, an electrically operated pump, steering column switch and associated hoses and nozzles.
2 On Hatchback models a larger reservoir is used incorporating a second pump which supplies water to the tailgate window via a small bore tube running inside the car. Control is by a separate switch mounted on the facia, which incorporates an intermittent wipe facility.
3 To renew a washer pump, disconnect the electrical plug and remove the pump by 'rocking it' out of its sealing grommet in the reservoir. Use a new seal when refitting.
4 A rubber connecting sleeve is used to join the plastic pipe to the pump nozzle as attempting to force the stiff plastic pipe onto the nozzle could cause it to fracture.
5 The windscreen washer nozzles should be adjusted so that the stream of fluid strikes the windscreen just above its centre. The tailgate nozzle should be adjusted so that the fluid strikes the tailgate window just to the left of centre.

39 Headlamp washer system – general

1 The headlamp washer system is a factory fitted option and consists of a reservoir located on the right-hand side of the engine compartment, high pressure pump and spray nozzles which are incorporated into the bumper overriders. The system is operated by the

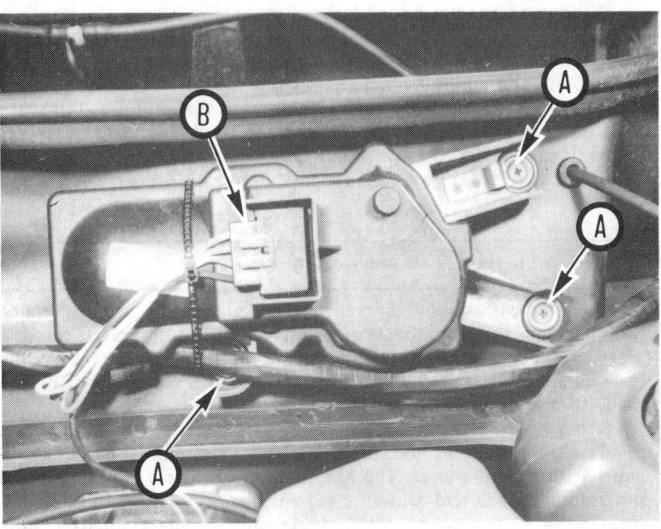

36.1 Windscreen wiper motor retaining screws (A) and wiring multi-plug (B)

36.2 Windscreen wiper motor linkage to spindle retaining nut (arrowed)

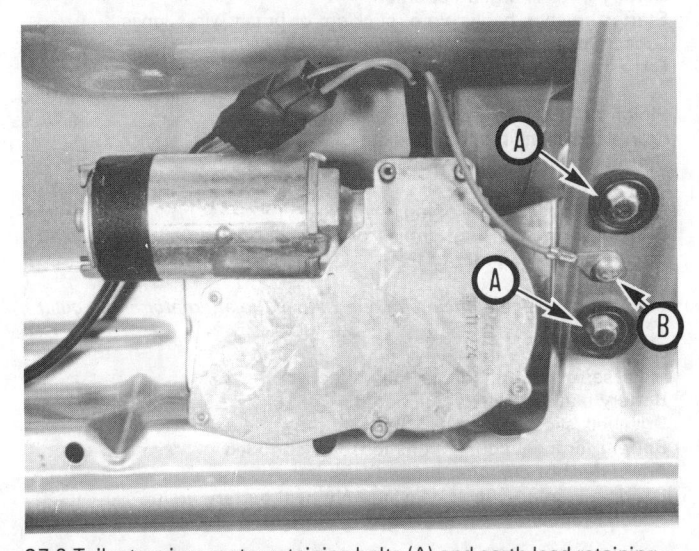

37.2 Tailgate wiper motor retaining bolts (A) and earth lead retaining screw (B)

windscreen washer switch, but only functions if the headlamps are switched on.
2 The procedure for renewal of the washer pump is the same as for the windscreen and tailgate units, described in the previous Section.
3 It is not possible to renew the washer nozzles separately and if this is necessary the complete overrider must be obtained. The procedure for removal and refitting the bumper overriders is contained in Chapter 12.
4 Adjust the nozzles when necessary so that the jet of fluid strikes the headlamp lens in the centre of the glass.

40 Washer fluid – general

1 It is recommended that only additives specially prepared for washer systems are used in the fluid reservoir. The use of household detergent or other cleaning agents is likely to damage the pump and rubber components of the system.
2 Never use cooling system antifreeze in a washer system or the paintwork will be damaged. In very cold weather, a small quantity of methylated spirit may be poured into the fluid to prevent freezing.

41 Radio or radio/cassette player (factory fitted unit) – removal and refitting

1 If the vehicle is equipped with a radio or radio/cassette player it may be removed in the following way for repair, replacement or as a preliminary operation to removal of the facia (refer to Chapter 12).
2 Disconnect the battery negative terminal.
3 Pull the knobs off the unit and then undo the nuts securing the trim panel to the front of the radio. Remove the trim panel.
4 Depress the spring clips securing the unit to its mounting frame and withdraw it sufficiently to gain access to the wiring at the rear.
5 Disconnect the aerial, speaker, earth and live feed electrical connections and remove the radio from the facia.
6 Refitting is the reverse sequence to removal.

42 Wiring diagram – explanatory notes

The following wiring diagram is laid out using a grid reference system, with the bottom line being the earth track. Using grid reference 79 at the bottom of the diagram as an example, follow the line upwards past the cigarette lighter (R3) to a connector (X2) and finally to a number in a box (18).
Referring back to grid reference 18 at the bottom of the diagram it will be seen that near the top of the diagram a number in a box (79) aligns with this reference. The line from this box is a continuation of grid reference 79 and shows the live feed to the cigarette lighter through the 20 amp fuse (F18) and originating from terminal 15 of the starter switch (S1).
The letters and numbers appearing adjacent to the lines are the wire identification. The first two letters represent the main wire colour and the second two are the tracer colour (where applicable). The number is the wire size (mm^2).
The wiring diagrams are applicable to all Nova model variants although some of the components shown are not fitted to UK models.

43 Fault diagnosis – electrical system

Symptom	Reason(s)
No voltage at starter motor	Battery discharged Battery defective internally Battery terminals loose or earth lead not securely attached to body Loose or broken connections in starter motor circuit Starter motor switch or solenoid faulty
Voltage at starter motor – faulty motor	Starter brushes badly worn, sticking, or brush wires loose Commutator dirty, worn or burnt Starter motor armature faulty Field coils earthed
Electrical defects	Battery in discharged condition Starter brushes badly worn, sticking, or brush wires loose Loose wires in starter motor circuit Dirt or oil on drivegear
Starter motor noisy or rough in engagement	Pinion or flywheel gear teeth broken or worn Starter drive main spring broken Starter motor retaining bolts loose
Alternator not charging*	Drivebelt loose and slipping, or broken Brushes worn, sticking, broken or dirty Brush springs weak or broken

*If all appears to be well but the alternator is still not charging, take the car to an automobile electrician for checking of the alternator and regulator

Battery will not hold charge for more than a few days	Battery defective internally Plate separators no longer fully effective Battery plates severely sulphated Drivebelt slipping Battery terminal connections loose or corroded Alternator not charging properly Short in lighting circuit causing continual battery drain Integral regulator unit not working correctly
Ignition light fails to go out, battery runs flat in a few days	Drivebelt loose and slipping, or broken Alternator faulty

Chapter 10 Electrical system

Symptom	Reason(s)

Failure of individual electrical equipment to function correctly is dealt with alphabetically below

Symptom	Reason(s)
Fuel or temperature gauge gives no reading	Temperature gauge transmitter faulty Fuel tank sender unit faulty Faulty gauge Broken or disconnected wiring connections Blown fuse
Horn operates all the time	Horn push either earthed or stuck down Horn cable to horn push earthed
Horn fails to operate	Blown fuse Cable or cable connection loose, broken or disconnected Horn has an internal fault
Horn emits intermittent or unsatisfactory noise	Cable connections loose
Lights do not come on	If engine running, battery discharged Blown fuse Wire connections loose, disconnected or broken Light switch shorting or otherwise faulty
Lights come on but fade out	If engine not running, battery discharged
Lights give very poor illumination	Lamp glasses dirty Reflector tarnished or dirty Lamps badly out of adjustment Incorrect bulb with too low wattage fitted Existing bulbs old and badly discoloured
Wiper motor fails to work	Blown fuse Wire connections loose, disconnected or broken Brushes badly worn Armature worn or faulty Field coils faulty
Wiper motor works very slowly and takes excessive current	Commutator dirty, greasy or burnt Linkage seized, bent or damaged Armature bearings dry or unaligned Armature badly worn or faulty
Wiper motor works slowly and takes little current	Brushes badly worn Commutator dirty, greasy or burnt Armature badly worn or faulty
Wiper motor works, but wiper blades remain static	Linkage disengaged or faulty Drive spindle damaged or worn Wiper motor gearbox parts badly worn

Key to wiring diagram for Vauxhall Nova – 1983 and 1984 models

Component	Description	Current track
E1	Right parking lamp	39
E2	Right tail lamp	40
E3	Number plate lamp	41
E4	Left parking lamp	37
E5	Left tail lamp	38
E6	Engine compartment lamp	43
E7	Main beam (right)	48
E8	Main beam (left)	47
E9	Dipped beam (right)	51
E10	Dipped beam (left)	50
E11	Instrument lights	43
E13	Luggage compartment lamp	71
E14	Passenger (interior) lamp	73
E15	Glovebox lamp	80
E16	Cigarette lighter lamp	78
E17	Reversing lamp	82
E19	Heated rear window	35
E20	Left fog lamp	118
E21	Right fog lamp	121
E24	Rear fog guard lamp	124
E32	Clock illumination lamp	77
E33	Ashtray lamp	77
E34	Heater control illumination lamp	43
F2,F3,F4,F5	Fuses (in fusebox)	47,48,50,51
F6,F12,F15	Fuses (in fusebox)	155,118, 152
F7,F8,F9,F10,F13,F16,F14,F17,F18	Fuses (in fusebox)	60,35,30,28, 38,71,42, 54,19
F20	Fuse (headlamp washer unit)	106
F21	Fuse (mixture preheating)	124
G1	Battery	1
G2	Alternator	13–15
H1	Radio	75
H2	Horn	58
H3	Turn signal lamp indicator light	67
H4	Oil pressure warning light	17
H5	Parking brake 'on'/low brake fluid warning light	25
H6	Hazard warning indicator light	64,65
H7	Charge warning light	15
H8	Main beam indicator light	49
H9	Right stop lamp	61
H10	Left stop lamp	60
H11	Front right turn signal lamp	68
H12	Rear right turn signal lamp	69
H13	Front left turn signal lamp	65
H14	Rear left turn signal lamp	66
H17	Trailer turn signal lamp indicator light	144
H19	Headlamps 'on' buzzer warning system	142,143
H20	Choke warning light	24
K1	Heated rear window relay	34,35
K2	Flasher unit	63
K5	Fog lamp relay	118,119
K8	Windscreen wiper time delay relay	98–101
K9	Headlamp washer unit time delay relay	104–106
K10	Trailer flasher unit	144,145
K28	Day running light relay	155,156
K30	Rear window wiper time delay relay	111–115
K34	Idle stabilization control unit	7–10
K44	Dashpot relay	161–163
K45	Mixture preheating relay	126,127
L1	Ignition coil	165,166
L2	Ignition coil (transistorized ignition system)	9,10
M1	Starter	6,7,162–164
M2	Windscreen wiper motor	54–57, 96–100
M3	Heater blower motor	30,31, 130–133
M4	Radiator cooling fan motor	28
M5	Windscreen washer pump	53,95
M8	Rear window wiper motor	86,113
M9	Rear window washer pump	85,110
M24	Headlamp washer pump	106
P1	Fuel gauge	22
P2	Temperature gauge	20
P3	Clock	76
P4	Fuel level sender unit	22
P5	Temperature sensor	20
P7	Tachometer	136
P8	Oil pressure gauge	138
P9	Voltmeter	137
P10	Oil pressure sensor	138
R1	Ballast resistor wire	165
R3	Cigarette lighter	79
R7	Mixture preheater	126
S1	Starter switch	6,7
S2.1	Light switch	41,42
S2.2	Passenger (interior) lamp switch	72
S3	Heated rear window and heater blower switch	130–132
S5.2	Main/dipped beam switch	49,50
S5.3	Turn signal switch	68,69
S6	Ignition distributor	166–168
S7	Reversing lamp switch	82
S8	Stop lamp switch	60
S9.1	Windscreen wiper switch	53–56
S9.2	Windscreen wiper time delay switch	95–99
S9.3	Rear window wiper switch	85–87, 110–113
S11	Low brake fluid switch	26
S12	Clutch switch	161
S13	Parking brake 'on' switch	25
S14	Oil pressure warning light switch	17
S15	Luggage compartment light switch	71
S16	Door switch (right)	73
S17	Door switch (left)	74
S18	Glovebox lamp switch	80
S21	Fog lamps switch	120–122
S22	Rear fog guard lamp switch	124,125
S29	Radiator cooling fan switch	28
S47	Door 'open' and headlamps 'on' warning system switch	140,141
S50	Choke warning light switch	24
S52	Hazard warning switch	63–66
S64	Horn switch	58
S65	Heated rear window and heater blower switch	30–33
S66	Vacuum switch	10
S73	Temperature switch	127
U1	Day running light voltage transformer	156–160
X1	Trailer socket	147–153
X2	Auxiliary connector	75,79, 80,143
Y10	Ignition distributor (transistorized ignition system)	12
Y15	Inductive sensor with ignition module	8,9
Y17	Idle cut-off solenoid valve	81
Y18	Dashpot solenoid valve	161

Note: diagram covers all models – therefore alternative wiring is shown for some components where differences occur within the range. Refer to Section 42 for explanation of use.

Colour code

BL	– Blue	GE	– Yellow	RT	– Red	LI	– Lilac
HBL	– Light blue	GR	– Grey	WS	– White	VI	– Violet
BR	– Brown	GN	– Green	SW	– Black		

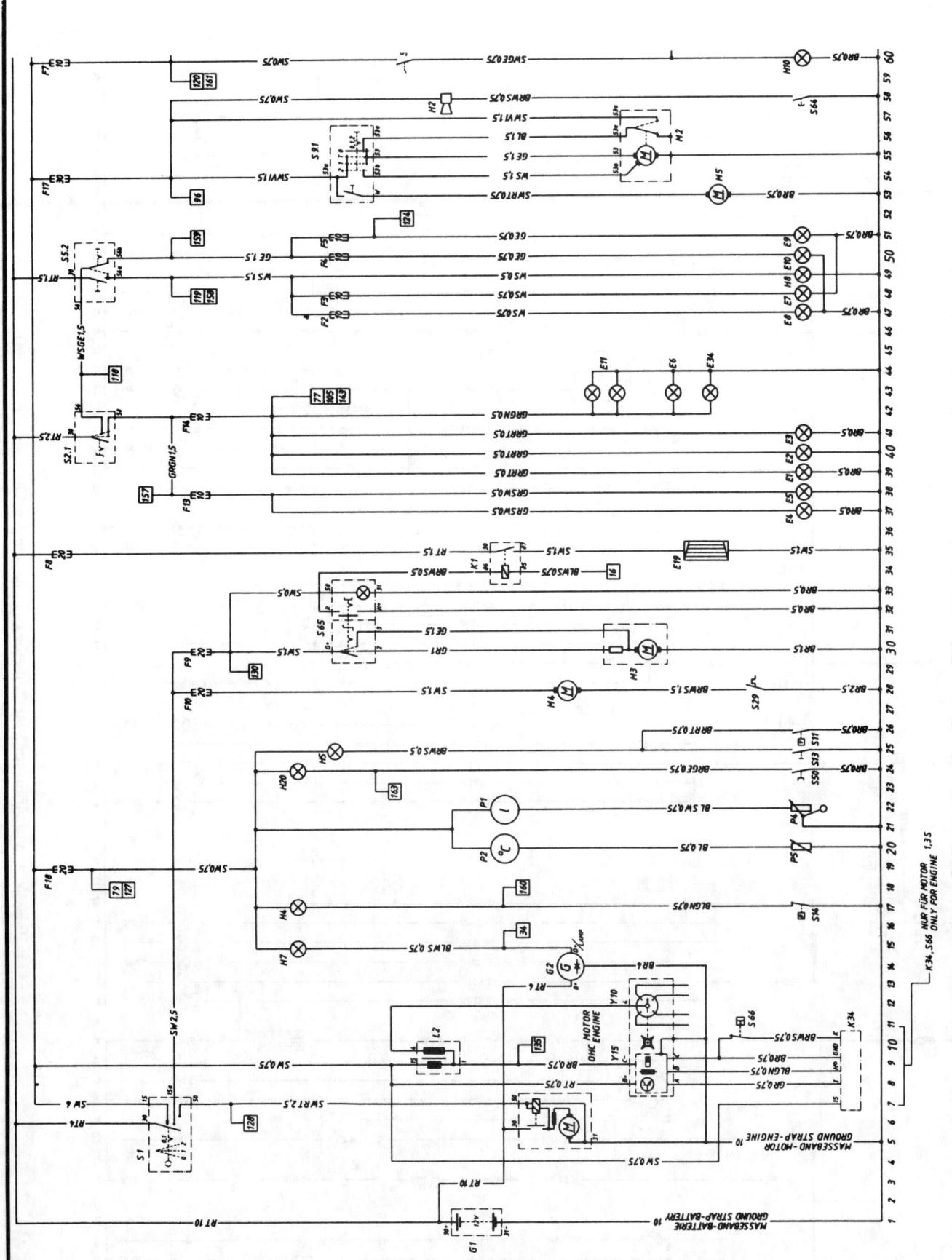

Fig. 10.30 Wiring diagram for Vauxhall Nova – 1983 and 1984 models

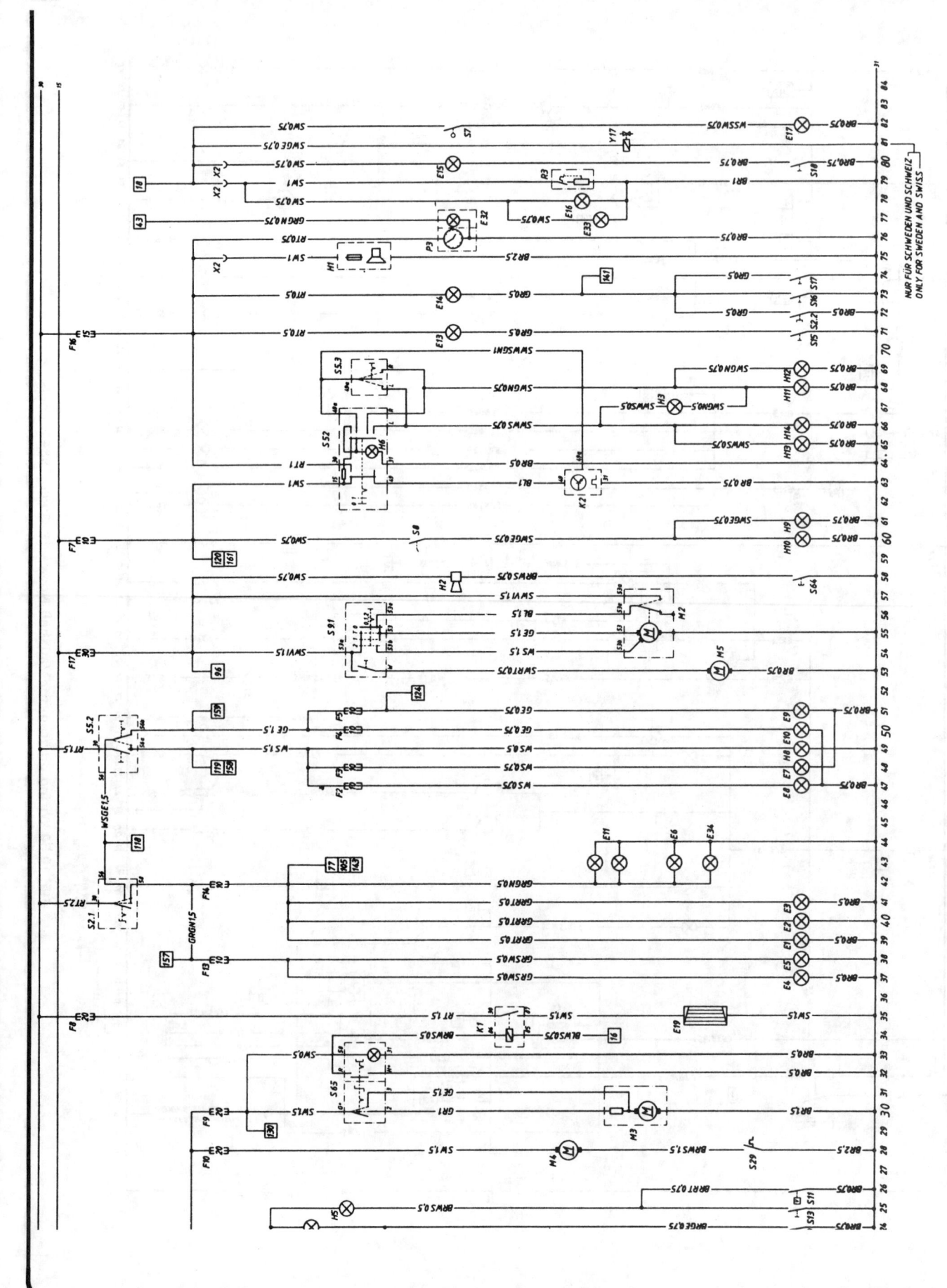

Fig. 10.30 Wiring diagram for Vauxhall Nova – 1983 and 1984 models (continued)

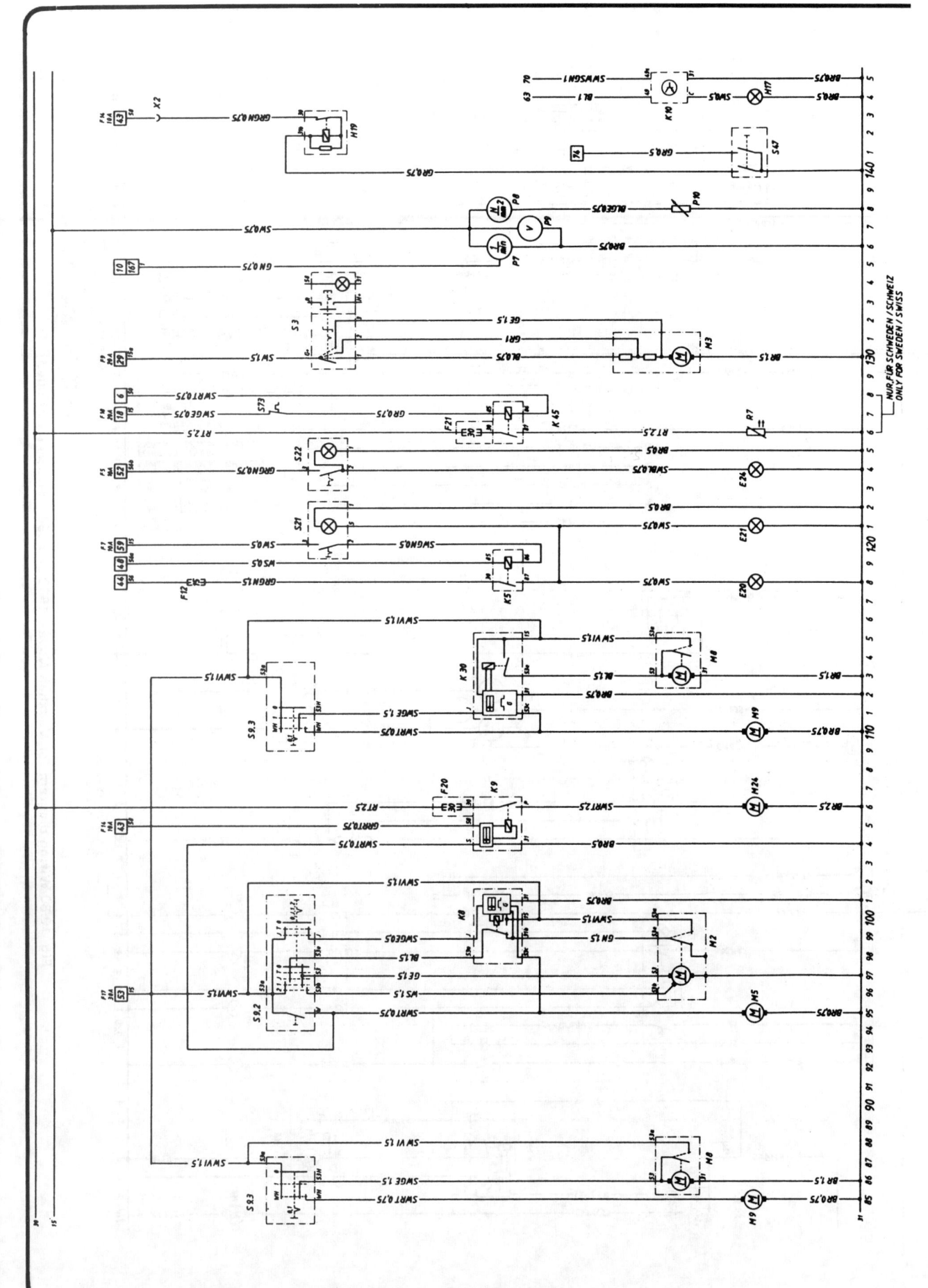

Fig. 10.30 Wiring diagram for Vauxhall Nova – 1983 and 1984 models (continued)

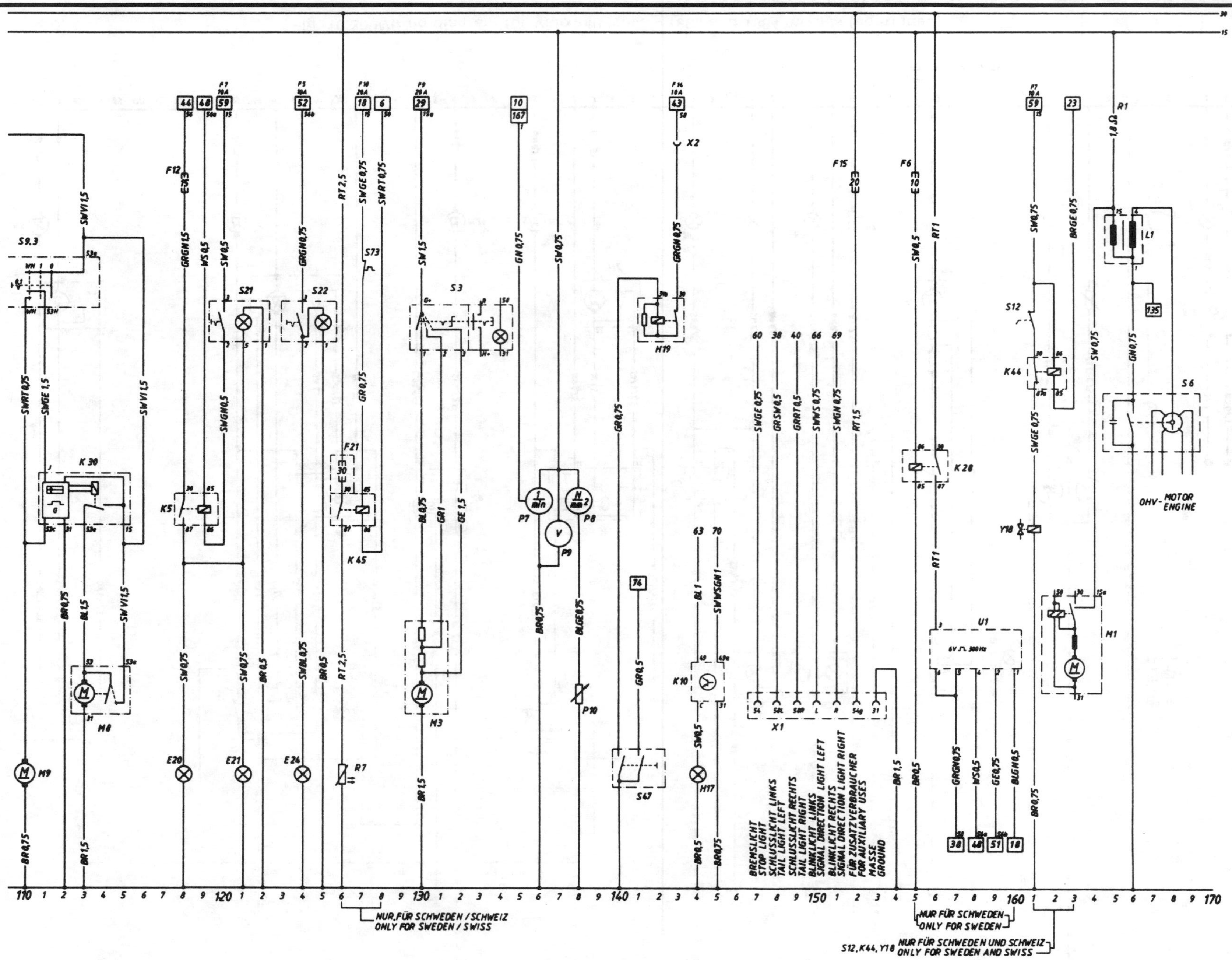

192

Fig. 10.30 Wiring diagram for Vauxhall Nova – 1983 and 1984 models (continued)

Chapter 11 Suspension

For modifications, and information applicable to later models, see Supplement at end of manual

Contents

Control arm balljoint – removal and refitting	8
Fault diagnosis – suspension	21
Front anti-roll bar – removal and refitting	10
Front hub bearings – removal and refitting	6
Front suspension control arm – removal and refitting	7
Front suspension strut – dismantling and reassembly	4
Front suspension strut – removal and refitting	3
Front suspension tie-bar – removal and refitting	9
General description	1
Maintenance and inspection	2
Rear anti-roll bar – removal and refitting	16
Rear axle – removal and refitting	17
Rear axle mounting/pivot bushes – removal and refitting	18
Rear coil spring – removal and refitting	14
Rear hub bearings – adjustment	11
Rear hub bearings – removal and refitting	12
Rear shock absorbers – removal and refitting	13
Rear suspension stub axle – removal and refitting	15
Steering knuckle – removal and refitting	5
Suspension geometry and wheel alignment – general	19
Wheels and tyres	20

Specifications

Front suspension
Type	Independent by MacPherson struts with coil springs and integral telescopic shock absorbers. Anti-roll bar on 1.2 and 1.3 models
Camber (laden)*	0°30′ negative to 1° positive
Maximum deviation from side to side	1°
Castor (laden)*	0°30′ positive to 2°30′ positive
Maximum deviation from side to side	1°

Rear suspension
Type	Semi-independent by trailing arms and transverse torsion member with coil springs and telescopic shock absorbers. Anti-roll bar on 1.2 and 1.3 models
Camber (laden)*	0° to -1°
Maximum deviation from side to side	0°30′
Toe setting	0 to 6 mm (0 to 0.2 in) toe-in

Laden indicates a vehicle containing two front seat occupants and a half-filled fuel tank

Roadwheels
Size:
Pressed steel type	4½J x 13
Aluminium alloy type	4½J x 14
Maximum permissible wheel rim run-out	1.0 mm (0.039 in)

Tyres
Type	Steel-braced radial ply
Tyre size (according to model)	135 SR 13
	145 SR 13
	155/70 SR 13
	175/70 SR 13
	165/65 SR 14

Tyre pressures (cold)

	bar	lbf/in²
135 SR 13:		
Up to 3 persons:		
Front	1.9	27
Rear	1.7	24
Fully loaded:		
Front	2.1	30
Rear	2.6	37

Chapter 11 Suspension

Tyre pressures (cold) – continued

	bar	lbf/in²
145 SR 13:		
Up to 3 persons:		
Front	1.6	23
Rear	1.6	23
Fully loaded:		
Front	1.8	26
Rear	2.4	34
155/70 SR 13:		
Up to 3 persons:		
Front	1.7	24
Rear	1.7	24
Fully loaded:		
Front	1.9	27
Rear	2.4	34
175/70 SR 13:		
Up to 3 persons:		
Front	1.5	22
Rear	1.5	22
Fully loaded:		
Front	1.7	24
Rear	2.3	33
165/65 SR 14:		
Up to 3 persons:		
Front	1.7	24
Rear	1.7	24
Fully loaded:		
Front	1.9	27
Rear	2.4	34

Torque wrench settings

	Nm	lbf ft
Front suspension		
Control arm mounting/pivot bolt nut	75	55
Tie-bar and balljoint-to-control arm nuts	100	74
Control arm balljoint-to-steering knuckle clamp bolt	30	22
Suspension strut-to-steering knuckle bolts	110	81
Tie-bar-to-front mounting bracket nut	90	66
Tie-bar front mounting bracket bolts	75	55
Anti-roll bar clamp nuts	20	15
Suspension strut upper mounting-to-turret nuts	30	22
Suspension strut piston rod nut	55	41
Rear suspension		
Rear axle mounting/pivot bolt nuts	105	77
Shock absorber lower mounting bolt	60	44
Anti-roll bar retaining bolts	80	59
Stub axle retaining bolts:		
Stage 1	60	44
Stage 2	Tighten further through 30°	Tighten further through 30°
Roadwheels		
Roadwheel bolts	90	66

1 General description

The independent front suspension is of the MacPherson strut type with coil springs and telescopic hydraulic shock absorbers. Each strut assembly consists of two parts, the upper part being a hollow tube into which the integral shock absorber locates, the coil spring and the strut upper mounting assembly. The lower part of the strut consists of the steering knuckle, which houses the front hub bearings and the hub flange. The two parts of the strut are bolted together and may be removed from the car independently of each other.

Lateral movement of each MacPherson strut is governed by a pressed-steel control arm. The control arm is supported in a rubber bush at its inner end and is attached to the steering knuckle at its outer end via a balljoint.

Fore-and-aft movement of each suspension assembly is controlled by a tie-bar bolted to the control arm at one end and supported in a rubber mounting at the other.

An eccentrically positioned anti-roll bar is fitted to 1.2 and 1.3 models.

The rear suspension is of semi-independent type consisting of trailing arms, connected by a transverse torsion member. Springing and damping of the rear suspension is by mini-block coil springs and telescopic hydraulic shock absorbers. A rear anti-roll bar is also fitted to 1.2 and 1.3 models.

Pressed-steel roadwheels and radial ply tyres are fitted as standard, with alloy sports roadwheels and low profile radial ply tyres available as options on SR versions.

2 Maintenance and inspection

1 Every 9000 miles (15 000 km) or 6 months, whichever comes first, a thorough inspection of the front and rear suspension components should be carried out using the following procedure as a guide.

Front suspension

2 Jack up the front of the car and securely support it on axle stands.
3 Visually inspect the control arm balljoint dust covers for splits or deterioration and renew the balljoint assembly, as described in Section 8, if any damage is apparent.
4 Grasp the roadwheel at the 12 o'clock and 6 o'clock positions and try to rock it. Very slight free play may be felt, but if the movement is

195

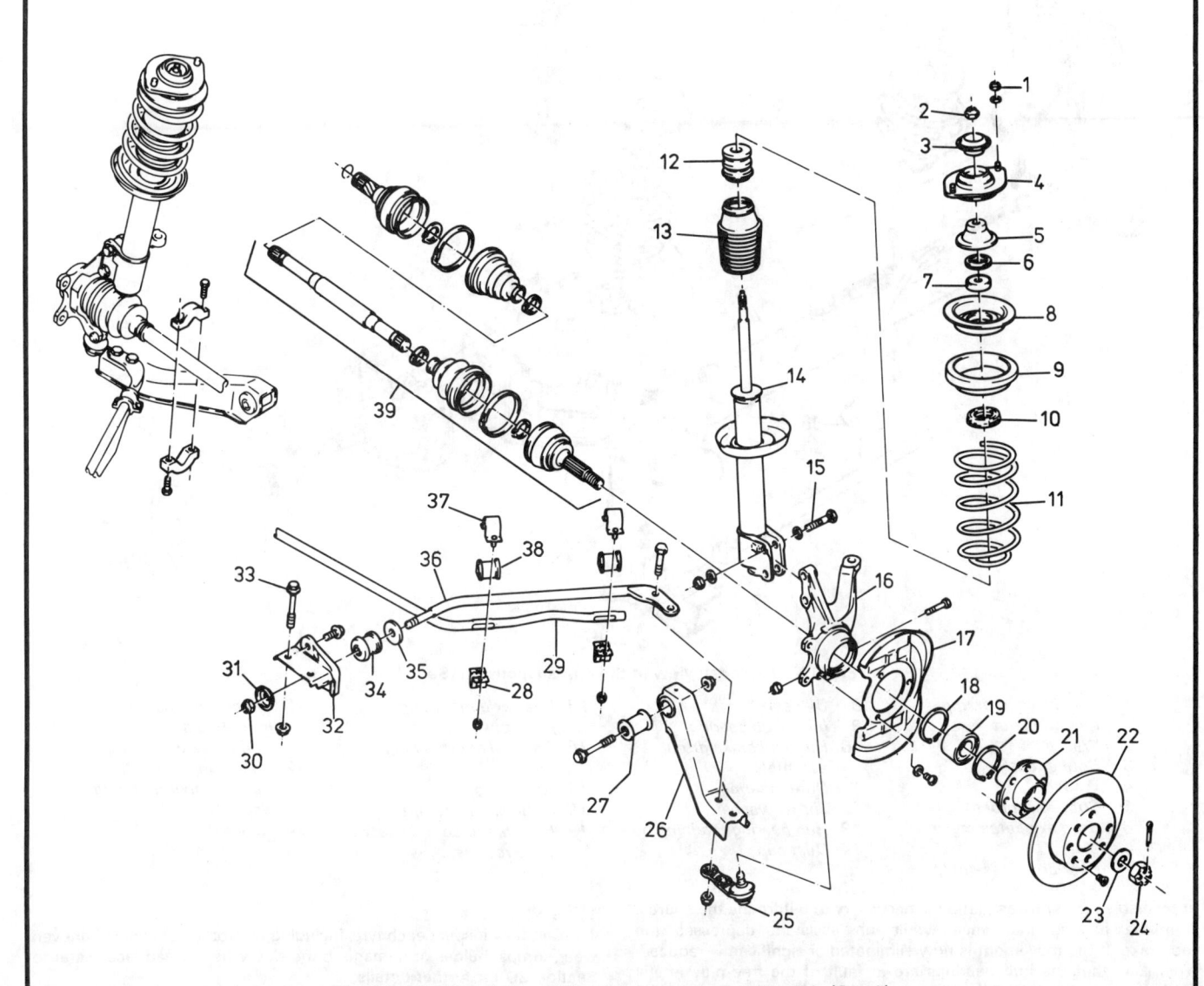

Fig. 11.1 Exploded view of the front suspension assembly (Sec 1)

1 Upper mounting retaining nut
2 Piston rod retaining nut
3 Support bearing nut
4 Upper mounting support bearing
5 Spacer plate
6 Seal ring
7 Ball-bearing
8 Upper spring plate
9 Damper pad
10 Support ring
11 Coil spring
12 Bump stop
13 Rubber gaiter
14 Suspension strut
15 Strut-to-steering knuckle retaining bolt
16 Steering knuckle
17 Disc shield
18 Inner circlip
19 Hub bearing
20 Outer circlip
21 Front hub flange
22 Brake disc
23 Thrust washer
24 Driveshaft retaining nut
25 Control arm balljoint
26 Control arm
27 Inner rubber mounting bush
28 Anti-roll bar lower clamp
29 Anti-roll bar
30 Tie-bar retaining nut
31 Thrust washer
32 Tie-bar bracket
33 Tie-bar bracket retaining bolt
34 Tie-bar front mounting bush
35 Thrust washer
36 Tie-bar
37 Anti-roll bar upper clamp
38 Anti-roll bar bush
39 Driveshaft assembly

Chapter 11 Suspension

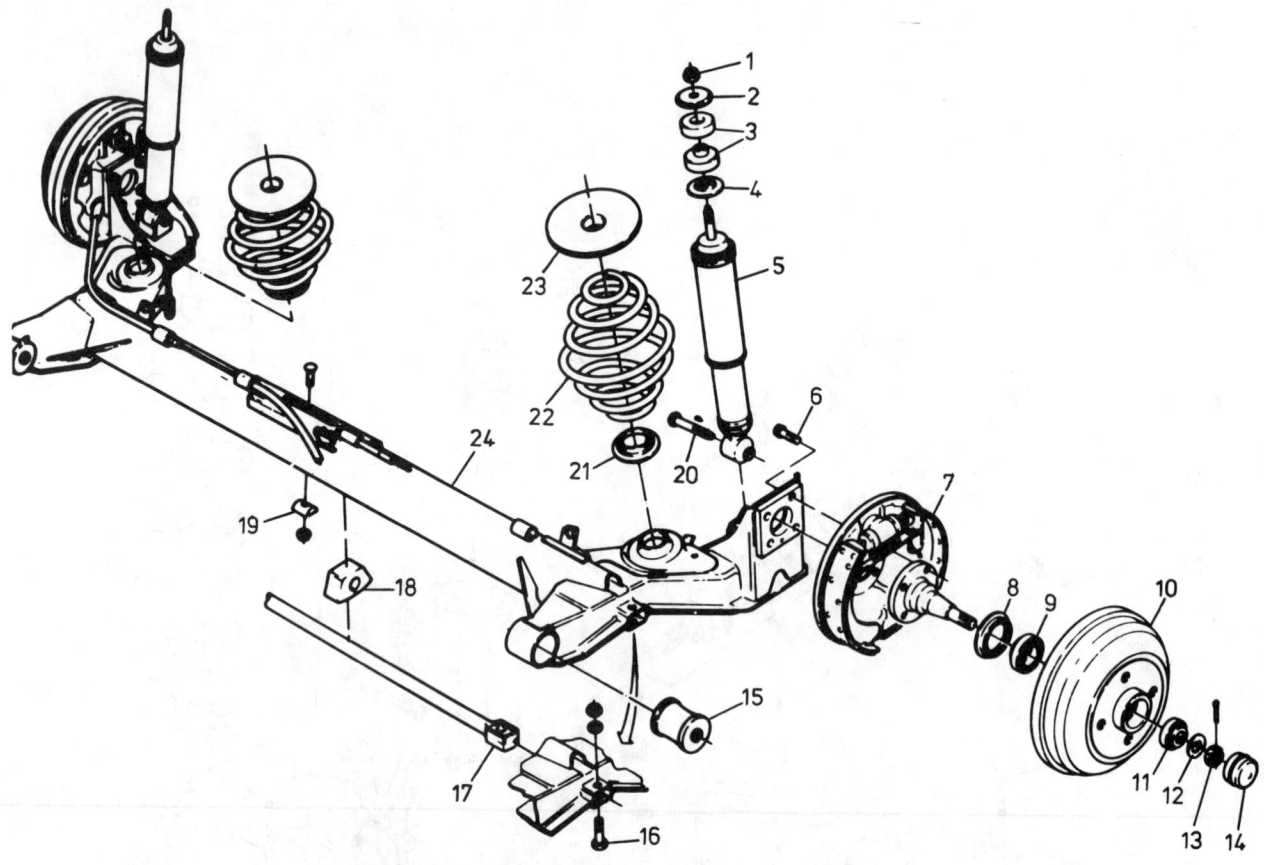

Fig. 11.2 Exploded view of the rear suspension (Sec 1)

1. Shock absorber upper mounting nut
2. Washer
3. Rubber cushions
4. Washer
5. Shock absorber
6. Stub axle retaining bolts
7. Rear brake assembly
8. Oil seal
9. Inner hub bearing
10. Hub and brake drum assembly
11. Outer bearing
12. Thrust washer
13. Hub bearing retaining nut
14. Dust cap
15. Rear axle mounting/pivot bush
16. Anti-roll bar retaining bolt
17. Anti-roll bar
18. Anti-roll bar damper
19. Handbrake compensating yoke retainer
20. Shock absorber lower mounting bot
21. Spring lower seating
22. Coil spring
23. Spring upper damping ring
24. Rear axle

appreciable further investigation is necessary to determine the source. Continue rocking the wheel while an assistant depresses the footbrake. If the movement is now eliminated or significantly reduced, it is likely that the hub bearings are at fault. If the free play is still evident with the footbrake depressed, then there is wear in the suspension joints or mountings. Pay close attention to the control arm balljoint and control arm inner mounting. Renew any worn components, as described in the appropriate Sections of this Chapter.

5 Using a large screwdriver or flat bar check for wear in the anti-roll bar mountings (where fitted) and control arm inner mountings by carefully levering against these components. Some movement is to be expected as the mountings are made of rubber, but excessive wear should be obvious. Renew any bushes that are worn.

6 Visually inspect the condition of the tie-bar front mounting bush for signs of cuts, swelling or other deterioration. Renew the bush if necessary, as described in Section 9.

Rear suspension

7 Jack up the rear of the car and securely support it on axle stands. Release the handbrake.

8 Visually inspect the rear suspension components, attachments and linkages for any obvious signs of wear or damage.

9 Grasp the roadwheel at the 12 o'clock and 6 o'clock positions and try to rock it. Any excess movement here indicates incorrect adjustment or wear in the rear hub bearings. Wear may also be accompanied by a rumbling sound when the wheel is spun or a noticeable roughness if the wheel is turned slowly. Adjustment and repair procedures are described in Sections 11 and 12.

Wheels and tyres

10 Carefully inspect each tyre, including the spare, for signs of uneven wear, lumps, bulges or damage to the sidewalls or tread face. Refer to Section 20 for further details.

11 Check the condition of the wheel rims for distortion, damage and excessive run-out. Also make sure that the balance weights are secure with no obvious signs that any are missing. Check the torque of the wheel bolts and also check the tyre pressures.

Shock absorbers

12 Check for any signs of fluid leakage around the front suspension strut or rear shock absorber body or from the rubber gaiter or cover around the piston rod. Should any fluid be noticed the shock absorber or strut is defective internally and renewal is necessary.

13 The efficiency of the shock absorber may be checked by bouncing the car at each corner. Generally speaking the body will return to its normal position and stop after being depressed. If it rises and returns on a rebound, the shock absorber or suspension strut is probably suspect. Examine also the upper and lower mountings for any sign of wear. Refer to the appropriate Sections of this Chapter for renewal procedures.

3 Front suspension strut – removal and refitting

1 Remove the wheel trim, slacken the wheel bolts and jack up the front of the car. Support the car on axle stands and remove the front roadwheel.

Chapter 11 Suspension

2 Undo the nuts and withdraw the two bolts securing the strut to the steering knuckle (photo).
3 From within the engine compartment undo the two nuts and washers securing the strut upper mounting assembly to the turret (photo).
4 Release the strut from the steering knuckle and withdraw the assembly from under the wheel arch (photo).
5 Refitting is the reverse sequence to removal, bearing in mind the following points:

(a) *The bolts securing the strut to the steering knuckle must be fitted with the bolt heads to the rear and secured with new locknuts*
(b) *Ensure that all nuts and bolts are tightened to the specified torque*

4 Front suspension strut – dismantling and reassembly

Note: *Before undertaking any dismantling work on the suspension strut it is necessary to obtain a suitable tool to compress the coil spring and safely hold it in compression while the work is carried out. This tool is vital and any attempt to dismantle the strut without it may result in personal injury.*

1 Begin by removing the suspension strut, as described in the previous Section.
2 With the strut assembly on the bench, attach suitable spring

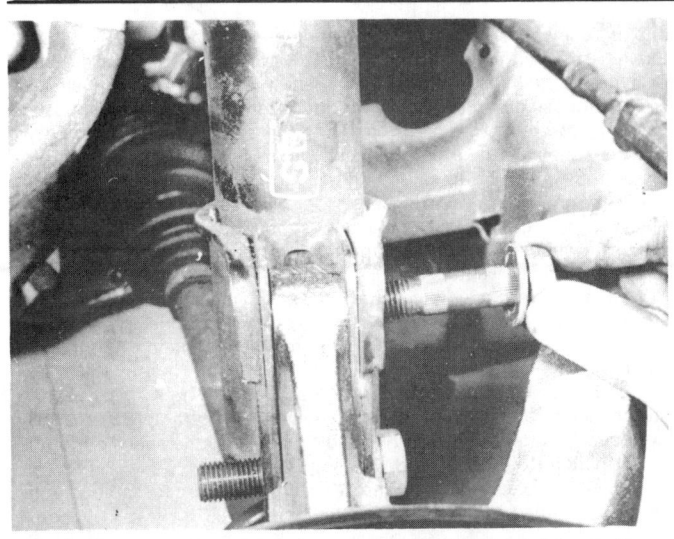

3.2 Remove the suspension strut-to-steering knuckle retaining bolts

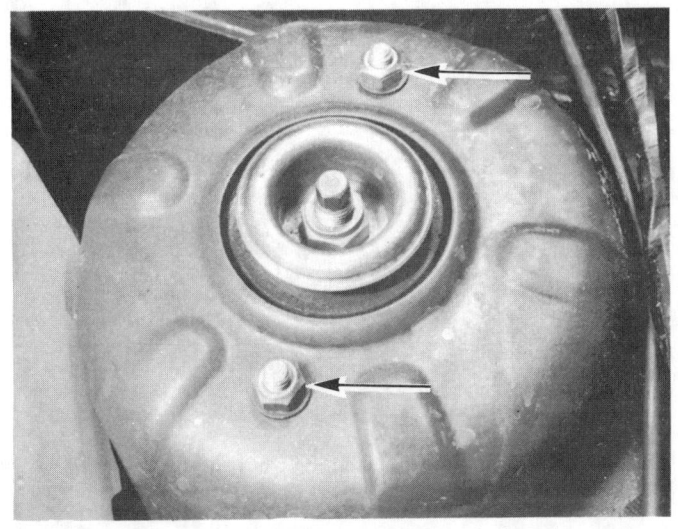

3.3 Undo the strut upper mounting retaining nuts (arrowed) ...

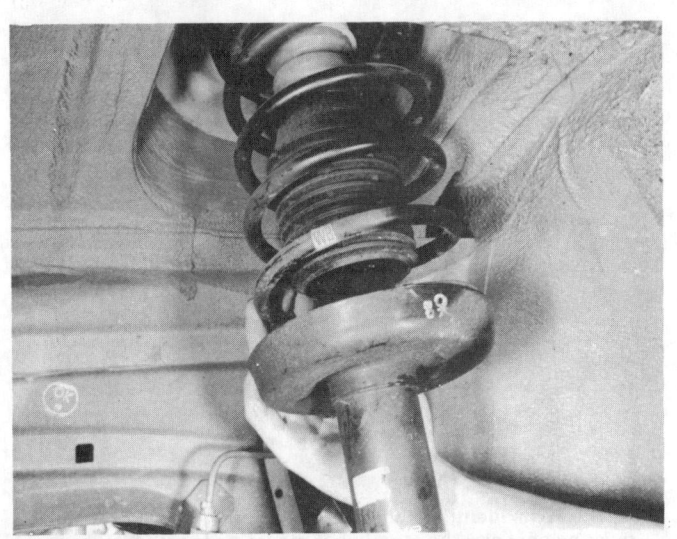

3.4 ... and withdraw the strut from under the wheel arch

Fig. 11.3 Removing the suspension strut upper mounting assembly (Sec 4)

4.2 Coil spring compressors in position on the front suspension strut

compressors to the coil spring and tighten them evenly until all tension is removed from the upper mounting assembly. Position the compressors opposite each other and make sure that they are located securely over the spring coils (photo).
3 With the spring suitably compressed, unscrew the piston rod retaining nut using a suitably cranked ring spanner, whilst at the same time holding the strut piston rod with a second spanner or socket. Remove the nut and lift off the upper mounting support bearing, support bearing stop, spacer plate and seal ring.
4 Lift off the upper spring plate, support ring and damper pad.
5 Remove the coil spring from the strut assembly leaving the compressors still attached.
6 Remove the rubber bump stop and the rubber gaiter from around the top of the strut.
7 With the suspension strut dismantled examine the components for signs of wear or damage. If the ball-bearing in the upper spring plate is in need of renewal, it can be driven out using a suitable mandrel. Tap in a new bearing using a soft-faced mallet, ensuring that the smaller internal diameter of the bearing is uppermost in relation to the installed position of the spring plate.
8 The front shock absorber is an integral part of the strut assembly and if renewal is necessary the complete strut must be obtained. Examine the unit for signs of damage to the body, distorted piston rod or hydraulic leakage and, if evident, renew the unit. Hold the strut in a vertical position and fully extend the piston rod. Apply firm pressure to the piston rod and push it back into the strut. After doing this two or three times an equal resistance should be felt over the full length of piston travel, with no trace of binding, tight spots or dead areas. If this is not the case renewal is necessary.
9 With new components obtained as required, begin reassembly by fitting the rubber gaiter and bump stop to the strut, ensuring that the gaiter is correctly seated.
10 Place the spring, with compressors still in place, in position on the strut and refit the damper pad, support ring and upper spring plate. Position the spring plate so that the small identification hole on the inner face lies opposite the strut lower mounting.
11 Refit the seal ring, spacer plate, upper mounting support bearing and the support bearing stop over the spring. With the piston rod fully extended fit a new retaining nut and tighten it to the specified torque whilst holding the piston rod.
12 Release the spring compressors and refit the strut assembly to the car, as described in the previous Section.

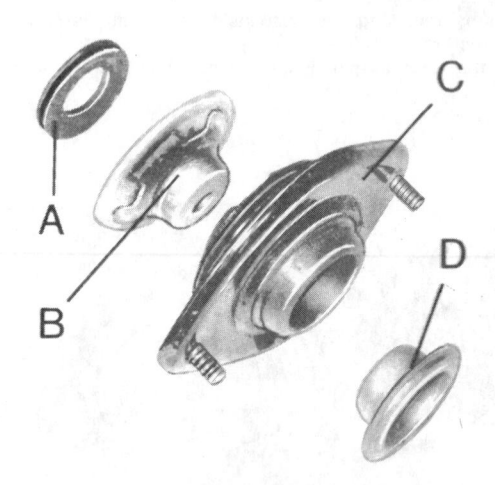

Fig. 11.4 Suspension strut upper mounting assembly components (Sec 4)

A Seal ring
B Spacer plate
C Upper mounting support bearing
D Support bearing stop

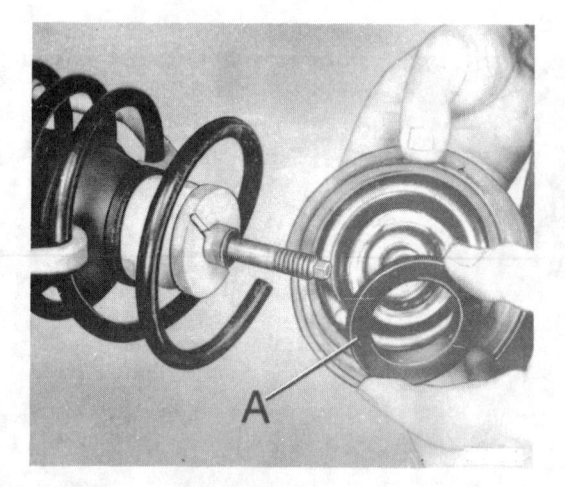

Fig. 11.5 Removing the upper spring plate and support ring (A) (Sec 4)

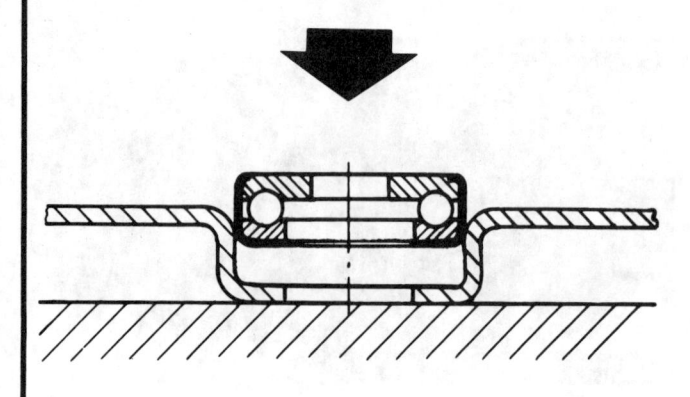

Fig. 11.6 The upper spring plate ball-bearing must be fitted with its smaller internal diameter uppermost (Sec 4)

Fig. 11.7 The identification hole in the spring plate (arrowed) must be opposite the strut lower mounting (arrowed) when fitted (Sec 4)

5.2A Remove the driveshaft retaining nut ..

5.2B ... and the thrust washer

5.5 Undo the retaining screw and lift off the brake disc

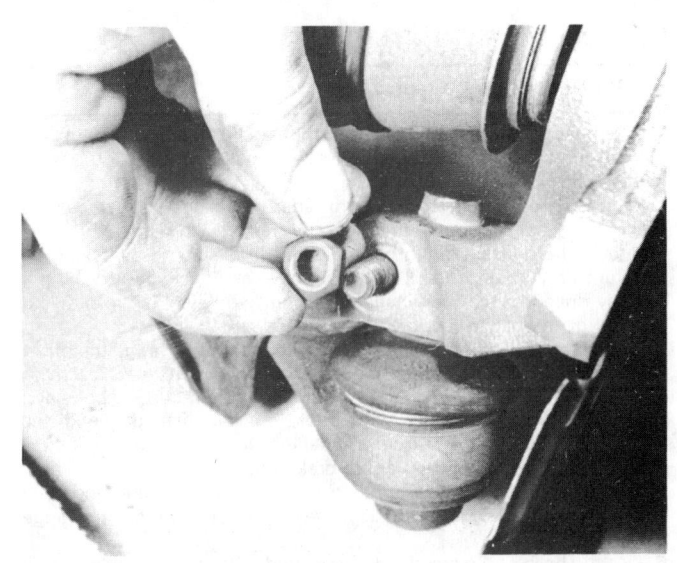

5.6 Undo the control arm balljoint retaining nut and remove the bolt

5.9 Remove the steering knuckle assembly from the balljoint, strut and driveshaft

5.17A Tighten the driveshaft retaining nut in three stages (see text) ...

5.17B ... and then secure the nut with a new split pin

Fig. 11.8 Using a chisel to release the brake caliper mounting bolt protective caps (Sec 5)

5 Steering knuckle – removal and refitting

1 Remove the wheel trim, slacken the wheel bolts and jack up the front of the car. Support the car on axle stands and remove the front roadwheel.
2 Extract the split pin from the driveshaft retaining nut, then using a socket and bar, unscrew the nut and recover the thrust washer (photos). The driveshaft can be prevented from turning by having an assistant firmly depress the footbrake, or alternatively by attaching a suitably drilled bar to the wheel hub using two wheel bolts, allowing the end of the bar to rest on an axle stand.
3 Using a small chisel, loosen the protective caps covering the brake caliper mounting bolts and then prise off the caps with a screwdriver.
4 Using an Allen key undo the two brake caliper mounting bolts and slide the caliper, complete with pads, off the disc. Suspend the caliper using string or wire from a suitable place under the wheel arch. Take care not to strain the flexible brake hose.
5 If the steering knuckle is being removed for overhaul, undo the small retaining screw and withdraw the brake disc from the hub flange (photo).
6 At the base of the steering knuckle, undo the nut and bolt securing the shank of the control arm balljoint to the knuckle (photo).
7 Undo the locknut securing the steering tie-rod outer balljoint to the steering knuckle arm. Release the joint using a universal balljoint separator.
8 Undo the two nuts and bolts securing the steering knuckle to the suspension strut.
9 Lever the control arm down to release the balljoint and then ease the steering knuckle out of its location in the suspension strut. Withdraw the steering knuckle from the driveshaft constant velocity joint and remove it from the car (photo). If the constant velocity joint is tight, tap it out with a soft-faced mallet, or in extreme cases use a hub puller.
10 To refit the steering knuckle, engage the constant velocity joint with the hub flange, locate the knuckle over the control arm balljoint shank, and position the upper part of the knuckle in the suspension strut.
11 Refit the suspension strut-to-steering knuckle retaining bolts, washers and new locknuts. Make sure that the bolts are fitted with their heads towards the rear of the car and tighten the nuts to the specified torque.
12 Refit the nut and bolt securing the control arm balljoint to the steering knuckle with the bolt head also to the rear and with the nut tightened to the specified torque.
13 Fit a new driveshaft retaining nut and thrust washer, but only tighten the nut finger tight at this stage.
14 Engage the steering tie-rod balljoint with the steering knuckle arm, refit the nut and tighten to the specified torque.
15 If the brake disc was previously removed, place it in position on the hub flange and refit the small retaining screw.
16 Locate the brake caliper over the disc and refit the two retaining bolts after applying a small quantity of thread locking compound. Tighten the bolts to the specified torque (see Chapter 9). Refit the protective caps to the bolt heads.
17 Refer to the Specifications of Chapter 7 and tighten the driveshaft retaining nut to the Stage 1 setting to seat the components. Back off the nut and retighten it to the Stage 2 setting. Finally tighten the nut by a further 90° and secure with a new split pin (photos). If the split pin hole in the constant velocity joint does not align with a slot in the nut, turn the nut back until the next slot is in alignment. *Do not tighten to align.*
18 Refit the roadwheel and lower the car to the ground.

6 Front hub bearings – removal and refitting

1 Remove the steering knuckle, as described in the previous Section.
2 If the brake disc is still in position, unscrew the small retaining screw and lift the disc off the hub flange.
3 The hub should now be removed from the steering knuckle using one of two methods. Either use a press or screw two roadwheel bolts into the hub flange and, using progressively thicker packing pieces, tighten the bolts to force off the hub. Note that whatever method is used it is quite likely that the bearing will come apart as the hub is removed, with the bearing inner race remaining in position on the hub. The bearing must be discarded if this happens and a new item obtained. *Under no circumstances should the bearing be reassembled and used again if it comes apart during hub removal.*
4 With the hub removed, undo the three small screws and lift off the disc shield.
5 Using circlip pliers extract the two bearing retaining circlips.
6 The bearing can now be removed from the steering knuckle by driving it out using a tube of suitable diameter in contact with the bearing outer race, or preferably by using a press and suitable mandrel.
7 If the bearing came apart during removal of the hub, remove the bearing inner race from the hub using a two or three-legged puller. A U-shaped exhaust clamp which is a snug fit in the groove of the inner race will provide useful purchase for the puller legs if the race is tight.
8 Before fitting the new bearing, insert the outer circlip into its groove in the steering knuckle. Make sure that the circlip is positively located and positioned so that the tabs are towards the bottom of the steering knuckle.
9 Press or drive in the new bearing until it contacts the circlip by applying pressure to the outer race.
10 With the bearing in place, fit the inner circlip with its tabs towards the bottom of the knuckle (photo).
11 Refit the disc shield and secure it with the three retaining screws.
12 Press or drive the hub into the bearing, ensuring that the inner race is well supported during this operation.
13 The steering knuckle can now be refitted to the car, as described in the previous Section.

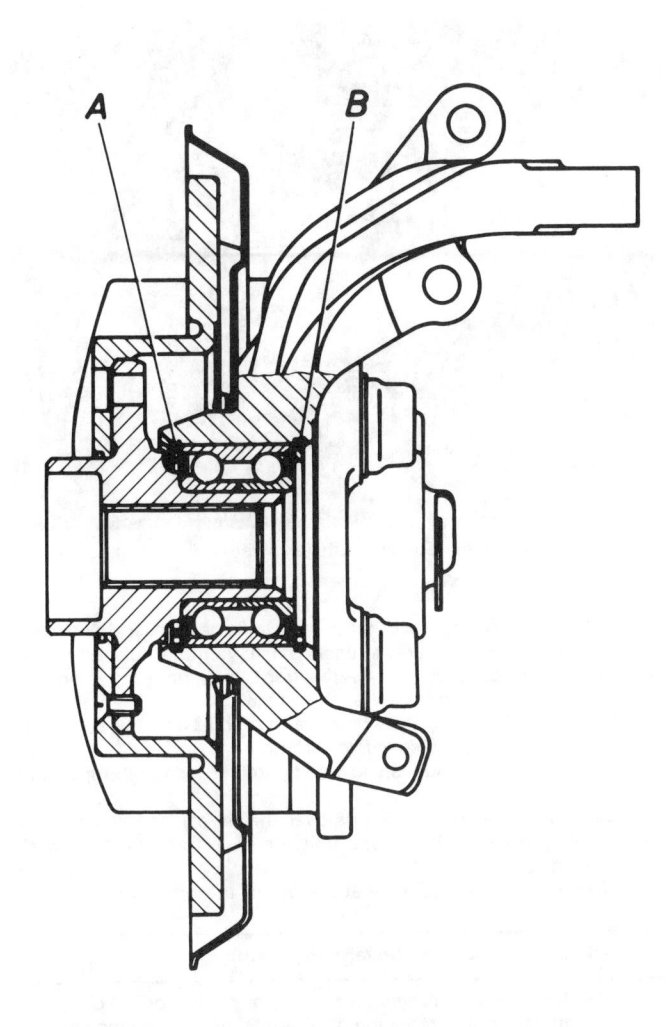

Chapter 11 Suspension

Fig. 11.9 Sectional view of the steering knuckle and hub assembly (Sec 6)

A and B are bearing retaining circlip locations

7.2 Control arm balljoint retaining nuts (arrowed)

6.10 The tabs of the bearing retaining circlips (arrowed) must be towards the bottom of the steering knuckle

7 Front suspension control arm – removal and refitting

1 Remove the wheel trim, slacken the wheel bolts and jack up the front of the car. Support the car on axle stands and remove the roadwheel.
2 Undo the two nuts and remove the bolts securing the control arm balljoint to the arm (photo).
3 At the inner end of the control arm, undo the mounting/pivot bolt retaining nut and withdraw the bolt from the control arm and mounting bracket.
4 Ease the arm from its inner mounting and remove it from under the car.
5 If necessary the inner rubber mounting bush may be renewed using a press or wide opening vice, suitable tubes and mandrels. The bushes are removed towards the rear facing side of the arm and refitted towards the front facing side. Lubricate the new bushes with soapy water before fitting and ensure that the three raised projections in the rubber part of the bush face to the rear when installed.
6 Refitting the control arm is the reverse sequence to removal bearing in mind the following points:

   (a) The head of the inner pivot/mounting bolt must be towards the front of the car
   (b) New balljoint retaining bolts and nuts must be used and fitted with the nuts below the control arm
   (c) Tighten all nuts and bolts to the specified torque
   (d) Jack up the control arm to a near horizontal position before tightening the inner mounting/pivot bolt

8 Control arm balljoint – removal and refitting

1 Remove the wheel trim, slacken the wheel bolts and jack up the front of the car. Support the car on axle stands and remove the roadwheel.
2 Undo the two nuts and remove the bolts securing the balljoint to the control arm.
3 Undo the nut and bolt securing the balljoint shank to the steering knuckle and remove the joint from the car.
4 The balljoint is a sealed unit and cannot be repaired or overhauled. If the unit is defective due to a split or perished dust cover, or due to excessive play in the joint itself, renewal is necessary.
5 Refitting is the reverse sequence to removal, bearing in mind the following points:

   (a) The balljoint shank-to-steering knuckle retaining bolt must be fitted with its head towards the rear of the car
   (b) New balljoint-to-control arm retaining bolts and nuts must be used and fitted with the nuts below the control arm
   (c) Tighten all nuts and bolts to the specified torque

Chapter 11 Suspension

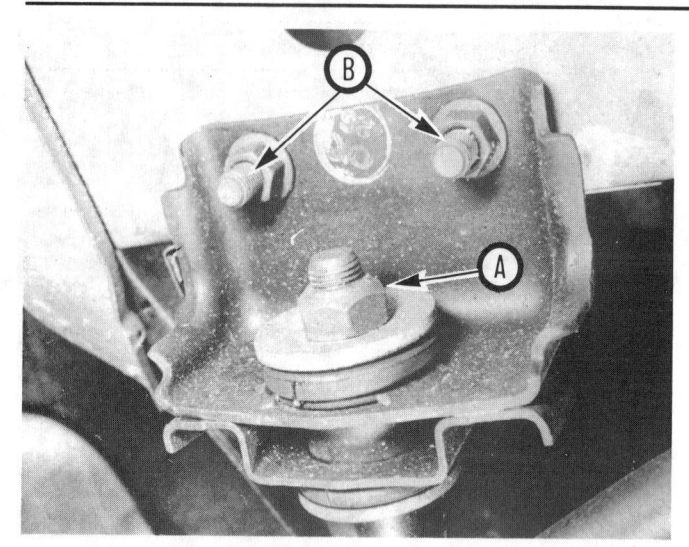

9.3 Tie-bar-to-mounting bracket retaining nut (A) and bracket-to-underbody retaining bolts (B)

10.2 Anti-roll bar clamp retaining nuts (arrowed)

9 Front suspension tie-bar – removal and refitting

1 Remove the wheel trim, slacken the wheel bolts and jack up the front of the car. Support the car on axle stands and remove the roadwheel.
2 On models fitted with an anti-roll bar, undo the clamp nuts and remove the clamps.
3 At the front of the tie-bar, undo the retaining nut and thrust washer securing the tie-bar to the mounting bracket (photo).
4 Undo the two nuts and remove the bolts securing the tie-bar and balljoint to the control arm.
5 Withdraw the tie-bar from its front mounting location and remove it from under the car. Recover the thrust washer.
6 If the tie-bar front mounting rubber bush is in need of renewal, undo the nuts and bolts and lift off the front mounting bracket.
7 Using a press, or wide opening vice, suitable tubes and mandrels, press the bush out of the bracket from rear to front. Lubricate the new bush in soapy water and press it into the bracket from front to rear. The narrow flange of the bush must be to the rear when installed, and both flanges of the bush must protrude equally.
8 Refit the mounting bracket and tighten the nuts and bolts to the specified torque.
9 Position the flat thrust washer on the tie-bar and insert the tie-bar end into the mounting bracket rubber bush. Refit the dished thrust washer with the concave side to the front, and then screw on the retaining nut – finger tight only at this stage.
10 Secure the tie-bar to the control arm using two new bolts and locknuts, tightened to the specified torque.
11 The front mounting nut can now be tightened, also to the specified torque.
12 Locate the anti-roll bar clamps over the tie-bar and anti-roll bar and secure with the two nuts and bolts on each clamp, tightened to the specified torque.
13 Refit the roadwheel and lower the car to the ground.

10 Front anti-roll bar – removal and refitting

1 Drive the car up on ramps, over a pit, or raise it on a hoist.
2 Undo the four clamp retaining nuts each side securing the anti-roll bar to the tie-bars (photo).
3 Lift off the clamps and remove the anti-roll bar from under the car.
4 If necessary remove the split rubber bushes from each tie-bar.
5 Refit the rubber bushes and then place the anti-roll bar in position with the large bow towards the right-hand side of the car.

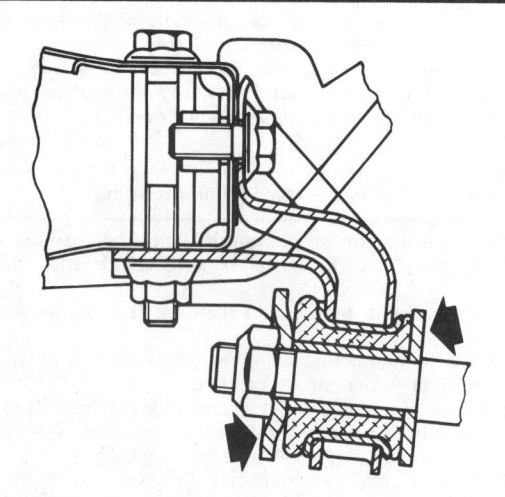

Fig. 11.10 Sectional view of the tie-bar and front mounting bracket (Sec 9)

The tie-bar bush must be fitted with its narrow flange to the rear and the thrust washer (arrowed) positioned as shown

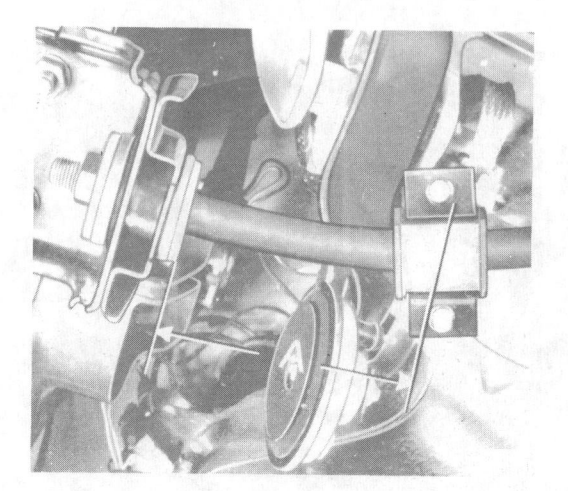

Fig. 11.11 Positioning dimension for anti-roll bar clamps (Sec 10)

$A = 115$ mm (4.53 in)

Chapter 11 Suspension

6 Secure the anti-roll bar with the clamps located over the flats in the bar and positioned so that the distance from the centre of the front clamp to the tie-bar thrust washer is as shown in Fig. 11.11.
7 Tighten the clamp retaining nuts to the specified torque.

11 Rear hub bearings – adjustment

1 Remove the wheel trim, slacken the wheel bolts and jack up the rear of the car. Support the car on axle stands and remove the roadwheel. Release the handbrake.
2 Using a stout screwdriver, prise off the dust cap from the centre of the wheel hub.
3 Extract the split pin from the bearing retaining nut on the stub axle.
4 Using a torque wrench, tighten the bearing retaining nut to 25 Nm (18 lbf ft) whilst turning the hub.
5 Slacken the nut until it is just possible to move the thrust washer, located behind the nut, using a screwdriver.
6 Insert a new split pin and bend its ends around the nut. If the slots in the nut do not align with the hole in the stub axle, tighten the nut slightly, but make sure that it is still possible to move the thrust washer. If the thrust washer now becomes tight, slacken the nut to the previous split pin hole.
7 Refit the dust cap and roadwheel, then lower the car to the ground.

12 Rear hub bearings – removal and refitting

1 Remove the wheel trim, slacken the wheel bolts and jack up the rear of the car. Support the car on axle stands and remove the roadwheel. Release the handbrake.
2 Refer to Chapter 9 if necessary and back off the rear brake adjusters.
3 Using a stout screwdriver, prise off the dust cap from the centre of the hub (photo).
4 Extract the split pin (photo) and unscrew the hub bearing retaining nut from the stub axle.
5 Withdraw the hub and brake drum assembly squarely off the stub axle. While doing this place a thumb over the bearing thrust washer to retain the thrust washer and outer bearing inner race in the hub.
6 With the hub assembly on the bench lift out the thrust washer and the outer bearing inner race.
7 Turn the hub over and prise out the oil seal using a stout screwdriver or tyre lever. Lift out the inner bearing inner race.
8 Suitably support the hub on blocks or on a vice and, with a soft metal drift, drive out the bearing outer races.
9 Thoroughly clean the centre of the hub with paraffin or a suitable solvent.
10 Before refitting the new bearings, remove any burrs that may be present in the bore of the hub with a fine file or scraper.
11 Fit the bearing outer races to the hub, with their larger internal

Fig. 11.12 Check the rear hub bearing adjustment by moving the thrustwasher with a screwdriver – arrowed (Sec 11)

Fig. 11.13 Removing the rear hub and brake drum assembly (Sec 12)

Fig. 11.14 The hub inner oil seal can be removed by prising it out with a screwdriver (Sec 12)

204   Chapter 11 Suspension

12.3 Prise off the rear hub dust cap using a screwdriver

12.4 Extract the split pin and undo the bearing retaining nut

12.14A With the hub assembly refitted, fit the outer bearing inner race ...

12.14B ... followed by the thrust washer ...

12.14C ... and bearing retaining nut

diameter facing outward. Press or tap the races into position using a tube of suitable diameter. Take care to keep the races square as they are installed otherwise they may jam in the hub bore and crack.
12 Pack the bearing inner races with a lithium based grease, and also apply a liberal quantity of grease to the space in the hub between the bearings.
13 Place the inner bearing inner race in position, lubricate the sealing lip of a new oil seal and tap the seal squarely into place.
14 Fit the hub assembly to the stub axle, followed by the outer bearing inner race, the thrust washer and retaining nut (photos).
15 Adjust the hub bearings, as described in the previous Section, and then adjust the brakes, as described in Chapter 9.
16 Refit the roadwheel and lower the car to the ground.

13 Rear shock absorber – removal and refitting

1   Drive the rear of the car up on ramps, over a pit, or raise it on a hoist.
2   Open the boot lid or tailgate and remove the protective cover over the shock absorber upper mounting.
3   Undo the upper mounting nut and lift off the washer and rubber cushion (photo).
4   Working under the car, undo the bolt securing the shock absorber lower mounting to the bracket on the rear axle trailing arm (photo).

Chapter 11 Suspension

5 Ease the lower mounting out of the bracket, lift the shock absorber clear of the trailing arm and remove it from the car. Remove the remaining flat washer and rubber cushion from the upper mounting.
6 Mount the unit vertically between vice jaws so that it is held by its lower mounting eye.
7 Fully extend and contract the unit two or three times over its full length of travel, by which time an equal resistance should be felt with no trace of binding, tight spots or dead areas. If this is not the case, or if there is any sign of fluid leakage or damage to the shock absorber body, renew the unit.
8 To refit the shock absorber place the washer and rubber cushion in position and insert the shock absorber upper mounting into its location.
9 Engage the lower mounting with the trailing arm bracket, refit the bolt and tighten it to the specified torque.
10 From inside the luggage compartment, refit the upper rubber cushion, flat washer and retaining nut. Tighten the nut until the dimension shown in Fig. 11.15 is obtained. Refit the protective cover over the upper mounting on completion.

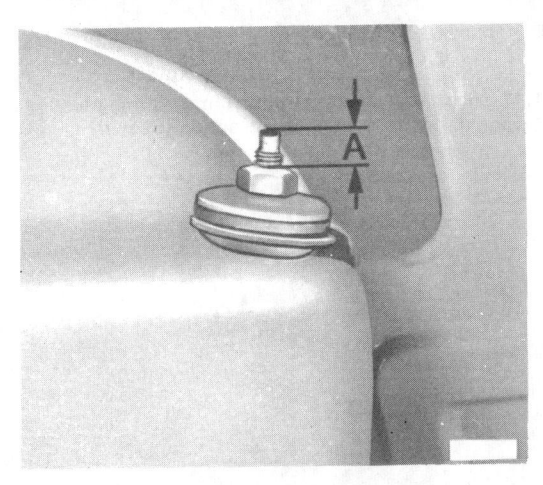

Fig. 11.15 Shock absorber upper mounting setting dimension (Sec 13)

A = 9 mm (0.35 in)

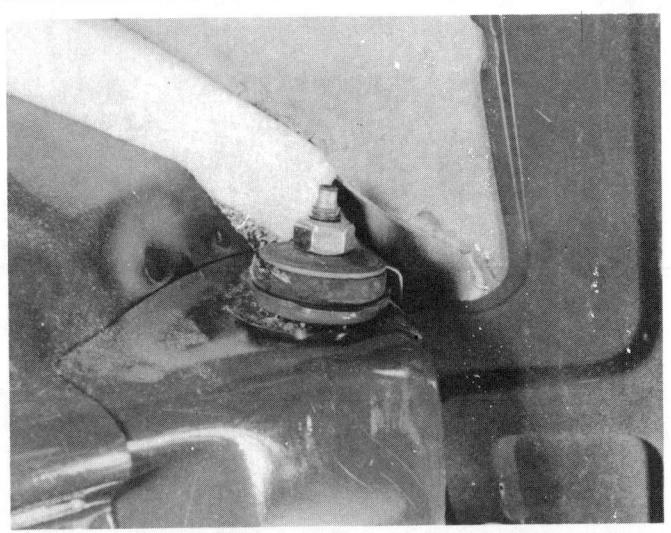

13.3 Shock absorber upper mounting nut, washer and rubber cushion

13.4 Shock absorber lower mounting bolt

14 Rear coil spring – removal and refitting

1 Jack up the rear of the car and support it on axle stands.
2 Place a jack beneath the trailing arm on the left-hand side of the car and raise the trailing arm slightly.
3 Undo the bolt securing the left-hand shock absorber lower mounting to the trailing arm bracket and ease the shock absorber out of the bracket.
4 Repeat the procedures in paragraphs 2 and 3 on the right-hand shock absorber.
5 Lower the jack and then pull down on the trailing arm until sufficient clearance exists to enable the spring to be lifted out.
6 Refitting the coil spring is the reverse sequence to removal, but make sure that the spring upper damping ring is in place and that the spring seats properly in the ring and trailing arm. Tighten the shock absorber mountings to the specified torque.

15 Rear suspension stub axle – removal and refitting

1 Remove the rear hub assembly, as described in Section 12, paragraphs 1 to 5 inclusive.
2 Place a jack beneath the trailing arm of the rear axle and raise the jack slightly.
3 Undo the bolt securing the shock absorber lower mounting to the trailing arm bracket. Ease the shock absorber out of the bracket to provide access to the stub axle retaining bolts.
4 Using a suitable Allen key or socket bit, undo the stub axle retaining bolts and withdraw the stub axle from the trailing arm.
5 Before refitting it will be necessary to obtain four new stub axle retaining bolts. The microencapsulated bolts are treated with a thread locking compound which becomes active once the bolt is fitted. After removal the bolts must not be re-used. Note also that the bolts have a shelf life of only three years and it is therefore necessary to inspect the date code stamped on the side of the bolt head to ensure that the bolts are still effective. The date code digit shown in Fig. 11.19 indicates the year of manufacture.
6 If the original stub axle is to be refitted it will be necessary to clear the retaining bolt threads of old sealant using an M10 x1.25 tap.
7 Refitting the stub axle is the reverse sequence to removal, bearing in mind the following points:

  (a) Tighten the stub axle retaining bolts in two stages as shown in the Specifications
  (b) Tighten the shock absorber lower mounting bolt to the specified torque
  (c) Refit the rear hub, as described in Section 12, and adjust the hub bearings, as described in Section 11

16 Rear anti-roll bar – removal and refitting

1 Remove the wheel trim, slacken the wheel bolts and jack up the rear of the car. Support the car on axle stands and remove one rear roadwheel.
2 Undo the retaining bolt securing the anti-roll bar to the underside of each rear axle trailing arm.

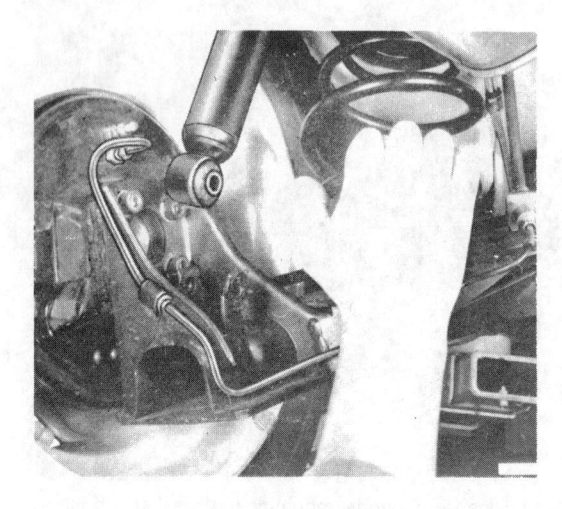

Fig. 11.16 Removing a rear coil spring (Sec 14)

Fig. 11.17 The rear coil spring upper damping ring must be in position before refitting the spring (Sec 14)

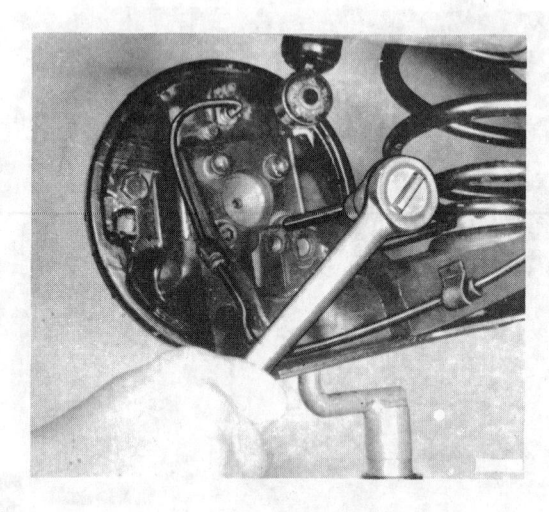

Fig. 11.18 Using a socket and Allen key bit to remove the stub axle retaining bolts (Sec 15)

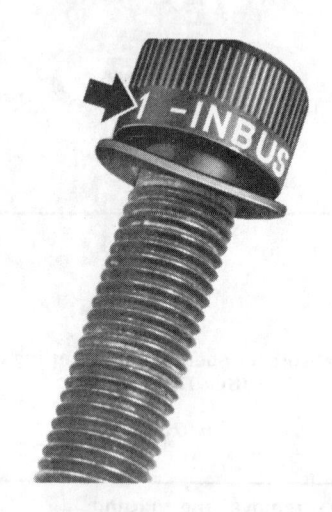

Fig. 11.19 The date code on the stub axle retaining bolts is stamped on the bolt head. The first digit (arrowed) indicates the year of the manufacture (Sec 15)

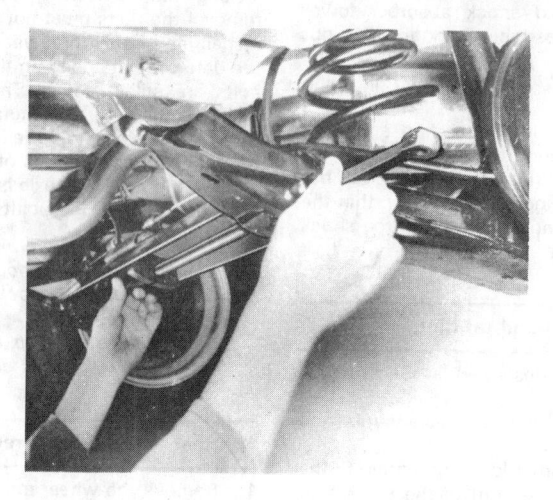

Fig. 11.20 Removing the rear anti-roll bar (Sec 16)

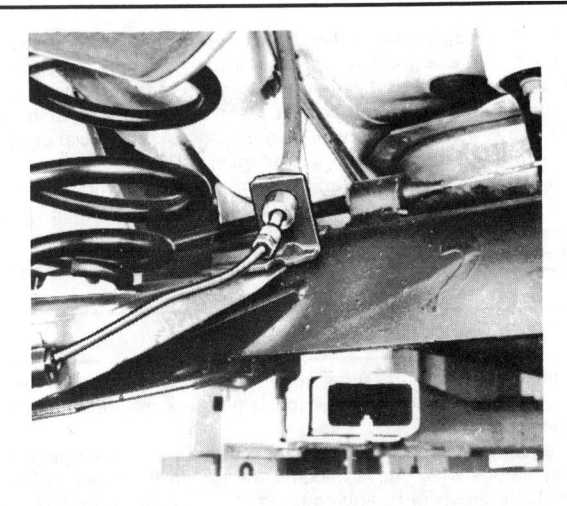

Fig. 11.21 Rear brake pipe disconnected at axle support bracket (Sec 17)

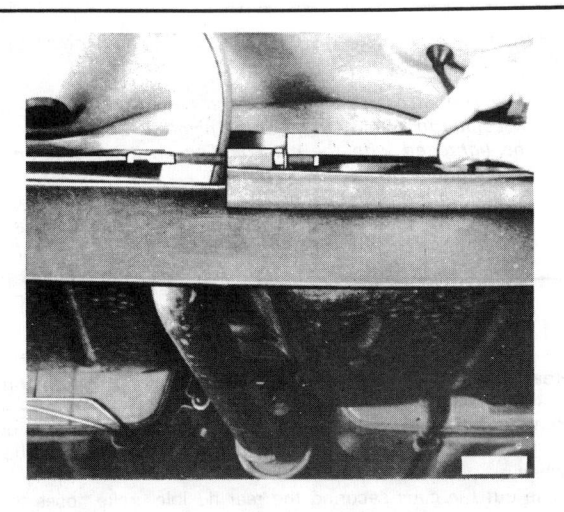

Fig. 11.22 Measure and record and handbrake cable exposed length before removal (Sec 17)

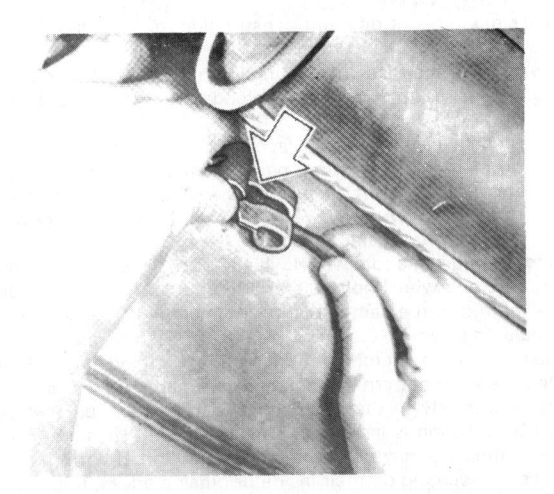

Fig. 11.23 Removing the handbrake cable from the connecting link (Sec 17)

17.14 Rear axle mounting/pivot

3 Slide the anti-roll bar sideways towards the side from which the roadwheel has been removed and withdraw the bar from the trailing arms and dampers.
4 Refitting is the reverse sequence to removal, but lubricate the anti-roll bar ends with medium grease to facilitate installation. Tighten the retaining bolts to the specified torque.

17 Rear axle – removal and refitting

1 Remove the wheel trim, slacken the wheel bolts and jack up the rear of the car. Support the car on axle stands and remove the rear roadwheels.
2 Using a brake hose clamp or self-locking wrench with suitably protected jaws, clamp the left-hand rear flexible brake hose located above the rear axle.
3 Wipe clean the brake hose to left-hand rear axle brake pipe union and then unscrew the union nut from the hose. Plug the pipe and hose ends to reduce brake fluid loss and prevent dirt entry.
4 Prise out the clip securing the brake hose to the support bracket on the axle and remove the hose from the bracket.
5 Repeat paragraphs 2 to 4 on the right-hand brake hose.
6 Measure and record the length of exposed thread protruding through the handbrake cable adjusting nut at the compensating yoke on the rear axle.
7 Hold the cable with pliers, unscrew the adjusting nut and remove

the cable end from the yoke.
8 Remove the retainer and detach the handbrake cable from the connecting link located just to the rear of the handbrake lever rod. Release the disconnected cable from the retaining clips on the underbody.
9 Place a jack under the left-hand rear axle trailing arm and raise the arm slightly.
10 Undo the bolt securing the left-hand shock absorber lower mounting to the bracket on the trailing arm. Ease the shock absorber out of the bracket and lower the jack.
11 Position the jack under the right-hand trailing arm and remove the right-hand shock absorber lower mounting in the same way.
12 With both shock absorber mountings released, pull down on the trailing arms and lift out the coil springs. Remove the damping rings located above the springs.
13 Position the jack under the centre of the axle and just take the axle weight.
14 Undo the nuts and remove both rear axle mounting/pivot bolts (photo).
15 Lower the jack slowly and at the same time ease the mountings out of their locations using a screwdriver.
16 Steady the axle assembly and lower it to the ground. Withdraw the rear axle from under the car.
17 With the axle removed the suspension and brake components can be removed if necessary by referring to the relevant Sections of this Chapter and of Chapter 9.

Chapter 11 Suspension

18 Refitting the rear axle is the reverse sequence to removal, bearing in mind the following points:

(a) *Ensure that all retaining nuts and bolts are tightened to the specified torque, noting that the pivot/mounting bolts should be tightened with the car standing on its wheels and two persons occupying the front seats*
(b) *When refitting the handbrake cable, initially set the adjusting nut so that the same amount of thread protrudes as was noted during removal. After refitting adjust the cable, as described in Chapter 9*
(c) *Bleed the hydraulic system, as described in Chapter 9, on completion of the installation*

18 Rear axle mounting/pivot bushes – removal and refitting

1 Remove the wheel trim, slacken the wheel bolts and jack up the rear of the car. Support the car on axle stands and remove the rear roadwheels.
2 Prise out the clips securing the rear flexible brake hoses to their support brackets on the rear axle and underbody. Also release the rigid brake pipes from their underbody clips.
3 Slacken, but do not remove, the bolts securing the shock absorber lower mountings to the trailing arm brackets.
4 Position a jack beneath the centre of the rear axle and just take the axle weight.
5 Undo the nuts and remove the axle pivot/mounting bolts.
6 Lower the jack slightly and ease the pivot/mountings out of their locations using a screwdriver. Continue lowering the jack until the mountings are clear of the underbody. Bend the brake pipes slightly if necessary to allow the axle to be lowered, but take care not to kink them or to strain the hoses.
7 Using a tube or pipe of suitable diameter, thick flat washers or packing pieces and a long bolt and nut, draw the old bushes out of their axle locations. It may be beneficial to cut off the rubber flange of the bush using a sharp knife to enable it to be drawn into the tube more easily.
8 Refit the new bushes in the same way, but lubricate them first with a soapy water solution. Ensure that the bushes are uniformly fitted with an equal amount of flange protruding on each side.
9 Refitting the axle is now a reverse of the removal sequence, ensuring that all nuts and bolts are tightened to the specified torque. *The pivot/mounting bolts should be tightened with the car standing on its wheels and two persons occupying the front seats.*

19 Suspension geometry and wheel alignment – general

With the exception of the front wheel toe setting all the front and rear suspension angles on Nova models are set during manufacture and are not adjustable. Unless the vehicle has suffered accident damage, or there is gross wear in the suspension mountings or joints, it can be assumed that these settings are correct. If for any reason it is believed that they are not correct, the task of checking them should be left to a GM dealer who will have the necessary special equipment needed to measure the small angles involved.

Front wheel toe setting adjustment procedures and further information on front wheel alignment will be found in Chapter 8. For reference purposes the front and rear suspension angles are given in the Specifications at the beginning of this Chapter.

20 Wheels and tyres

1 Check the tyre pressures regularly (see Routine Maintenance) when the tyres are cold, and periodically check the tread depth using a depth gauge.
2 Frequently inspect the tyre walls and treads for damage and pick out any stones which have become trapped in the tread pattern.
3 In the interests of extending tread life, the wheels and tyres can be moved between front and rear on the same side of the car and the spare incorporated in the rotational pattern. If the wheels have previously been balanced on the car it will be necessary to have them rebalanced after rotation.
4 Never mix tyres of different construction, or very dissimilar tread patterns.
5 Always keep the roadwheel bolts tightened to the specified torque and if the wheel bolt holes become elongated or flattened, renew the wheel.
6 Occasionally clean the inner faces of the roadwheels and, if there is any sign of rust or corrosion, paint them with metal preservative paint. **Note:** *Corrosion on alloy wheels may be evidence of a more serious problem which could lead to wheel failure. If corrosion is evident, consult your dealer for advice.*
7 Before removing a roadwheel which has been balanced on the car, always mark one wheel bolt hole and the hub, so that the roadwheel may be refitted in the same position, to maintain the balance.
8 Should unexpected excessive wear be noticed on any of the tyres its cause must be identified and rectified immediately. Generally speaking the wear pattern can be used as a guide to the case. If a tyre is worn excessively in the centre of the tread face, but not on the edges, over inflation is indicated. Similarly if the edges are worn, but not the centre, this may be due to under inflation. If both the front or rear tyres are wearing on their inside or outside edges, this is likely to be due to incorrect toe setting. If only one tyre is exhibiting this tendency then there may be a problem with the steering geometry, a worn steering or suspension component, or a faulty tyre. Wheel and tyre imbalance is indicated by irregular and uneven wear patches appearing periodically around the tread face.

21 Fault diagnosis – suspension

Note: *Before diagnosing suspension faults, be sure that the trouble is not due to incorrect tyre pressures, a mixture of tyre types or binding brakes*

Symptom	Reason(s)
Vehicle pulls to one side	Incorrect front or rear wheel alignment Faulty front tyre Accident damage to steering or suspension components
Vehicle wanders	Excessive wear in suspension mountings, joints or components Incorrect front or rear wheel alignment Wear in steering components (see Chapter 8)
Wheel wobble or vibration	Roadwheels out of balance Roadwheels buckled Lump or bulge in tyre Excessive wear in suspension mountings, joints or components Faulty shock absorbers or front struts
Excessive pitching or rolling on corners or during braking	Faulty shock absorbers or front struts Worn anti-roll bar rubber bushes or loose mounting clamps Worn front tie-bar mounting bushes

276 Page Panel control.

1) Oil and Filter ✓
2) Brake Fluid and check the Brakes (Pads) ✓
3) Change antifreeze ✓
4) Air Filter, Spark plugs, drive belt tension condition Screen wipers?
5) Lubrication, controls, hinges, locks,

Hydraulic Fluid annually regardless of mileage. ✓

Miles 9548   Anti Freeze
6-10-94

Serviced Nooo Lux 23-11-95
New Front Brake required for next Service.

Chapter 12 Bodywork

For modifications, and information applicable to later models, see Supplement at end of manual

Contents

Bonnet – removal and refitting	6
Bonnet release cable – removal and refitting	7
Boot lid – removal and refitting	27
Boot lid lock – removal and refitting	28
Centre console – removal and refitting	33
Exterior rear view mirror – removal and refitting	32
Facia – removal and refitting	36
Front bumper – removal and refitting	9
Front door – removal and refitting	13
Front door exterior handle – removal and refitting	15
Front door fixed quarter-light – removal and refitting	20
Front door lock – removal and refitting	14
Front door lock cylinder – removal, dismantling, reassembly and refitting	17
Front door remote control handle – removal and refitting	16
Front door trim panel – removal and refitting	12
Front door window – removal and refitting	19
Front door window regulator – removal and refitting	18
Front seat – removal and refitting	29
Front wing – removal and refitting	11
General description	1
Glove compartment – removal and refitting	35
Heater – description	37
Heater components – removal and refitting	38
Maintenance – bodywork and underframe	2
Maintenance – hinges and locks	5
Maintenance – upholstery and carpets	3
Minor body damage – repair	4
Parcel shelf – removal and refitting	34
Radiator grille – removal and refitting	8
Rear bumper – removal and refitting	10
Rear fixed quarter-light – removal and refitting	21
Rear seat – removal and refitting	30
Rear window or tailgate glass – removal and refitting	26
Sunroof – operation	31
Tailgate – removal and refitting	23
Tailgate lock – removal and refitting	24
Tailgate support strut – removal and refitting	25
Vents and grilles – removal and refitting	39
Windscreen – removal and refitting	22

Specifications

General
Type .... All-steel unitary construction
Model code:
- 2-door Saloon .... 91 and 92
- 3-door Hatchback .... 93 and 94
- 4-door Saloon .... 96 and 97
- 5-door Hatchback .... 98 and 99

For vehicle dimensions and weights refer to the introductory section at the beginning of this manual

Torque wrench settings

	Nm	lbf ft
Bumper mounting bolts	12	9
Tailgate strut pivot pegs	20	15
Bonnet-to-hinge bolts	20	15
Boot lid-to-hinge bolts	20	15
Front seat U-shaped clip bolts	20	15
Seat belt anchorages	35	26

1 General description

The vehicle body structure is a welded fabrication of many individually shaped panels which form a 'monocoque' bodyshell. Certain areas are strengthened locally to provide for suspension, steering and engine anchorages and load distribution. The resultant structure is extremely strong and rigid.

The front wings are bolted in position and are detachable should renewal be necessary.

2 Maintenance – bodywork and underframe

The general condition of a vehicle's bodywork is the one thing that significantly affects its value. Maintenance is easy but needs to be regular. Neglect, particularly after minor damage, can lead quickly to further deterioration and costly repair bills. It is important also to keep watch on those parts of the vehicle not immediately visible, for instance the underside, inside all the wheel arches and the lower part of the engine compartment.

The basic maintenance routine for the bodywork is washing – preferably with a lot of water, from a hose. This will remove all the loose solids which may have stuck to the vehicle. It is important to flush these off in such a way as to prevent grit from scratching the finish. The wheel arches and underframe need washing in the same way to remove any accumulated mud which will retain moisture and tend to encourage rust. Paradoxically enough, the best time to clean the underframe and wheel arches is in wet weather when the mud is thoroughly wet and soft. In very wet weather the underframe is usually cleaned of large accumulations automatically and this is a good time for inspection.

Periodically, except on vehicles with a wax-based underbody protective coating, it is a good idea to have the whole of the underframe of the vehicle steam cleaned, engine compartment included, so that a thorough inspection can be carried out to see what minor repairs and renovations are necessary. Steam cleaning is available at many garages and is necessary for removal of the accumulation of oily grime which sometimes is allowed to become thick in certain areas. If steam cleaning facilities are not available, there are one or two excellent grease solvents available, such as Holts Engine Cleaner or Holts Foambrite, which can be brush applied. The dirt can then be simply hosed off. Note that these methods should not be used on vehicles with wax-based underbody protective coating or the coating will be removed. Such vehicles should be inspected annually, preferably just prior to winter, when the underbody should be washed down and any damage to the wax coating repaired using Holts Undershield. Ideally, a completely fresh coat should be applied. It would also be worth considering the use of such wax-based protection for injection into door panels, sills, box sections, etc, as an additional safeguard against rust damage where such protection is not provided by the vehicle manufacturer.

After washing paintwork, wipe off with a chamois leather to give an unspotted clear finish. A coat of clear protective wax polish, like the many excellent Turtle Wax polishes, will give added protection against chemical pollutants in the air. If the paintwork sheen has dulled or oxidised, use a cleaner/polisher combination such as Turtle Extra to restore the brilliance of the shine. This requires a little effort, but such dulling is usually caused because regular washing has been neglected. Care needs to be taken with metallic paintwork, as special non-abrasive cleaner/polisher is required to avoid damage to the finish. Always check that the door and ventilator opening drain holes and pipes are completely clear so that water can be drained out. Bright work should be treated in the same way as paint work. Windscreens and windows can be kept clear of the smeary film which often appears by the use of a proprietary glass cleaner like Holts Mixra. Never use any form of wax or other body or chromium polish on glass.

3 Maintenance – upholstery and carpets

Mats and carpets should be brushed or vacuum cleaned regularly to keep them free of grit. If they are badly stained remove them from the vehicle for scrubbing or sponging and make quite sure they are dry before refitting. Seats and interior trim panels can be kept clean by wiping with a damp cloth and Turtle Wax Carisma. If they do become stained (which can be more apparent on light coloured upholstery) use a little liquid detergent and a soft nail brush to scour the grime out of the grain of the material. Do not forget to keep the headlining clean in the same way as the upholstery. When using liquid cleaners inside the vehicle do not over-wet the surfaces being cleaned. Excessive damp could get into the seams and padded interior causing stains, offensive odours or even rot. If the inside of the vehicle gets wet accidentally it is worthwhile taking some trouble to dry it out properly, particularly where carpets are involved. *Do not leave oil or electric heaters inside the vehicle for this purpose.*

4 Minor body damage – repair

The colour bodywork repair photographic sequences between pages 32 and 33 illustrate the operations detailed in the following sub-sections.
Note: *For more detailed information about bodywork repair, the Haynes Publishing Group publish a book by Lindsay Porter called The Car Bodywork Repair Manual. This incorporates information on such aspects as rust treatment, painting and glass fibre repairs, as well as details on more ambitious repairs involving welding and panel beating.*

Repair of minor scratches in bodywork

If the scratch is very superficial, and does not penetrate to the metal of the bodywork, repair is very simple. Lightly rub the area of the scratch with a paintwork renovator like Turtle Wax New Color Back, or a very fine cutting paste like Holts Body + Plus Rubbing Compound to remove loose paint from the scratch and to clear the surrounding bodywork of wax polish. Rinse the area with clean water.

Apply touch-up paint, such as Holts Dupli-Color Color Touch or a paint film like Holts Autofilm, to the scratch using a fine paint brush; continue to apply fine layers of paint until the surface of the paint in the scratch is level with the surrounding paintwork. Allow the new paint at least two weeks to harden; then blend it into the surrounding paintwork by rubbing the scratch area with a paintwork renovator or a very fine cutting paste, such as Holts Body + Plus Rubbing Compound or Turtle Wax New Color Back. Finally, apply wax polish from one of the Turtle Wax range of wax polishes.

Where the scratch has penetrated right through to the metal of the bodywork, causing the metal to rust, a different repair technique is required. Remove any loose rust from the bottom of the scratch with a penknife, then apply rust inhibiting paint, such as Turtle Wax Rust Master, to prevent the formation of rust in the future. Using a rubber or nylon applicator fill the scratch with bodystopper paste like Holts Body + Plus Knifing Putty. If required, this paste can be mixed with cellulose thinners, such as Holts Body + Plus Cellulose Thinners, to provide a very thin paste which is ideal for filling narrow scratches. Before the stopper-paste in the scratch hardens, wrap a piece of smooth cotton rag around the top of a finger. Dip the finger in cellulose thinners, such as Holts Body + Plus Cellulose Thinners, and then quickly sweep it across the surface of the stopper-paste in the scratch; this will ensure that the surface of the stopper-paste is slightly hollowed. The scratch can now be painted over as described earlier in this Section.

Repair of dents in bodywork

When deep denting of the vehicle's bodywork has taken place, the first task is to pull the dent out, until the affected bodywork almost attains its original shape. There is little point in trying to restore the original shape completely, as the metal in the damaged area will have stretched on impact and cannot be reshaped fully to its original contour. It is better to bring the level of the dent up to a point which is about ⅛ in (3 mm) below the level of the surrounding bodywork. In cases where the dent is very shallow anyway, it is not worth trying to pull it out at all. If the underside of the dent is accessible, it can be hammered out gently from behind, using a mallet with a wooden or plastic head. Whilst doing this, hold a suitable block of wood firmly against the outside of the panel to absorb the impact from the hammer blows and thus prevent a large area of the bodywork from being 'belled-out'.

Should the dent be in a section of the bodywork which has a double skin or some other factor making it inaccessible from behind, a different technique is called for. Drill several small holes through the metal inside the area – particulary in the deeper section. Then screw long self-tapping screws into the holes just sufficiently for them to gain a good purchase in the metal. Now the dent can be pulled out by pulling on the protruding heads of the screws with a pair of pliers.

The next stage of the repair is the removal of the paint from the damaged area, and from an inch or so of the surrounding 'sound' bodywork. This is accomplished most easily by using a wire brush or abrasive pad on a power drill, although it can be done just as effectively by hand using sheets of abrasive paper. To complete the preparation for filling, score the surface of the bare metal with a screwdriver or the tang of a file, or alternatively, drill small holes in the affected area. This will provide a really good 'key' for the filler paste.

To complete the repair see the Section on filling and re-spraying.

Repair of rust holes or gashes in bodywork

Remove all paint from the affected area and from an inch or so of the surrounding 'sound' bodywork, using an abrasive pad or a wire brush on a power drill. If these are not available a few sheets of abrasive paper will do the job just as effectively. With the paint removed you will be able to gauge the severity of the corrosion and therefore decide whether to renew the whole panel (if this is possible) or to repair the affected area. New body panels are not as expensive as most people think and it is often quicker and more satisfactory to fit a new panel than to attempt to repair large areas of corrosion.

Remove all fittings from the affected area except those which will act as a guide to the original shape of the damaged bodywork (eg headlamp shells etc). Then, using tin snips or a hacksaw blade, remove all loose metal and any other metal badly affected by corrosion. Hammer the edges of the hole inwards in order to create a slight depression for the filler paste.

Wire brush the affected area to remove the powdery rust from the surface of the remaining metal. Paint the affected area with rust inhibiting paint like Turtle Rust Master; if the back of the rusted area is accessible treat this also.

Before filling can take place it will be necessary to block the hole in some way. This can be achieved by the use of aluminium or plastic mesh, or aluminium tape.

Aluminium or plastic mesh or glass fibre matting, such as the Holts Body + Plus Glass Fibre Matting, is probably the best material to use for a large hole. Cut a piece to the approximate size and shape of the hole to be filled, then position it in the hole so that its edges are below the level of the surrounding bodywork. It can be retained in position by several blobs of filler paste around its periphery.

Aluminium tape should be used for small or very narrow holes. Pull a piece off the roll and trim it to the approximate size and shape required, then pull off the backing paper (if used) and stick the tape over the hole; it can be overlapped if the thickness of one piece is insufficient. Burnish down the edges of the tape with the handle of a screwdriver or similar, to ensure that the tape is securely attached to the metal underneath.

Bodywork repairs – filling and re-spraying

Before using this Section, see the Sections on dent, deep scratch, rust holes and gash repairs.

Chapter 12 Bodywork

Many types of bodyfiller are available, but generally speaking those proprietary kits which contain a tin of filler paste and a tube of resin hardener are best for this type of repair, like Holts Body + Plus or Holts No Mix which can be used directly from the tube. A wide, flexible plastic or nylon applicator will be found invaluable for imparting a smooth and well contoured finish to the surface of the filler.

Mix up a little filler on a clean piece of card or board – measure the hardener carefully (follow the maker's instructions on the pack) otherwise the filler will set too rapidly or too slowly. Alternatively, Holts No Mix can be used straight from the tube without mixing, but daylight is required to cure it. Using the applicator apply the filler paste to the prepared area; draw the applicator across the surface of the filler to achieve the correct contour and to level the filler surface. As soon as a contour that approximates to the correct one is achieved, stop working the paste – if you carry on too long the paste will become sticky and begin to 'pick up' on the applicator. Continue to add thin layers of filler paste at twenty-minute intervals until the level of the filler is just proud of the surrounding bodywork.

Once the filler has hardened, excess can be removed using a metal plane or file. From then on, progressively finer grades of abrasive paper should be used, starting with a 40 grade production paper and finishing with 400 grade wet-and-dry paper. Always wrap the abrasive paper around a flat rubber, cork, or wooden block – otherwise the surface of the filler will not be completely flat. During the smoothing of the filler surface the wet-and-dry paper should be periodically rinsed in water. This will ensure that a very smooth finish is imparted to the filler at the final stage.

At this stage the 'dent' should be surrounded by a ring of bare metal, which in turn should be encircled by the finely 'feathered' edge of the good paintwork. Rinse the repair area with clean water, until all of the dust produced by the rubbing-down operation has gone.

Spray the whole repair area with a light coat of primer, either Holts Body + Plus Grey or Red Oxide Primer – this will show up any imperfections in the surface of the filler. Repair these imperfections with fresh filler paste or bodystopper, and once more smooth the surface with abrasive paper. If bodystopper is used, it can be mixed with cellulose thinners to form a really thin paste which is ideal for filling small holes. Repeat this spray and repair procedure until you are satisfied that the surface of the filler, and the feathered edge of the paintwork are perfect. Clean the repair area with clean water and allow to dry fully.

The repair area is now ready for final spraying. Paint spraying must be carried out in a warm, dry, windless and dust free atmosphere. This condition can be created artificially if you have access to a large indoor working area, but if you are forced to work in the open, you will have to pick your day very carefully. If you are working indoors, dousing the floor in the work area with water will help to settle the dust which would otherwise be in the atmosphere. If the repair area is confined to one body panel, mask off the surrounding panels; this will help to minimise the effects of a slight mis-match in paint colours. Bodywork fittings (eg chrome strips, door handles etc) will also need to be masked off. Use genuine masking tape and several thicknesses of newspaper for the masking operations.

Before commencing to spray, agitate the aerosol can thoroughly, then spray a test area (an old tin, or similar) until the technique is mastered. Cover the repair area with a thick coat of primer; the thickness should be built up using several thin layers of paint rather than one thick one. Using 400 grade wet-and-dry paper, rub down the surface of the primer until it is really smooth. While doing this, the work area should be thoroughly doused with water, and the wet-and-dry paper periodically rinsed in water. Allow to dry before spraying on more paint.

Spray on the top coat using Holts Dupli-Color Autospray, again building up the thickness by using several thin layers of paint. Start spraying in the centre of the repair area and then, with a single side-to-side motion, work outwards until the whole repair area and about 2 inches of the surrounding original paintwork is covered. Remove all masking material 10 to 15 minutes after spraying on the final coat of paint.

Allow the new paint at least two weeks to harden, then, using a paintwork renovator or a very fine cutting paste such as Turtle Wax New Color Back or Holts Body + Plus Rubbing Compound, blend the edges of the paint into the existing paintwork. Finally, apply wax polish.

5 Maintenance – hinges and locks

1 Oil the hinges of the bonnet, boot and doors with a drop or two of light oil, periodically. A good time is after the car has been washed.

2 Oil the bonnet release catch pivot pin and safety catch pivot pin, periodically.
3 Do not over-lubricate door locks and strikers. Normally a little oil on the lock dovetail and a thin smear of high melting-point grease on the striker is adequate. Make sure that before lubrication they are wiped thoroughly clean and correctly adjusted.

6 Bonnet – removal and refitting

1 Open the bonnet and support it in the fully open position.
2 Mark the position of the hinges on the underside of the bonnet.
3 With the help of an assistant, support the weight of the bonnet and unbolt and remove it from the vehicle.
4 Refitting is a reversal of removal, but check the bonnet alignment (even gap between edge of bonnet and wing) before finally tightening the hinge bolts.
5 When closing the bonnet, it should close smoothly and positively, with moderate hand pressure. If it does not, align the lockpin and plate and adjust the projection of the lockpin by releasing its locknut and turning it by using a screwdriver in its end slot. The standard setting is as shown in Fig. 12.1.

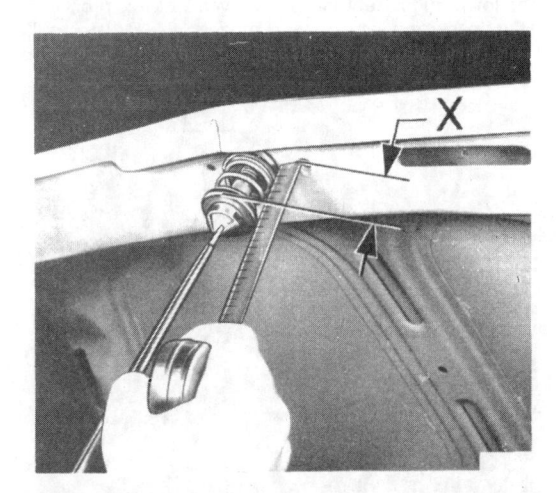

Fig. 12.1 Bonnet lockpin standard setting dimension (Sec 6)

X = 40 to 42 mm (1.58 to 1.65 in)

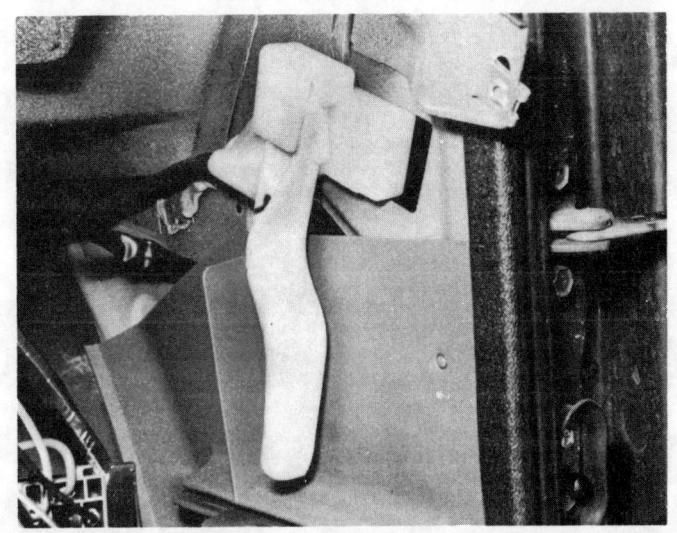

7.3 Bonnet release handle fitting above right-hand footwell (facia removed for photographic access)

7 Bonnet release cable – removal and refitting

1 Remove the cable clip from the top of the front body panel.
2 Using a screwdriver, prise the cable end fitting out of the release slide.
3 Working inside the vehicle, pull the bonnet release handle out of its retainer (photo).
4 Release the cable end fitting from the handle.
5 Pull the cable assembly through its grommet in the engine compartment rear bulkhead into the engine compartment.
6 Fit the new cable by reversing the removal operations, then adjust the cable to remove any slackness by altering the cable setting at the clip on the body panel.
7 The cable should release the bonnet with a gentle pull on the control handle. If it is stiff, check the setting of the catch lockpin, as described in the preceding Section.
8 Apply some grease to the bonnet release slide and to the lockpin.

8 Radiator grille – removal and refitting

1 On early Saloon models the radiator grille is removed by simply easing its upper edge away from the body panel and then lifting it to release the lower lugs. Refitting is the reverse of this procedure.
2 On Hatchback and later Saloon models, the radiator grille is an integral part of the front bumper assembly and reference should be made to the following Section.

Fig. 12.2 Radiator grille attachments on early Saloon models (Sec 8)

A Lower lugs   B Clip locations in body panel

9.2 Bumper/radiator grille retaining screws on Hatchback models – arrowed

9 Front bumper – removal and refitting

1 Working under the front wheel arches, undo the two bolts each side securing the bumper bracket to the chassis member, and the single bolt each side securing the bumper to the front wing.
2 On Hatchback and later Saloon models, unscrew the three screws securing the integral radiator grille to the body panel (photo).
3 On vehicles equipped with a headlamp washer system, remove the fluid hoses from the bumper overriders by releasing the hose connection detent with a screwdriver.
4 Carefully withdraw the bumper assembly from the car (photo).
5 Refitting is the reverse sequence to removal.

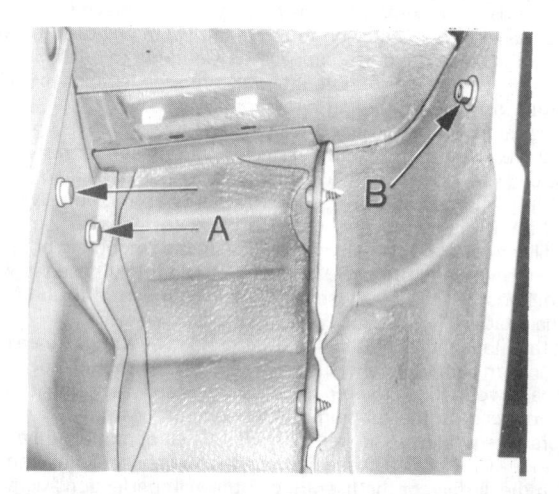

Fig. 12.3 Front bumpers mounting bolt locations viewed from under the wheel arch (Sec 9)

A Bumper bracket-to-chassis member retaining bolts
B Bumper-to-front wing retaining bolt

10 Rear bumper – removal and refitting

1 Remove the rear number plate lamp, as described in Chapter 10, and then disconnect the electrical lead from it.
2 Open the boot lid or tailgate and remove the trim on the rear body panel
3 Undo and remove the nut and washer securing the bumper to the body panel (photo).
4 Withdraw the bumper rearwards to disengage the end slides on each side of the bumper from the brackets on the body (photos). **Note:**

9.4 Removing the front bumper and bracket from the body aperture

Chapter 12 Bodywork

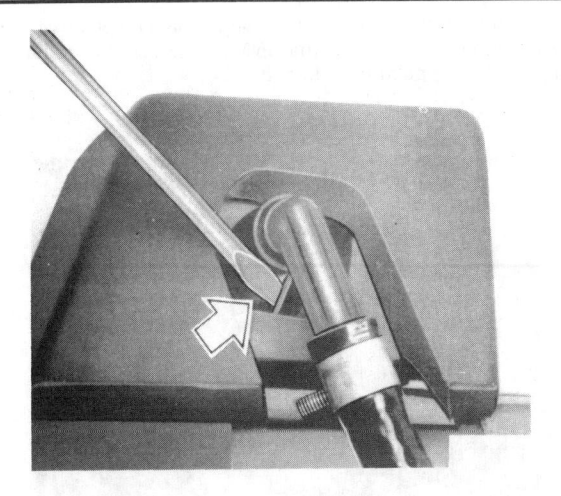

Fig. 12.4 Headlamp washer hose connection detent (arrowed) on bumper overrider (Sec 9)

Fig. 12.5 Rear bumper centre retaining catch (arrowed) on Hatchback models (Sec 10)

10.3 Rear bumper-to-body retaining nut in luggage compartment

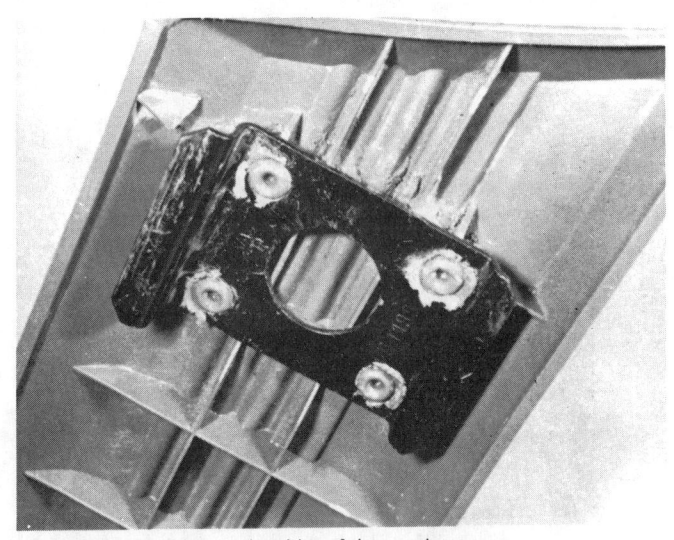

10.4A The end slides on the sides of the rear bumper ...

10.4B ... engage with brackets on the side of the body

On Hatchback models there is an additional catch below the centre of the bumper which engages with a bracket on the underbody. Lift the catch with a screwdriver to release it as the bumper is withdrawn.
5   Refitting is the reverse sequence to removal.

11 Front wing – removal and refitting

1   Remove the front bumper, as described in Section 9.
2   Remove the parcel shelf, as described in Section 34.
3   From inside the car, remove the trim panel from the side of the footwell to expose the front wing-to-A-pillar mounting bolts.
4   Undo the two bolts securing the front wing to the A-pillar.
5   From inside the wheel arch undo the two bolts securing the wing to the front panel.
6   Undo the six bolts securing the top edge of the wing to the engine compartment inner panel and the single bolt securing the wing to the sill.
7   Cut around the seams of the wing with a sharp knife to release the mastic and then lift the wing away.
8   Clean away all old mastic from the body mating flange and apply a thick bead of new sealant.
9   Place the wing in position and screw in the bolts finger tight.

10 Align the wing with the adjacent body panels and then tighten all bolts and screws.
11 Apply protective coating to the underside of the wing and refinish the outer surface to match the body colour.
12 Refit the remaining components using the reverse sequence to removal.

12 Front door trim panel – removal and refitting

1 Remove the securing clip for the window regulator handle. To do this force back the trim bezel and pull out the clip using a length of wire with a hooked end. Remove the bezel (photo).
2 Undo the two screws and remove the armrest (photo).
3 Carefully prise off the escutcheon plate around the door remote control handle (photo).
4 Insert the fingers or a broad blade between the trim panel and the door and release the panel from the door. Use a jerking action to do this in order to free the plastic securing buttons from the holes in the door (photo). The buttons are rather brittle and it is a wise precaution to purchase some spare ones to replace any broken during this operation.
5 If the trim panel has been removed for access to the door internal components, carefully peel back the polythene condensation barrier as necessary (photo).
6 Refitting is the reverse sequence to removal, but note that the window regulator handle securing clip is fitted to the handle before the handle is fitted to the door (photo). The handle complete with clip is then simply driven out onto the splined shaft of the regulator by striking it with the palm of the hand.

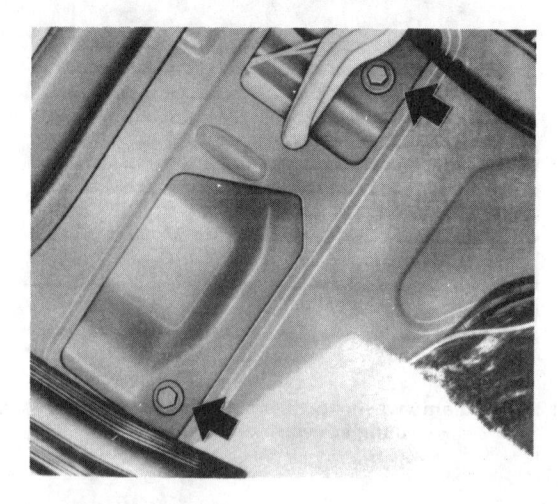

Fig. 12.6 Front wing-to-A-pillar mounting bolts – arrowed (Sec 11)

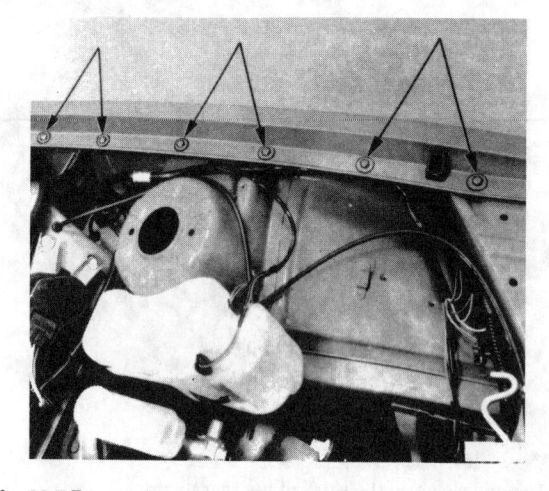

Fig. 12.7 Front wing upper retaining bolts — arrowed (Sec 11)

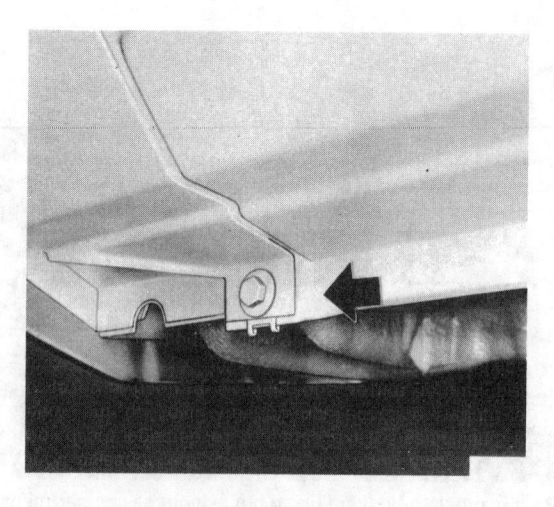

Fig. 12.8 Front wing-to-sill panel retaining bolt – arrowed (Sec 11)

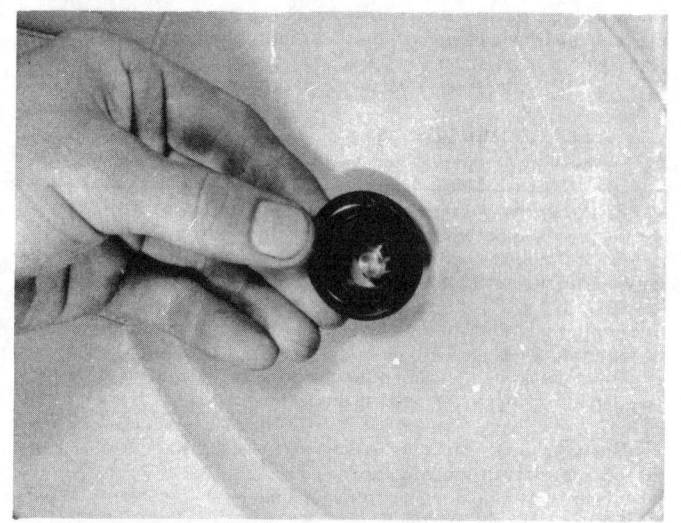

12.1 After removing the window regulator handle lift off the bezel

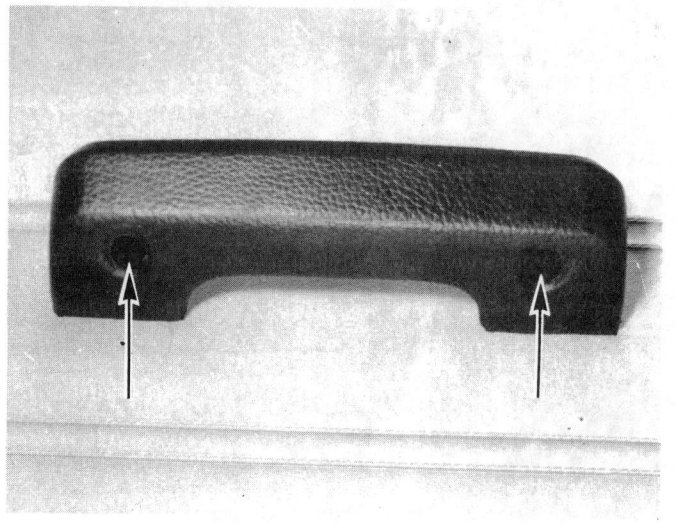

12.2 Undo the two screws (arrowed) and remove the armrest

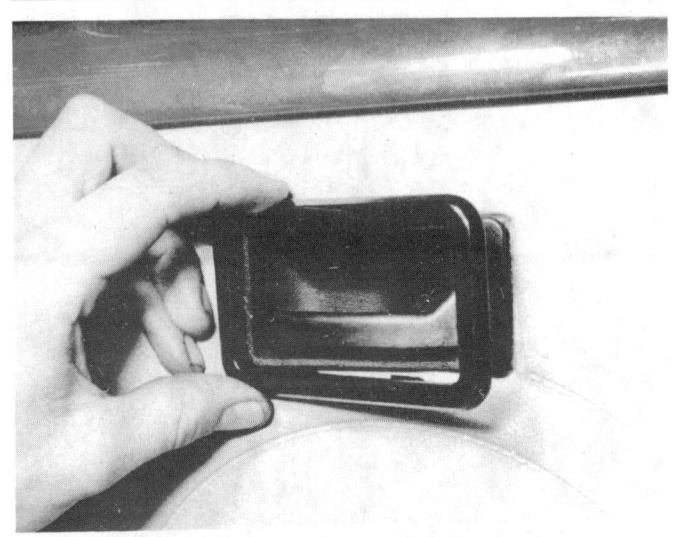

Chapter 12 Bodywork

12.3 Carefully prise off the remote control handle escutcheon

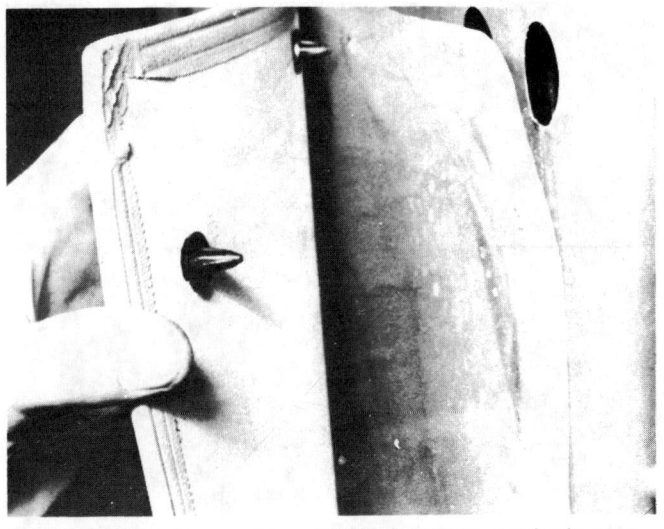

12.4 Remove the trim panel by releasing the plastic securing buttons using a jerking action

12.5 Peel back the condensation barrier for access to the door internal components

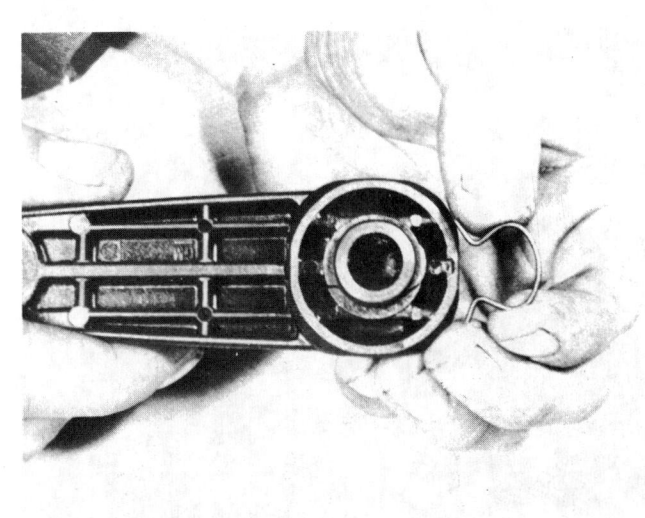

12.6 Fit the regulator handle retaining clip to the handle before fitting the handle to the regulator shaft

13.3 Front door hinge pin (arrowed)

13 Front door – removal and refitting

1 The door hinges are welded onto the door frame and the body pillar so that there is no provision for adjustment or alignment.
2 To remove a door, open it fully and support it under its lower edge on blocks covered with pads of rag.
3 Disconnect the door check and drive out the hinge pins (photo). Remove the door.
4 If the door can be moved up and down on its hinge due to wear in the pivot pins or holes, it may be possible to drill out the holes and fit slightly oversize pins.
5 Door closure may be adjusted by moving the socket-headed striker.

14 Front door lock – removal and refitting

1 Remove the door trim panel, as described in Section 12.
2 Temporarily place the handle on the regulator and wind the window fully up.
3 Undo the bolt securing the window rear guide channel to the door (photo) and move the channel aside.
4 Disconnect the three control rods from the lock by prising the clip collars off the rods (photo) and then withdrawing the rods.

14.3 Window rear guide channel retaining bolt

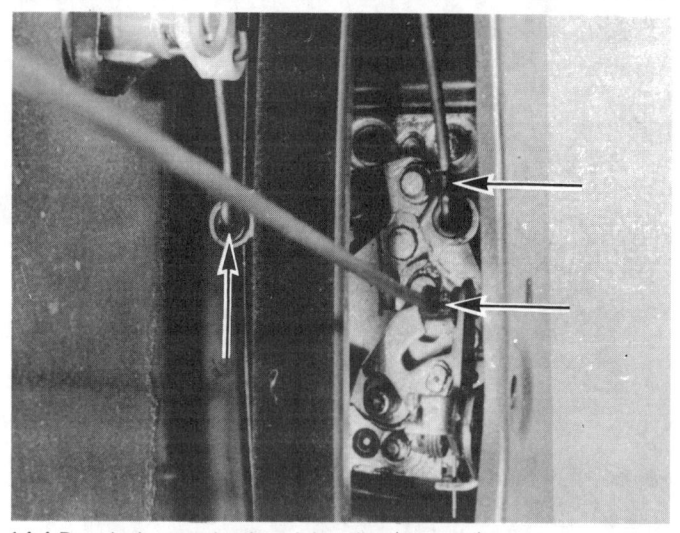

14.4 Door lock control rod retaining clips (arrowed)

14.5 Door lock retaining screws

15.4 The door exterior handle retaining nuts are accessible using a socket through two holes in the door inner panel

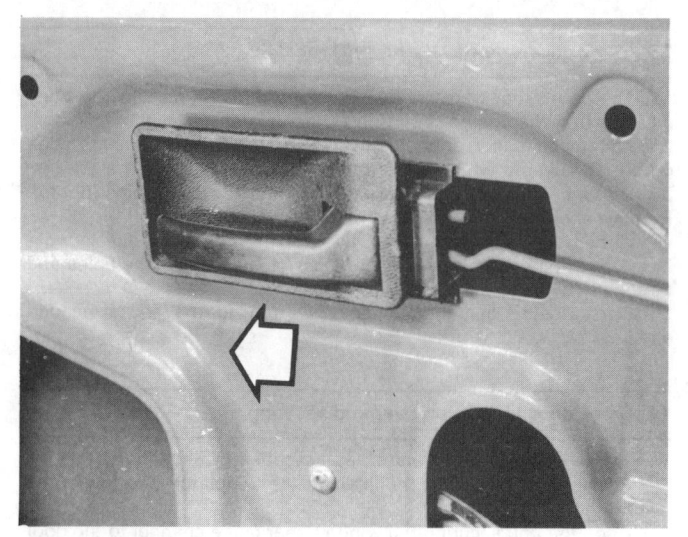

5 Undo the three screws securing the lock assembly to the door (photo) and withdraw the lock from inside the door.
6 Refitting is the reverse sequence to removal.

15 Front door exterior handle – removal and refitting

1 Remove the door trim panel, as described in Section 12.
2 Undo the bolt securing the window rear guide channel to the door and move the channel aside.
3 Disconnect the exterior handle control rod at the door lock assembly by prising the retaining clip collar off the rod and withdrawing the rod.
4 Using a socket and extension bar through the two holes in the door panel, undo the two exterior handle retaining nuts and recover the washers (photo).
5 Manipulate the handle as necessary and withdraw it, complete with control rod, from the door.
6 Refitting is the reverse sequence to removal.

16.3 Disengage the door remote control handle by moving it forwards

16 Front door remote control handle – removal and refitting

1 Remove the door trim panel, as described in Section 12.
2 Disconnect the remote control handle control rod at the door lock

assembly by prising the retaining clip collar off the rod and withdrawing the rod.
3 Move the handle forward to disengage it from its retaining slot in the door and remove the handle complete with control rod (photo).
4 Refitting is the reverse sequence to removal.

17 Front door lock cylinder – removal, dismantling, reassembly and refitting

1 Remove the door trim panel, as described in Section 12.
2 Undo the bolt securing the window rear guide channel to the door and move the channel aside.
3 Disconnect the lock cylinder control rod at the door lock assembly by prising the retaining clip collar off the rod and withdrawing the rod.
4 Remove the lock cylinder retaining clip by prising it forward using a screwdriver (photo).
5 Withdraw the cylinder and control rod from the door.
6 To dismantle, insert the ignition key in the lock and then force a screwdriver between the housing and the arm to force off the retaining circlip.
7 Separate the components, noting their fitted sequence.
8 Reassembly is a reversal of dismantling, but note that the ends of the spring must be crossed as shown (Fig. 12.11) and also the relative position of the arm to the housing must be observed. Use a new circlip.

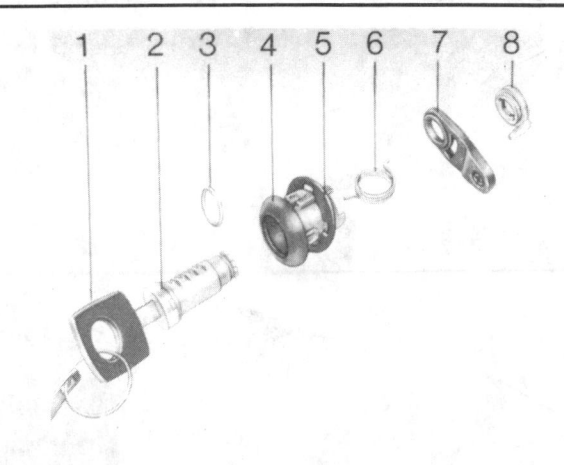

Fig. 12.10 Front door lock cylinder components (Sec 17)

1 Ignition key   5 Seal
2 Cylinder   6 Spring
3 Circlip   7 Arm
4 Housing   8 End piece

17.4 Release the lock cylinder retaining clip (arrowed) by prising it forwards

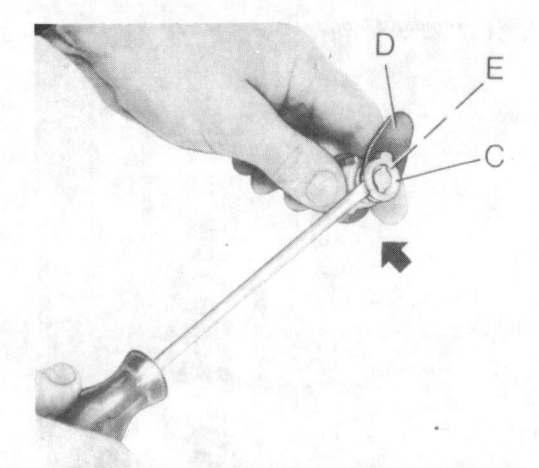

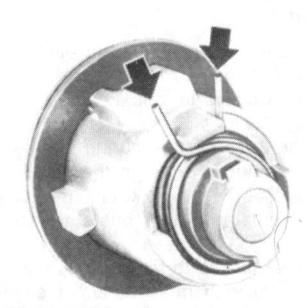

Fig. 12.11 Lock cylinder spring ends (arrowed) correctly located (Sec 17)

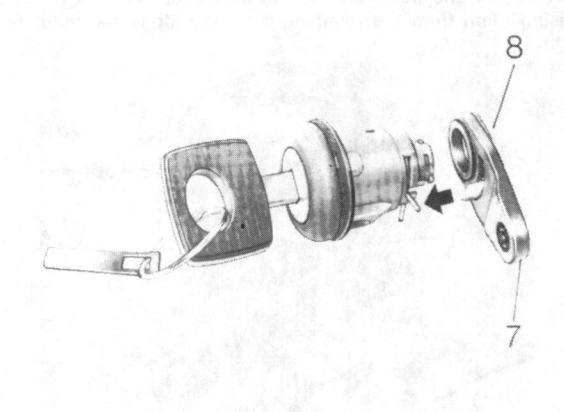

Fig. 12.9 Separating front door lock cylinder arm from housing (Sec 17)

C Housing   E Circlip
D Arm

Fig. 12.12 Relationship of cylinder arm to housing (Sec 17)

7 Arm   8 End piece

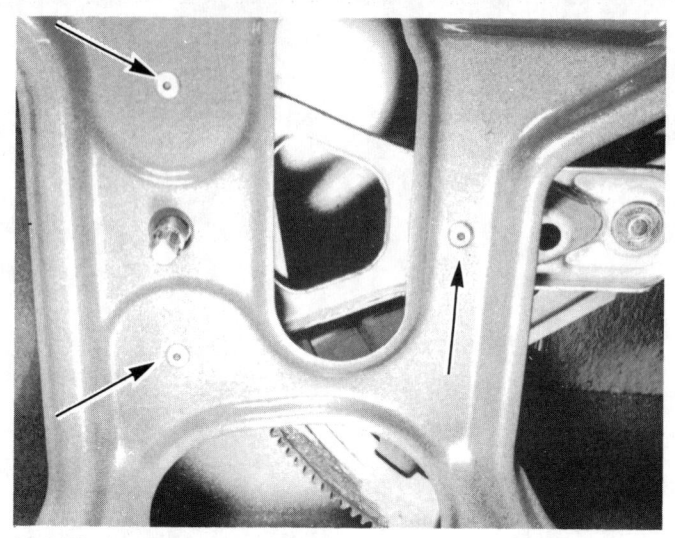

Chapter 12 Bodywork

toward the rear to remove.
4   Lower the window fully, move the window upper rear corner forwards and lift the window up and out of the door at an angle. As the window is lifted, disengage the lower channel from the regulator lifting arm.
5   Refitting is the reverse sequence to removal.

18.3 Window regulator retaining rivets (arrowed)

18 Front door window regulator – removal and refitting

Note: *This procedure applies to two-door models. On four-door models a cable type regulator is fitted, similar to that fitted to the rear doors (Chapter 13, Section 12)*

1   Remove the door trim panel, as described in Section 12.
2   Temporarily refit the regulator handle and position the window halfway open. Retain the window in this position using wooden wedges.
3   Drill off the heads of the three rivets securing the regulator to the door panel (photo). Take care not to drill into the panel.
4   Disengage the regulator lifting arm from the window lower channel and withdraw the regulator through one of the large apertures in the door panel.
5   Refitting is the reverse sequence to removal. Lubricate the lifting arm and regulator mechanism with molybdenum disulphide grease and secure with three new 4.8 x 11 mm (0.19 x 0.43 in) blind steel rivets.

19 Front door window – removal and refitting

1   Remove the door trim panel, as described in Section 12.
2   Undo the bolt securing the window rear guide channel to the door and remove the channel by sliding it down and out of the window rubber seal.
3   Withdraw the inner and outer waist seal finishers by carefully releasing them from the retaining clips. Withdraw the outer finisher

Fig. 12.14 Removal of the front door outer waist seal finisher (Sec 19)

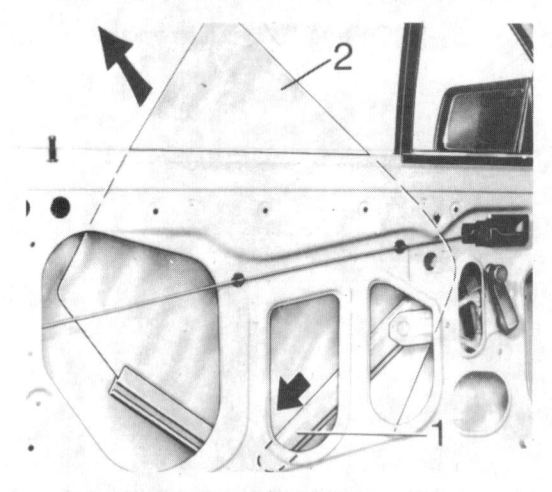

Fig. 12.15 Front door window removal (Sec 19)

1   Regulator lifting arm   2   Window

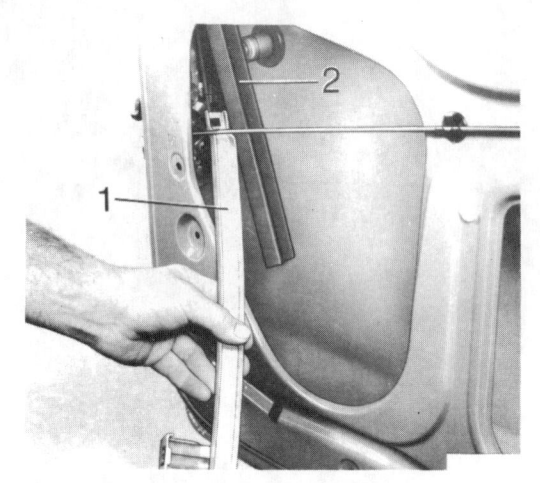

Fig. 12.13 Removing the front door window rear guide channel (Sec 19)

1   Guide channel   2   Window rubber seal

Fig. 12.16 Front door quarter-light retaining screw locations – arrowed (Sec 20)

Chapter 12 Bodywork

20 Front door fixed quarter-light – removal and refitting

1 Remove the front door window, as described in the previous Section.
2 Undo and remove the screws securing the quarter-light to the upper door frame and inner panel.
3 Move the quarter-light to the rear to disengage the rubber seal from the frame and withdraw the quarter-light upwards out of the door.
4 Refitting is the reverse sequence to removal.

21 Rear fixed quarter-light – removal and refitting

1 The glass can be pushed out from inside the car. Start at one corner and push the window with the palms of the hands until it is released from the rubber seal. Have an assistant outside the car to support the glass as it is released.
2 With the quarter-light glass removed, the rubber seal can be removed from the aperture.
3 Thoroughly clean the sealing edge of the glass and the aperture flange and then position a new rubber seal around the glass. *Do not use any sealant or adhesive.*
4 Cut a length of strong cord, greater in length than the periphery of the glass, and insert it into the body aperture flange channel of the rubber seal. The ends of the cord should meet and overlap at the front lower corner of the glass.
5 Position the glass and rubber seal in the body aperture from the outside. From inside the car pull the sealing lip of the rubber seal over the body aperture flange by pulling on the cord. As the cord lifts the rubber over the flange, have your assistant tap the glass with the palm of his hand from the outside.
6 Continue pulling out the cord until the rubber seal is positioned over the complete periphery of the flange.

22 Windscreen – removal and refitting

1 If the windscreen is to be renewed it is a job which is better left to a windscreen replacement specialist. They will do the job in half the time and, most important, ensure that it is correctly fitted with no leakage around the sealing weatherstrip. However, if you wish to do it yourself, proceed as follows.
2 Where a windscreen is to be replaced due to cracking, the facia air vents should be covered before attempting removal.
3 First release the rubber weatherstrip from the body by running a small blunt screwdriver around and under the rubber, both inside and outside the car. This operation will break the adhesive bond of the sealant originally used. Take care not to damage the paintwork or catch the rubber with the screwdriver. Pivot up the wiper arms, remove the interior mirror and place a protective cover over the bonnet. Salvage your tax disc.
4 Have an assistant push the inner lip of the rubber surround off the flange of the windscreen body aperture. Once the rubber surround starts to peel off the flange, the screen may be forced gently outward by careful hand pressure. The second person should support and remove the screen complete with rubber surround.
5 Fit a new rubber weatherseal to the glass and ensure that all old sealant is removed from the body flange. Scrape it away and then clean it off with a fuel-soaked cloth.
6 Apply a bead of sealant to the body flange all round the windscreen aperture.
7 Cut a piece of strong cord greater in length than the periphery of the glass and insert it into the body flange locating channel of the rubber surround.
8 Offer the windscreen to the body aperture and pass the ends of the cord, previously fitted and located at bottom centre, into the vehicle interior.
9 Press the windscreen into place, at the same time have an assistant pull the cords to engage the lip of the rubber channel over the body flange.
10 Remove any excess sealant with a paraffin-soaked rag.
11 Refit the interior mirror and the tax disc then lower the wiper arms into position.

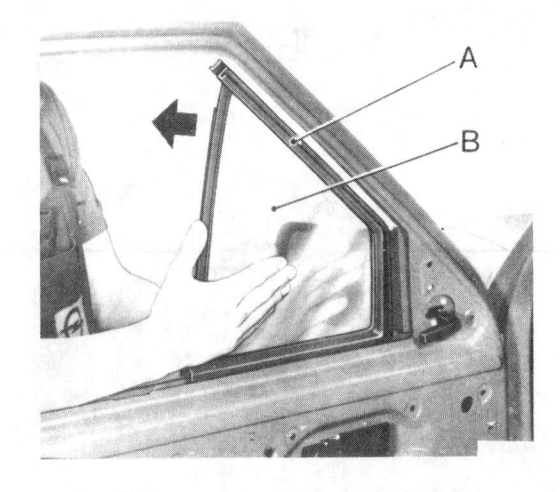

Fig. 12.17 Front door fixed quarter-light removal (Sec 20)

A Rubber seal   B Quarter-light

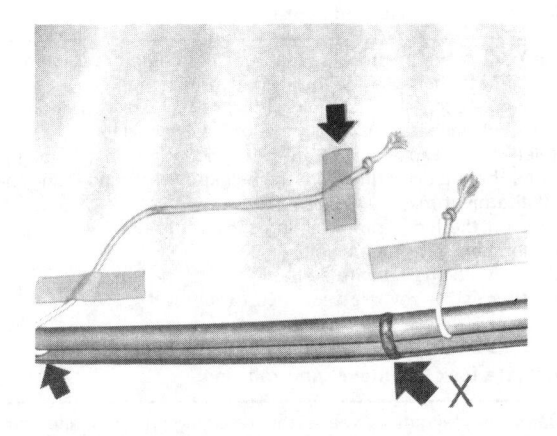

Fig. 12.18 Windscreen with rubber weatherstrip and draw cords (arrowed) in position (Sec 22)

X Rubber collar position at bottom centre of windscreen

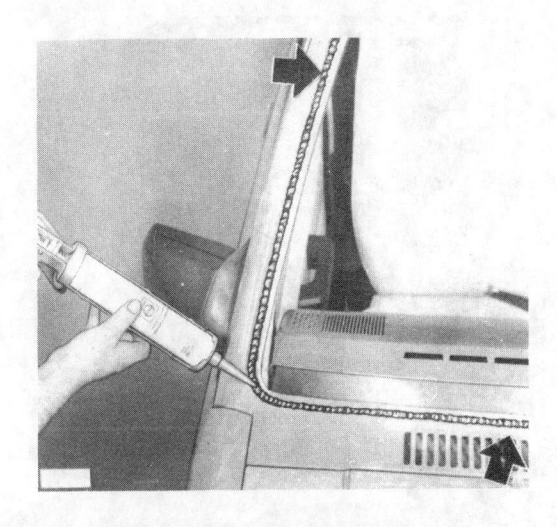

Fig. 12.19 Applying sealant to windscreen aperture body flange (Sec 22)

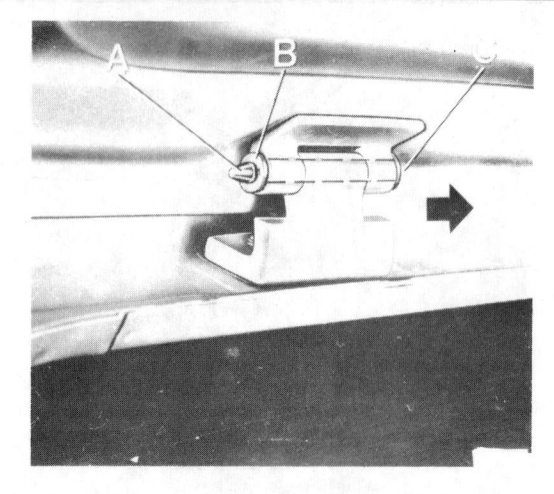

Chapter 12 Bodywork

Fig. 12.20 Tailgate hinge arrangement (Sec 23)

A Hinge pin
B Retaining ring
C Hinge pin collar faces centre of car

23 Tailgate – removal and refitting

1 Disconnect the electrical leads from the heated rear window.
2 Disconnect the hose for the tailgate washer jet.
3 Disconnect the electrical wiring for the tailgate wiper motor.
4 Open the tailgate fully and have an assistant support it.
5 Release the support strut from its mounting on the tailgate by extracting the retaining ring from the ball end fitting and lifting the ball end off the pivot peg.
6 Prise off the hinge pin retaining clips and then drive the pins out toward the centre of the vehicle.
7 Carefully lift the tailgate away.
8 Refitting is the reverse sequence to removal.

24 Tailgate lock – removal and refitting

1 Open the tailgate, release the retaining buttons and carefully remove the tailgate trim panel.
2 Unscrew the tailgate lock retaining screws (photo), release the control rod from the lock lever and withdraw the lock.
3 To remove the lock cylinder, prise out the retaining clip (photo) and withdraw the cylinder complete with control rod.
4 Dismantling and reassembly of the lock cylinder is the same as described in Section 17 for the front door lock.
5 Refitting is the reverse sequence to removal. Adjust the position of the striker if necessary to give positive clearance of the tailgate without slamming.

25 Tailgate support strut – removal and refitting

1 Open the tailgate and have an assistant support it in the open position.
2 Extract the retaining rings from the ball end fittings and lift the ball ends off the pivot pegs on the tailgate and body (photo).
3 Refitting is the reverse sequence to removal.
4 The struts are gas pressurised and should not be punctured, cut open or subjected to heat.

26 Rear window or tailgate glass – removal and refitting

The procedure is basically the same as described for the windscreen in Section 22, but first remember to disconnect the leads for the heater element and, on Hatchback models, also disconnect the washer hose and wiper arm. On SR models it will be necessary to remove the tailgate spoiler to gain access to the glass rubber seal.

24.2 Tailgate lock and retaining screws

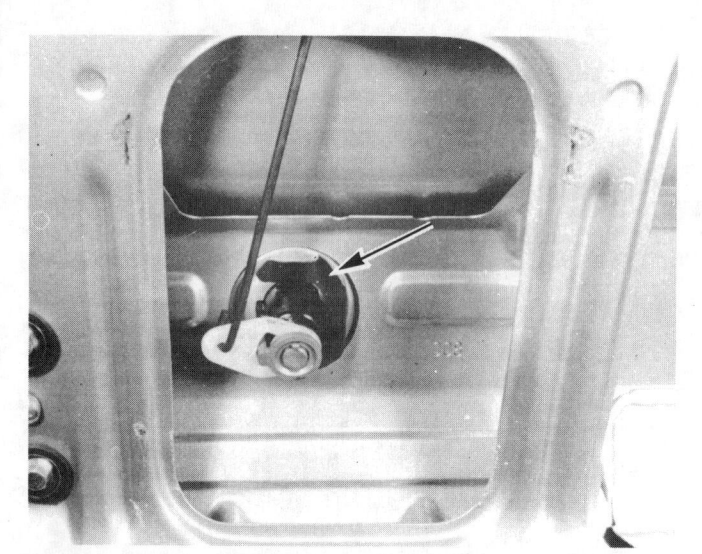

24.3 Tailgate lock cylinder retaining clip (arrowed)

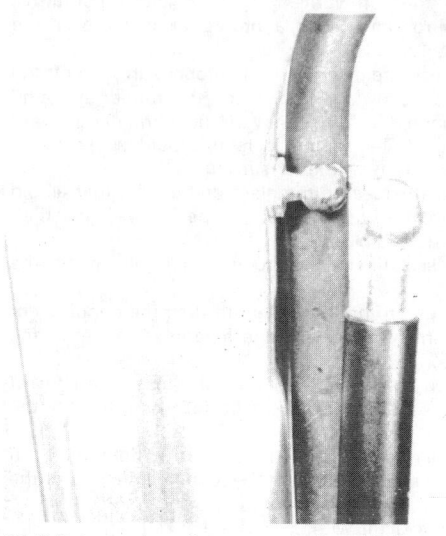

25.2 Release the tailgate strut ball end fittings after extracting the retaining rings

28.1 Boot lid lock and retaining screws

28.4 Boot lid lock striker

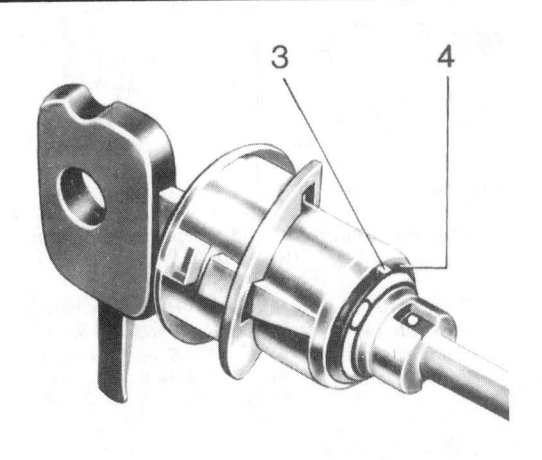

Fig. 12.21 Boot lid lock cylinder washer (3) and snap-ring (4) (Sec 28)

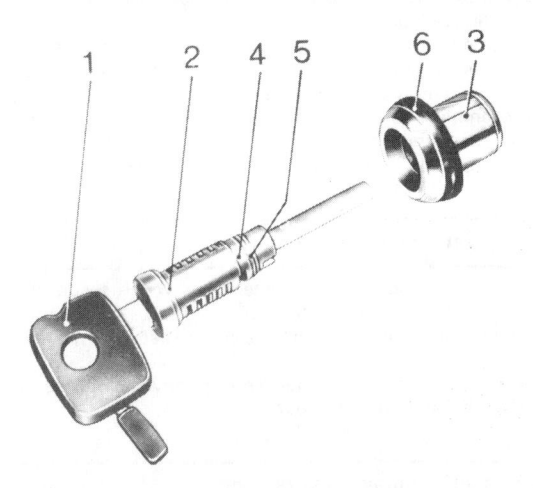

Fig. 12.22 Boot lid lock cylinder components (Sec 28)

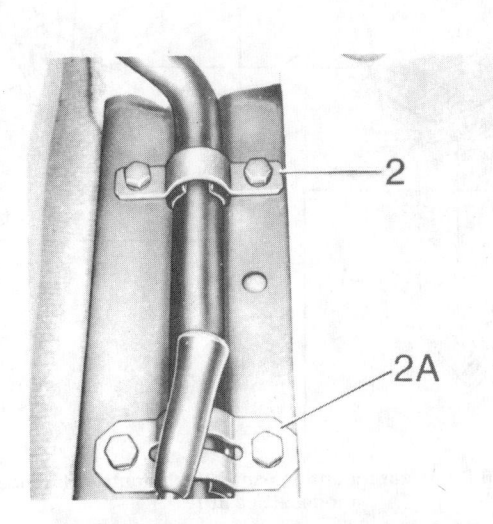

1 Ignition key
2 Lock barrel
3 Housing
4 Washer
5 Snap-ring
6 Seal

Fig. 12.23 Front seat mountings (Sec 29)

2 Narrow U-shaped clip   2A Wide U-shaped clip

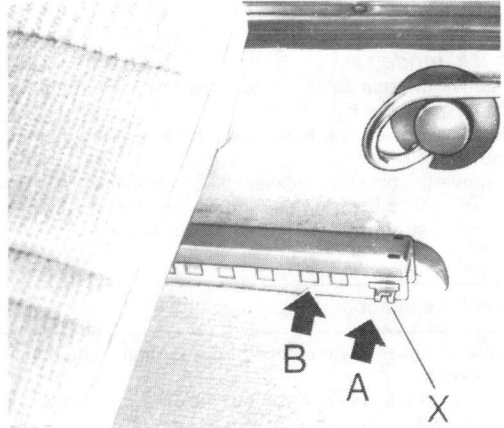

Fig. 12.24 Front seat slide rail atachments (Sec 29)

X Retaining clip   A and B Alternative retaining clip locations

27 Boot lid – removal and refitting

1 Open the lid and mark the position of the hinges on the underside of the lid.
2 With the help of an assistant, unbolt the hinges from the lid and lift the lid away.
3 The hinges themselves should not normally be disturbed, but if they must be removed then the counterbalance springs will have to be released and removed.
4 Unhook them using a lever of sufficient length to be able to counteract the tension of the spring rods.
5 If the hinges are unbolted from the body, fit new sealing washers in order to prevent water seepage into the luggage compartment.
6 Refitting is the reverse sequence to removal. Align the hinges with the previously made marks and, if necessary, adjust the position of the striker so that the boot lid will shut and lock without slamming.

28 Boot lid lock – removal and refitting

1 Open the boot lid, undo and remove the retaining screws (photo) and withdraw the lock.
2 To remove the lock cylinder, prise off the retaining clip and withdraw the cylinder.
3 The cylinder may be dismantled by extracting the retaining snap-ring and washer and, with the key inserted, withdrawing the key and lock barrel assembly from the housing.
4 Reassembly and refitting is the reverse of the dismantling and removal procedures. If necessary adjust the position of the striker (photo) so that the boot lid will shut and lock without slamming.

29 Front seat – removal and refitting

1 Unbolt and remove the U-shaped clips which secure the tubular crossmember of the seats to the floor.
2 Remove the spring clip from the outer seat slide rail and remove the seat.
3 Refitting is the reverse sequence to removal. Note that the wider U-shaped clip is fitted to the outside.

30 Rear seat – removal and refitting

Saloon models
1 Pull the two handles at the base of the seat cushion to release the retainers.
2 Remove the cushion.
3 Prise up the metal tabs at the base of the seat back, then lift the seat back upwards off the securing hooks.
4 Refitting is a reversal of removal.

Hatchback models
5 To remove the seat cushion unclip the trim from the front edge of the seat to expose the hinges.
6 Undo the screw each side securing the cushion to the hinge loops and lift out the cushion.
7 To remove the backrest, release the catches and fold the backrest fully forward. Now slide it up and out of the pivot guides.
8 Refitting is the reverse sequence to removal.

31 Sunroof – operation

1 To raise or lower this type of sunroof (optional equipment) rotate the control knob.
2 To remove the sunroof, set it in the closed position and then use a coin to turn the screw which is located in the centre of the control knob.
3 Slightly raise the sunroof panel and press the release lever to disengage the safety catch.
4 From outside the vehicle, pivot the sunroof panel and lift it from its retainers at the front edge.
5 Rotate the control knob to the fully closed position.

6 A storage bag is provided on the back of the rear seat. Store the sunroof glass panel so that its curved surface is towards the seat back and the hinge tongues pointing upward.
7 To refit the sunroof, set the control knob to 'open' and engage the panel in the retainers and with the safety catch.

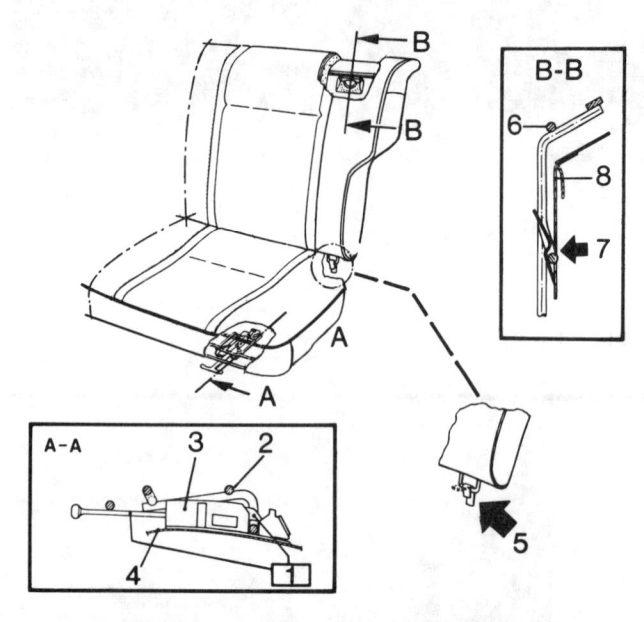

Fig. 12.25 Rear seat mountings and attachments – Saloon models (Sec 30)

1 Release handles
2 Seat retainers
3 Handle bracket
4 Vehicle floor
5 Retaining metal tab
6 Retaining wire
7 Securing hooks
8 Rear body panel

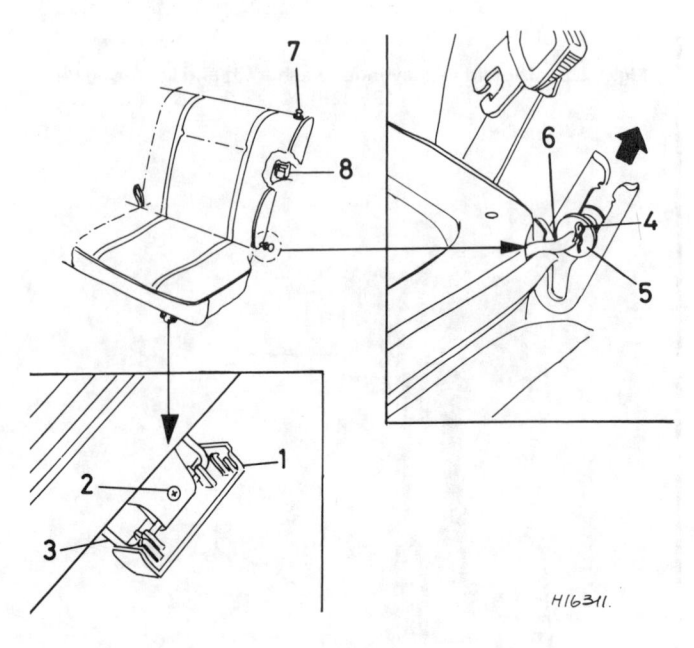

Fig. 12.26 Rear seat mountings and attachments – Hatchback models (Sec 30)

1 Hinge trim
2 Retaining screw
3 Hinge loop
4 Clip
5 Bushing
6 Pivot
7 Release button
8 Backrest catch

Chapter 12 Bodywork

32 Exterior rear view mirror – removal and refitting

1 Working inside the vehicle, prise off the mirror remote control handle cover.
2 Prise off the triangular shaped plate.
3 Extract the three mirror mounting screws which are now exposed and have an assistant support the mirror head on the outside of the vehicle.
4 Refitting is a reversal of removal.

33 Centre console – removal and refitting

1 The centre console is secured by self-tapping screws which are hidden under square cover plates on the top surface of the console.
2 Prise out the plates with a small blade and then extract the screws (photo).
3 Release the gear lever rubber boot from around the console and remove the console from the car.
4 Refitting is the reverse sequence to removal.

34 Parcel shelf – removal and refitting

1 Undo the three screws, one at each end of the shelf and one above the centre console, and withdraw the parcel shelf from its guide rails towards the centre of the car (photo).
2 Refitting is the reverse sequence to removal.

35 Glove compartment – removal and refitting

1 Open the glove compartment door and undo the three screws inside and the two screws below the door hinges.
2 Withdraw the glove compartment from the facia.
3 On models equipped with a courtesy lamp in the glove compartment, disconnect the leads to the lamp and switch before removing the glove compartment.
4 Refitting is the reverse sequence to removal.

36 Facia – removal and refitting

1 Disconnect the battery negative terminal.
2 Remove the steering wheel, as described in Chapter 8.
3 Undo the four retaining screws and lift off the steering column shroud lower half.
4 Insert the ignition key into the ignition/steering lock cylinder and turn it to position 'II'.
5 Using a short length of small diameter wire inserted into the hole in the lock housing, depress the detent spring and withdraw the key and lock cylinder.
6 By judicious manipulation, the upper half of the column shroud can now be removed. To do this it is necessary to carefully bend the shroud outward slightly to clear the switch stalks.
7 Refer to Chapter 10 and remove the instrument panel.
8 Refer to Section 34 and remove the parcel shelf.
9 Remove the ashtray.
10 Undo the two screws beneath the face level vent deflectors and withdraw the central switch and vent panel from the facia. Make a careful note of the electrical connections on the switches at the rear of the panel, disconnect the wires and remove the panel (photo).
11 Undo the screw at the base of the heater control unit housing and the additional screw directly below that. Release the housing from the facia, slide the control cable ends out of the lever mechanism and remove the illumination bulb holder. Lift away the control unit housing.
12 Remove the remaining facia switches and the radio, as described in Chapter 10.
13 Lift off the fusebox cover, detach the fusebox from its location by depressing the retaining catches and move the fusebox to one side.
14 Tap out the retaining pin securing the choke knob to the cable end fitting and unscrew the knob (photo).
15 Undo the retaining ring securing the choke cable to the facia (photo) and push the cable through to the inside.
16 Undo the two screws, one each side securing the ends of the facia to the mounting brackets (photo).

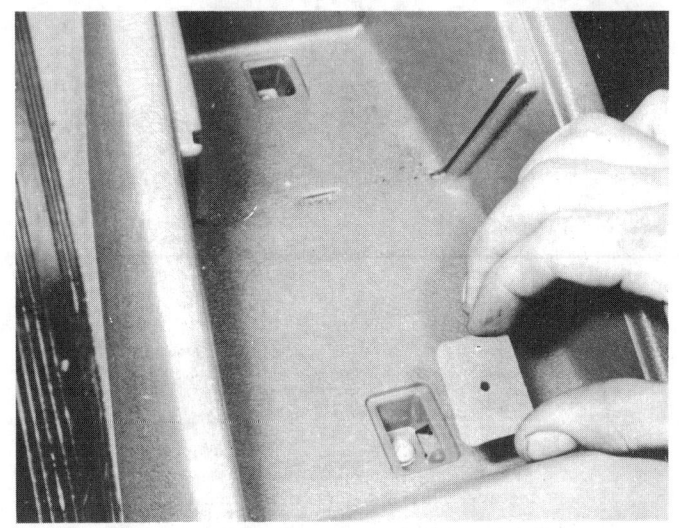

33.2 The centre console retaining screws are located under cover plates

34.1 Slide the parcel shelf out of its guide rails after undoing the securing screws

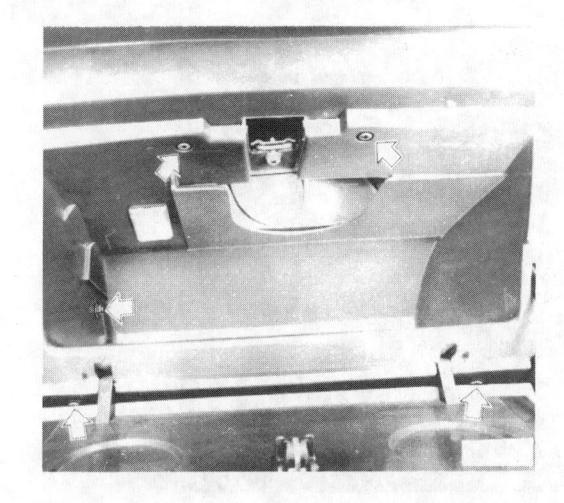

Fig. 12.27 Glove compartment retaining screw locations (arrowed) – left-hand drive models shown (Sec 35)

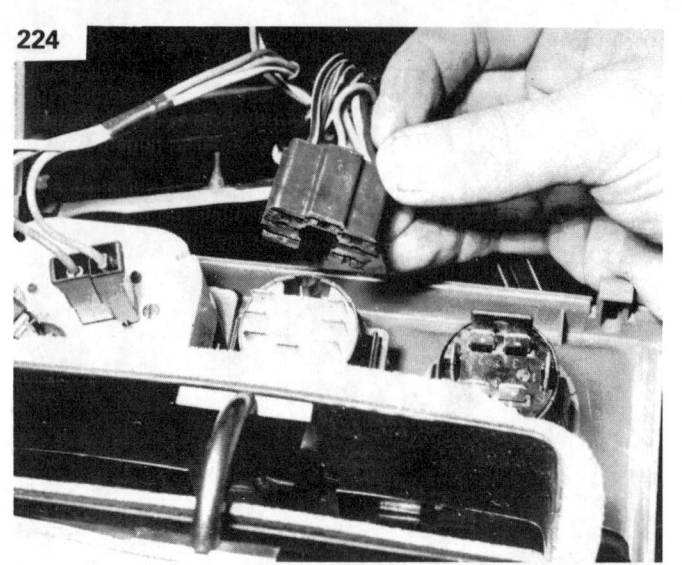

36.10 Disconnect the wiring at the rear of the facia switch and vent panel

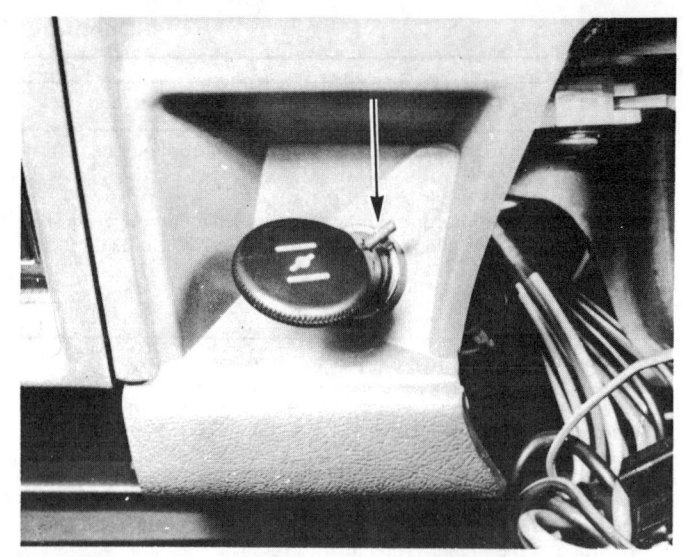

36.14 Remove the choke knob retaining pin (arrowed) and unscrew the knob

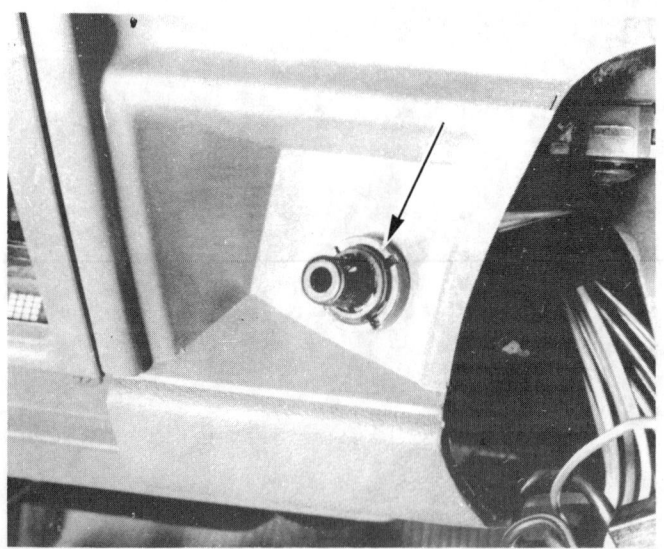

36.15 Unscrew the choke cable end fitting retaining ring (arrowed) to release the cable from the facia

36.16 Undo the facia end retaining screws

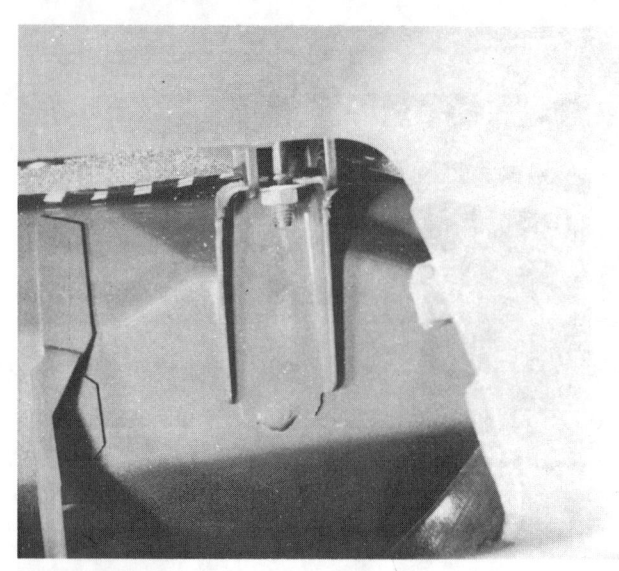

36.17A Undo the nuts in the instrument panel aperture ...

36.17B ... and glove compartment aperture

Chapter 12 Bodywork

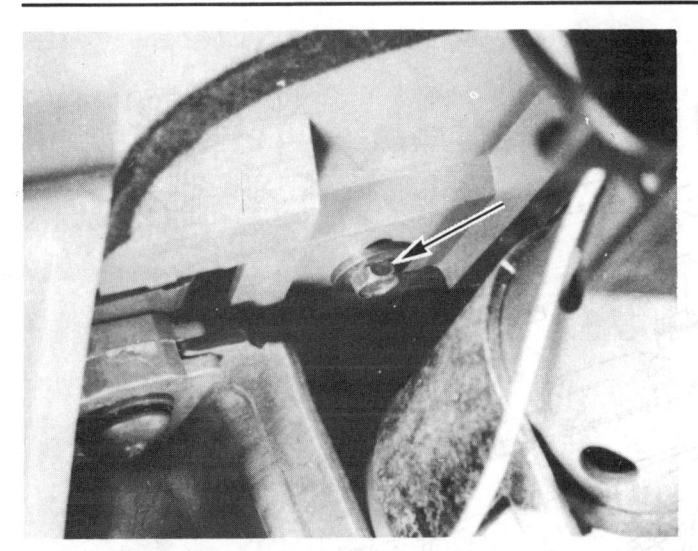

36.18 Undo the bolt above the steering column (arrowed) ...

36.20 ... and remove the facia

38.5 Control cable attachments at the heater control unit

38.27 Heater air distribution housing retaining nut location

17 Undo the two nuts securing the facia to the bulkhead. One is accessible through the instrument panel aperture (photo) and the other through the glove compartment aperture (photo).
18 Undo the bolt directly below the steering column, securing the facia to the column mounting bracket (photo).
19 Undo the two remaining screws, one behind the ashtray aperture and one below it.
20 Carefully withdraw the complete facia panel away from its location, while at the same time moving aside the disconnected wiring cables and air ducting as necessary (photo). Remove the facia through the driver's door opening.
21 Refitting is the reverse sequence to removal, but it is advisable to engage the help of an assistant to guide the cables and wiring as the facia is placed in position.

37 Heater – description

1 The heater system depends upon fresh air being drawn into the grille at the base of the windscreen and passed through a matrix which is heated from the engine cooling system.
2 Temperature regulation is controlled by mixing cold intake air with warm air, using flap valves both for this function and for the direction of air to the interior, windscreen or side air outlets.
3 An electric booster fan is mounted within the engine compartment to supplement the normal ram effect provided when the vehicle is in forward motion.
4 An independent fresh air ventilation system provides a supply of unheated fresh air at the nozzles on the instrument panel.
5 Stale air is exhausted from the vehicle interior through a vent in the central body pillar.

38 Heater components – removal and refitting

Control unit
1 Disconnect the battery negative terminal.
2 Undo the two screws beneath the face level vent deflectors and withdraw the central switch and vent panel from the facia.
3 Make a careful note of the electrical connections on the switches at the rear of the panel, disconnect the wires and remove the panel.
4 Remove the ashtray, undo the control unit retaining screw and the cable clamp retaining screw just below it.
5 Withdraw the control unit slightly and slide the cable ends out of the lever mechanism (photo).
6 Withdraw the illumination bulb holder and remove the control unit.
7 Refitting is the reverse sequence to removal.

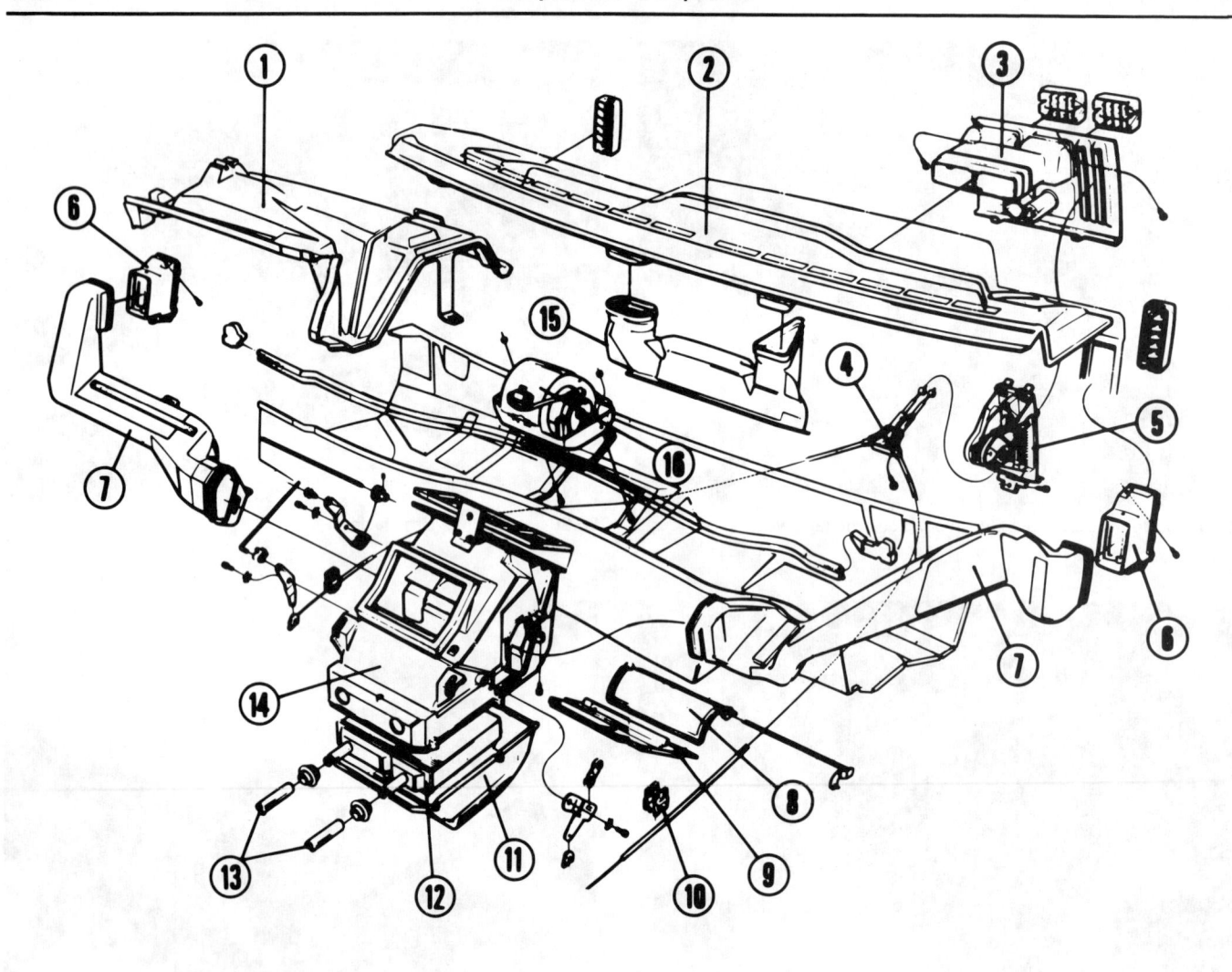

Fig. 12.28 Exploded view of the heater assembly and related components (Sec 38)

1 Blower motor water deflector	5 Control unit	9 Mixer flap	13 Heater hoses
2 Facia	6 Side heater vent housing	10 Clip retainer	14 Air distributor housing
3 Central switch and vent panel	7 Ducting	11 Matrix cover	15 Ducting
4 Control cables	8 Air distribution flap	12 Matrix	16 Blower motor

Blower motor

8 Open the bonnet and remove the water deflectors from the opening on the top of the engine compartment bulkhead.
9 Release the motor wiring plug from its retaining clip using a screwdriver and disconnect the plug.
10 Prise off the clips which retain the cover over the motor and lift off the cover.
11 Undo the two motor retaining screws, noting the earth lead under one of the screws, and remove the motor.
12 Refitting is the reverse sequence to removal.

Heater matrix

13 Clamp the heater hoses close to the engine compartment bulkhead using self-locking grips, or alternatively drain the cooling system as described in Chapter 2.
14 Undo the hose clips and disconnect the heater hoses from the matrix pipe stubs. Plug the pipe stubs to prevent coolant spillage into the car interior during removal of the matrix.
15 Remove the parcel shelf, as described in Section 34.
16 Undo the centre console retaining screws and move the console to one side. Refer to Section 33 if necessary.
17 Undo the retaining screws and remove the cover from the base of the matrix housing.
18 Undo the bolts securing the heater matrix front mounting bracket and release the matrix from its rear mounting clips.
19 Withdraw the matrix from the bulkhead and heater assembly and remove it from the car.
20 If the matrix is blocked, try reverse flushing, as described for the radiator in Chapter 2. If this fails, try a radiator cleansing agent, but use it strictly in accordance with the manufacturer's instructions.
21 If the matrix is leaking, have it professionally repaired by a radiator repairer, or purchase a new one. Temporary repairs are not worth the trouble if the unit has to be removed again after a short period of service.
22 Refitting is the reverse sequence to removal. Refill or top up the cooling system, as described in Chapter 2.

Air distribution housing

23 Remove the facia, as described in Section 36.
24 Remove the heater matrix, as described earlier in this Section.

227

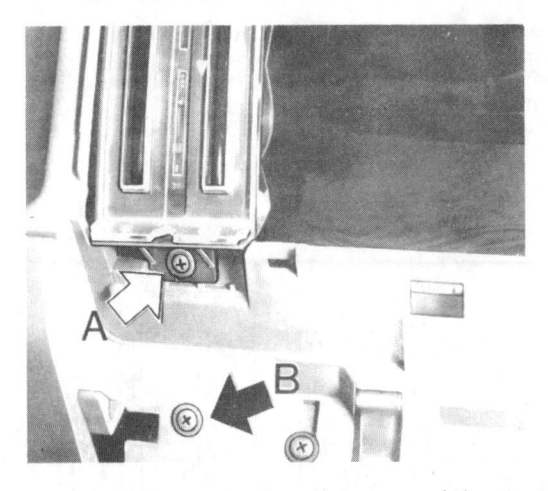

Fig. 12.29 Heater control unit retaining screw (A) and cable clamp retaining screw (B) (Sec 38)

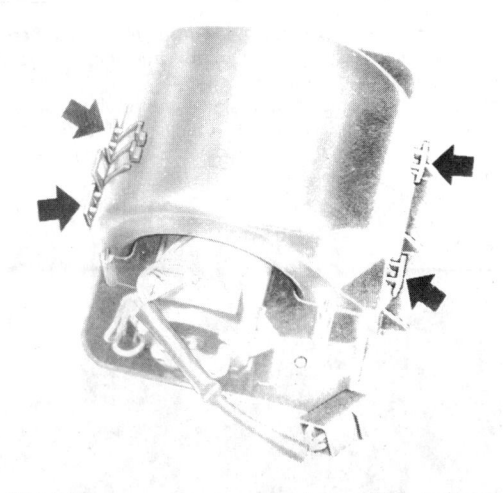

Fig. 12.30 Blower motor cover retaining clip locations – arrowed (Sec 38)

Fig. 12.31 Blower motor retaining screws (arrowed) and earth lead (A) (Sec 38)

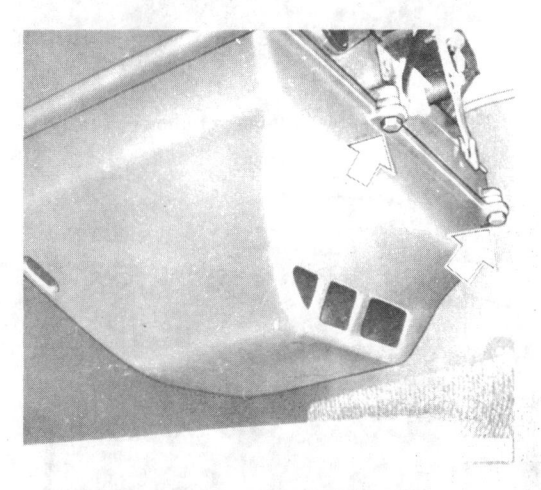

Fig. 12.32 Heater matrix cover retaining screws – arrowed (Sec 38)

Fig. 12.33 Heater matrix removal (Sec 38)

Small arrows indicate matrix mounting points

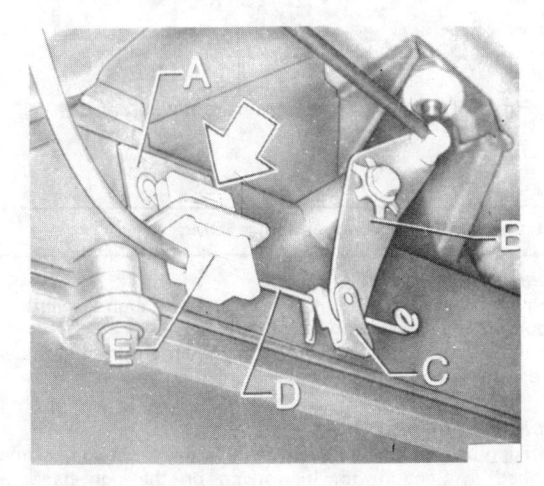

Fig. 12.34 Control cable attachments on air distribution housing (Sec 38)

A   Clip retainer bracket
B   Flap lever
C   Inner cable retaining clip
D   Inner cable
E   Outer cable clip
      Arrow indicates clip retainer catches

25 Disconnect the control cable ends from the mixer flap and air distribution flap on the side of the matrix housing, and release the outer cables from the clip retainers.
26 Undo the retaining screws and release the retaining clips securing the heater ducting to the bulkhead. Detach the ducting from the air distribution housing and remove it from the car.
27 Undo the two nuts, one each side, and the single bolt at the top securing the air distribution housing to the bulkhead (photo). Remove the unit from the car.
28 Refitting is the reverse sequence to removal.

Fig. 12.35 Side heater vent insert removal (Sec 39)

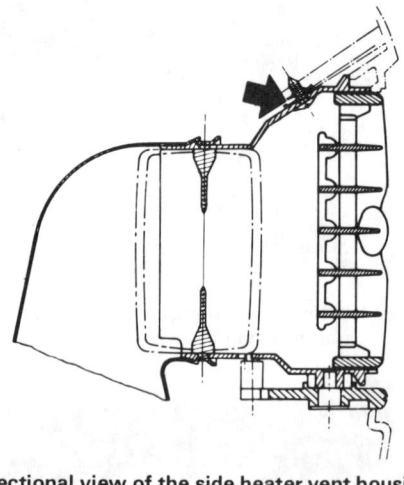

Fig. 12.36 Sectional view of the side heater vent housing (Sec 39)

Arrow indicates retaining screw location

Fig.12.37 Face level vent insert removal (Sec 39)

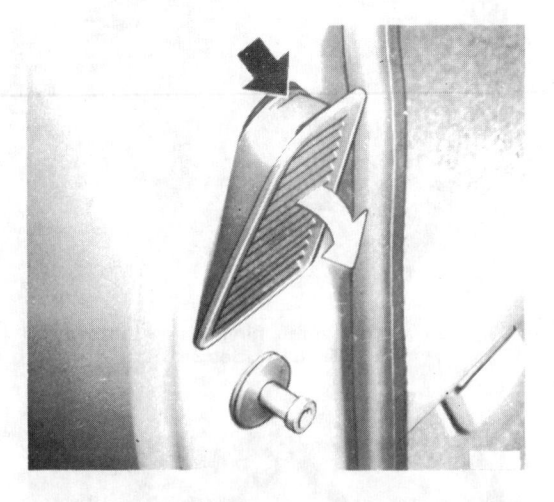

Fig. 12.38 Stale air exhaust grille removal (Sec 39)

39 Vents and grilles – removal and refitting

Side heater vent insert
1 Using a hooked length of stiff wire inserted through the vent grille at the top, firmly pull the vent out of its location.
2 Refit the vent by pushing it into position, but ensure that the vent cams and recesses engage properly.

Side heater vent housing
3 If working on the left-hand vent remove the glove compartment, as described in Section 35. If working on the right-hand vent disconnect the battery negative terminal, remove the fusebox cover and withdraw the fusebox by releasing the retaining catches. Move the fusebox to one side to gain access to the vent housing.
4 Undo the screw securing the vent housing to the facia, detach the housing from the facia and ducting and remove it.
5 Refitting is the reverse sequence to removal.

Face level vent insert
6 Rotate the insert downwards as far as possible, and pull it out of the housing.
7 Refit the vent in the same way.

Stale air exhaust grille
8 Depress the upper catch on the grille with a screwdriver, tip the grille forwards and remove it from the body pillar.
9 Refitting is the reverse sequence to removal.

Chapter 13 Supplement:
Revisions and information on later models

Contents

Introduction	1
Specifications	2
Routine maintenance – September 1986 on	3
Engine (1.0 ohv)	4
Engine identification	
Hydraulic timing chain tensioner	
Engine (1.2 and 1.3 ohc)	5
Engine identification	
Camshaft renewal	
Oil filter – thread change (1985 on)	
Crankshaft pulley bolt – modification	
Camshaft toothed belt automatic tensioner	
Engine (1.4 ohc)	6
General information	
Camshaft toothed belt – removal, refitting and tensioning (June 1990-on engines)	
Camshaft toothed belt tensioner – removal and refitting	
Cylinder head – removal and refitting	
Oil pump – removal and refitting	
Pistons and connecting rods – removal and refitting	
Crankshaft front oil seal – removal and refitting	
Crankshaft and main bearings – removal and refitting	
Engine – removal and refitting	
Crankshaft pulley bolt – modification	
Engine (1.6 ohc)	7
Engine procedures – general	
Camshaft toothed belt automatic tensioner	
Crankshaft pulley bolt – modification	
Cooling system (1.4 ohc models)	8
Cooling system filling	
Thermostat – removal, testing and refitting	
Water pump – removal and refitting	
Drivebelt – removal, refitting and adjustment	
Radiator electric cooling fan – removal and refitting	
Cooling fan thermal switch – testing, removal and refitting	
Fuel system (carburettor)	9
Weber 32 TL carburettor – adjustments	
Pierburg 1B1 carburettor – adjustments	
Pierburg 2E3 carburettor – description	
Pierburg 2E3 carburettor – adjustments	
Pierburg 2E3 carburettor – removal and refitting	
Pierburg 2E3 carburettor – overhaul	
Unleaded fuel	
Exhaust gas recirculation (EGR) system – description and maintenance	
Fuel system (Multec central fuel injection)	10
General description	
Fuel tank – removal and refitting	
Fuel filler neck – removal and refitting	
Fuel pump – removal and refitting	
Fuel pump control relay – renewal	
Fuel filter – renewal	
Fuel pressure – check	
Air cleaner – removal and refitting	
Air cleaner element – renewal	
Fuel injector – removal and refitting	
Fuel pressure regulator – removal and refitting	
Idle air control stepper motor – removal and refitting	
Throttle position sensor – removal and refitting	
Throttle valve housing – removal and refitting	
Electronic Control Unit (ECU) – removal and refitting	
Octane number plug and unleaded fuel – general	
Exhaust gas oxygen sensor – removal and refitting	
Coolant temperature sensor – removal and refitting	
Manifold absolute pressure (MAP) sensor – removal and refitting	
Road speed sensor – removal and refitting	
Exhaust system and catalytic converter – general	
Active carbon canister – renewal	
Fuel system (Bosch L3.1-Jetronic)	11
General description	
Fuel pump – removal and refitting	
Fuel pump diaphragm damper – removal and refitting	
Fuel filter – renewal	
Fuel gauge sender unit – removal and refitting	
Air cleaner element – renewal	
Air cleaner – removal and refitting	
Fuel injectors – removal, testing and refitting	
Airflow sensor unit – removal and refitting	
Airflow sensor output unit – removal and refitting	
Throttle valve switch – checking, removal and refitting	
Throttle housing – removal and refitting	
Fuel pump control relay – renewal	
Coolant temperature sensor – removal, testing and refitting	
Oil temperature switch (up to 1990 model year) – removal, testing and refitting	
Oil temperature sensor (from 1990 model year) – renewal	
Air intake temperature switch (up to model year 1990) – removal, testing and refitting	
Fuel pressure regulator – removal and refitting	
Supplementary air valve – removal, testing and refitting	
Coasting air valve (vacuum control) – adjustment	
Idle speed and mixture adjustment	
Exhaust gas recirculation (EGR) system – description and maintenance	
EGR valve – removal, cleaning and refitting	
Exhaust system – general	
Ignition system	12
Bosch distributor (1.0 models) – description	
Condenser (Bosch distributor) – removal and refitting	
Contact breaker points (Bosch distributor) – removal, refitting and adjustment	
Bosch distributor – removal and refitting	
Bosch distributor – overhaul	
Ignition system (1.4 models) – general	
Ignition coil and signal amplifier/module (1.4 models) – removal, checking and refitting	
Crankshaft pulley markings – 1.4 models	
Distributor (all models) – removal and refitting	
Distributor (1.4 models) – removal and refitting	
Ignition timing (1.4 models with Multec central fuel injection) – adjustment	

Chapter 13 Supplement: Revisions and information on later models

Ignition system (1.6 GTE/GSi) – description
Distributor (1.6 GTE/GSi) – removal and refitting
Ignition/fuel injection control unit (1.6 GTE/GSi) – removal and refitting
Ignition timing (1.6 GTE/GSi) – adjustment
Ignition coil (1.6 GTE/GSi) – checking
Retarding the ignition timing (for unleaded fuel)
Clutch and manual transmission 13
Gearchange linkage (five-speed) – adjustment
Transmission – modifications
Driveshafts 14
Driveshaft – overhaul
Braking system 15
Brake pedal relay lever – modification
Rear brake shoes (self-adjusting type) – renewal
Rear brake shoes – modifications
Rear wheel cylinder – modifications
Handbrake (self-adjusting brakes) – adjustment
Handbrake lever and long (main) cable (vehicles fitted with catalytic converter) – removal and refitting
Brake disc (all models) – refinishing
Brake master cylinder (later models) – overhaul
Brake pedal free play – checking and adjustment
Brake discs (1.6 GTE/GSi)
Front brake pads (from 1992 model year) – inspection and renewal
Front brake caliper (from 1992 model year) – removal, overhaul and refitting
Front brake disc (from 1992 model year) – removal and refitting
Electrical system 16
Fuses and relays
Heated rear window – precautions and repair
Fan and heated rear window switch (from 1991 model year) – removal and refitting
Electrically-operated exterior mirrors
Dim-dip headlamps
Headlamps-on warning buzzer
Headlamp switch (from 1991 model year) – removal and refitting
Headlamp (from 1991 model year) – removal and refitting
Headlamp range-control motor – removal and refitting
Headlamp range-control motor switch – removal and refitting
Headlamp washer system
Front direction indicator lamp (from 1991 model year) – removal and refitting
Hazard warning lamp switch (from 1991 model year) – removal and refitting
Rear foglamp switch (from 1991 model year) – removal and refitting
Cigarette lighter (from 1991 model year) – removal and refitting
Clock (from 1991 model year) – removal and refitting
Instrument panel and instruments (from 1991 model year) – removal and refitting

Heater controls (from 1991 model year) – removal and refitting
Underbonnet lamp
Luggage compartment lamp
Central door locking system
Electric windows
Front seat heaters
Radio equipment (factory-fitted) – removal
Direction indicator side repeater lamp – bulb renewal
Suspension 17
Modified components
Front wheel camber
Bodywork 18
Exterior mirror glass – renewal
Front seat head restraints – removal and refitting
Passenger grab handles – removal and refitting
Rear door – dismantling and reassembly
Rear quarter trim panel and window – removal and refitting
Front door armrest/doorpull/doorpocket – removal and refitting
Split type rear seats
Heater/ventilation – four- and five-door models
Radiator grille (from 1991 model year) – removal and refitting
Front bumper (from 1991 model year) – removal and refitting
Rear bumper (from 1991 model year) – removal and refitting
Front wing – removal and refitting
Centre console (from 1991 model year) – removal and refitting
Parcel shelf (from 1991 model year) – removal and refitting
Glovebox and cover (from 1991 model year) – removal and refitting
Facia (from 1991 model year) – removal and refitting
Sliding sunroof (up to August 1989) – removal and refitting
Sliding sunroof (from August 1989) – removal and refitting
Tailgate lock (from August 1988)
Seat belts – maintenance, removal and refitting
Rear seat belts – after-market fitting

1 Introduction

This Supplement contains information which is additional to, or a revision of, the material contained in the preceding twelve Chapters of this manual. Since this manual was first written, the Nova range has undergone significant changes and revisions. Whilst primarily intended to cover models produced from 1984 onwards, additional or revised information is included on earlier models. In order to use the Supplement to its best advantage, it is therefore recommended that it is referred to before the main Chapters of this manual.

Project vehicles

The vehicles used in the preparation of this Supplement, and appearing in many of the photographic sequences, were a 1986 model Nova 1.3 five-door Hatchback, a 1989 model Nova 1.6 GTE, and a 1991 model Nova 1.4 Merit with a catalytic converter.

2 Specifications

The specifications listed below are revisions of or supplementary to the specifications listed at the beginning of each Chapter

Engine – 1.2 and 1.3 litre ohc
Lubrication system
See Section 5

Oil filter (1.2 and 1.3 UNF thread)	Champion C103
Oil filter (1.2 and 1.3 Metric)	Champion G102

Torque wrench settings

	Nm	lbf ft
Crankshaft pulley:		
23 mm thread length (M10)	55	40
30 mm thread length (M12):		
Stage 1	55	40
Stage 2	Tighten through a further 45 to 60°	

Chapter 13 Supplement: Revisions and information on later models

Engine – 1.4 litre ohc
Details as for 1.2 and 1.3 litre engines, except for the following:

General
Designation:
  Carburettor models .... 14 NV
  Multec central fuel injection models .... C 14 NZ
Bore .... 77.6 mm
Stroke .... 73.4 mm
Capacity .... 1389 cc
Compression ratio 9.4:1
Output (DIN):
  Carburettor models .... 56 kW (75 bhp) at 5600 rpm
  Multec central fuel injection models .... No information available

Camshaft
Identification code:
  14 NV engine .... F
  C 14 NZ engine .... G
Camlift:
  14 NV engine .... 6.12 mm (inlet and exhaust)
  C 14 NZ engine .... 5.20 mm (inlet), 5.60 mm (exhaust)

Gudgeon pin
Length .... 55.0 mm
Diameter .... 18.0 mm
Clearance in piston .... 0.007 to 0.010 mm

Pistons
Standard diameter .... 77.535 to 77.605 mm

Torque wrench settings

	Nm	lbf ft
Big-end cap bolts:		
15 mm thread length	28	21
40 mm thread length:		
Stage 1	25	18
Stage 2	Tighten through a further 30°	
Crankshaft pulley:		
23 mm thread length (M10)	55	40
30 mm thread length (M12):		
Stage 1	55	40
Stage 2	Tighten through a further 45 to 60°	
Toothed belt tensioner bolt	20	15

Engine – 1.6 litre ohc
Details as for 1.2 and 1.3 litre engines, except for the following:

General
Designation E16SE
Bore .... 79.0 mm
Stroke .... 81.5 mm
Capacity .... 1598 cc
Compression ratio .... 10.0 : 1
Output (DIN) .... 74 kW (100 bhp) at 5600 rpm

Camshaft
Identification code .... D
Endfloat .... 0.09 to 0.21 mm
Camlift .... 5.61 mm (inlet), 6.12 mm (exhaust)

Gudgeon pin
Length .... 55.0 mm
Diameter .... 18.0 mm
Clearance in piston .... 0.007 to 0.010 mm

Valves
Valve stem diameter:
  Inlet .... 6.998 to 7.012 mm
  Exhaust .... 6.978 to 6.992 mm

Lubrication system
Oil capacity with filter change .... 3.5 litres (6.2 pts)
Dipstick MIN to MAX quantity .... 1.0 litre (1.8 pts)
Oil filter .... Champion G102

Chapter 13 Supplement: Revisions and information on later models

Torque wrench settings

	Nm	lbf ft
Crankshaft pulley:		
23 mm thread length (M10)	55	40
30 mm thread length (M12):		
Stage 1	55	40
Stage 2	Tighten through a further 45 to 60°	

Cooling system

Thermostat (all later models except 1.0 litre models)
Identification 92
Starts to open .... 92°C
Fully open .... 107°C

Expansion tank filler cap (all later models)
Opening pressure .... 1.20 to 1.35 bars

Cooling system capacity
1.4 models .... 5.7 litres (10.0 pints)
1.6 models .... 6.1 litres (10.7 pints)

Fuel system (carburettor)

Air filter element
1.2 litre (September 1975 on) and 1.4 litre models .... Champion W103
1.6 litre models .... Champion U559

Carburettor type
1.2 and 1.3 (13 SB engines to August 1985) litre models .... Pierburg 1B1
1.3 (13 SB engines from August 1985) and 1.4 litre models .... Pierburg 2E3

Weber 32 TL carburettor specification
Float level .... 23.50 to 24.00 mm (0.93 to 0.95 in)

Pierburg 1B1 carburettor specification
Float level .... 26.0 to 28.0 mm (1.02 to 1.10 in)
Accelerator pump injection quantity:
  1.2 models .... 0.55 to 0.85 cc/stroke
  1.3 models .... 0.45 to 0.75 cc/stroke

Pierburg 2E3 carburettor specification

Application .... 1.3 litre (from 1985 model year) and 1.4 litre models
Type .... Twin barrel downdraught, automatic choke

Calibration – 13 SB engine:	Primary	Secondary
Venturi diameter	20 mm	24 mm
Main jet	X97.5	X112.5
Air correction jet	80	100
Emulsion tube code number	88	60
Partial load enrichment orifice	0.5 mm	–
Pre-atomizer diameter	8 mm	7 mm
Mixture outlet orifice	2.5 mm	3.0 mm
Idle fuel jet	37.5	–
Idle air jet	130	–
Fuel load enrichment jet	–	85 to 105

Adjustment data – 13 SB engine (see text):
  Idle speed .... 900 to 950 rpm
  CO% at idle .... 1.0 to 1.5
  Choke valve pull-down gap .... 1.7 to 2.1 mm
  Fast idle speed .... 2100 to 2500 rpm
  Throttle valve fast idle gap .... 0.8 to 0.9 mm
  Accelerator pump delivery .... 1.03 to 1.27 cc/stroke
  Float level .... 19 to 39 (0.75 to 1.54 in)

Calibration – 14 NV engine:	Primary	Secondary
Main jet	95	110
Air correction jet	117.5	90
Idle fuel jet	45	–
Idle air jet	130	–

Adjustment data – 14 NV engine (see text):
  Idle speed .... 900 to 950 rpm
  CO% at idle .... 0.5 to 1.0
  Choke valve pull-down gap .... 1.9 ± 0.2 mm
  Fast idle speed .... 2200 to 2600 rpm
  Throttle valve fast idle gap .... 0.85 ± 0.05 mm
  Accelerator pump delivery .... 1.2 to 1.35 cc/stroke
  Float level .... 29 ± 1 mm

Chapter 13 Supplement: Revisions and information on later models

Fuel injection system (Multec central fuel injection)
General
Application.. C 1.4 NZ engine
Type.. Central (single-point) fuel injection – Rochester TBI 700 type
Idle speed.. 830 to 990 rpm
CO% at idle.. Less than 0.4
Fuel pump delivery (line) pressure............... 0.76 bar
Air filter element... Champion W103

Torque wrench settings
	Nm	lbf ft
Fuel tank strap nuts	20	15
Fuel filler neck clampbolt:		
2-door models	15	11
4-door models	10	7
Fuel inlet and return unions (on throttle valve housing)	35	26
Coolant temperature sensor	10	7.4
Oxygen sensor	30	22
Catalytic converter flange joints	20	15
Throttle valve housing	22	16

Fuel injection system (Bosch L3.1-Jetronic)
General
Application.. 1.6 GTE/GSi
Type.. Bosch L3.1-Jetronic with rear-mounted electric fuel pump
Idle speed.. 900 to 950 rpm
CO% at idle.. 0.2 to 0.4
Fuel pump delivery..................................... 1 litre/min at 12 volts
Normal fuel pressure with intake manifold connection open to atmospheric pressure.. 2.5 bar (36.3 lbf/in^2)
Air filter element... Champion U559
Fuel filter element....................................... Champion L201

Torque wrench settings
	Nm	lbf ft
Oil temperature switch	20	15
Air intake temperature switch	6	4

Ignition system
Distributor
Type:
    1.0 litre engines (adjustment data as for Delco-Remy 3470269 distributor in Chapter 4)................... Bosch 0231170340
    1.2 and 1.3 litre engines............................ AC Delco or Bosch
    1.4 litre engines... Bosch or Lucas
    1.6 litre engines... Bosch
Direction of rotation (except 1.0 litre engines)..... Anti-clockwise
Direction of rotation (1.0 litre engines).............. Clockwise

Ignition coil
Type.. AC Delco or Bosch (according to engine type)
Primary resistance:
    AC Delco (later 1.2 and 1.3 litre engines)..... 0.43 ohms
    Bosch (14 NV engine)................................ 0.82 ± 0.08 ohms
    Bosch (1.6 litre engine).............................. 0.72 ± 0.07 ohms
Secondary resistance:
    AC Delco (later 1.2 and 1.3 litre engines)..... 5000 ohms
    Bosch (14 NV engine)................................ 7450 to 9050 ohms
    Bosch (1.6 litre engine).............................. 6900 to 8500 ohms

Ignition timing
All except 1.4 litre engines........................... 10 ± 2° BTDC at 700 to 1000 rpm
1.4 litre engines... 5 ± 2° BTDC at 700 to 1000 rpm
Timing mark location................................... Crankshaft pulley notch (see text)

Spark plugs
Type:
    1.4 and 1.6 litre models............................. Champion RN7YCC or RN7YC
Electrode gap:
    RN7YCC... 0.8 mm (0.032 in)
    RN7YC... 0.7 mm (0.028 in)

Torque wrench setting
	Nm	lbf ft
Spark plugs (1.4 and 1.6 litre models)	20	15

Clutch – 1.6 GTE/GSi models
Driven plate diameter .... 200 mm (7.9 in)

Transmission
General
Designation and application:
- F 10/5 W .... 1.4 litre models except SR
- F 13/5 CR .... 1.4 litre SR models and 1.6 GTE/GSi model

Reverse gear ratio (all models)
- Transmissions with an 'A' prefix 3.31:1
- Transmissions without an 'A' prefix 3.18:1

Gear ratios (F 13/5 CR transmission)
- 1st 3.55:1
- 2nd 2.24:1
- 3rd 1.48:1
- 4th 1.12:1
- 5th 0.89:1
- Reverse .... 3.31:1 (3.18:1 on transmission with 'A' prefix)

Final drive ratio (F 13/5 CR transmission)
- 1.4 litre models 3.94:1
- 1.6 litre models 3.74:1

Braking system – 1.6 GTE/GSi models
Front ventilated discs
- Disc diameter .... 236.0 mm (9.3 in)
- Disc thickness .... 20.0 mm (0.79 in)
- Minimum thickness after refinishing .... 18.0 mm (0.71 in)

Braking system – from 1992 model year

Torque wrench settings	Nm	lbf ft
Front caliper guide bolts	30	22
Front caliper-to-mounting bracket bolts	30	22
Front caliper bracket-to-steering knuckle bolts	95	70

Front suspension
General
- Camber (laden*) – 1.6 GTE/GSi models .... 0°45' negative to 0°45' positive
- Castor (laden*) – from 1988 model year .... 0°45' positive to 2°45' positive

Laden indicates a vehicle containing two front seat occupants and a half-filled fuel tank

Rear suspension
General
- Camber (laden*) – from 1987 model year .... 0°40'' negative to 1°35' negative
- Toe setting from 1987 model year (laden*):
  - All models except 1.6 GTE/GSi .... 0.5 mm (0.002 in) toe-out to 4.0 mm (0.16in) toe-in
  - 1.6 GTE/GSi models .... 0 to 4.0 mm (0.16 in) toe-in

Laden indicates a vehicle running two front seat occupants and a half-filled fuel tank

Roadwheels and tyres
Roadwheels (1.6 GTE/GSi models)
- Type .... Pressed steel or aluminium alloy
- Size .... 5J x 14

Tyres (1.6 GTE/GSi models)
- Size .... 175/65 HR 14

Tyre pressures (cold) – revisions and additions for later models

145 SR 13:	Bar	lbf/in^2
Up to 3 persons:		
Front	1.7	24
Rear	1.7	24
Fully loaded:		
Front	2.0	29
Rear	2.4	34

Chapter 13 Supplement: Revisions and information on later models

Tyre pressures (cold) – revisions and additions for later models (continued)

	Bar	lbf/in²
175/70 SR 13:		
Up to 3 persons:		
Front	1.7	24
Rear	1.7	24
Fully loaded:		
Front	2.0	29
Rear	2.4	34
165/65 SR 14:		
Fully loaded:		
Front	2.0	29
Rear	2.4	34
175/65 HR 14:		
Up to 3 persons:		
Front	1.8	26
Rear	1.6	23
Fully loaded:		
Front	2.0	29
Rear	2.2	31

Dimensions from 1991 model year
Overall length:
- Saloon .... 3986 mm (159.9 in)
- Hatchback ...... 3652 mm (143.8 in)

Overall width:
- Saloon .... 1542 mm (60.7 in)
- Hatchback ...... 1535 mm (60.4 in)

Rear track .... 1300 mm (51.2 in)

General weights (later models)
Kerb weight
Two-door Saloon:
- Nova .... 745 kg (1643 lb)
- Nova Merit ..... 760 kg (1676 lb)
- Nova L .... 775 kg (1709 lb)

Four-door Saloon:
- Nova Merit ..... 780 kg (1720 lb)
- Nova L 1.2 ..... 795 kg (1753 lb)
- Nova L 1.3 ..... 800 kg (1764 lb)
- Nova GL .... 820 kg (1808 lb)

Three-door Hatchback:
- Nova .... 740 kg (1632 lb)
- Nova Merit 1.0 ... 740 kg (1632 lb)
- Nova Merit 1.2 ... 750 kg (1654 lb)
- Nova L .... 765 kg (1687 lb)
- Nova SR .... 775 kg (1709 lb)
- Nova GTE .... 834 kg (1839 lb)

Five-door Hatchback:
- Nova Merit ..... 770 kg (1698 lb)
- Nova L 1.2 ..... 785 kg (1731 lb)
- Nova L 1.3 ..... 795 kg (1753 lb)
- Nova GL .... 825 kg (1819 lb)

Maximum towing weights

	4-speed	5-speed
Trailer with brakes:		
1.0 models	500 kg (1103 lb)	550 kg (1213 lb)
1.2 models	600 kg (1323 lb)	600 kg (1323 lb)
1.3, 1.4 and 1.6 models	–	800 kg (1764 lb)
Trailer without brakes:		
All models	400 kg (882 lb)	

Maximum roof rack load
All models .... 80 kg (176 lb)

3 Routine maintenance – September 1986 on

1 From September 1986, the manufacturers have extended some of the routine maintenance time intervals as follows:
Previously: 9000 miles (15 000 km) or six months
Now: 9000 miles (15 000 km) or twelve months
Previously: 18 000 miles (30 000 km) or twelve months
Now: 18 000 miles (30 000 km or two years)

Note: *On vehicles with self-adjusting rear brakes, lining wear need only be checked every 18 000 miles, or two years*

2 If reducing servicing expenses is of paramount importance, then follow these extended service periods. Where economy is not essential, it is recommended that the original maintenance schedule featured in the introductory Section of this manual is adhered to.

3 On ohv engines, the valve clearances must be checked and adjusted after the first 600 miles (1000 km) on new vehicles and those fitted with reconditioned engines.

Underbonnet view on 1.6 GTE model

1 Battery
2 Supplementary air valve
3 Throttle housing
4 Throttle valve switch
5 Fuel distribution tube
6 Inlet manifold and air box
7 Exhaust gas recirculation valve
8 Fuel pressure regulator
9 Ignition/fuel injection control unit
10 Coasting air valve
11 Brake vacuum servo unit
12 Fuel pump control relay
13 Windscreen wiper motor
14 Windscreen washer reservoir
15 Front suspension strut upper mounting
16 Brake fluid reservoir
17 Cooling system expansion tank
18 Ignition coil
19 Electric cooling fan
20 Distributor
21 Engine oil level dipstick
22 Engine oil filler cap
23 Exhaust manifold
24 Fuel injectors
25 Air inlet duct
26 Airflow sensor unit
27 Multi-tone horn relay and fuse

Front underside view on 1.6 GTE model

1. Differential cover plate
2. Engine/transmission rear mounting
3. Driveshaft inner constant velocity joint
4. Intermediate exhaust pipe
5. Steering gear
6. Twin-branch exhaust downpipe
7. Front suspension control arm
8. Right-hand driveshaft
9. Anti-roll bar
10. Tie-bar front mounting
11. Engine oil drain plug
12. Oil filter
13. Clutch access cover
14. Horn
15. Left-hand driveshaft
16. Transmission oil level plug

Underbonnet view on 1.4 model with Multec central fuel injection (air cleaner removed)

1. Battery
2. Alternator
3. Fuel pump control relay
4. Manifold absolute pressure (MAP) sensor
5. Throttle position sensor
6. Octane number plug
7. Fuel injector
8. Throttle valve housing
9. Fuel pressure regulator
10. Idle air control stepper motor
11. Brake vacuum servo unit
12. Engine oil filler cap
13. Engine oil level dipstick
14. Distributor
15. Ignition coil
16. Cooling system expansion tank
17. Windscreen washer reservoir
18. Brake fluid reservoir
19. Windscreen wiper motor
20. Accelerator (throttle) cable

Chapter 13 Supplement: Revisions and information on later models

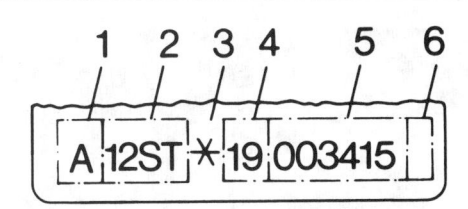

Fig. 13.1 Engine identification number (Sec 4)

1 Build letter for special export markets (applicable only to Austria/Sweden/ Switzerland)
2 Engine type identification; example shown indicates: (12) engine capacity 1.2 litres; (S) high-compression engine requiring four-star fuel; (T) power output (randomly selected letter)
3 Star indicates capability to run on low-lead petrol
4 Manufacturing plant identification
5 Engine serial number
6 Blank

4 The brake hydraulic fluid must be renewed every twelve months, regardless of mileage.
5 Except in abnormally dusty conditions, the air cleaner element need only be renewed every 18 000 miles (30 000 km), or two years.
6 The carburettor fuel filter (10S engine only) also needs to be renewed every 18 000 miles (30 000 km), or two years.
7 On models where the coolant cannot be drained completely (because of the lack of a block drain plug, or because the radiator bottom hose is not at the lowest point of the radiator), it is recommended that the cooling system by flushed whenever the coolant is renewed.
8 Owners of later 1.2, 1.4 or 1.6 litre models should note that the manufacturer no longer calls for routine maintenance of the toothed belt. However, due to the importance of the belt to the engine's reliability, owners are advised to check the belt's condition at the interval given in the *'Routine maintenance'* Section at the front of this manual (noting the new time intervals given in paragraph 1 above).
9 Unbolt (or unclip, as applicable) the toothed belt cover. Using a spanner applied to the crankshaft pulley bolt, rotate the crankshaft so that the full length of the belt is checked. Check the belt carefully for any signs of uneven wear, splitting, cracks (especially at the roots of the belt teeth) or oil contamination, and renew it if there is the slightest doubt about its condition; refer to Section 6 of this Chapter and to Section 31 of Chapter 1 for details. Obviously, the belt's tension, once set on installation, does **not** require checking or adjustment.

4 Engine (1.0 ohv)

Engine identification
1 The engine capacity/type may be identified by reference to the engine number. The second block of characters ('2' in Fig. 13.1) indicates the engine type.

Hydraulic timing chain tensioner
2 On some engines, a hydraulic type of timing chain tensioner is used.
3 Before removing the timing chain or the tensioner, fully depress the slipper to lock it in the fully retracted position.
4 When refitting the tensioner, have it in the retracted position and, after the timing chain has been installed, lightly depress the slipper. This will automatically set it in its extended state.

5 Engine (1.2 and 1.3 ohc)

Engine identification
1 Refer to Section 4 of this Chapter.

Camshaft renewal
2 It is possible for the camshaft to become excessively worn on 1.3 engines built up to 1982 (engine number 13S - 814742).
3 Where this is the case, renew the camshaft, as described in Chapter 1, Sections 32 and 33, with the improved 'TIG' type camshaft.
4 Before installing the new camshaft, an oil channel must be drilled in the camshaft housing in the following way.
5 Locate the defective camshaft in the camshaft housing, so that the end face of the camshaft is flush with the inner face of the camshaft

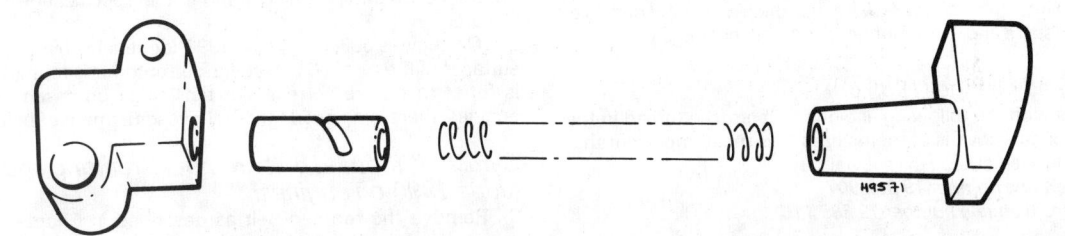

Fig. 13.2 Hydraulic type timing chain tensioner components (Sec 4)

Fig. 13.3 Cast-in identification lug (arrowed) between 4th exhaust cam and rear bearing on 1.3 engine TIG camshaft (Sec 5)

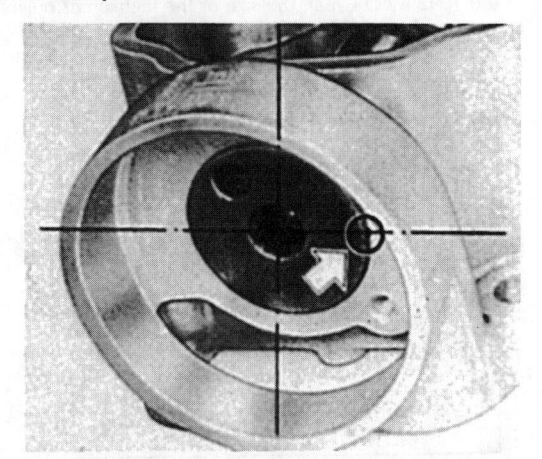

Fig. 13.4 Oil channel drilling point in camshaft bearing – arrowed (Sec 5)

Fig. 13.5 Drilling camshaft bearing oil channel (Sec 5)

Fig. 13.6 Oil channel drilled in camshaft bearing (Sec 5)

bearing with the distributor recess. Lock the camshaft to prevent it moving, using a pair of self-locking pliers.
6 Centre-punch the edge of the bearing as shown in Fig. 13.4.
7 Now drill a hole 4.0 mm (0.165 in) diameter at the punch mark, 16.0 mm (0.63 in) deep. On no account exceed the specified depth. Accurate drilling will result in an oil channel being formed in the camshaft housing bearing surface.
8 Remove all swarf, and de-burr the channel.
9 If a new TIG type camshaft has been fitted, it is imperative that when the engine is first started, it is **not** allowed to idle. Instead, the initial four minutes of running should be carried out in the following way; using a tachometer connected in accordance with its manufacturer's instructions:

   Stage 1 – one minute at 2000 rpm
   Stage 2 – one minute at 1500 rpm
   Stage 3 – one minute at 3000 rpm
   Stage 4 – one minute at 2000 rpm

10 Failure to follow this procedure will cause initial wear and damage, due to low oil pressure and lack of lubrication at the bearings.

Oil filter – thread change (1985 on)
11 On vehicles with the following engine numbers, the thread in the oil filter and its fixing sleeve is of metric type. It is most important that the correct filter is used at time of renewal.
   12 ST: from engine number 19 382 907
   13 S/13 SB: from engine number 19 389 115

Crankshaft pulley bolt – modification
12 From the 1992 model year, the size of the crankshaft pulley bolt has been increased from M10 to M12. When refitting a later M12 type bolt, the revised torque wrench setting given in the Specifications at the start of this Chapter must be used.

Camshaft toothed belt automatic tensioner
13 1.2 litre engines built from June 1990, identifiable by the squared-off top surfaces of the toothed belt covers, are fitted with a toothed belt automatic tensioner. This means that there is no need for regular adjustment of belt tension, and this requires a modified procedure on refitting the belt; refer to Section 6 of this Chapter.

6 Engine (1.4 ohc)

General information
1 The 1.4 litre engines are mechanically similar to the 1.2/1.3 units described in Part B of Chapter 1. Refer to Chapter 1, noting the differences mentioned below and in the Specifications Section of this Chapter.
2 On engines built from June 1990, identifiable by the squared-off top surfaces of the toothed belt covers, an automatic toothed belt tensioner is fitted to ensure correct belt tensioning on assembly. With these engines, there is no need for regular readjustment of belt tension.

Camshaft toothed belt – removal, refitting and tensioning (June 1990-on engines)
3 Remove the toothed belt as described in Chapter 1, Section 31, noting the following points. It may be necessary to remove the air cleaner and intake duct to improve access. The toothed belt cover is in two parts, the lower cover being fitted over the water pump; either may be secured by clips or hexagon-headed screws may be used. It is only necessary to drain the cooling system if a leak results from slackening the pump bolts, or from moving the pump (see paragraph 7 below). Lock the tensioner in its slackest position (paragraph 12 below), then withdraw the belt as described.
4 Whenever the toothed belt is disturbed, its tension must be set as follows – note that this procedure must **only** be carried out on a **cold** engine.
5 Refit the toothed belt so that the front run is taut, then remove the tensioner locking rod, and check that the index mark on the crankshaft sprocket and the stamped line on the camshaft sprocket are aligned with their respective oil pump housing/timing belt rear cover notches (photo).
6 Temporarily refit the crankshaft pulley bolt, and remove the spark plugs so that the crankshaft can be rotated easily. Turning the water pump is easier if a special spanner (Kent-Moore part number KM-421-A) is used; alternatives are available such as Sykes-Pickavant part number 031300. If the pump cannot be turned to adjust the belt tension, it can be freed by lightly striking it from below using a hammer and a long drift. If a coolant leak results from disturbing the pump, drain the cooling system, remove the pump and renew its O-ring, lubricating it with grease or petroleum jelly (Chapter 2, Section 8).

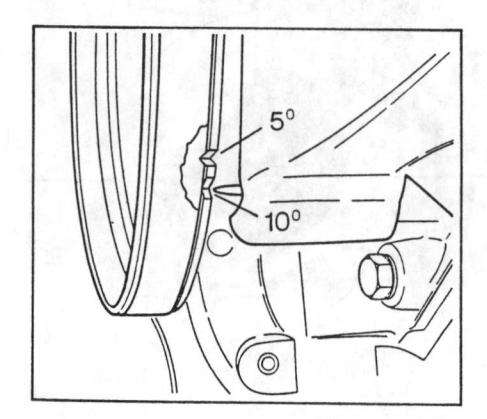

Fig. 13.7 Ignition timing pointer aligned with the crankshaft pulley 10° BTDC notch – 1.4 engines (Sec 6)

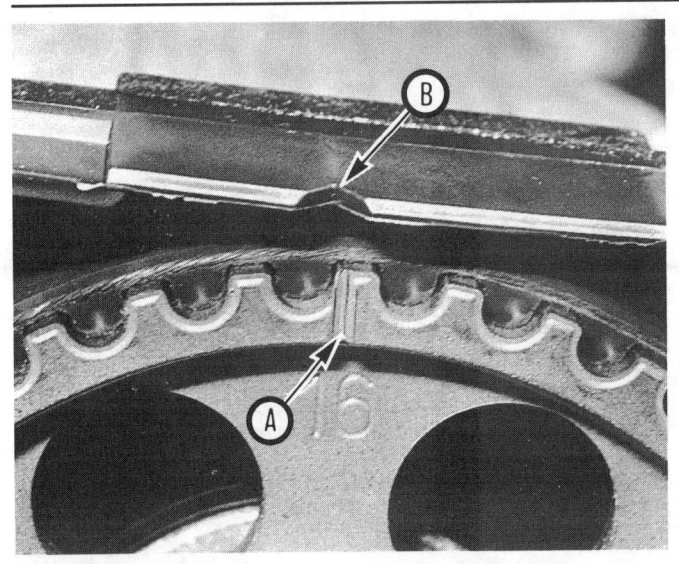

Chapter 13 Supplement: Revisions and information on later models

6.5 Camshaft sprocket mark (A) aligned with toothed belt rear cover marking (B) – 1.4 engine

6.11 Toothed belt tension is correct when tensioner indicator pointer aligns with baseplate notch centre (A). Note crankshaft sprocket index mark aligned with pump housing/rear cover notches (B) – 1.4 engine

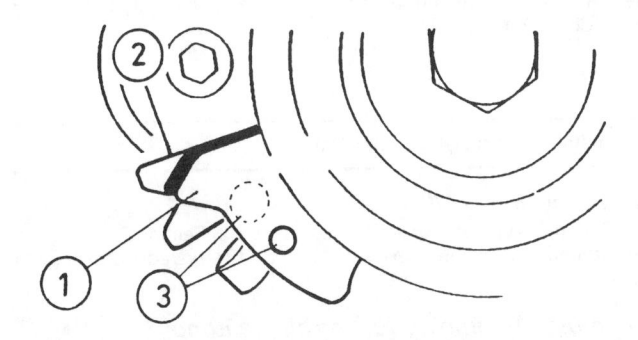

Fig. 13.8 To lock tensioner for removal/refitting, move indicator arm (1) clockwise until locking rod can be passed through holes (3) in arm and baseplate (2) (Sec 6)

15 When working on a fuel injection engine, refer to Section 10 of this Chapter for details of removal and refitting of the fuel injection components.
16 To enable removal of the cylinder head, it will be necessary to remove the main toothed belt rear cover, which is secured with Allen screws. To remove the main toothed belt rear cover, it will be necessary to remove the camshaft sprocket, the crankshaft sprocket and the toothed belt tensioner (where applicable).
17 When refitting the cylinder head, new securing bolts must be used.
18 The procedure for cylinder head bolt tightening is as described in Chapter 1, Section 32.

Oil pump – removal and refitting

19 The procedure is as described in Chapter 1, Section 37, with the following differences.
20 To allow removal of the main toothed belt rear cover, the camshaft

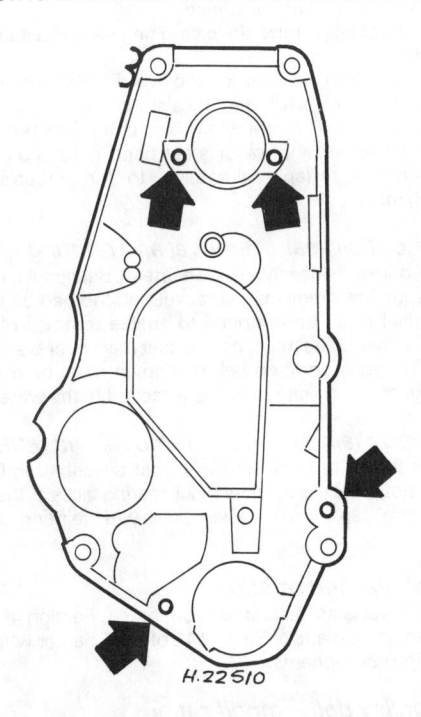

7 Tighten the belt by slackening the three water pump bolts, and turning the pump clockwise until the holes in the tensioner indicator arm and baseplate align (the tensioner indicator arm will then have moved fully clockwise to its stop).
8 Lightly tighten the pump securing bolts, just sufficiently to prevent the pump from moving.
9 Turn the crankshaft smoothly (or else the belt may jump a tooth), without jerking it or moving the water pump, through two complete revolutions clockwise until the timing marks are again aligned.
10 Slacken the toothed belt by turning the water pump anti-clockwise until the tensioner's indicator pointer is in the centre of its baseplate notch; the belt tension is then correct.
11 Tighten the pump bolts, turn the crankshaft through two turns clockwise, and recheck the setting (photo). If the pointer and notch are not aligned, the operation must be repeated.

Camshaft toothed belt tensioner – removal and refitting

12 To lock the tensioner in its slackest position for removal and refitting, move the tensioner indicator arm clockwise until the holes align in the baseplate and arm, and a close-fitting rod can be passed through them (Fig. 13.8); unbolt the tensioner.
13 On refitting, engage the tensioner baseplate lug in the hole in the oil pump housing, tighten the tensioner bolt securely and remove the locking rod; the tensioner should be quite free to move. Set the belt tension as described above.

Cylinder head – removal and refitting

14 The procedure is broadly similar to that described in Chapter 1, Section 32, but the following differences should be noted.

Fig. 13.9 Camshaft toothed belt rear cover fixing screws (arrowed) – engines with toothed belt tensioner (Sec 6)

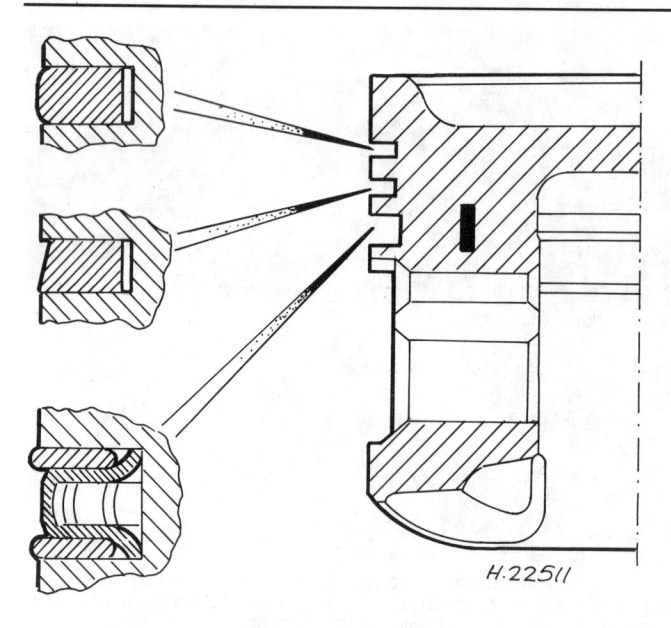

Fig. 13.10 A cross-sectional view of the piston ring orientation – 1.4 engine (Sec 6)

sprocket, crankshaft sprocket and toothed belt tensioner (where applicable) must be removed. The main toothed belt rear cover is secured with Allen screws.
21 An oil pressure or oil temperature sensor may be fitted to the oil pump, in which case the wiring must be disconnected before removing the pump. If the sensor is removed form the pump, it must be refitted using a new gasket.
22 When refitting the oil pick-up pipe, apply suitable thread-locking compound to the oil pick-up pipe securing bolts before inserting and tightening them.

Pistons and connecting rods – removal and refitting
23 On later models, arrow direction markings may be found on the piston crowns. Where this is the case, the arrows point towards the camshaft toothed belt end of the engine.
24 When refitting the pistons, the piston ring gaps must be offset by 180° from each other.
25 The second piston ring carries the word 'TOP' which must be uppermost when the ring is fitted to the piston.
26 In the case of the oil control ring (which comprises three separate components), the upper and lower steel band gaps must be offset between 25 and 50 mm (approximately 1 to 2 in) either side of the expander ring gap.

Crankshaft front oil seal – removal and refitting
27 The procedure is as described in Chapter 1, Section 41, noting that the camshaft sprocket, crankshaft sprocket and toothed belt tensioner (where applicable) must be removed to enable removal of the main toothed belt rear cover. The main toothed belt rear cover is secured with Allen screws. The main toothed belt rear cover must be manipulated from the smaller toothed belt rear cover attached to the water pump.

Crankshaft and main bearings – removal and refitting
28 The procedure is broadly similar to that described in Chapter 1, Section 44, but note that the cylinder block mating faces of the front and rear main bearing caps must be smeared with jointing compound before fitting.

Engine – removal and refitting
29 Engine removal is as described in Chapter 1, Section 46, but note that reference must be made to Section 10 of this Chapter when dealing with fuel injection components.

Crankshaft pulley bolt – modification
30 From the 1992 model year, the size of the crankshaft pulley bolt has been increased from M10 to M12. When refitting a later M12 type bolt, the revised torque wrench setting given in the Specifications at the start of this Chapter must be used.

7 Engine (1.6 ohc)

Engine procedures – general
1 Overhaul and repair procedures for the 1.6 engine are identical to those for the 1.2 and 1.3 engines, except where references are made to the fuel system components. Where this occurs, refer to Section 11 of this Chapter for details of the fuel injection components fitted to the 1.6 engine.

Camshaft toothed belt automatic tensioner
2 1.6 litre engines built from June 1990, identifiable by the squared-off top surfaces of the toothed belt covers, are fitted with a toothed belt automatic tensioner. This means that there is no need for regular readjustment of belt tension, and this requires a modified procedure on refitting the belt; refer to Section 6 of this Chapter.

Crankshaft pulley bolt – modification
3 From the 1992 model year, the size of the crankshaft pulley bolt has been increased from M10 to M12. When refitting a later M12 type bolt, the revised torque wrench setting given in the Specifications at the start of this Chapter must be used.

8 Cooling system (1.4 ohc models)

Cooling system – filling
1 The procedure is as described in Chapter 2, Section 5, noting that the temperature gauge transmitter must be removed from the inlet manifold.

Thermostat – removal, testing and refitting
2 The procedure is as described for 1.2 and 1.3 litre models in Chapter 2, Section 7, but note that the camshaft toothed belt, camshaft sprocket, crankshaft sprocket and toothed belt tensioner (where applicable) must be removed to enable removal of the main toothed belt rear cover. The main toothed belt rear cover is secured with Allen screws. The main toothed belt rear cover must be manipulated from the smaller toothed belt rear cover attached to the water pump.
3 Refer to Section 6 of this Chapter for details of camshaft toothed belt removal and refitting.

Water pump – removal and refitting
4 The water pump removal and refitting procedure is as described for 1.2 and 1.3 models in Chapter 2, Section 8, noting the following.
5 Refer to Section 6 of this Chapter for details of camshaft toothed belt removal and refitting.
6 Before removing the water pump, it will be necessary to remove the securing screw from the smaller toothed belt rear cover attached to the water pump. The pump can then be removed complete with the smaller toothed belt rear cover. If desired, the toothed belt rear cover can be removed from the pump by rotating the cover to release it from the flange on the pump.

Drivebelt – removal, refitting and adjustment
7 The procedure is as described for 1.2 and 1.3 models in Chapter 2, Section 9.

Radiator electric cooling fan – removal and refitting
8 Although the procedure basically follows that given in Chapter 2, Section 10, it may also be necessary to remove the distributor and withdraw the engine oil dipstick to improve access.

Cooling fan thermal switch – testing, removal and refitting
9 The procedure is as described in Chapter 2, Section 11, but note that a new sealing ring must be used when refitting the switch.

Chapter 13 Supplement: Revisions and information on later models

Fig. 13.11 Clamp screw (arrowed) for accelerator pump adjustment on Pierburg 1B1 carburettor (Sec 9)

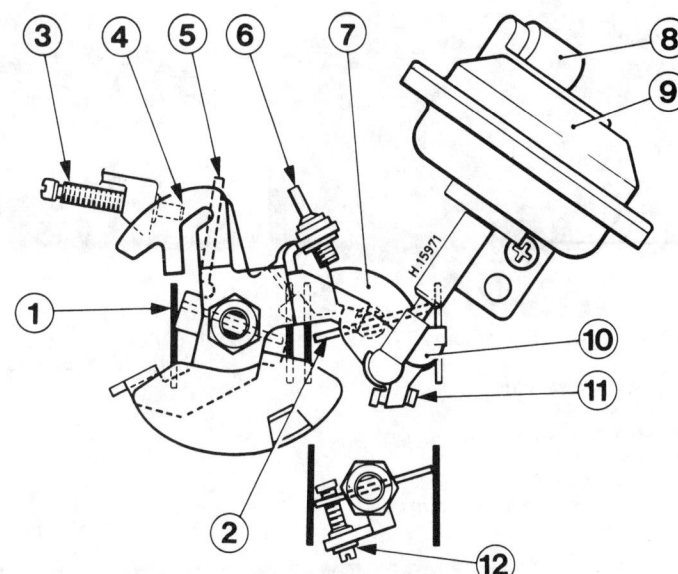

Fig. 13.12 Throttle linkage – Pierburg 2E3 carburettor (Sec 9)

1. Primary throttle valve
2. Secondary throttle valve
3. Throttle stop screw
4. Segment lever
5. Throttle lever
6. Fast idle adjustment screw
7. Secondary interlock lever
8. Mounting bracket
9. Vacuum unit
10. Spring
11. Fork lever
12. Secondary stop screw

9 Fuel system (carburettor)

Weber 32 TL carburettor – adjustments

1 It is rare for the following adjustments to be required, except after new components have been fitted, or as the result of a fault having been diagnosed attributable to maladjustment of the following assemblies.

Float level

2 Refer to Chapter 3, and remove the carburettor top cover and flange gasket. Hold the cover vertically so that the floats hang to their fullest extent from it.

3 Measure from the surface of the cover flange to the tips of the floats nearest to the cover. Compare the measurement with the dimension specified in this Chapter. If it does not match, then bend the float hinge tab as necessary.

Accelerator pump stroke

4 If incorrect accelerator adjustment is suspected, remove the carburettor, and hold it over a suitable container.

5 Operate the pump lever ten times, and then measure the ejected fuel in a measuring glass. Divide the total volume collected by ten, and compare the result with the quantity specified in Chapter 3. If the figures do not agree, check the roller on the accelerator pump arm for wear, or the arm for distortion. Bending the pump arm very carefully will vary the fuel volume ejected. Make sure that the float chamber is kept filled during the time that the pump lever is operated. Use a piece of flexible tubing and a small funnel attached to the fuel inlet nozzle on the carburettor through which to pour the fuel.

Pierburg 1B1 carburettor – adjustments

6 As for the Weber 32 TL carburettor, it is rare for the following adjustments to be required. (Refer to paragraph 1 of this Section).

Float level

7 The operations are very similar to those described for the Weber 32 TL carburettor (paragraphs 2 and 3).

Accelerator pump stroke

8 Checking is carried out as described for the Weber 32 TL carburettor, but if adjustment is needed, proceed as follows.

9 Release the clamp screw (Fig. 13.11) and turn the cam plate slightly.

10 Refill the carburettor bowl with fuel, and recheck the volume ejected as previously described. As the adjustment is based upon trial and error, further adjustment of the cam plate may be required to obtain the correct setting.

Pierburg 2E3 carburettor – description

11 The Pierburg 2E3 carburettor is a dual-barrel downdraught instrument with automatic choke. It is fitted to 1.3 litre models after the 1985 model year and to the 1.4 litre carburettor models introduced in late 1989 (photos).

12 The automatic choke is of the usual strangler type, controlled by a bi-metallic spring; the choke cover is heated electrically, and also by the engine coolant, ensuring a rapid response to changing engine and ambient temperature. Over-choking is avoided by the eccentric mounting of the choke valve plate, by a vacuum pull-down system, and by a mechanical linkage with the throttle mechanism.

13 With the engine at operating temperature, idling mixture is supplied via a bypass system. Although an idle cut-off valve is shown in some of the illustrations, this is not normally fitted to the Nova. Unusually, idle speed adjustment takes place at the throttle stop screw.

14 Opening of the throttle valves is sequential. The primary throttle valve is opened mechanically; the secondary throttle valve is opened by vacuum developed in both venturis, but is prevented from so doing until the primary valve is at least halfway open. For safety reasons, both throttle valves are closed mechanically.

15 Efficient operation under all speed and load conditions is ensured by a part-load enrichment valve, and by primary and secondary transition systems. An accelerator pump provides extra fuel needed for rapid acceleration.

Pierburg 2E3 carburettor – adjustment

Idle speed and mixture

16 The engine must be at normal operating temperature (but the radiator fan must not be working), and all other factors which may affect the idle speed (spark plug conditions and gaps, ignition timing, etc) must be correctly set. The air cleaner element must be clean, and the hot air pick-up system (where applicable) must be operating correctly. Electrical loads (lights, heated rear windows, etc) must be switched off. The crankcase ventilation hose must be disconnected, and its connection in the air filter plugged.

17 Connect a tachometer and an exhaust gas analyser to the engine in

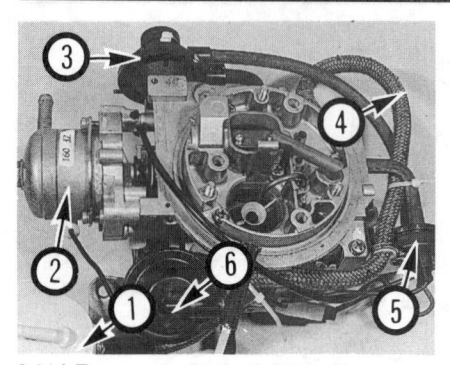

9.11A Top cover – Pierburg 2E3 carburettor

1 Vapour separator
2 Choke cover
3 Choke pull-down unit
4 Fuel hose
5 Vacuum switch
6 Secondary throttle vacuum unit

9.11B View of choke cover side – Pierburg 2E3 carburettor

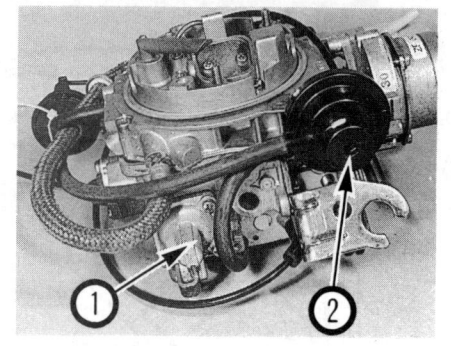

9.11C View of accelerator pump (1) and choke pull-down unit (2) – Pierburg 2E3 carburettor

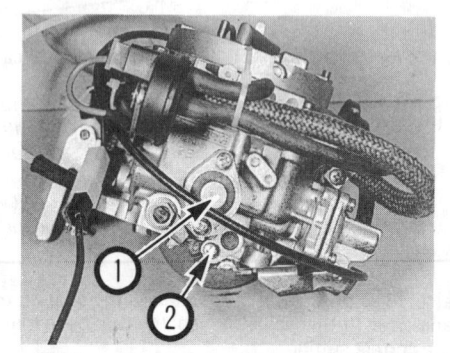

9.11D View of part-load enrichment valve (1) and accelerator pump cam (2) – Pierburg 2E3 carburettor

9.18 Throttle stop screw (arrowed) – Pierburg 2E3 carburettor

accordance with their maker's instructions. If an exhaust gas analyser is not available, a proprietary mixture indicating device may be used instead.
18 Start the engine and allow it to idle. If the idle speed is outside the specified limits, adjust by means of the throttle stop screw (photo).
19 When the idle speed is correct, connect up an exhaust gas analyser, and check the CO level in the exhaust gas. If it is outside the specified limits, adjust by means of the idle mixture adjustment screw. In production, the screw is covered by a tamperproof plug which must first be removed (photo).
20 With the idle mixture correct, readjust the idle speed if necessary.
21 When both speed and mixture are correct, stop the engine and disconnect the test equipment. Fit a new tamperproof plug to the idle mixture adjustment screw.

Fast idle
22 The engine must be at operating temperature, and the idle speed and mixture must be correctly adjusted. Remove the air cleaner to improve access.
23 Position the fast idle adjustment screw on the second-highest step of the fast idle cam. Connect a tachometer to the engine. Make sure that the choke plate is fully open.
24 Start the engine without touching the throttle pedal, and compare the engine speed with that given on the Specifications. If adjustment is necessary, remove the tamperpoof cap from the head of the fast idle screw by crushing it with pliers, and adjust by means of the screw (photo).
25 When adjustment is correct, stop the engine and disconnect the tachometer. Fit a new tamperproof cap.

Choke pull-down
26 Remove the air cleaner.
27 Remove the choke cover by removing the three screws and the securing ring. There is no need to disconnect the coolant hoses, just move the cover aside. Notice how the loop in the end of the bi-metallic spring engages in the choke drive lever (photo).
28 Move the choke drive lever to close the choke valve completely. Position the fast idle screw on the highest step of the cam.
29 Apply vacuum to the choke pull-down unit (at the hose nearest the carburettor body) using a modified bicycle pump or similar item. Apply light pressure to the choke drive lever in a clockwise direction (as if to close the choke valve), and check the choke valve gap by inserting a gauge rod or twist drill of the specified size (photo).
30 If adjustment is necessary, turn the socket-headed adjusting screw on the side of the choke housing (photo).
31 Refit the choke cover, making sure that the spring loop engages in the choke drive lever. Align the notches in the choke cover and housing when tightening the screws (photo).

Vacuum unit – leak checking
32 If a vacuum source incorporating a gauge is available, apply approximately 300 mbar (9 in Hg) to the choke pull-down unit, at the hose nearest the carburettor body. Close off the vacuum source, and check that the vacuum is held. If there is a leak, rectify or renew the leaking component.
33 Similarly check the secondary throttle vacuum unit.
34 If a suitable vacuum source is not available, testing of suspect vacuum units must be by substitution of a known good item.

Pierburg 2E3 carburettor – removal and refitting
35 Remove the air cleaner.
36 Disconnect the battery earth lead.
37 Disconnect and plug the coolant hoses from the automatic choke, noting which hose goes to which connection. Be prepared for some coolant spillage.
38 Disconnect and plug the fuel supply and return hoses from the

Chapter 13 Supplement: Revisions and information on later models

9.19 Idle mixture adjustment screw under tamperproof plug (arrowed) – Pierburg 2E3 carburettor

9.24 Fast idle adjustment screw under tamperproof cap (arrowed) – Pierburg 2E3 carburettor

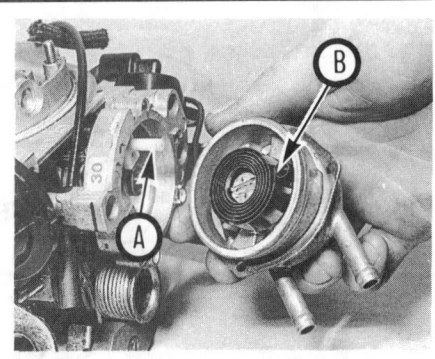

9.27 Choke drive lever (A) must engage with spring loop (B) – Pierburg 2E3 carburettor

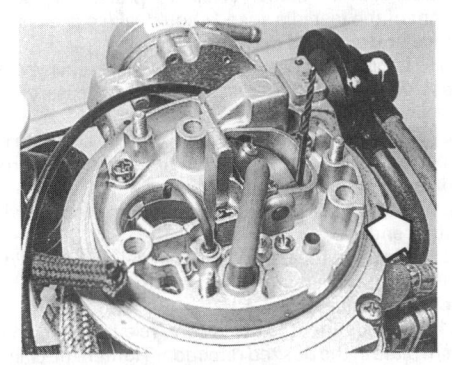

9.29 Using a twist drill to check choke pull-down gap, vacuum applied to hose (arrowed) – Pierburg 2E3 carburettor

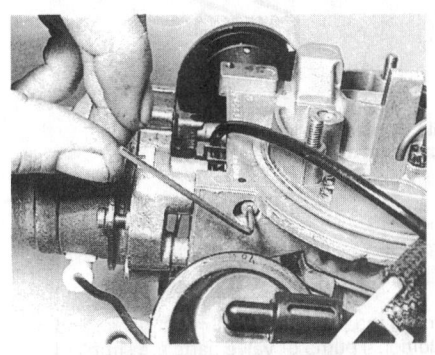

9.30 Choke pull-down socket-headed adjusting screw – Pierburg 2E3 carburettor

carburettor or vapour separator, as applicable. Be prepared for some fuel spillage.
39 Disconnect the throttle cable outer by pulling it out of its bracket. Unhook the inner cable from the throttle operating plate.
40 Disconnect the distributor vacuum hose at the distributor.
41 Disconnect the carburettor electrical supply at the wiring harness connector near the bulkhead.
42 Remove the three nuts from the top of the carburettor, and lift the carburettor off its studs.
43 Refit in the reverse order to removal. Use a new carburettor-to-manifold gasket if the old one was damaged. If much coolant was lost, check the coolant level after the engine has been run, and top up if necessary.

Pierburg 2E3 carburettor – overhaul

44 With the carburettor removed from the vehicle, drain the fuel from the float chamber and vapour separator (where applicable). Clean the outside of the carburettor.

45 Remove the hoses and wires from the carburettor, making identifying marks or notes to avoid confusion on reassembly.
46 Access to the jets and float chamber is obtained by removing the top half of the carburettor, which is secured by screws; note the fitted position of each screw, as their lengths differ. Blow through the jets and drillings with compressed air, or air from a foot pump – do not probe them with wire. If it is wished to remove the jets, unscrew them carefully with well-fitting tools.
47 Remove the fuel strainer from the inlet pipe by hooking it out with a small screwdriver, or by snaring ti with a long thin screw. Remove the strainer (photo).
48 Clean any foreign matter from the float chamber. Renew the inlet needle valve and seat if wear is evident, or if high mileage has been covered. Renew the float if it is punctured or otherwise damaged.
49 No procedure has been specified for float level adjustment, in any case the tolerance allowed is so wide that precision setting is clearly unnecessary. Simply check that the inlet needle valve is closed completely before the float reaches the top of its stroke.

9.31 Choke cover and housing alignment marks (arrowed) – Pierburg 2E3 carburettor

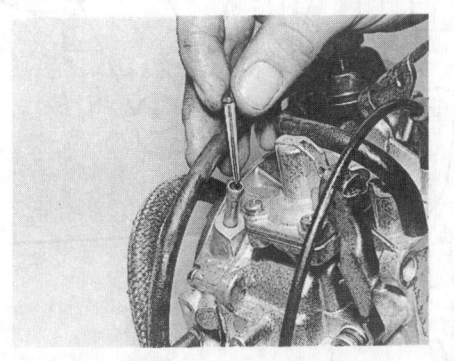

9.47 Fuel inlet strainer – Pierburg 2E3 carburettor

9.64 Accelerator pump cam adjustment – Pierburg 2E3 carburettor. Clockwise (+) to increase fuel delivery, anti-clockwise (–) to reduce fuel delivery

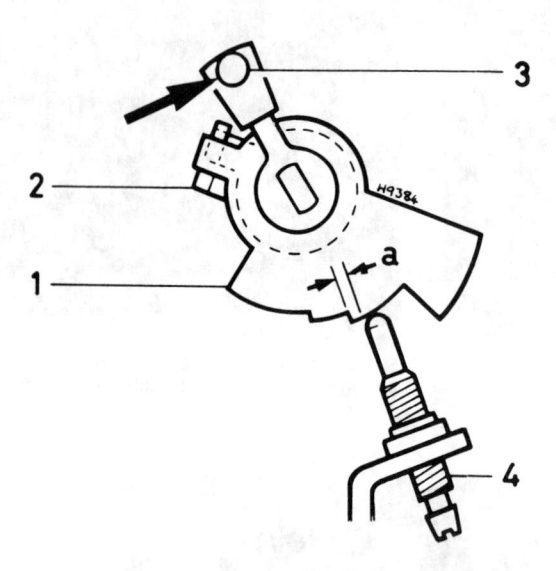

Chapter 13 Supplement: Revisions and information on later models

Fig. 13.13 Fast idle cam adjustment – Pierburg 2E3 carburettor (Sec 9)

1 Fast idle cam
2 Adjustment lever
3 Choke drive lever (press in direction arrowed)
4 Fast idle adjustment screw
a = 0.2 to 0.8 mm (0.008 to 0.032 in)

50 Renew the diaphragms in the part-load enrichment valve, and in the accelerator pump. If additional pump or valve parts are supplied in the overhaul kit, renew these parts also.

51 Further dismantling is not recommended. Pay particular attention to the throttle opening mechanism if it is decided to dismantle it; the interlocking arrangement is important.

52 Reassemble in the reverse order to dismantling. Use new gaskets and seals throughout; lubricate linkages with a smear or molybdenum-based grease.

53 Before refitting the carburettor, carry out the checks and adjustments described in the following paragraphs.

Fast idle cam position

54 The choke pull-down adjustment described previously must be correct.

55 If not already done, remove the choke cover.

56 Open the throttle, then close the choke valve by light finger pressure on the choke drive lever. Release the throttle.

57 Check that the fast idle adjustment screw is resting on the second highest step of the fast idle cam, in the position shown in Fig. 13.13. If not, first check that the choke return spring is correctly positioned, then adjust by bending the lever '2' (Fig. 13.13).

58 Refit and secure the choke cover, observing the alignment marks.

Throttle valve fast idle gap

59 Position the fast idle adjustment screw on the highest step of the fast idle cam.

60 Use a gauge rod or twist drill of the specified diameter to measure the opening of the primary throttle valve. Adjust if necessary at the fast idle adjustment screw. (This is a preliminary adjustment; final adjustment of the fast idle speed should take place with the engine running.)

Accelerator pump delivery

61 It will be necessary to feed the float chamber with fuel from a small reservoir during this test. Take all necessary fire precautions when dealing with fuel and fuel vapour.

62 Position the primary barrel over an accurate measuring glass. Fully open and close the throttle ten times, taking approximately one second for each opening, and pausing for three seconds after each return stroke. Make sure that the fast idle cam is not restricting throttle travel at either end.

63 Measure the quantity of fuel delivered, and divide by ten to obtain the quantity per stroke. Compare the quantity per stroke with the specified value.

64 If adjustment is necessary, release the clamp screw and turn the cam plate in the desired direction. Tighten the clamp screw, and recheck the pump delivery (photo).

Unleaded fuel

65 Unleaded fuel can be used in all carburettor-equipped 1.2, 1.3 and 1.4 models.

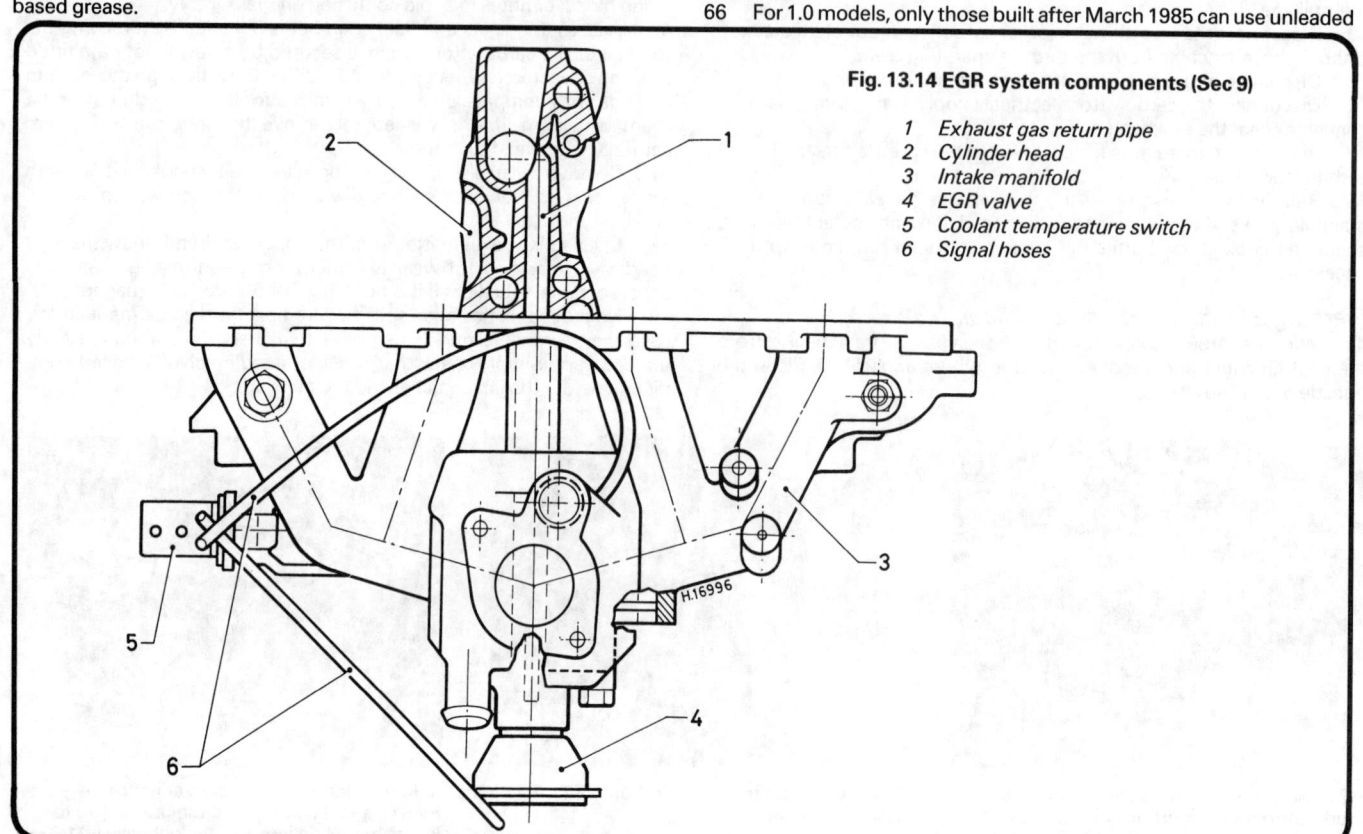

66 For 1.0 models, only those built after March 1985 can use unleaded

Fig. 13.14 EGR system components (Sec 9)

1 Exhaust gas return pipe
2 Cylinder head
3 Intake manifold
4 EGR valve
5 Coolant temperature switch
6 Signal hoses

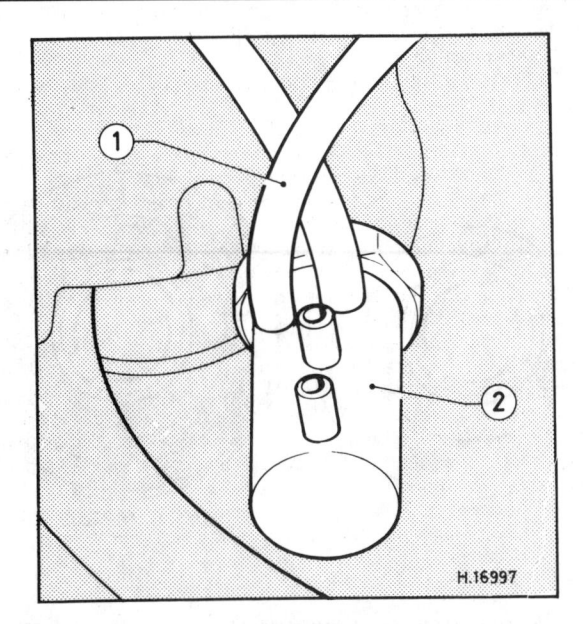

Chapter 13 Supplement: Revisions and information on later models

Fig. 13.15 EGR system signal hose (1) and coolant temperature switch (2) (Sec 9)

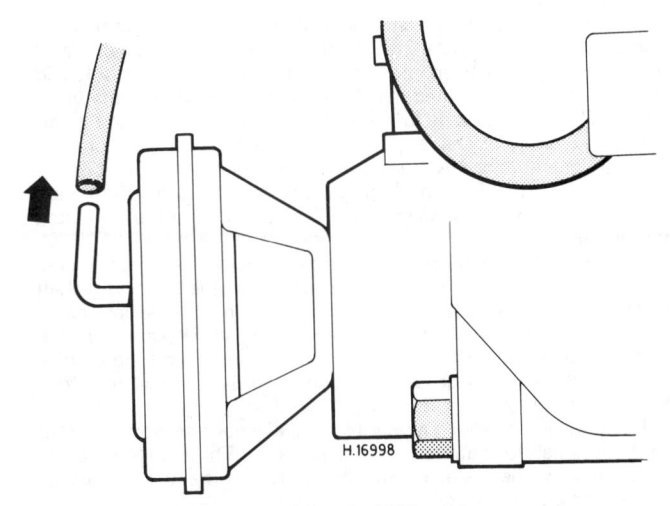

Fig. 13.16 Disconnecting the EGR valve vacuum hose (arrowed) (Sec 9)

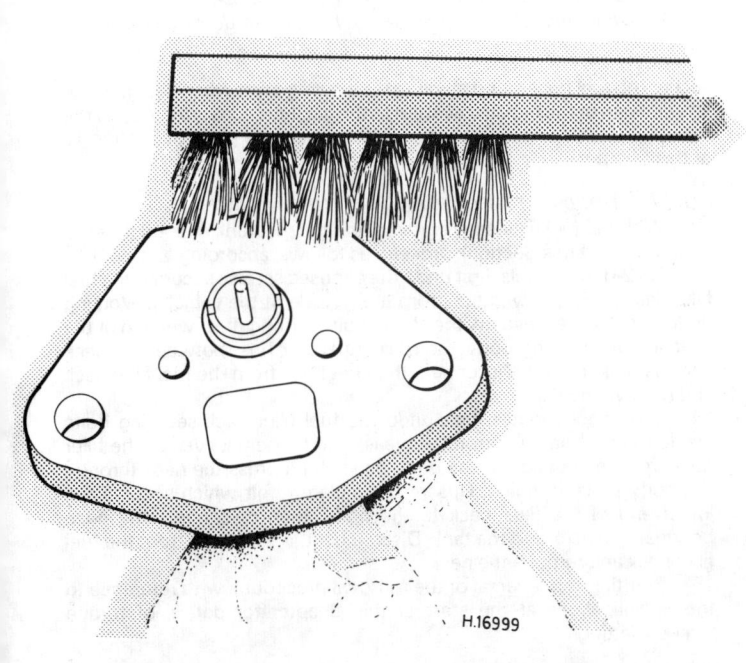

Fig. 13.17 Cleaning EGR valve with a wire brush (Sec 9)

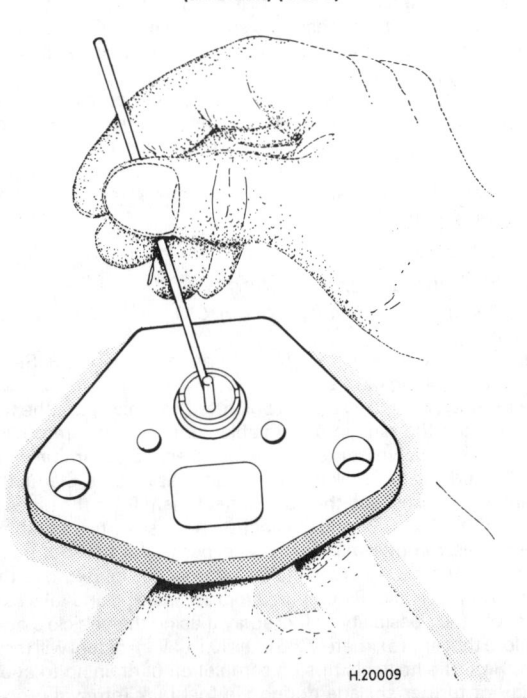

Fig. 13.18 Cleaning EGR valve with a pointed tool (Sec 9)

fuel freely. The use of unleaded fuel in earlier 1.0 models is permissible provided that one tankful of leaded fuel is used after every five of unleaded fuel.

67 All 1.4 models have the ignition timing adjusted for use with unleaded fuel before they leave the factory, and no further adjustment is required. Leaded petrol can be used if desired, with no adverse affects.
68 On all except 1.4 models, if there is any evidence of 'pinking' when running on unleaded fuel, retard the ignition timing as specified in Section 12 of this Chapter.

Exhaust gas recirculation (EGR) system – description and maintenance

69 This system is fitted to models operating in certain territories which have very stringent anti-pollution regulations.
70 The system is designed to reintroduce small amounts of exhaust gas into the combustion cycle, to reduce the generation of oxides of nitrogen. The volume of exhaust gas reintroduced is governed by manifold vacuum and coolant temperature.
71 The main components comprise an exhaust gas return pipe (1) (Fig 13.14) which is located in the cylinder head (2), intake manifold (3) and the EGR valve (4).
72 At coolant temperatures below 17°C (62°F) a coolant temperature switch (5) prevents gas recirculation by holding the EGR valve closed.
73 It is recommended that the EGR valve is cleaned annually in the following way, and the connecting hoses inspected for condition.
74 Pull off the hose from the valve, then unbolt the valve and remove it. Clean away all carbon using a wire brush and a pointed tool, but take care not to damage the valve seat.
75 Refit the valve and reconnect the hose.

10 Fuel system (Multec central fuel injection)

Warning: *Unleaded fuel must* **always** *be used in vehicles equipped with a catalytic converter. Leaded fuel* **must not** *be used, as it will 'poison' the catalyst.*

General description

1 The Multec central fuel injection system fitted to 1.4 models is a very simple method of fuel metering when compared to a conventional carburettor. Fuel is injected into the inlet manifold by a single solenoid valve (fuel injector) mounted centrally in the top of the throttle valve housing. The length of time for which the injector remains open determines the quantity of fuel reaching the cylinders for combustion. The electrical signals which determine the fuel injector opening duration are calculated by the electronic control unit (ECU) from information supplied by its network of sensors. The fuel pressure is regulated mechanically.

2 The signals fed to the ECU include inlet manifold vacuum from the manifold absolute pressure (MAP) sensor; engine speed and crankshaft position from the distributor; road speed from a sensor at the base of the speedometer cable; the position of the throttle valve plate from the throttle position sensor; engine coolant temperature; and the oxygen content in the exhaust gases, via a sensor in the exhaust manifold. Battery voltage is also monitored by the ECU.

3 Using the information gathered from the various sensors, the ECU sends out signals to control the system actuators. The actuators include the fuel injector, the idle air control stepper motor, the fuel pump relay and the ignition control unit.

4 The ECU also has a diagnostic function which can be used in conjunction with special Vauxhall test equipment for fault diagnosis. With the exception of basic checks to ensure that all relevant wiring and hoses are in good condition and securely connected, fault diagnosis should be entrusted to a Vauxhall dealer.

5 The system incorporates a three-way catalytic converter to reduce exhaust gas pollutants, and closed-loop fuel mixture control (by means of the exhaust gas oxygen sensor) is used. The mixture control remains in an open-loop mode (using pre-programmed values stored in the ECU memory) until the exhaust gas oxygen sensor reaches its normal operating temperature.

Fuel tank – removal and refitting

Warning: *Refer to the warning note in Chapter 3, Section 1 before proceeding.*

6 The procedure is similar to that described in Chapter 3, Section 7, but note the following differences.

7 The fuel tank may have a drain bolt fitted, in which case the fuel can be drained from the tank into a suitable container after removing the bolt. A non-return valve is fitted in the fuel filler neck, therefore if a drain bolt is not fitted, the fuel tank must be drained as follows. Clamp the fuel return and feed hoses, and then disconnect them from the pipes under the floor of the vehicle. Extend the ends of the hoses into a suitable clean container, then remove the clamps and pump the fuel out using the vehicle's own fuel pump or a suitable external pump. Remove the fuel tank cap to aid fuel flow. Be prepared for fuel spillage, and take extreme care to avoid the possibility of fuel spray if using the vehicle's own fuel pump. Note that approximately 2.5 litres (0.5 gallon) of fuel will remain in the tank. Refit the hose clamps on completion of draining to avoid the possibility of further spillage during the fuel tank removal procedure. After removing the tank, remember that it still contains fuel.

8 To remove the handbrake cable, so that the fuel tank can be lowered from its location, first detach the short handbrake cable from the compensator then, after the exhaust system and the central heatshield have been removed, disconnect the long handbrake cable at its 'S'-shaped connector. Further information regarding handbrake cable removal and refitting can be found in Section 15 of this Chapter. Move the handbrake cable as far away from the working area as possible. If the support bracket at the rear edge of the fuel tank cannot be released at this stage, it may be released as the fuel tank is lowered.

9 The wiring must be disconnected from the fuel gauge sender unit and the fuel pump before the tank is released from its location. The fuel pump wiring plug is accessible from inside the vehicle, under the rear seats (refer to the fuel pump removal and refitting procedure later in this Section for details). Also disconnect or remove the fuel filler neck from the body, as necessary (refer to the removal and refitting procedure later in this Section for details).

10 Where applicable, disconnect the active carbon canister hose from the fuel tank as the fuel tank is lowered. If not already done, detach the handbrake cable support bracket at the rear edge of the fuel tank.

11 During refitting, tighten the fuel tank mounting strap retaining nuts to their specified torque wrench setting and use new bolts to attach the central heatshield (where applicable). If necessary, use new clips to secure the fuel hoses, and always use a new gasket when securing the forward end of the exhaust system to the catalytic converter. Adjust the handbrake cable in accordance with the procedure given in Section 15 of this Chapter.

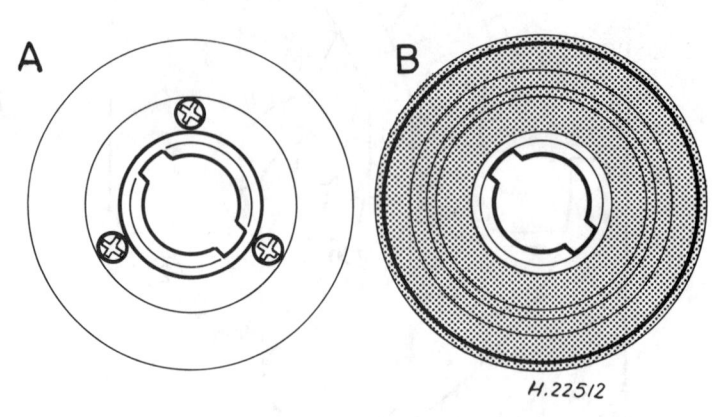

Fig. 13.19 Fuel filler neck attachment to body panel (Sec 10)

A Secured by screws on 2-door models
B Filler neck sleeve fits over flange on 4-door models

Fuel filler neck – removal and refitting

12 With the fuel filler cap removed and the fuel tank drained (refer to paragraph 7 of this Section), proceed as follows, according to model.

13 On 2-door models, first undo the exposed screws securing the fuel filler neck to the body panel (from the outside of the vehicle). Working under the vehicle, unscrew the clamp bolt securing the lower end of the filler neck to the underbody, then disconnect the hose between the neck and the tank (photo). Disconnect the vent hose from the fuel filler neck and remove the neck.

14 On 4-door models, first undo the fuel filler neck securing bolts inside the wheelarch. Carefully release the rubber sleeves of the filler neck from the outside of the body panel, then press the neck through the body panel opening. Unscrew the clamp bolt which secures the lower end of the filler neck to the body, then disconnect the hose between the neck and the tank. Disconnect the vent hose from the fuel filler neck and remove the neck.

15 Refitting is a reversal of the removal procedure, with reference to the Specifications at the start of this Chapter for details of torque wrench settings.

Fuel pump – removal and refitting

16 Fold up the rear seat cushion, and move floor covering beneath it to one side. Expose the top surface of the fuel pump mounting bracket by removing the cover plug in the vehicle floorpan.

17 Disconnect the multiplug from the pump (ensure that the battery negative lead has been disconnected).

18 Disconnect the fuel hose(s) from the pump after releasing the securing clips. Clamp the hoses to minimise the possibility of fuel loss.

19 Undo the bolts securing the fuel pump mounting bracket to the top of the fuel tank, then withdraw the pump mounting bracket and pump from the tank. Note the position of the sealing ring.

20 Press off the pre-filter at the base of the pump assembly.

21 Press the connecting hose (between the mounting bracket and the pump) towards the top of the mounting bracket, then remove the fuel pump and its rubber mounting from the mounting bracket.

22 Refitting is a reversal of the removal procedure, noting the following. A new rubber mounting, pre-filter, and pump mounting bracket-to-tank sealing ring must be used. Suitable thread-locking compound must be applied to the pump mounting bracket bolt threads before insertion and tightening.

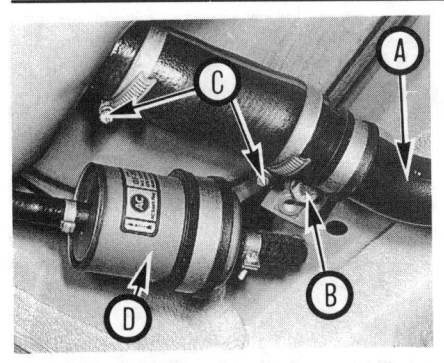

Chapter 13 Supplement: Revisions and information on later models

10.13 Fuel filler neck (A), its lower end clamp bolt (B), connecting hose clips (C) and the fuel filter (D) – 2-door model

10.23 Fuel pump control relay mounted on the engine compartment bulkhead

10.26 Removing the air cleaner assembly from the throttle valve housing

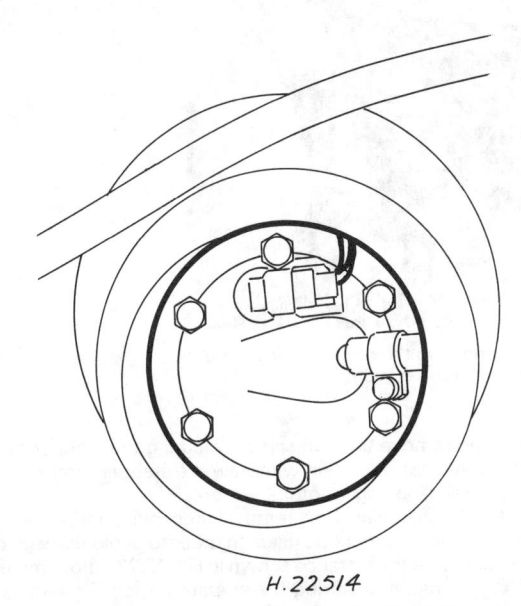

Fig. 13.20 Fuel pump viewed through access hole in the vehicle floorpan (Sec 10)

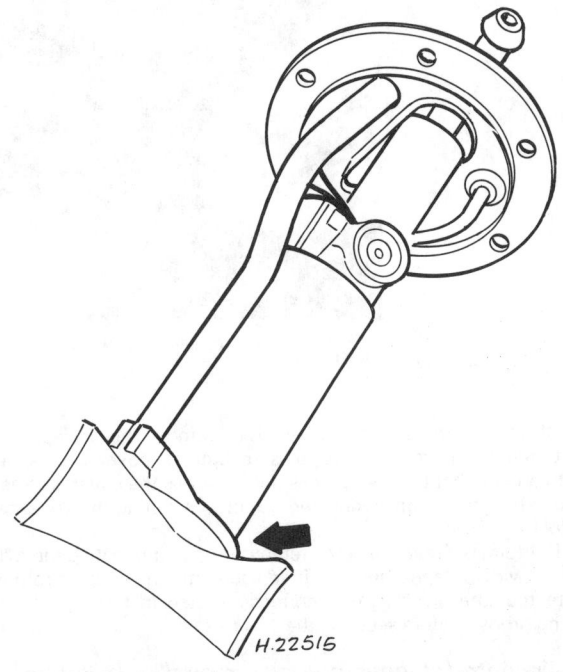

Fig. 13.21 A new rubber mounting (arrowed) must be used when refitting the fuel pump (Sec 10)

Fuel pump control relay – renewal

23 The procedure is as described in Section 11 of this Chapter for models with the Bosch L3.1-Jetronic fuel injection system, but note that the relay is located on the right-hand side of the bulkhead (photo).

Fuel filter – renewal

24 The filter is located to the right-hand side of the fuel tank, and the renewal procedure is as described in Section 11 of this Chapter for models with the Bosch L3.1-Jetronic fuel injection system.

Fuel pressure – check

25 Fuel pressure checking must be entrusted to a Vauxhall dealer, or other suitable specialist, who has the necessary special equipment.

Air cleaner – removal and refitting

26 Remove the two large screws from the upper surface of the air cleaner lid, disconnect the hoses (having noted their fitted positions), and lift the air cleaner from the throttle valve housing. Note the position of the gasket on the throttle valve housing (photo).

27 Refitting is a reversal of the removal procedure, ensuring that the intake air tube (at the front end of the air cleaner assembly) locates correctly.

Air cleaner element – renewal

28 The air cleaner element can be renewed after removing the lid of the air cleaner assembly. To remove the lid, first remove the two large screws from the upper surface of the lid, then remove the small guide screw (by the intake pipe), and release the clips around the circumference of the lid.

Fuel injector – removal and refitting

29 With the battery earth lead disconnected and the air cleaner removed, disconnect the multiplug from the fuel injector.

30 Undo the screw securing the fuel injector clamp to the top of the throttle valve housing, then lift the fuel injector from its location (with the careful use of a screwdriver if necessary) (photos).

31 Refitting is a reversal of the removal procedure, using new seals. Apply a light smear of grease to the seals before fitting, and use thread-locking compound on the clamp screw threads.

Fuel pressure regulator – removal and refitting

32 As the fuel system operates under high pressure, this operation requires depressurisation of the fuel system before dismantling begins. The procedure is described in the following paragraph.

33 Disconnect the fuel pump control relay multiplug and the oil pressure switch multiplug (where applicable). Start the engine and run it for a minimum of five seconds to reduce the fuel pressure. Reconnect the fuel pump control relay and oil pressure switch multiplugs (as applicable) after switching off the ignition.

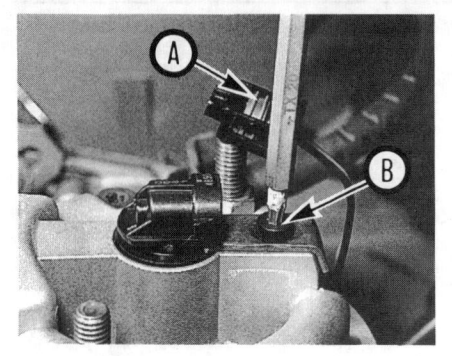

Chapter 13 Supplement: Revisions and information on later models

10.30A Disconnect the multiplug (A), then remove the clamp retaining screw (B) ...

10.30B ... and remove the injector

10.35A Undo the fuel pressure regulator cover screws (arrowed) ...

10.35B ... then carefully remove the cover

10.37 Disconnecting the multiplug from the idle air control stepper motor

34 Disconnect the battery earth lead and remove the air cleaner.
35 Carefully unscrew the fuel pressure regulator cover screws, taking care to ensure that the spring does not fly out as the cover is released. Remove the cover, cup, spring and diaphragm, noting their orientation (photos).
36 Refitting is a reversal of the removal procedure, noting the following. A new diaphragm must be fitted each time the cover is removed. Ensure that the diaphragm is correctly seated in the throttle valve housing groove before securing the cover.

Idle air control stepper motor – removal and refitting

37 With the battery earth lead disconnected and the air cleaner removed, disconnect the multiplug from the stepper motor (photo).

10.38 Withdrawing the idle air control stepper motor from the throttle valve housing. Note the sealing ring location (arrowed)

38 Undo and remove the two screws securing the stepper motor to the throttle valve housing, then withdraw the stepper motor. Note the position of the sealing ring (photo).
39 Refitting is a reversal of the removal procedure, noting the following. A new sealing ring must be used. In order to avoid damage during refitting, ensure that the distance shown in Fig. 13.22 is not greater than 28 mm (1.10 in) before inserting the stepper motor. If the distance is greater than the maximum allowed, carefully push the stepper motor cone back in to its stop. Suitable thread-locking compound must be applied to the threads of the stepper motor securing screws before insertion and tightening.

Throttle position sensor – removal and refitting

40 With the battery earth lead disconnected and the air cleaner removed, disconnect the multiplug from the throttle position sensor.
41 Unscrew the two securing screws and withdraw the sensor (photo).
42 Refitting is a reversal of the removal procedure, noting the following. Install the sensor with the throttle valve in the closed position, and ensure that the sensor seats correctly onto its shaft. Suitable thread-

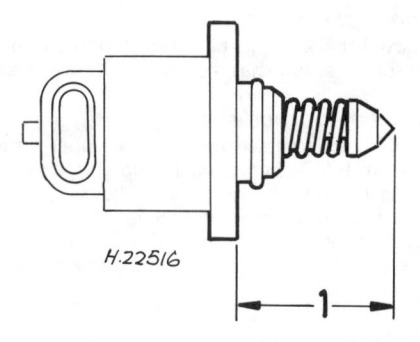

Fig. 13.22 Distance (1) must not be greater than 28 mm (1.10 in) when refitting the idle air control stepper motor (Sec 10)

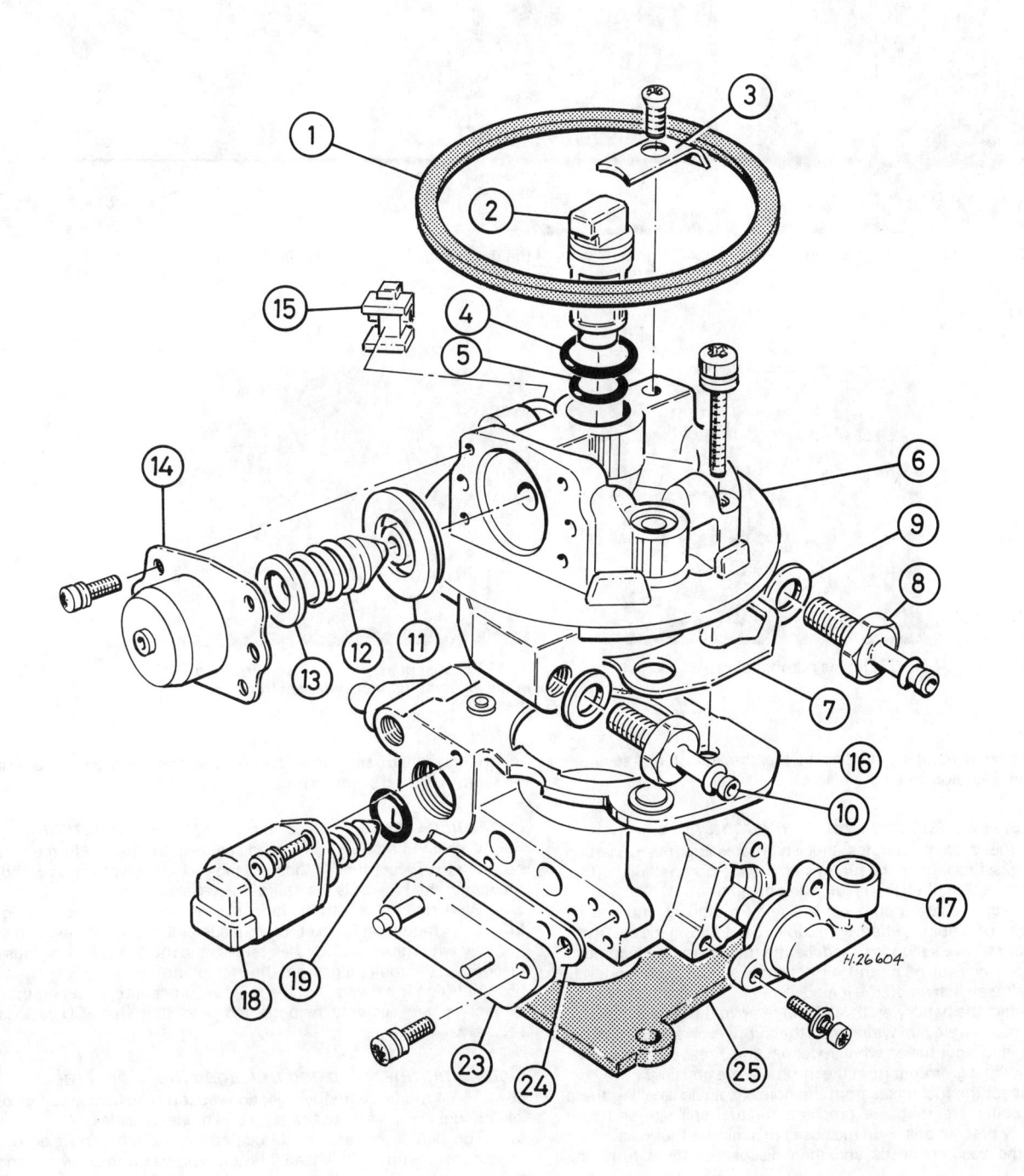

Fig. 13.23 An exploded view of the throttle valve housing (Sec 10)

1 Gasket
2 Fuel injector
3 Clamp
4 Upper seal
5 Lower seal
6 Throttle valve housing upper section
7 Gasket
8 Fuel inlet union
9 Seal
10 Fuel return union
11 Fuel pressure regulator diaphragm
12 Fuel pressure regulator spring
13 Fuel pressure regulator cup (spring seat)
14 Fuel pressure regulator cover
15 Rubber fuel injector wiring grommet
16 Throttle valve housing lower section
17 Throttle position sensor
18 Idle air control stepper motor
19 Sealing ring
23 Vacuum connection flange
24 Gasket
25 Throttle valve housing-to-inlet manifold gasket

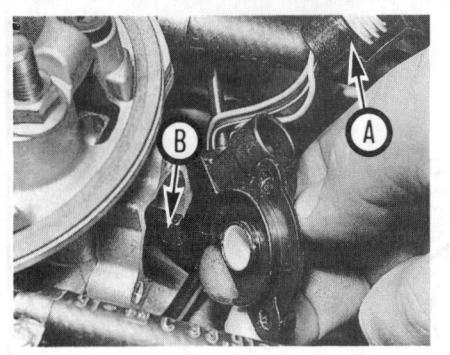

10.41 Withdrawing the throttle position sensor

A Multiplug
B Throttle valve shaft

10.49 A general view of the throttle valve operating linkage

10.54 Withdrawing the ECU

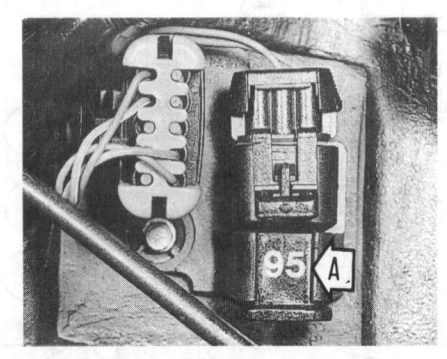

10.57 Octane number plug (A)

10.61 Exhaust gas oxygen sensor (arrowed) in the exhaust manifold (viewed from under the vehicle)

locking compound must be applied to the threads of the securing screws before insertion and tightening.

Throttle valve housing – removal and refitting

43 As the fuel system operates under high pressure, this operation requires depressurisation of the fuel system before dismantling begins. The procedure is described in the following paragraph.
44 Disconnect the fuel pump control relay multiplug and the oil pressure switch multiplug (where applicable). Start the engine and run it for a minimum of five seconds to reduce the fuel pressure. Reconnect the fuel pump control relay and oil pressure switch multiplugs (as applicable) after switching off the ignition.
45 Disconnect the battery earth lead and remove the air cleaner.
46 Disconnect all relevant wiring from the throttle valve housing; label the wiring to avoid confusion when reconnecting. Press out the rubber fuel injector wiring grommet from the throttle valve housing.
47 Disconnect the fuel hoses from the housing, again labelling them to aid subsequent refitting. Be prepared for fuel spillage, and take suitable safety precautions. Plug the hoses to minimise fuel loss.
48 Label the vacuum hoses and then disconnect them from the housing.
49 Disconnect the throttle valve operating linkage at the throttle valve housing (photo).
50 Undo the two nuts securing the throttle valve housing to the inlet manifold (recover the washers which fit under the nuts), then carefully lift off the housing. The gasket can be discarded.
51 If required, the throttle valve housing may now be split into its upper and lower sections by removing the two securing screws, but note that a new gasket will be required during reassembly, along with suitable thread-locking compound for the screw threads. The fuel inlet and return unions may also be removed, but note that new sealing rings must be used during refitting, and the unions must be tightened to their specified torque wrench setting (see Specifications at the start of this Chapter).
52 Refitting is a reversal of the removal procedure, noting the following. Always use new gaskets and seals, use suitable thread-locking compound where applicable, and ensure that the washers are fitted under the housing securing nuts.

Electronic Control Unit (ECU) – removal and refitting

53 With the battery earth lead disconnected, first undo the screws securing the parcel shelf in the passenger side footwell, then withdraw the parcel shelf towards the rear of the vehicle.
54 Working underneath the facia, slacken the two nuts clamping the ECU to its mounting bracket, then release the ECU from its guide lug. Partially withdraw the ECU, then release the multiplug locking lugs and disconnect the multiplugs from the unit (photo).
55 Refitting is a reversal of the removal procedure, ensuring that the multiplugs are securely reconnected, and that the ECU is located securely.

Octane number plug and unleaded fuel – general

56 All 1.4 models with fuel injection **must** run on unleaded fuel **only**; leaded fuel will 'poison' the catalyst and render it useless.
57 The octane number plug, located on the left-hand side of the engine compartment bulkhead (photo), is provided to allow the engine to be adapted to run efficiently on other grades of unleaded fuel, if premium unleaded fuel (95 RON) is not available. All UK specification 1.4 fuel injection models are adjusted to run on premium unleaded fuel (95 RON) before leaving the factory, and before changing fuel octane grade a Vauxhall dealer should be consulted (a change of fuel octane grade may be necessary if travelling abroad).
58 The octane number plug may be changed from 95 RON (premium unleaded fuel) to 91 RON (normal unleaded fuel) by removing the plug, turning it through 180° then re-inserting it so that the number '91' can be read off the side with the locking clip. The procedure is the same when changing back to 95 RON. The change of octane grade at the octane number plug should be carried out after the fuel tank has been filled with the appropriate grade of unleaded fuel.

Exhaust gas oxygen sensor – removal and refitting

59 This operation must be performed with the engine at its normal

Chapter 13 Supplement: Revisions and information on later models 253

operating temperature, therefore great care must be taken to avoid burns from the hot exhaust.

60 Disconnect the battery earth lead.

61 Disconnect the sensor multiplug, then remove the sensor by unscrewing it from its location in the exhaust manifold. Note the position of the sealing ring as the sensor is removed (photo). Take care not to drop the sensor and do not allow it to contact fuel or silicone substances.

62 Refitting is a reversal of the removal procedure, noting the following. If the original oxygen sensor is to be refitted, it must be coated with special grease – Vauxhall part number 19 48 602 (5 613 695). A new sealing ring must be used, and the sensor must be tightened to the specified torque.

Coolant temperature sensor – removal and refitting

63 Disconnect the battery earth lead.

64 Partially drain the cooling system to allow the temperature sensor to be removed without excessive coolant loss (refer to Chapter 2).

65 Disconnect the multiplug from the temperature sensor, then unscrew the sensor. Note the position of the sealing ring.

66 Refitting is a reversal of the removal procedure, using a new sealing ring and tightening the sensor to the specified torque.

Manifold absolute pressure (MAP) sensor – removal and refitting

67 Disconnect the battery earth lead.

68 Disconnect the vacuum hose and the multiplug from the MAP sensor (photo).

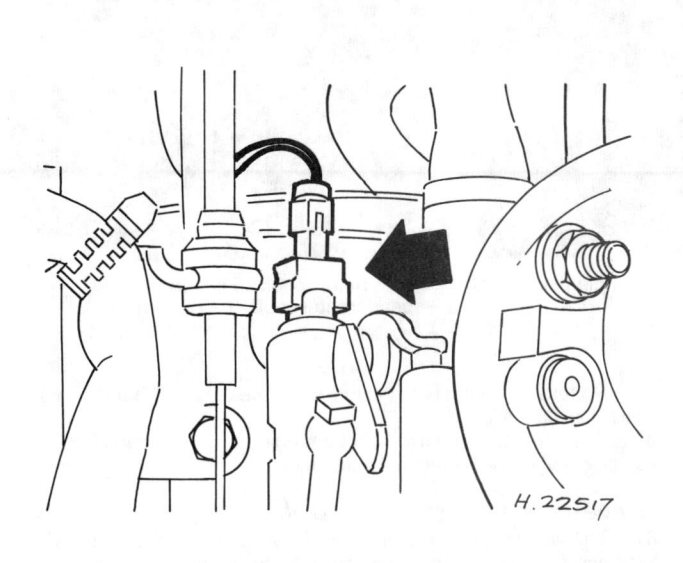

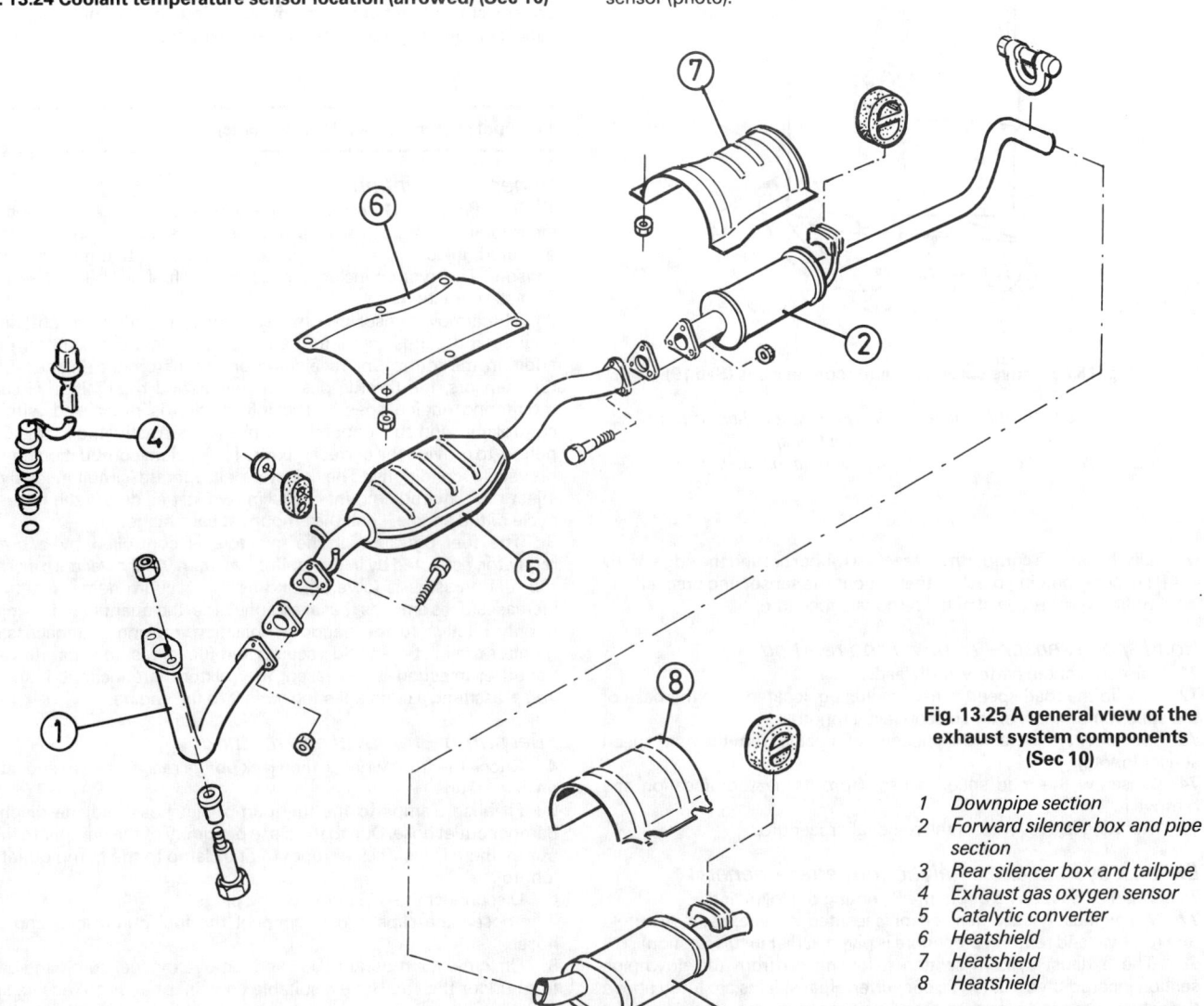

Fig. 13.24 Coolant temperature sensor location (arrowed) (Sec 10)

Fig. 13.25 A general view of the exhaust system components (Sec 10)

1 Downpipe section
2 Forward silencer box and pipe section
3 Rear silencer box and tailpipe
4 Exhaust gas oxygen sensor
5 Catalytic converter
6 Heatshield
7 Heatshield
8 Heatshield

Chapter 13 Supplement: Revisions and information on later models

10.68 Manifold Absolute Pressure (MAP) sensor multiplug (A) and vacuum hose (B)

10.72 Road speed sensor multiplug (arrowed) held in its clip

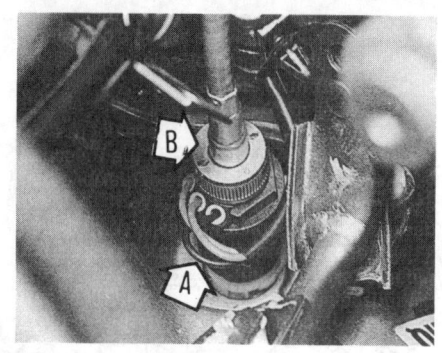

10.73 Road speed sensor (A), and speedometer cable-to-sensor fixing (B)

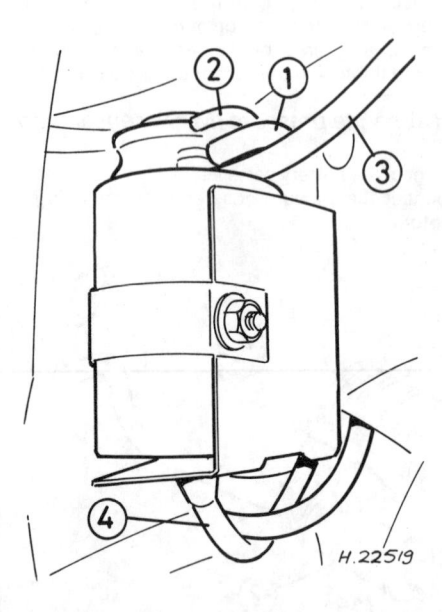

Fig. 13.26 Active carbon canister connections (Sec 10)

1 Hose to throttle valve housing
2 Vacuum hose
3 Fuel vapour hose (from fuel tank)
4 Ventilation hose

69 Lift the engine compartment sealing rubber around the edge of the MAP sensor mounting bracket, then lift out the sensor and bracket.
70 Refitting is a reversal of the removal procedure.

Road speed sensor – removal and refitting
71 Disconnect the battery earth lead.
72 Unclip the road speed sensor multiplug, located near the base of the speedometer cable, and disconnect it (photo).
73 Unscrew and detach the speedometer cable from the road speed sensor (photo).
74 Unscrew the road speed sensor from its gearbox location and remove it.
75 Refitting is a reversal of the removal procedure.

Exhaust system and catalytic converter – general
76 Refer to Section 22 of Chapter 3, noting the following.
77 An exhaust gas oxygen sensor is located in the exhaust manifold (the removal and refitting procedure is given earlier in this Section).
78 The exhaust system section following on from the downpipe section incorporates a catalytic converter. Heat shields are fitted above the catalytic converter and the silencer boxes to the rear, to prevent the extremely high temperatures encountered reaching the passenger compartment. If removed, these heat shields must always be refitted.

79 No tension spring type flexible joints are featured, but gaskets are used at all flange joints.
80 The catalytic converter will be rendered useless if leaded fuel is used instead of the correct unleaded fuel.

Active carbon canister – renewal
81 The active carbon canister is located under the front of the left-hand front wheelarch. The canister can be removed after disconnecting its vacuum and fuel vapour hoses (having labelled them to aid correct subsequent refitting), by releasing the locknut securing the canister mounting bracket to the vehicle body.

11 Fuel system (Bosch L3.1-Jetronic)

General description
1 The Bosch L3.1-Jetronic fuel injection system is fully electronic, incorporating an electronic air-flow sensor, a control unit, electrically-operated injection, an electric fuel pump, and engine monitoring sensors. The system includes deceleration fuel cut-off and automatic cold start functions.
2 The airflow sensor unit measures the volume of air entering the engine, and sends this information to the electronic control unit. In addition, the control unit receives information from the engine temperature sensors, the throttle position switch, and the ignition circuit (for monitoring engine speed). The information is processed within the control unit, and subsequently the injectors are activated for the correct period, to provide the correct mixture for efficient combustion under the prevailing conditions. The injectors all operate simultaneously, and inject the required amount of fuel in two stages during the four-stroke cycle of the engine – half the amount at each stage.
3 The fuel pressure at the injectors is controlled by a pressure regulator, activated by inlet manifold vacuum. As the vacuum increases, the fuel pressure is lowered, and as the vacuum decreases, the fuel increases. This ensures that at any one time, the quantity of fuel injected is only relative to the period of injection, making it unnecessary to monitor the inlet manifold vacuum and fuel pressure separately. If the period of injection is increased, the mixture is enrichened, and *vice-versa*, assuring a constant supply of air to the engine.

Fuel pump – removal and refitting
4 Chock the front wheels, then jack up the rear of the car and support on axle stands.
5 Fit hose clamps to the fuel pump inlet hose and the diaphragm damper outlet hose. Due to the close proximity of the damper to the fuel pump, there is insufficient room to fit a clamp to the pump outlet hose (photo).
6 Disconnect the fuel pump wiring plugs.
7 Loosen the clips and disconnect the fuel pump inlet and outlet hoses.
8 Unscrew the mounting bolt and remove the fuel pump and bracket from under the car. Have a suitable container handy to receive spilled fuel.
9 Remove the fuel pump and rubber sleeve from the bracket.
10 Refitting is a reversal of removal.

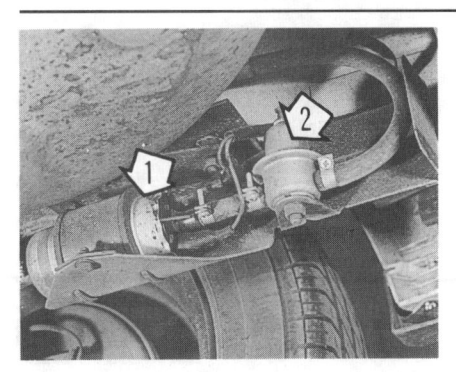

Chapter 13 Supplement: Revisions and information on later models

11.5 Fuel pump (1), and diaphragm damper (2)

11.13 Fuel filter

11.15 Fuel flow direction arrow on the fuel filter

11.18A Release the spring clips ...

11.18B ... and lift off the air cleaner cover and airflow sensor unit

11.19 Removing the air cleaner element

11.24A Unscrew the mounting bolts (arrowed) ...

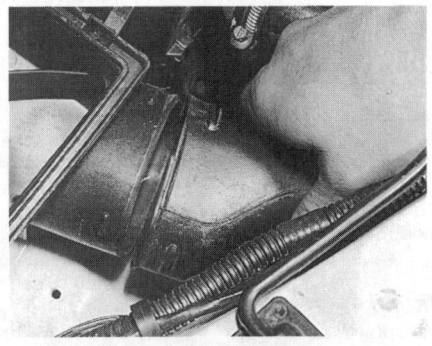

11.24B ... and disconnect the air cleaner body from the air inlet

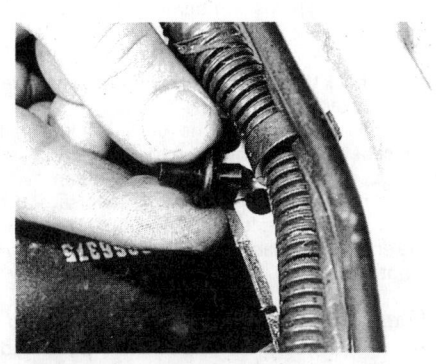

11.25A Remove the plastic clip ...

11.25B ... and remove the air inlet

11.27A Release the wire clips ...

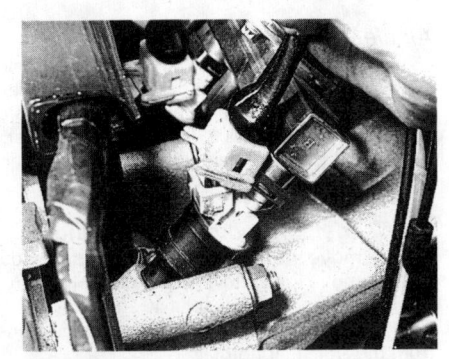

11.27B ... and disconnect the wiring plugs from the injectors

11.28 Unscrewing the fuel distribution tube bolts

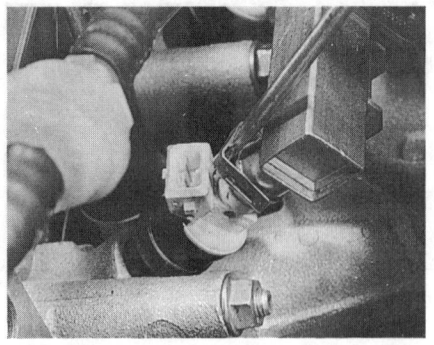

11.29A Using a screwdriver, prise out ...

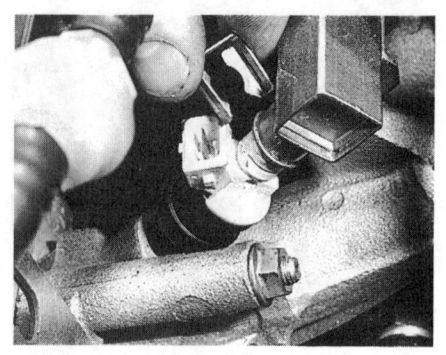

11.29B ... and remove the injector clips

11.30 Disconnecting the vacuum pipe from the fuel pressure regulator

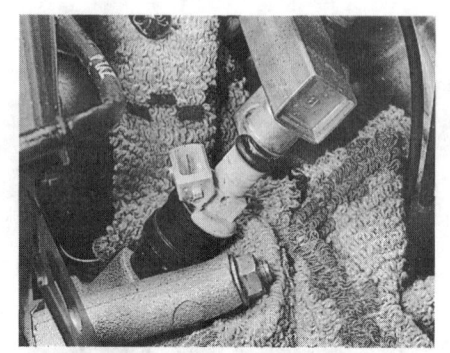

11.32 Separating the fuel distribution tube from the injectors

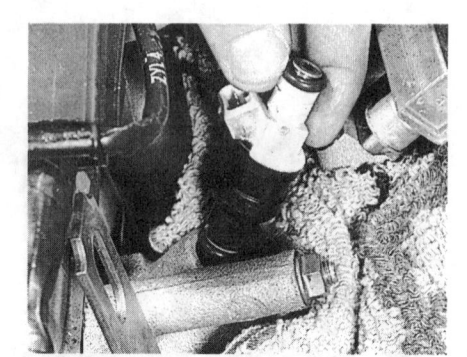

11.33 Removing an injector

Fuel pump diaphragm damper – removal and refitting
11 The procedure is similar to that for the fuel pump, however there are no wiring plugs or mounting bracket.

Fuel filter – renewal
12 Chock the front wheels, then jack up the rear of the car and support on axle stands.
13 Fit hose clamps to the fuel filter inlet and outlet hoses, then loosen the clips and disconnect the hoses (photo).
14 Loosen the mounting clamp and remove the fuel filter.
15 Fit the new fuel filter using a reversal of the removal procedure, but make sure that the fuel flow direction arrow points to the hose leading to the engine compartment (ie it should point to the centre-line of the car) (photo).

Fuel gauge sender unit – removal and refitting
16 The procedure is similar to that described in Chapter 3, Section 8, but the unit is attached to the fuel tank by six screws, and a gasket is fitted.
17 When refitting the fuel gauge sender unit, fit a new gasket, and apply a suitable sealing compound to the threads of the screws before inserting and tightening them.

Air cleaner element – renewal
18 Release the spring clips, and lift the air cleaner cover and airflow sensor unit from the lower half of the air cleaner (photos).
19 Extract the air cleaner element, and discard it (photo).
20 Wipe clean the air cleaner body and cover.
21 Insert the new element, refit the cover, and snap the spring clips into place to secure.

Air cleaner – removal and refitting
22 Remove the air cleaner element as previously described.
23 Remove the airflow sensor control unit from the air cleaner cover, with reference to paragraphs 40 and 41.
24 Unscrew the mounting bolts and disconnect the air cleaner body from the air inlet (photos).
25 If necessary, the air inlet may be removed by pushing the centre dowel from the plastic clip (photos).

26 Refitting is a reversal of removal.

Fuel injectors – removal, testing and refitting
27 Disconnect the wiring plugs from the injectors, after depressing and unhooking the wire clips (photos).
28 Unscrew the bolts securing the fuel distribution tube to the inlet manifold (photo).
29 Using a screwdriver, prise off the clips retaining the injectors to the fuel distribution tube (photos).
30 Disconnect the vacuum pipe from the fuel pressure regulator, and disconnect the brake servo and crankcase ventilation hoses (photo).
31 Place some cloth rags around each injector to soak up spilled fuel.
32 Pull the fuel distribution tube from the injectors (photo).
33 Pull the injectors from the inlet manifold (photo).
34 Carefully clean the injectors.
35 Connect an ohmmeter between the two terminals of each injector, and check that the resistance is 16.0 ± 1.0 ohms (photo). Renew all four injectors if any one is faulty.
36 Refitting is a reversal of removal, but fir new O-ring seals to the injectors, and lubricate the seals with a little fuel before reassembly (photos).

Airflow sensor unit – removal and refitting
37 Disconnect the wiring multi-plug from the airflow sensor unit by levering the spring clip outwards and unhooking the multi-plug (photos).
38 Loosen the clip and disconnect the air inlet duct leading to the throttle housing (photo).
39 Release the air cleaner spring clips, and withdraw the airflow sensor unit and upper air cleaner cover.
40 Unscrew the outer bolt securing the airflow sensor unit to the air cleaner cover (photo).
41 Slide the air intake funnel from inside the air cleaner cover, then unscrew the bolts and separate the airflow sensor unit (photos).
42 With the unit removed, check that the air valve moves freely, without any sign of sticking (photo). Wipe away any dirt from the valve and surrounding area, using a non-fluffy cloth.
43 Refitting is a reversal of removal, but make sure that the air inlet duct is securely located, and the clip tightened.

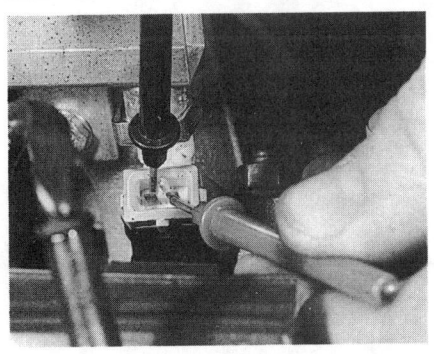

Chapter 13 Supplement: Revisions and information on later models

11.35 Checking the resistance of an injector

11.36A Fuel injector and O-ring seals

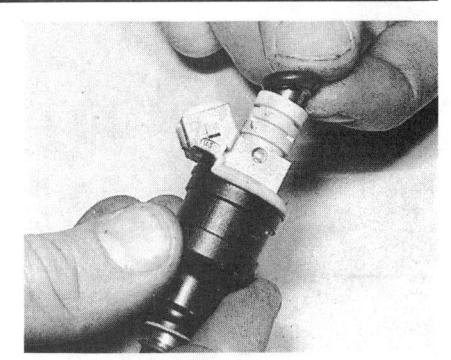

11.36B Fitting an O-ring seal

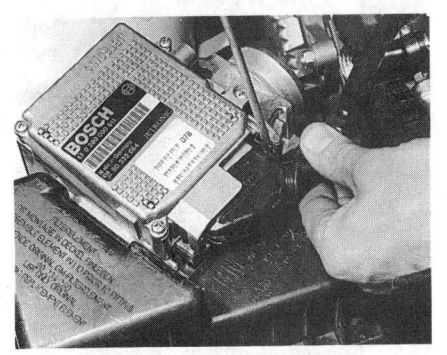

11.37A Lever the spring clip outwards ...

11.37B ... and unhook the multi-plug

11.38 Disconnecting the air inlet duct from the airflow sensor unit

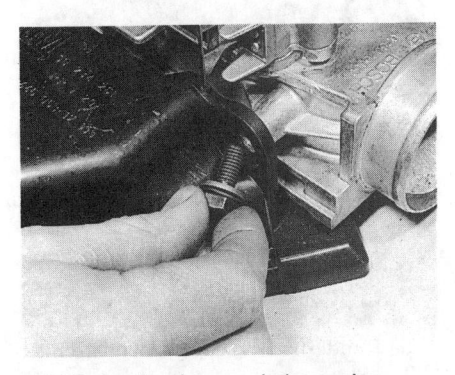

11.40 Removing the outer bolt securing airflow sensor unit to the air cleaner cover

11.41A Slide out the air intake funnel ...

11.41B ... and unbolt the airflow sensor unit

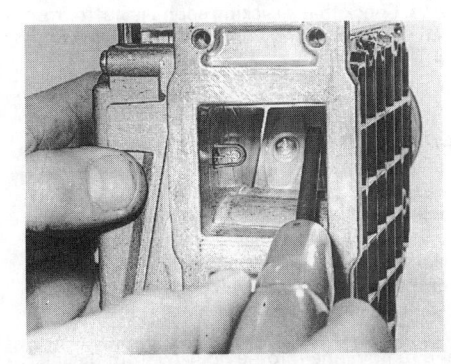

11.42 Checking the air valve movement

11.45A Undo the screws ...

11.45B ... and withdraw the airflow sensor cover

11.48A Remove the wire clip ...

11.48B ... and disconnect the wiring plug from the throttle valve switch

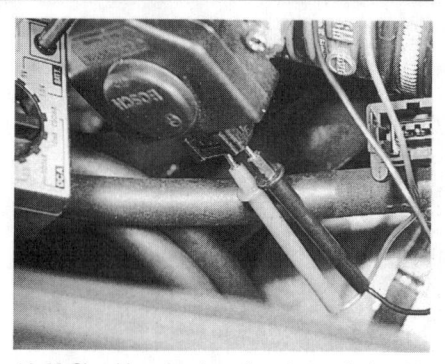

11.49 Checking the throttle valve switch with an ohmmeter

11.55A Disconnecting the air inlet duct from the throttle housing

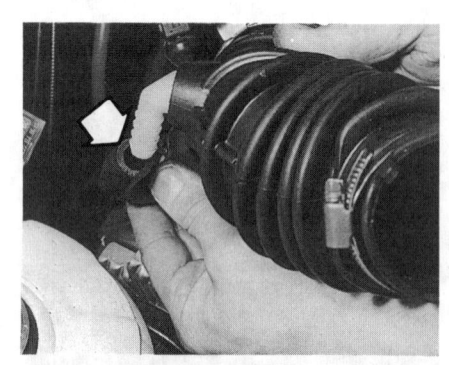

11.55B Disconnecting the hose (arrowed) for the supplementary air device

11.55C Crankcase ventilation hose connection (arrowed) to the valve cover

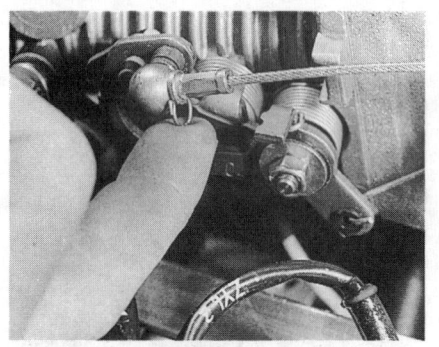

11.56A Release the spring clip ...

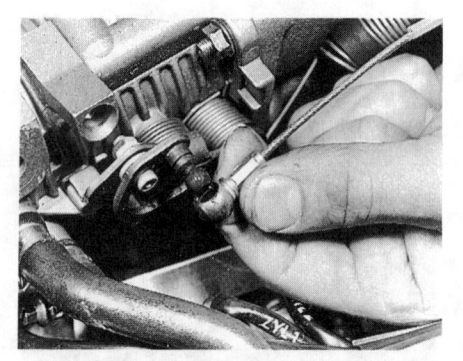

11.56B ... and disconnect the accelerator cable socket

11.58 Throttle housing upper mounting nuts (arrowed)

Airflow sensor output unit – removal and refitting

44 The output unit is located in the airflow sensor cover. First disconnect the wiring multi-plug.
45 Unscrew the four screws, and withdraw the cover from the airflow sensor (photos).
46 It is not possible to separate the output unit from the cover, so if faulty, both parts must be obtained as an assembly.
47 Refitting is a reversal of removal.

Throttle valve switch – checking, removal and refitting

48 With the engine stopped, disconnect the wiring plug from the throttle valve switch by depressing and unhooking the wire clip (photos).
49 Connect an ohmmeter between terminals 2 and 18 on the switch, and check that the resistance is zero (photo). Repeat the check with the ohmmeter connected between terminals 3 and 18. The ohmmeter should now read infinity.
50 To remove the switch, unscrew the two mounting screws and detach the switch from the throttle valve shaft.
51 To refit the switch, engage it with the throttle valve shaft, and insert the mounting screws loosely.
52 Turn the switch anti-clockwise until a slight resistance is felt, then tighten the mounting screws. Check that a clicking sound is heard from the switch when the throttle valve is slightly opened, and again when the throttle valve is closed.
53 Refit the wiring plug.

Throttle housing – removal and refitting

54 Drain the cooling system (Chapter 2).
55 Disconnect the air inlet duct (and the hose for the supplementary air device), crankcase ventilation hose, and coolant hoses. If necessary, completely remove the air inlet duct and ventilation hose (photos).
56 Disconnect the accelerator cable by releasing the spring clip and pulling the inner cable socket from the throttle valve lever (photos).
57 Disconnect the throttle valve switch wiring plug and the vacuum pipe.
58 Unscrew the nuts and remove the throttle housing from the inlet manifold air box (photo). Remove the gasket.

Chapter 13 Supplement: Revisions and information on later models

11.60 Fuel pump control relay

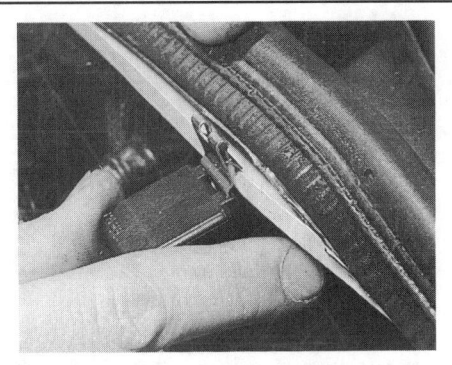

11.61 Remove the relay from the bulkhead ...

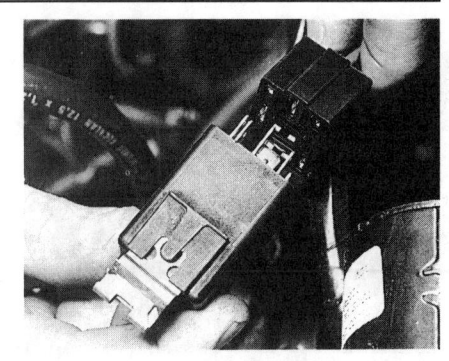

11.62 ... and disconnect the wiring plug

11.64 Coolant temperature sensor (arrowed) – shown with distributor removed

11.71 Oil temperature switch viewed from underneath the vehicle

11.80 Air intake temperature switch (arrowed)

59 Refitting is a reversal of removal, but fit a new gasket. Refill the cooling system with reference to Chapter 2.

Fuel pump control relay – renewal

60 The fuel pump control relay is located on the left-hand side of the bulkhead, above the brake vacuum servo unit (photo).
61 Pull up the moulding, and prise the relay clamp from the bulkhead (photo).
62 Disconnect the wiring plug (photo).
63 Refitting is a reversal of removal.

Coolant temperature sensor – removal, testing and refitting

64 The coolant temperature sensor is located on the left-hand end of the cylinder head, beneath the distributor (photo).
65 Refer to Chapter 2, and drain the cooling system.
66 Disconnect the wiring plug (blue), and unscrew the sensor.
67 Connect an ohmmeter between the sensor terminals, and position the sensor capsule in a container of water which is being heated. Use a thermometer to check the water temperature. At 20°C (68°F), the resistance should be between 2100 and 2900 ohms. At 80°C (176°F), the resistance should be between 270 and 390 ohms.
68 If possible, check the sensor at –10°C (14°F) using an antifreeze solution chilled in a deep freezer. The resistance should be between 7000 and 11 600 ohms.
69 Renew the sensor if it is faulty.
70 Refitting is a reversal of removal, using a new sealing ring, but refer to Chapter 2 when refilling the cooling system.

Oil temperature switch (up to 1990 model year) – removal, testing and refitting

71 The oil temperature switch is located on the right-hand rear of the cylinder block, and is wired in series with the air intake temperature switch (photo).
72 Apply the handbrake, then jack up the front of the car and support on axle stands.
73 Disconnect the switch wiring.
74 Unscrew the switch from the cylinder block.
75 Connect an ohmmeter between the switch terminals, and position the switch capsule in a container of water which is being heated. Use a thermometer to check the water temperature.
76 Below approximately 65°C (149°F), the ohmmeter should indicate infinity, showing that the circuit is switched off. Above approximately 65°C (149°F), there should be no resistance, showing that the circuit is switched on. Renew the switch if it fails to function correctly.
77 Insert and tighten the switch to the specified torque.
78 Reconnect the wiring, then lower the car to the ground.

Oil temperature sensor (from 1990 model year) – renewal

79 An oil temperature sensor is fitted from model year 1990, in place of the straightforward switch used previously. The sensor can be renewed by disconnecting the wiring plug, then unscrewing the sensor from the cylinder block. Ensure that the new sensor is fully tightened after refitting.

Air intake temperature switch (up to model year 1990) – removal, testing and refitting

80 The air intake temperature switch is located on the left-hand underside of the inlet manifold (photo). It is wired in series with the oil temperature switch.
81 Disconnect the wiring, and unscrew the switch from the inlet manifold.
82 Connect an ohmmeter between the switch terminals, and position the switch capsule in a container of water which is being heated. Use a thermometer to check the water temperature.
83 Below approximately 17°C (63°F), there should be zero resistance, indicating that the circuit is switched on. Above approximately 17°C (63°F), there should be infinity resistance, indicating that the circuit is switched off. Renew the switch if it fails to function correctly.
84 Refit and tighten the switch to the specified torque.
85 Reconnect the wiring.

Fuel pressure regulator – removal and refitting

86 The fuel pressure regulator is located on the left-hand end of the fuel distribution tube (photo). First disconnect the vacuum pipe.

Chapter 13 Supplement: Revisions and information on later models

11.86 Fuel pressure regulator

11.91 Disconnecting the supplementary air valve wiring plug

11.97 Coasting air valve

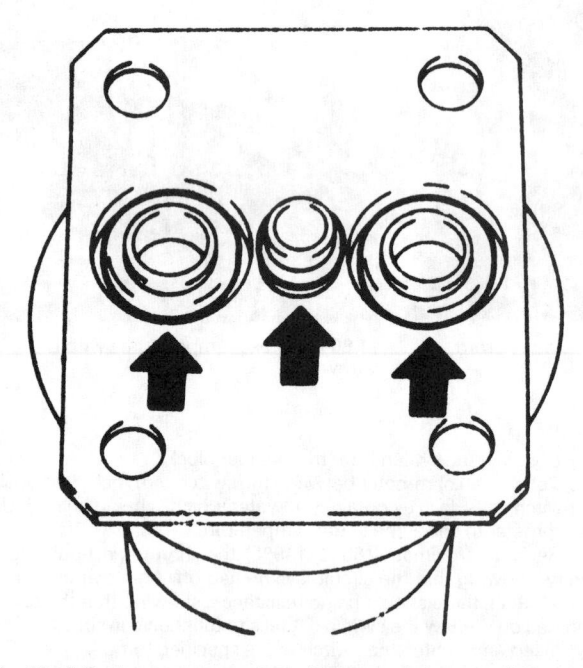

Fig. 13.27 Fuel pressure regulator O-ring seals (arrowed) (Sec 11)

87 Place some cloth rags around the regulator to soak up spilled fuel.
88 Unscrew the four screws, and remove the regulator from the distribution tube. Remove and discard the O-ring seals.
89 Refitting is a reversal of removal, but use new O-ring seals.

Supplementary air valve – removal, testing and refitting
90 The supplementary air valve is located on the right-hand side of the inlet manifold. It provides additional air for cold starting, to supplement the extra fuel supplied by the injectors. The valve gradually closes after the ignition is switched on, and remains closed while the engine is at normal temperature.
91 To remove the valve, first disconnect the wiring plug (photo).
92 Loose the clips and disconnect the two air hoses.
93 Unbolt the supplementary air valve from the inlet manifold.
94 Look through the air valve ports, and check that the regulator disc is open. This position corresponds to the setting with the engine cold, when additional air is required.
95 Apply 12 volts to the two terminals, and check that the regulator disc gradually closes. This position corresponds to the setting with the engine at normal operating temperature, when no additional air is required. Renew the supplementary air valve if it fails to function correctly.
96 Refitting is a reversal of removal, but make sure that the hoses are secure, and the clips tightened.

Coasting air valve (vacuum control) – adjustment
97 The coasting air valve is located on the left-hand side of the inlet manifold air box (photo), and its purpose is to weaken the fuel/air mixture during coasting. It does this by allowing air to bypass the airflow sensor unit. It is operated by vacuum taken from the vacuum spark port (also used by the exhaust gas recirculation valve).
98 To adjust the valve, disconnect the vacuum pipe from the bottom of the valve, and connect a vacuum pump instead. Apply vacuum, and check that the internal valve audibly switches at a vacuum of 630 mbar (18.6 in Hg). Any adjustment necessary may be made by resetting the adjustment screw.
99 On completion, reconnect the vacuum pipe.

Idle speed and mixture adjustment
100 Tamperproof plugs may be fitted over the idle speed and mixture adjustment screws. Under current UK legislation, the plugs may be removed and discarded if necessary. In some countries, only authorised persons may do this, and new plugs must be fitted after making adjustments.
101 Before making an adjustment, run the engine to normal operating temperature, and ensure that all electrical loads are switched off.
102 Connect a tachometer in accordance with the equipment manufacturer's instructions.
103 Allow the engine to idle, and check that the idle speed is as specified. If adjustment is necessary, turn the idle speed adjustment screw on the throttle housing anti-clockwise to increase the speed, or clockwise to decrease the speed (photo).
104 To check the idle mixture (CO level), connect an exhaust gas analyser in accordance with the equipment manufacturer's instructions.
105 With the engine idling, check that the CO level is as specified. If

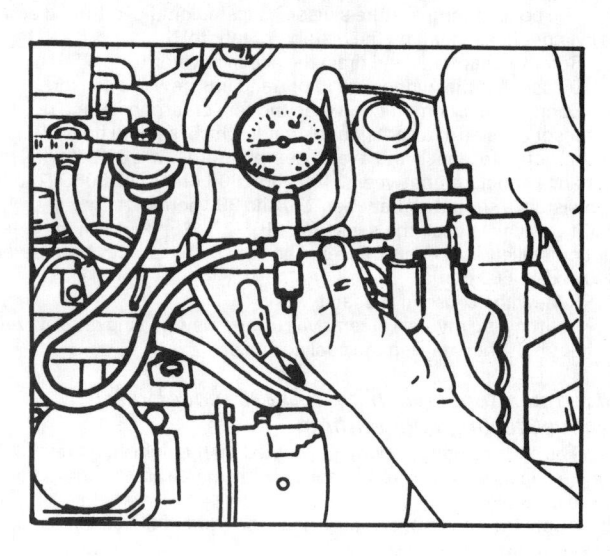

Fig. 13.28 Checking the coasting air valve (Sec 11)

Chapter 13 Supplement: Revisions and information on later models 261

11.103 Idle speed adjustment screw (arrowed)

11.105 Adjusting the CO level

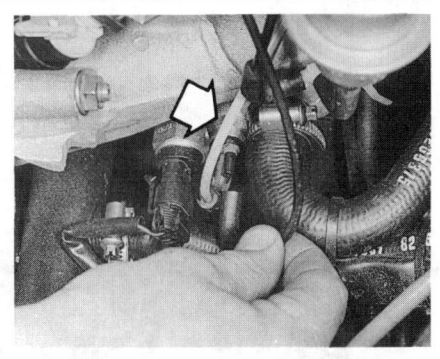

11.108 EGR ported vacuum switch (arrowed)

11.110 EGR valve and mounting bolts (arrowed)

11.111 Removing the EGR valve

11.112 EGR valve and sleeve

11.115A The twin-branch exhaust downpipe

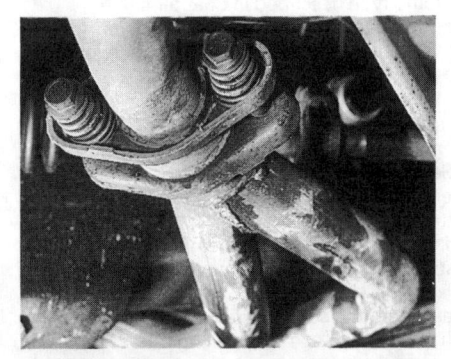

11.115B Exhaust downpipe-to-intermediate pipe connection

adjustment is necessary, turn the adjustment screw on the airflow sensor unit as required (photo).
106 On completion, disconnect the exhaust gas analyser and tachometer.

Exhaust gas recirculation (EGR) system – description and maintenance
107 An exhaust gas recirculation valve is fitted over the two centre branches of the inlet manifold. Internal channels within the cylinder head direct the exhaust gases to the valve, and when the valve is open, further channels direct the gases to the inlet ports of all four cylinders.
108 The valve is operated by vacuum taken from the vacuum spark port, through a temperature-controlled ported vacuum switch (PVS) (photo). Operation of the valve occurs only when the engine has reached normal operating temperature, and maximum opening is obtained at part-throttle openings. At idle and maximum-throttle positions, the valve is closed.
109 Further information on the EGR system is given in Section 9.

After cleaning the valve, it is recommended that the gasket is renewed.

EGR valve – removal, cleaning and refitting
110 Unscrew the mounting bolts securing the EGR valve to the inlet manifold (photo). Note that the right-hand bolt is located in the valve by a small sleeve, and care must be taken not to allow the sleeve to drop into the inlet manifold.
111 Remove the valve, disconnect the vacuum pipe, and remove the gasket, which will probably be stuck to the inlet manifold (photo).
112 Use a small wire brush to clean the valve, then scrape the carbon build-up from the ports (photo).
113 Refitting is a reversal of removal, but fit a new gasket.

Exhaust system – general
114 Refer to Chapter 3, Section 22.
115 A twin-branch downpipe is fitted. The connection between the downpipe and intermediate pipe consists of a central cone, and the bolts are fitted with springs, to allow slight flexing of the joint (photos).

Chapter 13 Supplement: Revisions and information on later models

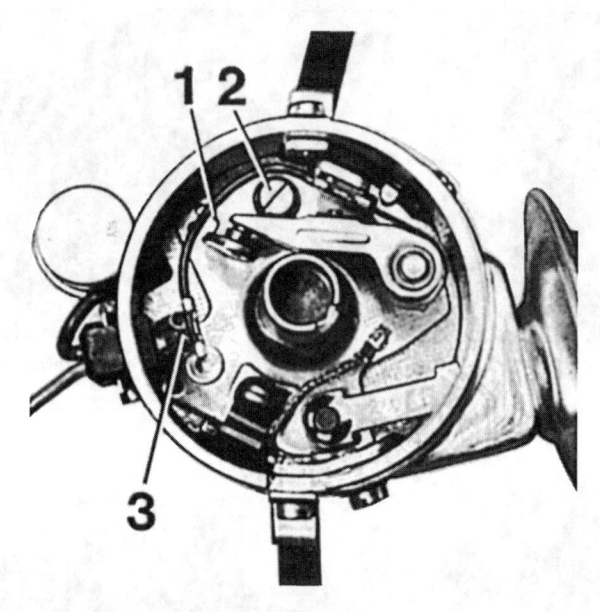

Fig. 13.29 Bosch distributor contact breaker points assembly (Sec 12)

1 Notch for adjustment of gap
2 Contact breaker securing screw
3 LT spade terminal

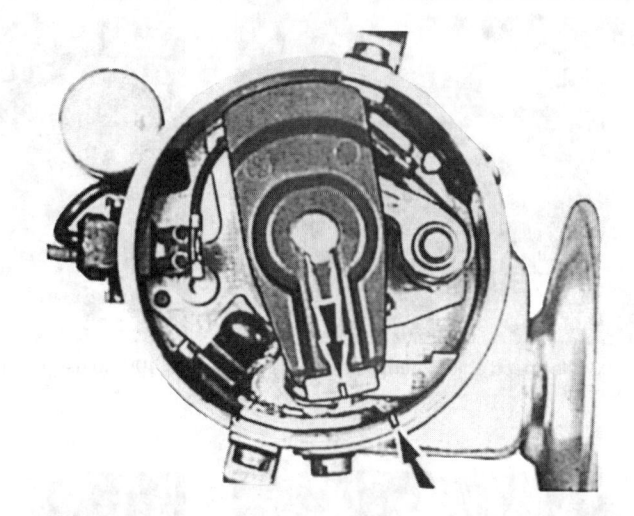

Fig. 13.30 Rotor position prior to fitting Bosch distributor (Sec 12)
Notches should be positioned as arrowed; with distributor fully engaged, notches should align

Fig. 13.31 Removing vacuum unit control rod C-clip from Bosch distributor (Sec 12)

12 Ignition system

Bosch distributor (1.0 models) – description
1 On some later models, a Bosch distributor may be fitted to an alternative to the Delco-Remy type described in Chapter 4.
2 Adjustment data and general specifications are as given for the Delco-Remy distributor. However, all relevant procedures relating to the Bosch distributor are given in the following paragraphs.

Condenser (Bosch distributor) – removal and refitting
3 The general comments regarding the function and testing of the condenser are given in Section 5 of Chapter 4. However, on the Bosch distributor, the condenser is fitted to the outside of the distributor body, and is removed and refitted as follows.
4 Release the two spring clips to remove the distributor cap. Pull the rotor arm from the distributor shaft, then remove the plastic anti-condensation shield.
5 Disconnect the LT lead from the spade terminal located inside the distributor, and disconnect the LT lead connecting the coil to the distributor.
6 Undo and remove the screw securing the condenser and LT lead assembly to the distributor body. Note that the condenser is supplied complete with LT lead to the coil, and LT spade terminal and grommet.
7 Refitting is the reverse of the removal procedure.

Contact breaker points (Bosch distributor) – removal, refitting and adjustment
8 Adjustment of the contact breaker points on the Bosch distributor is essentially the same as that described for the Delco-Remy distributor in Chapter 4, Section 3. It should be noted that the distributor cap is secured by two spring clips, and that the plastic anti-condensation shield must first be removed for access to the points.
9 Removal and refitting of the contact breaker points is similar to the procedure described in Chapter 5 for the Delco-Remy distributor. The Bosch points, however, are of one-piece construction, and are removed by disconnecting the LT lead from the spade terminal and removing the single securing screw.

Bosch distributor – removal and refitting
10 Removal and refitting is as given in Section 6 of Chapter 4. However, to assist in correctly aligning the rotor arm with the notch on the distributor body, the rotor should be placed in the position shown in Fig. 13.30 prior to fitting the distributor. Once the skew gear and oil pump drive are correctly engaged, the notch in the rotor arm contact should align with the notch in the rim of the distributor body.

Bosch distributor – overhaul
11 With the distributor removed from the engine, remove the distributor cap, rotor arm and plastic cover (if not already done).
12 Disconnect the LT lead from the spade terminal inside the distributor, then remove the condenser and contact breaker points as previously described.
13 Remove the C-clip from the vacuum unit control rod-to-baseplate pivot, and disengage the control rod. Remove the two vacuum unit securing screws from the outside of the distributor body, and remove the vacuum unit.
14 Remove the two screws securing the distributor cap retaining clips, noting that these also secure the contact breaker baseplate assembly. Remove the clips and baseplate.
15 This is the practical limit of dismantling. The distributor components should now be examined as detailed in Chapter 4, Section 7.
16 Reassembly is a reverse of the dismantling procedure.

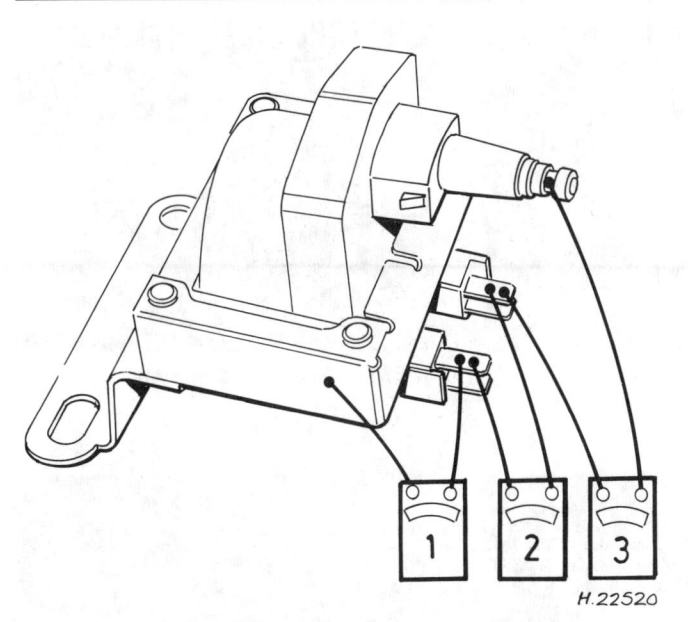

Chapter 13 Supplement: Revisions and information on later models

Fig. 13.32 Ignition coil testing – 1.4 models (Sec 12)

1 Short circuit to earth
2 Primary resistance
3 Secondary resistance

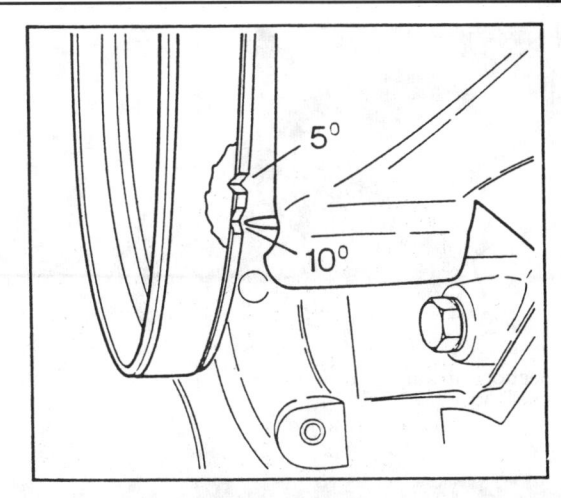

Fig. 13.33 Crankshaft pulley timing marks – 1.4 models (Sec 12)

Ignition system (1.4 models) – general

17 To avoid the risk of electric shock, always disconnect the battery earth lead or switch the ignition off before working on the ignition system or connecting/disconnecting test equipment.
18 The ignition system is broadly similar to that described for the 1.6 GTE/GSi models later in this Section, but the following major differences should be noted.
19 The electronic control unit (microprocessor) controlling the fuel and ignition functions is located under the left-hand side of the facia in the passenger compartment. Removal and refitting procedures are given in Section 10 of this Chapter.
20 A compact ignition coil is used, with a separate signal amplifier located below it, at the front left-hand side of the engine compartment.
21 The system uses a variety of sensors to calculate the optimum ignition and fuel system settings for the prevailing engine operating conditions. Further details can be found in Section 10 of this Chapter. The signals provided by the sensors include engine coolant temperature; engine load from the manifold absolute pressure (MAP) sensor; and engine speed and crankshaft position from the distributor.

Ignition coil and signal amplifier/module (1.4 models) – removal, checking and refitting

22 With the battery earth lead disconnected, first remove the HT lead from the coil tower.
23 Disconnect the multiplug from the base of the ignition coil (photo).
24 Undo the two ignition coil securing bolts and remove the coil (photo). This will expose the signal amplifier on its cooling plate below.
25 The signal amplifier and cooling plate can be removed by disconnecting the signal amplifier multiplug, then lifting out the assembly. If desired, the signal amplifier can be removed from the cooling plate by undoing the two securing bolts (photo).
26 Check the ignition coil primary and secondary resistances using an ohmmeter as described for 1.6 GTE/GSi later in this Section.
27 Refitting is a reversal of the removal procedure.

Crankshaft pulley markings – 1.4 models

28 Where two timing marks are found on the crankshaft pulley, the first notch represents 10° BTDC and the second notch represents 5° BTDC (refer to Fig. 13.33).

Distributor (all models) – removal and refitting

29 On models fitted with a distributor with offset-lug drive, or peg drive, instead of the skew-gear type drive, note the following.
30 Where reference is made in procedures to checking that the rotor arm points to a given reference point on the distributor body (with the crankshaft pulley notch aligned with its timing mark), note that a number of different types of distributor may be fitted, each having its own particular markings.
31 The alignment of the rotor arm with the notch in the distributor body rim applies to Bosch distributors. For Lucas and Delco distributors, the rotor arm should align with the arrow or mark on the outside of the distributor housing.

Distributor (1.4 models) – removal and refitting

32 The procedure is as described later in this Section for 1.6 GTE/GSi models, but before removing the distributor, make alignment marks between the distributor flange and the camshaft housing to obtain a basic static ignition timing setting when refitting.

12.23 Disconnecting the ignition coil multiplug

12.24 Ignition coil securing bolts (arrowed)

12.25 Signal amplifier-to-cooling plate securing bolts (arrowed)

12.34 Electronic module (arrowed) on the ignition coil mounting bracket

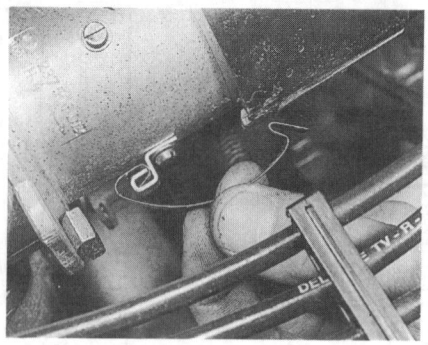

12.37A Release the spring clips ...

12.37B ... and remove the distributor cap ...

12.37C ... rotor arm ...

12.37D ... and plastic dust cover

12.38 Rotor arm aligned with the notch on the distributor body rim

Ignition timing (1.4 models with Multec central fuel injection) – adjustment

33 If the distributor has been removed, a basic static ignition timing setting can be obtained by re-aligning the marks made on the distributor flange and the camshaft housing before removal. Accurate adjustment of the ignition timing must be entrusted to a Vauxhall dealer, as dedicated equipment is required to electronically disable the automatic advance/retard functions.

Ignition system (1.6 GTE/GSi) – description

34 The ignition system fitted to the 1.6 fuel-injected engine is of the fully-electronic type, employing a microprocessor ignition timing control unit. The distributor is fitted with a Hall-effect trigger unit, and does not include centrifugal weights or a vacuum advance unit, as these functions are carried out electronically. Output from the Hall-effect trigger unit is fed to an electronic module located on the ignition coil mounting bracket (photo). This amplifies the signals, in order to operate the coil primary circuit.

35 The microprocessor for the ignition timing is located, together with the fuel injection microprocessor, in a sealed control unit on the left-hand side of the bulkhead. In practice, both microprocessors receive information from the engine temperature sensors and throttle position switch. This information is processed within the ignition timing processor, and the ignition timing is advanced or retarded according to predetermined parameters within the microprocessor memory.

Distributor (1.6 GTE/GSi) – removal and refitting

36 Remove the spark plugs (Chapter 4).
37 Release the spring clips, and withdraw the distributor cap. Pull off the rotor arm, remove the plastic dust cover, then refit the rotor arm (photos).
38 With the transmission in 4th gear and the handbrake released, pull the car forwards until, with a finger over the plug hole, compression can be felt in No 1 cylinder (nearest the crankshaft pulley). Continue turning the engine forwards until the notch on the crankshaft pulley is aligned with the timing pointer. Check also that the centre of the metal segment on the rotor arm is aligned with the notch on the distributor body rim (photo).

39 Disconnect the wiring plug from the distributor. To do this, depress the wire clip and unhook its ends (photos).
40 Unscrew the clamp nut, remove the clamp plate, and withdraw the distributor from the camshaft housing (photos).
41 Check the condition of the O-ring seal on the base of the distributor, and renew it if necessary (photo).
42 To refit the distributor, first check that the notch on the crankshaft pulley is still aligned with the timing pointer.
43 Turn the rotor arm so that it is aligned with the notch in the distributor body rim, then insert the distributor so that the drive peg is aligned with the hole in the end of the camshaft (photo). On Delco distributors, the offset-lug on the distributor drive coupling engages with a similarly offset slot in the end of the camshaft. With the distributor fully entered, re-align the rotor arm and notch.
44 Fit the clamp plate, and tighten the nut.
45 Fit the wire clip, and reconnect the wiring plug.
46 Remove the rotor arm and refit the plastic dust cover, then refit the rotor arm.
47 Refit the distributor cap and spark plugs.
48 Adjust the ignition timing, as described in paragraphs 53 to 58.

Ignition/fuel injection control unit (1.6 GTE/GSi) – removal and refitting

49 The control unit is located on the left-hand side of the bulkhead (photo), and consists of integrated circuits for the control of the ignition timing and fuel injectors.
50 To remove the control unit, first unscrew the upper mounting screw, and unhook the unit from the bottom bracket (photos).
51 Prise the spring clip to one side with a screwdriver, pull out the bottom of the wiring plug and unhook it at its top end (photo).
52 Refitting is a reversal of removal.

Ignition timing (1.6 GTE/GSI) – adjustment

53 Refer to Chapter 4, Section 10, paragraphs 1 to 5. The timing marks are the same as those on the 1.2 and 1.3 engines. Ignore the reference to the vacuum pipe.

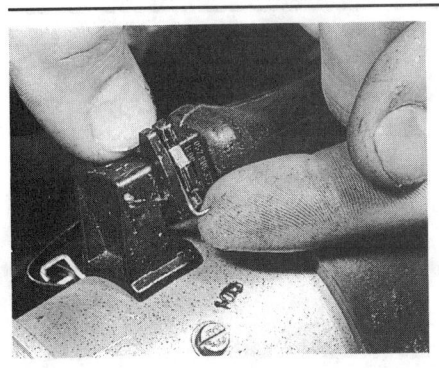

Chapter 13 Supplement: Revisions and information on later models

12.39A Release the wire clip ...

12.39B ... and disconnect the distributor wiring plug

12.40A Remove the clamp plate ...

12.40B ... and withdraw the distributor

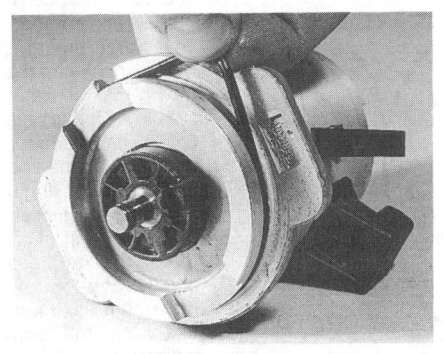

12.41 Removing the distributor O-ring seal

12.43 Hole in the camshaft (arrowed) must align with distributor drive peg

12.49 Ignition/fuel injection control unit

12.50A Unscrew the upper mounting screw ...

12.50B ... and unhook the control unit from the bracket

12.51 Removing the control unit wiring plug

12.55 Using split pins to interconnect the multiplug sockets

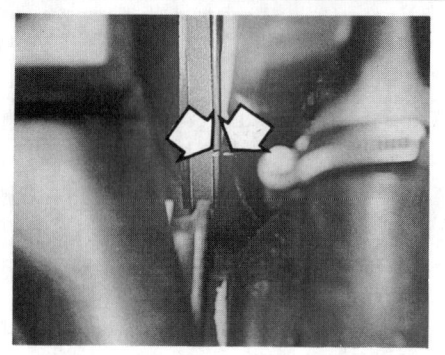

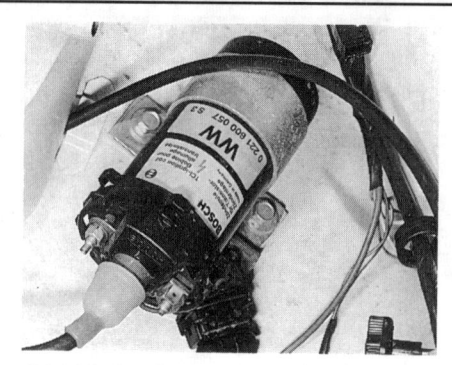

266 Chapter 13 Supplement: Revisions and information on later models

12.57 Crankshaft pulley timing marks aligned (arrowed)

12.59A Ignition coil

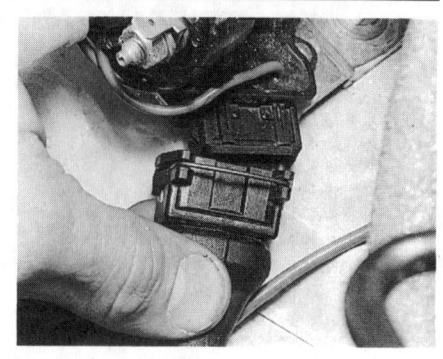

12.59B Disconnecting the ignition coil wiring plug

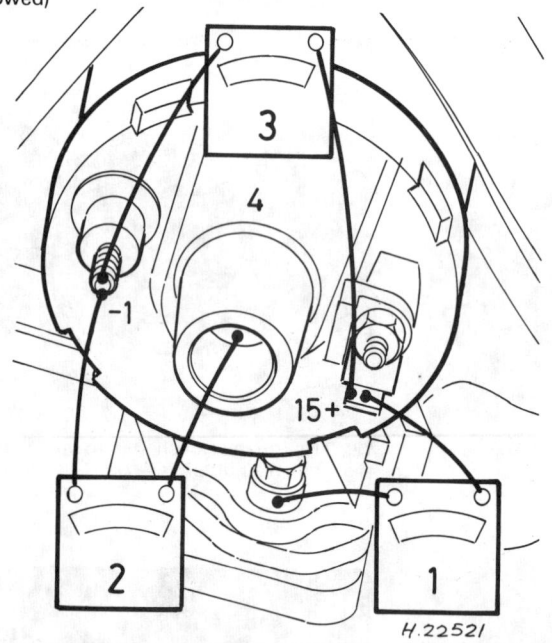

Fig. 13.34 Ignition coil testing – 1.6 GTE/GSi models (Sec 12)

1 Short circuit to earth
2 Secondary resistance
3 Primary resistance

54 Switch off all electrical equipment.
55 With the engine stopped, disconnect the wiring plug from the throttle valve switch, and use a suitable tool or wiring to interconnect the three pin sockets inside the plug. Split pins are ideal for this, as they do not damage the pin sockets (photo).
56 Start the engine and allow it to idle. The actual idle speed will be lower than normal, due to the throttle valve switch being disconnected.
57 Point the timing light at the timing marks, and check that they are aligned (photo). If they are not, loosen the distributor clamp nut and turn the distributor body clockwise to advance the timing, or anti-clockwise to retard the timing. Tighten the clamp nut after making the adjustment.
58 Stop the engine, and connect the wiring plug to the throttle valve switch. Disconnect the timing light.

Ignition coil (1.6 GTE/GSi) – checking

59 Remove the protective cap (where fitted) from the ignition coil, and disconnect the HT lead and primary wiring plug (photos).
60 Connect an ohmmeter between the primary terminals (1 and 15), and check that the primary winding resistance is as given in the Specifications.
61 Connect the ohmmeter between the HT terminal (4) and the primary negative terminal (1), and check that the secondary winding resistance is as given in the Specifications.
62 Refit the protective cap and wiring.

Retarding the ignition timing (for unleaded fuel)

63 When using unleaded fuel (on all except 1.4 models), the ignition timing may require retarding to prevent 'pinking'. The original recom-

mendation was to retard the ignition timing by 5°, but as from January 1989, this has been changed to 3°, and this applies retrospectively to all models (except the 1.4 which requires no adjustment).
64 Some timing lights have a facility for checking and altering the ignition timing in relation to a given setting (in this case the 10° BTDC timing marks). Retarding the timing by 3° will therefore be relatively simple with this type.
65 Where a standard timing light is being used, the crankshaft pulley must first be marked using the following procedure.
66 Turn the engine by means of the crankshaft pulley bolt, or by engaging top gear and pulling the car forward, until No 1 piston is at TDC on the firing stroke. This can be felt by removing No 1 spark plug and feeling for compression with your finger as the engine is turned. Use a blunt probe (eg a knitting needle) inserted through the plug hole to determine accurately the precise TDC point.
67 Mark the crankshaft pulley in line with the timing pointer to indicate the TDC position. Now divide the distance between the TDC and 10° BTDC marks by ten to determine a 1° increment, then multiply this by three to determine a 3° increment. Mark the crankshaft pulley at 7° BTDC for the new ignition timing, and use this mark to adjust the ignition timing as previously described.

13 Clutch and manual transmission

Gearchange linkage (five-speed) – adjustment

Note: *Special tools may be required for this operation*

1 Set the gearchange lever in neutral.
2 Remove the centre console, if fitted.
3 Loosen the guide lever clamp bolt (see Fig. 13.35).

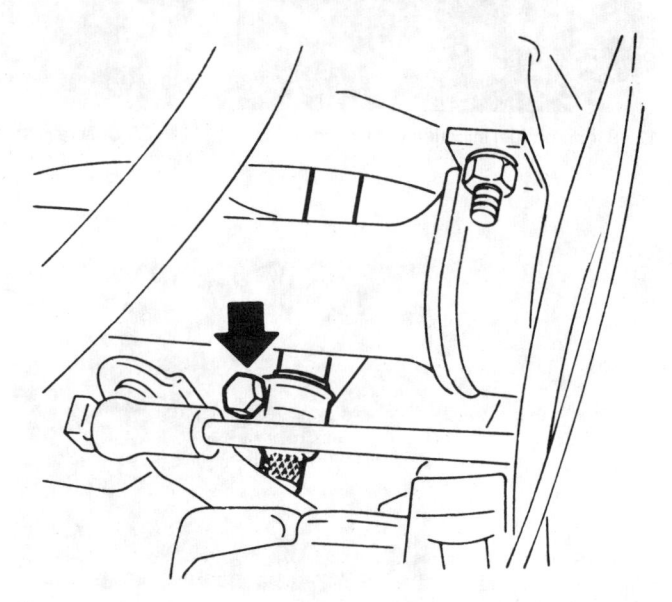

Fig. 13.35 Gearchange linkage (five-speed) guide lever clamp bolt (arrowed) (Sec 13)

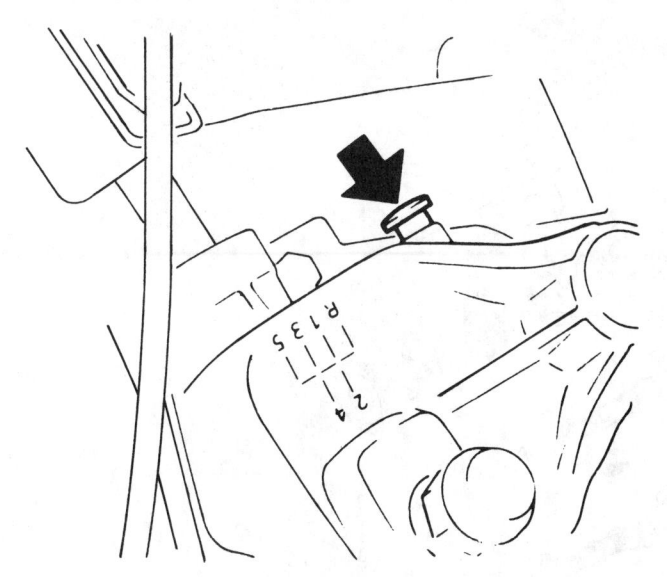

Chapter 13 Supplement: Revisions and information on later models

Fig. 13.36 Transmission cover adjustment hole plug (arrowed) (Sec 13)

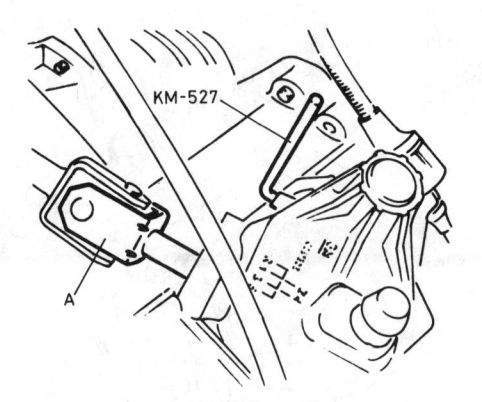

Fig. 13.37 Gearchange remote control rod (A) (Sec 13)

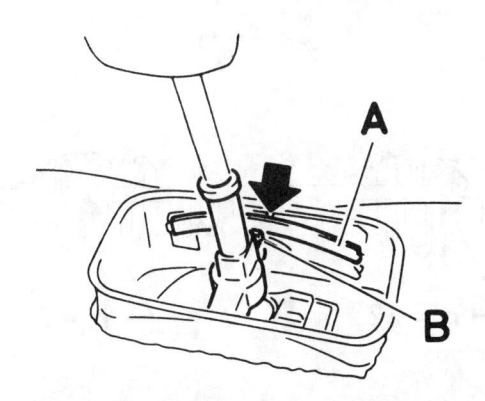

Fig. 13.38 Gearchange linkage – basic adjustment (Sec 13)

A Stop shoulder   B Front edge of gearshift lever hook

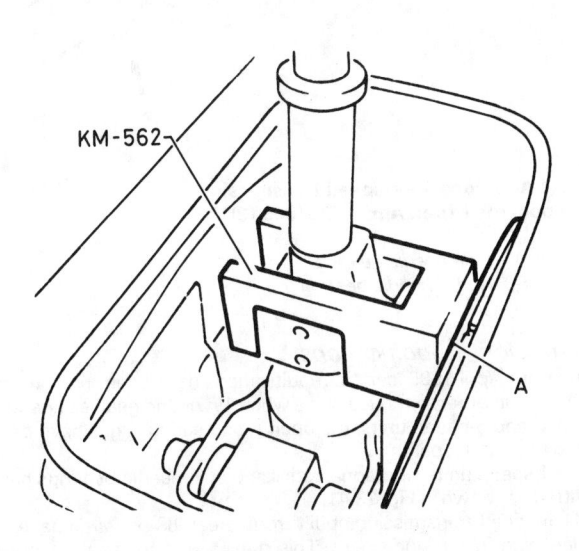

Fig. 13.39 Checking 5th gear selection fine adjustment (Sec 13)

A See text

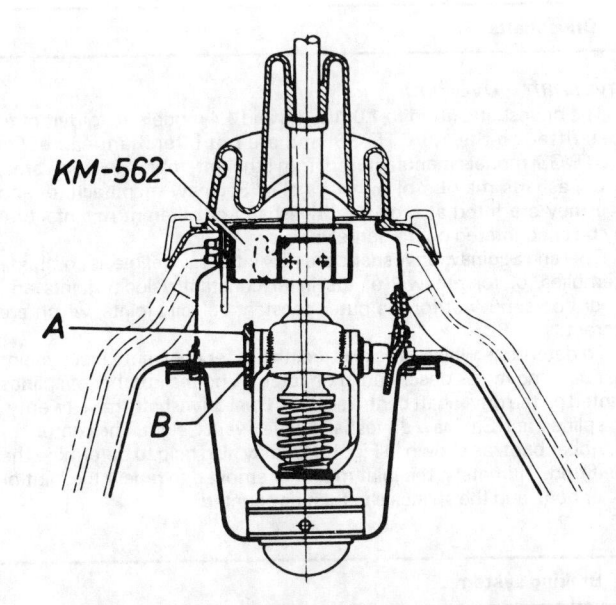

Fig. 13.40 5th gear selection fine adjustment (Sec 13)

A Lockplate tabs   B Adjusting nut

4 Prise out the plug from the adjustment hole in the transmission cover.
5 Turn the remote control rod ('A' in Fig. 13.37) in an anti-clockwise direction (viewed from the rear of the car) until special tool KM 527 can be inserted into the adjustment hole. If the tool is not available, a substitute rod can be used.
6 Pull the boot up the gearchange lever, and at the same time have an assistant hold the lever over, in line with the 1st/2nd speed gate. The hook on the lever must be up against the stop shoulder without any play (see Fig. 13.38).
7 The forward edge of the hook must lie within the marked notch in the transmission tunnel cut-out.
8 Holding the lever in its set position, tighten the guide lever clamp bolt.
9 Remove the special tool or rod and refit the plug.
10 To check 5th gear fine adjustment, another special tool will be required – KM 562. Locate this on the gearchange lever as shown (lever in neutral) and check that the play ('A' in Fig. 13.39) on the left stop shoulder does not exceed 0.4 mm (0.016 in).
11 If the play is excessive, flatten the lockplate tabs (A) (Fig. 13.40) and turn the adjusting nut (B) Lock the nut by bending up the lockplate tabs.
12 Check the engagement of all gears with the engine idling and clutch pedal depressed.
13 Refit the boot and centre console.

268 Chapter 13 Supplement: Revisions and information on later models

Fig. 13.41 Additional four-speed transmission components – from April 1986 (Sec 13)

A Split needle bearing
B Radial needle bearing

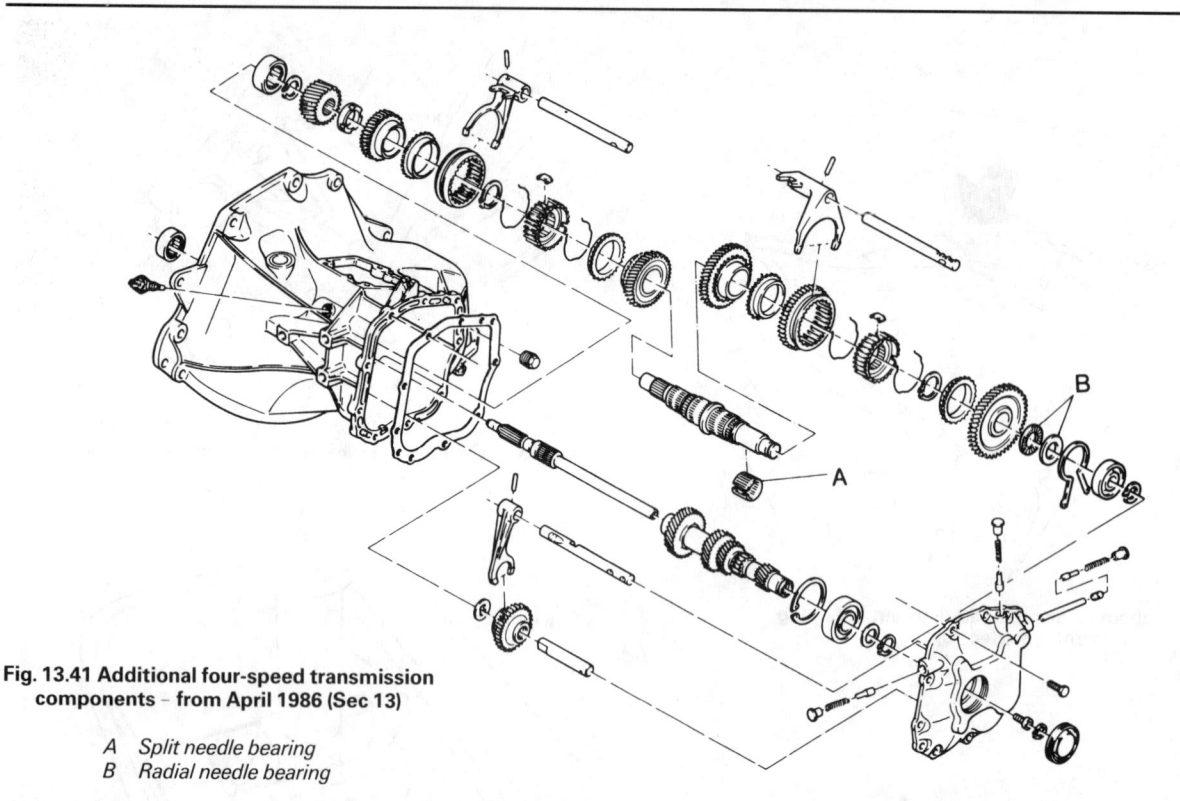

Transmission – modifications

14 As from April 1986, certain modifications have been made to the transmission in order to reduce noise when 1st or 2nd gear is engaged.
15 On 4- and 5-speed units, the open type bearing cages have been changed to sealed type.
16 On 4-speed units, additional radial and split needle bearings have been fitted as shown in Fig. 13.41.
17 All 5-speed transmissions built from late-1988 onwards have no threaded plug in the end cover. This means that for all procedures described in Chapters 5 and 6 of this manual which require the removal of this plug, the complete end cover must be unbolted from the transmission casing.

14 Driveshafts

Driveshaft – overhaul

1 The driveshafts fitted to 1.0, 1.2, 1.3 and 1.4 models are different to those fitted to the 1.6 GTE/GSi and are not interchangeable. On 1.6 GTE/GSi models manufactured from 5th September 1988 onwards, the driveshafts may be of either Opel or Saginaw manufacture, and since they are fitted at random, driveshafts of different manufacture may be encountered on the same vehicle.
2 Opel and Saginaw driveshafts are fully interchangeable as complete assemblies, or for renewal of the inner constant velocity joints and rubber boots, but not for the outer constant velocity joints, which are different.
3 To determine which type is fitted, the outer constant velocity joint must be removed as described in Chapter 7, and the number of splines counted on the driveshaft centre section. Opel driveshafts have twenty-two splines, and Saginaw driveshafts have twenty-eight. The contour of the rubber boot, as shown in Fig. 13.42, may also help to determine the type fitted. Ultimately, the joint must be removed to renew the joint or rubber boot, and the splines can then be counted.

15 Braking system

Brake pedal relay lever – modification

1 After 1984, the brake pedal relay lever on right-hand drive models has been modified. To remove the relay lever, first withdraw the parcel shelf (three screws) from under the facia panel (photo).
2 Working inside the car under the facia panel, disconnect the brake pedal and the servo pushrods by extracting the spring clips and clevis pins.
3 Disconnect the pedal return spring.
4 Unscrew the nuts which secure the relay lever support brackets to the bulkhead (photo).
5 Slide the relay lever assembly towards the left-hand side of the car, and withdraw it from under the heater casing.
6 Refitting is a reversal of removal. Check the brake pedal free play on completion, as described later in this Section.

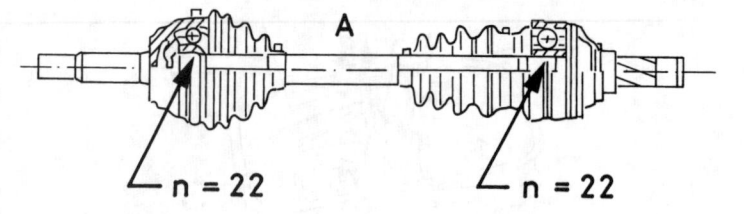

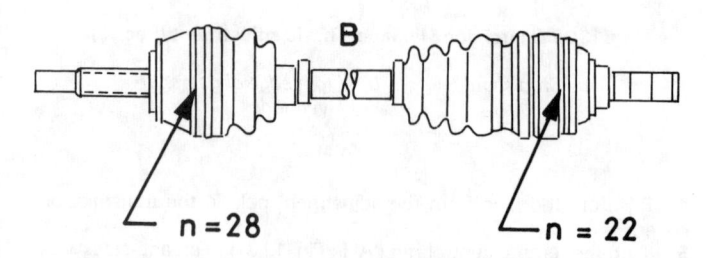

Fig. 13.42 Number of constant velocity joint splines (n) on Opel (A) and Saginaw (B) driveshafts – 1.6 GTE/GSi models from September 1988 (Sec 14)

Rear brake shoes (self-adjusting type) – renewal

7 Vehicles produced for the 1984 model year and later are equipped

Chapter 13 Supplement: Revisions and information on later models 269

15.1 Extracting a parcel shelf screw

15.4 Brake pedal relay lever right-hand bracket (arrowed)

15.9 Self-adjusting rear brake (right-hand) with brake drum removed. Later versions have a modified anchor block spring

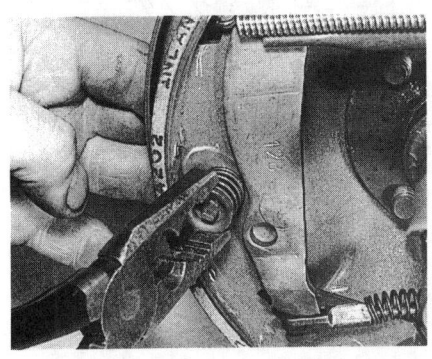

15.10 Removing a brake shoe steady spring cup

15.12 Releasing handbrake cable from shoe handbrake lever

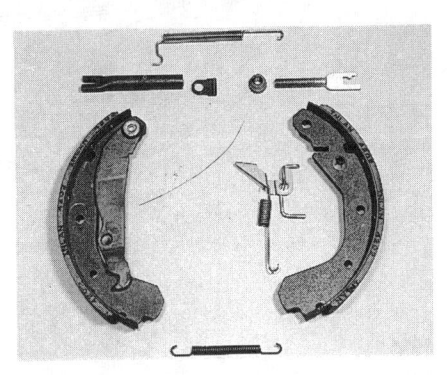

15.13 Self-adjusting brake shoe components. Modified anchor block spring on later versions

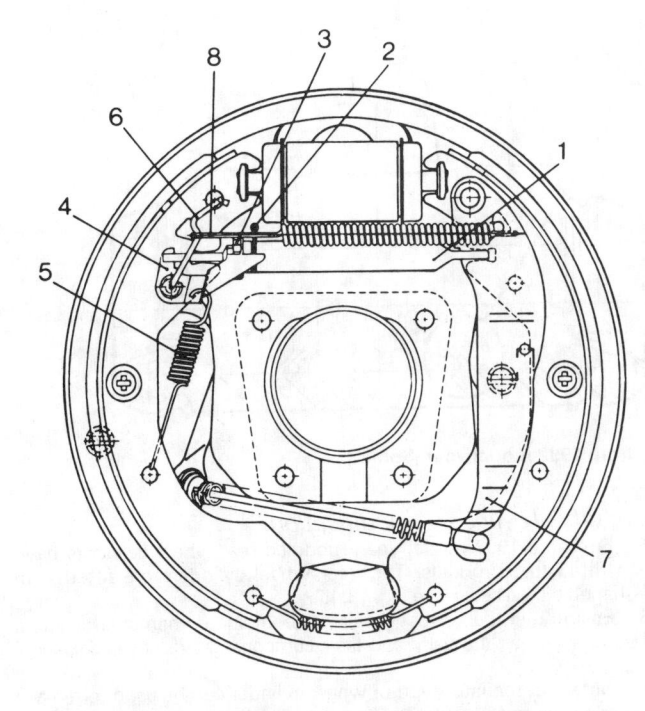

Fig. 13.43 Self-adjusting type rear brake assembly (Sec 15)

1 Adjuster strut
2 Thermoclip
3 Star wheel
4 Adjuster lever
5 Tension spring
6 Torsion spring
7 Shoe handbrake lever
8 Upper return spring

with self-adjusting rear brakes. The external adjusters described in Chapter 9 will not be found on the brake backplate, and periodic adjustment of the shoes will not be necessary.

8 Raise the rear of the car, support it securely and remove the roadwheels. Fully release the handbrake.
9 Prise off the grease cap, extract the split pin, unscrew the castellated nut, remove the thrustwasher and pull off the brake drum (photo).
10 Brush away all dust and dirt, taking care not to inhale it, as it may be injurious to health. Remove the shoe steady cups by gripping their edges with a pair of pliers, depressing them and turning them through 90° (photo).
11 Remove the shoe steady springs and pins.
12 Expand the shoes, using hand pressure to release them from the cylinder and the lower anchor block. Pull the shoes towards you, unhook the handbrake cable from the lever on the shoe (photo) and remove the complete shoe assembly.
13 Lay the assembly on the bench, and place the new shoes beside the old assembly in the correct leading and trailing positions. Always use new factory re-lined shoes, which will have been ground to contour, and come complete with a new handbrake lever riveted to the trailing shoe (photo).
14 Transfer the shoe return springs to their correct holes in the new shoes, and locate the self-adjuster components. The threaded strut should be clean and lightly greased, and fully retracted by turning the star wheel before fitting it between the shoes. Note that the shorter leg of the strut fork should be towards you (photos).
15 Offer the complete brake shoe assembly to the backplate, which should have a smear of grease applied to its shoe contact high spots.
16 Expand the shoes against the tension of their return springs, and engage the handbrake cable with the shoe lever, then engage the shoe ends with the pistons of the hydraulic cylinder and the anchor block.
17 Make sure that the shoes are concentric within the rim of the brake backplate, and fit the shoe steady pins, springs and cups.
18 Fit the brake drum.
19 Adjust the rear hub bearings as described in Chapter 11, Section 11.
20 Renew the shoes on the opposite side of the car in a similar way.

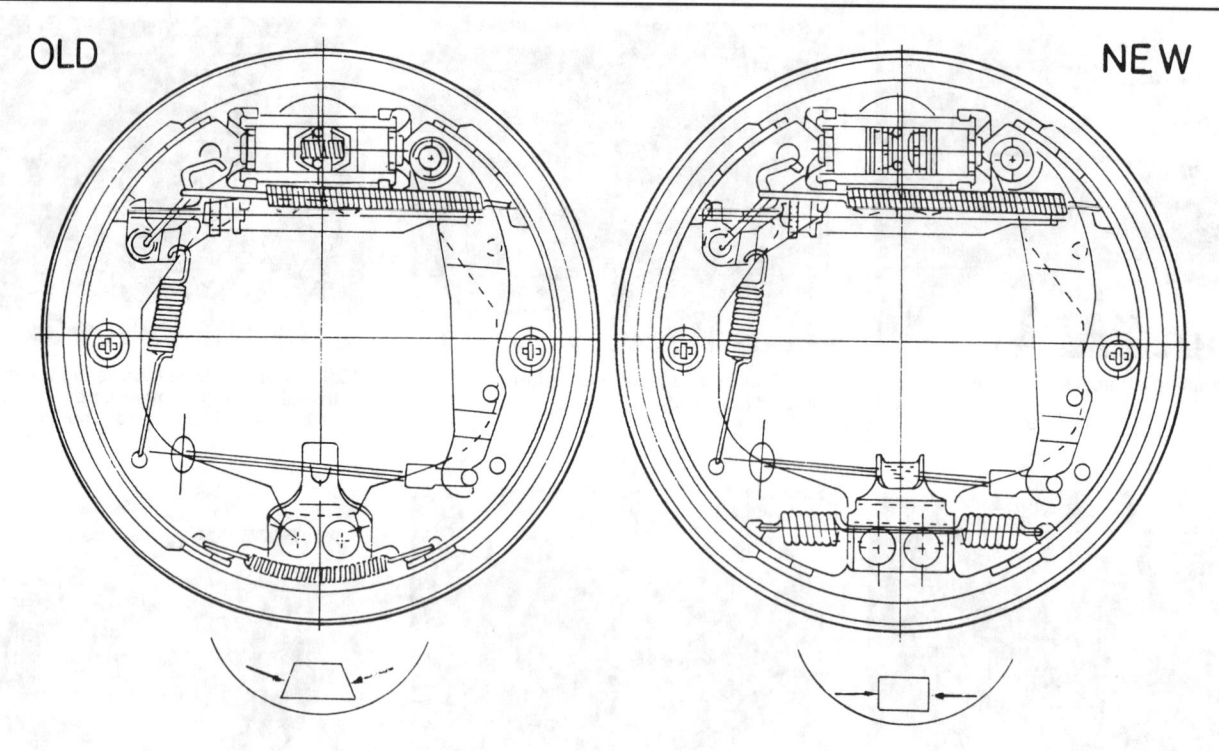

Fig. 13.44 Modified rear brake assembly – from 1992 model year (Sec 15)

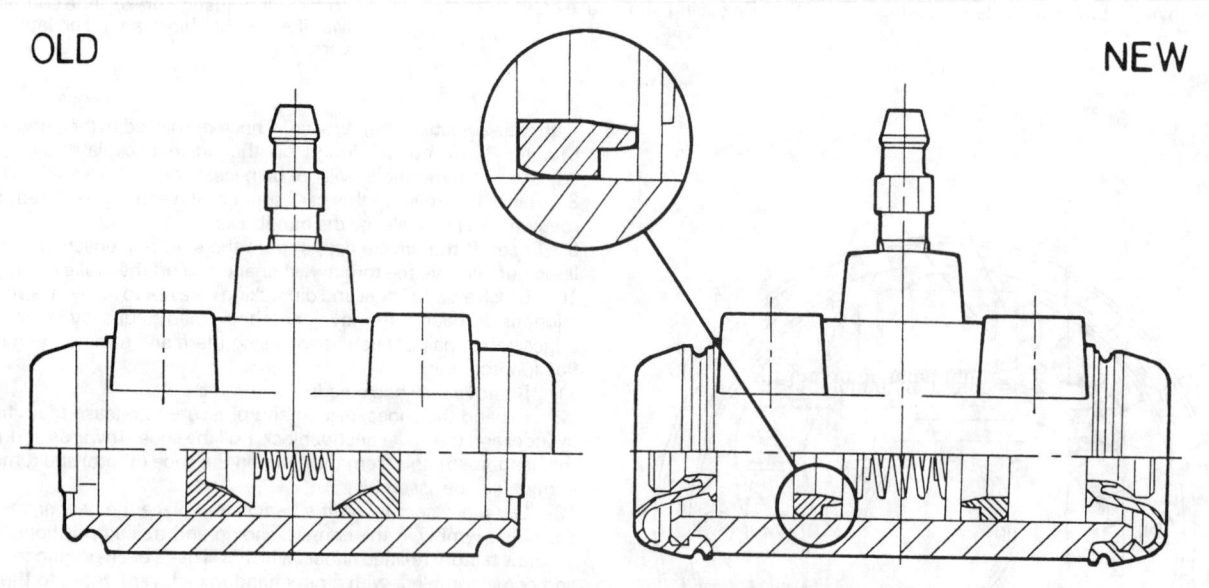

Fig. 13.45 Modified rear wheel cylinder – from 1992 model year (Sec 15)

21 Refit the roadwheels, lower the car, and then apply the footbrake several times to set the shoes at the closest adjustment to the drum. Check the operation of the handbrake and, if necessary, adjust as described in the following paragraphs.

Rear brake shoes – modifications

22 From the 1992 model year, the brake shoe lower anchor block has been modified.
23 To accommodate this modification, modified shoes and a modified lower shoe return spring have been fitted, and the location of the spring differs from earlier models as shown in Fig. 13.44.
24 Overhaul procedures are unaffected, but note that the later type shoes and return spring must not be used on models fitted with the earlier type of anchor block.

Rear wheel cylinder – modifications

25 From the 1992 model year, modified rear wheel cylinders have been fitted to all models. The later wheel cylinders are fitted with 'L'-shaped piston seals (see Fig. 13.45).
26 Ensure that the correct repair kit is obtained when overhauling a wheel cylinder, as the early and later components are not interchangeable.
27 Note that the later type of wheel cylinder can be used to replace the early type as a complete unit.

Handbrake (self-adjusting brakes) – adjustment

28 Adjustment of the handbrake is usually required due to cable stretch, or the fitting of new components. On cars with self-adjusting rear brakes, this should be carried out in the following way.

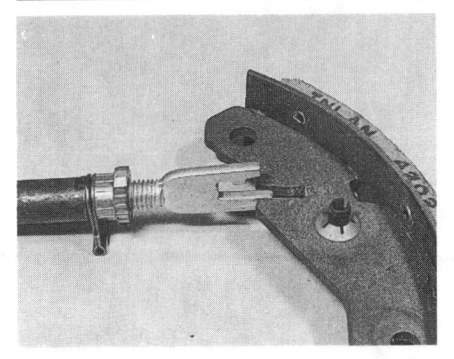

Chapter 13 Supplement: Revisions and information on later models

15.14A Shoe adjuster strut and star wheel

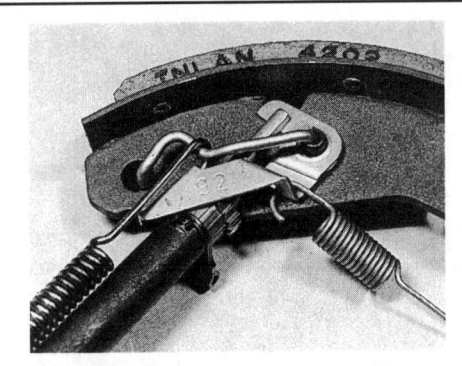

15.14B Shoe self-adjuster lever and spring arrangement

15.31 Handbrake cable yoke adjuster nut (arrowed)

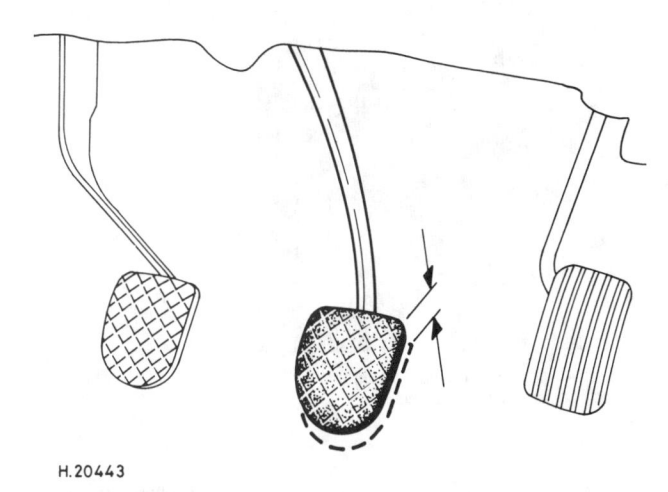

Fig. 13.46 Brake pedal free play (between the arrows). For dimensions, see text (Sec 15)

29 Adjust the shoes by giving at least ten applications of the brake pedal.
30 Raise the rear wheels off the floor, securely support the vehicle, and fully release the handbrake.
31 Turn the adjuster nut on the cable yoke on the rear axle until the roadwheels are locked. Now unscrew the nut until the wheels can be turned without any binding (photo).
32 Lower the car, and check that the handbrake starts to apply the brakes at the second notch of the lever ratchet.

Handbrake lever and long (main) cable (vehicles fitted with catalytic converter) – removal and refitting
33 The procedure is similar to that described in Chapter 9, Section 24, but note that it will be necessary to remove the heat shield(s) for access.

Brake disc (all models) – refinishing
34 A brake disc which has had its thickness reduced to the minimum permitted by refinishing (Chapter 9, Specifications) may only be used for the life of one set of new pads. When these pads have worn out, a new disc must be fitted.

Brake master cylinder (later models) – overhaul
35 The maker's overhaul kit for later-type master cylinders (early 1984 onwards) consists of the two pistons, with seals already fitted, in an assembly tube. The tube is used to aid insertion of the pistons.
36 Overhaul procedures are otherwise as described in Chapter 9, Section 15.

Brake pedal free play – checking and adjustment
37 This is not a routine operation. It should only be necessary after renewal of the vacuum servo or the relay lever.
38 With the engine stopped, press the brake pedal firmly several times to destroy any residual vacuum in the servo. Measure the pedal free play: it should be 6 to 9 mm (0.24 to 0.36 in).
39 If adjustment is necessary, this is carried out by disconnecting the vacuum servo clevis, and screwing the clevis up or down the pushrod. Reconnect the clevis and recheck the free play.

Brake discs (1.6 GTE/GSi)
40 On the 1.6 GTE/GSi the front brake discs are of the ventilated type, incorporating internal cavities for the dissipation of heat.

Front brake pads (from 1992 model year) – inspection and renewal
41 From the 1992 model year, a new type of front brake caliper has been fitted to all models. The pads can be inspected for wear as described in Chapter 9, Section 5, but if the pads require renewal, proceed as follows.
42 Using a screwdriver, prise the pad retaining clip from the outboard edge of the caliper, noting how it is located (photo).
43 Prise out the two guide bolt dust caps from the inboard edge of the caliper, then using a suitable Allen key, unscrew the guide bolts, and lift the caliper and inboard pad from the bracket (photos). Recover the outboard brake pad. Suspend the caliper body with wire or string, to avoid straining the brake fluid hose.
44 Pull the inboard pad from the caliper piston, noting that it is retained by a clip attached to the pad backing plate (photo).
45 Brush the dust and dirt from the caliper, but take care not to inhale it. Carefully remove any rust from the edge of the brake disc.
46 In order to accommodate the new thicker pads, the caliper piston must be depressed fully into its cylinder bore, using a flat bar of metal such as a tyre lever. The action of depressing the piston will cause the fluid level in the reservoir to rise, so to avoid spillage, syphon out some fluid using an old hydrometer or a teat pipette. Do not lever between the piston and disc to depress the piston.
47 Check that the cutaway recesses in the piston are positioned vertically. If necessary, carefully turn the piston to its correct position (photo).
48 Apply a little brake grease to the contact surfaces of the new brake pads.
49 Fit the new inboard pad to the caliper piston, ensuring that the piston is correctly located.
50 Locate the outboard pad on the caliper bracket, with the friction material facing the disc.
51 Refit the caliper to the bracket, and tighten the guide bolts to the specified torque.
52 Refit the guide bolt dust caps.
53 Refit the pad retaining clip, locating it as noted before removal.
54 Repeat the operations on the remaining side of the vehicle.
55 Refit the road wheels and lower the vehicle to the ground. Do not fully tighten the roadwheel bolts until the vehicle is resting on its wheels.
56 Apply the footbrake hard several times to position the pads against the discs.
57 Check and if necessary top up the brake fluid level.
58 New pads should be carefully bedded in and, where possible, heavy braking should be avoided during the first 100 miles (160 km) or so after fitting new pads.

15.42 Prising out the brake pad retaining clip

15.43A Removing a caliper guide bolt dust cap

15.43B Withdrawing the caliper and inboard pad, and the outboard pad

15.44 Removing the inboard pad from the caliper piston

15.47 Caliper piston cutaway recess (arrowed) correctly positioned

Front brake caliper (from 1992 model year) – removal, overhaul and refitting

Note: Before dismantling a caliper, check that suitable replacement parts can be obtained, and retain the old components to compare them with the new ones. New sealing rings must be used on the fluid hose union bolt on refitting.

59 Where applicable, remove the wheel trims, then loosen the relevant front roadwheels and apply the handbrake. Jack up the front of the vehicle, and support safely on axle stands positioned under the body side members. Remove the roadwheel.
60 Remove the brake pads, as described earlier in this Section.
61 Working under the bonnet, remove the brake fluid reservoir cap, and secure a piece of polythene over the filler neck with a rubber band, or by fitting the cap. This will reduce the loss of fluid during the following procedure.
62 Unscrew the brake fluid hose union bolt from the rear of the caliper, and disconnect the hose. Recover the two sealing rings from the union bolt (one on either side of the hose and end fitting). Be prepared for fluid spillage, and plug the open ends to prevent dirt ingress and further fluid loss.
63 Withdraw the caliper body from the vehicle.
64 If desired, the caliper bracket can be removed from the steering knuckle by unscrewing the two securing bolts.
65 To overhaul the caliper, proceed as follows, otherwise proceed to paragraph 80 for details of refitting.
66 Brush the dirt and dust from the caliper, but take care not to inhale it.
67 Using a screwdriver, carefully prise the dust seal from the end of piston and the caliper body, and remove it.
68 Place a thin piece of wood in front of the piston to prevent it from falling out of its bore and sustaining damage, then apply low air pressure – *eg* from a foot pump – to the hydraulic fluid union hole in the rear of the caliper body, to eject the piston from its bore.
69 Remove the wood and carefully withdraw the piston.
70 Carefully prise the seal from its groove in the caliper piston bore, using a plastic or wooden instrument.
71 Inspect the surfaces of the piston and its bore in the caliper for scoring or evidence of metal-to-metal contact. If evident, renew the complete caliper assembly.
72 If the piston and bore are in good condition, discard the seals and obtain a repair kit, which will contain all the necessary renewable items.
73 Clean the piston and cylinder bore with brake fluid or methylated spirit – nothing else.
74 Commence reassembly by fitting the seal into the caliper bore.
75 Locate the dust seal in its groove in the piston. Dip the piston in clean brake fluid and insert it squarely into the cylinder. Check that the cutaway recesses in the piston are positioned vertically. If necessary, carefully turn the piston to its correct position.
76 When the piston has been partially depressed, engage the dust seal with the rim of the caliper bore.
77 Push the piston further into its bore, but not as far as the stop, ensuring that it does not jam.
78 If desired, the guide bolt sleeves can be renewed. Extract the nylon compression sleeve from within each rubber, then carefully compress the rubber shoulder, and push the rubber through the hole in the caliper body to remove it from the inboard end.
79 Fit the new sleeves using a reversal of the removal procedure.
80 Where applicable, refit the caliper bracket to the hub carrier, and tighten the securing bolts to the specified torque.
81 Reconnect the brake fluid hose union, using new sealing rings on the union bolt.
82 Refit the brake pads, as described earlier in this Section.
83 Remove the polythene from the brake fluid reservoir filler neck, and bleed the relevant brake hydraulic circuit as described in Chapter 9, Section 4.
84 Refit the roadwheel and lower the vehicle to the ground. Do not fully tighten the roadwheel bolts until the vehicle is resting on its wheels.

Front brake disc (from 1992 model year) – removal and refitting

85 The brake discs can be examined for wear as described in Chapter 9, Section 7. To remove a brake disc, proceed as follows.
86 Remove the brake pads as described previously in this Section.

Chapter 13 Supplement: Revisions and information on later models

87 Unscrew the two securing bolts and remove the caliper bracket.
88 Remove the securing screw and withdraw the disc from the hub, where applicable tilting it to clear the brake caliper.
89 Refitting is a reversal of removal, but make sure that the mating faces of the disc and hub are perfectly clean, and apply a little locking fluid to the threads of the securing screw. Refit the brake pads as described earlier in this Section.

16 Electrical system

Fuses and relays
1 On 1986 and later models, additional equipment circuits are protected by the following fuses:

No 1: Electric fuel pump (where applicable)
No 4: Headlamp range-control motor (left-hand)
No 5: Headlamp range-control motor (right-hand)
No 11: Electrically-operated exterior mirrors and electrically-operated front windows
No 14: Headlamp-on warning buzzer
No 16: Central door locking system

2 The fuse ratings on later models have been rationalized to include only 10A (red) 15A (blue), 20A (yellow) and 30A (green) fuses.
3 On some later models fitted with a multi-tone horn, a relay is located on the left-hand side of the bulkhead, and a fuse is attached to the base of the relay (photo).

Heated rear window – precautions and repair
4 Care should be taken to avoid damage to the element for the heated rear window. Avoid scratching with rings on the fingers when cleaning, and do not allow luggage to rub against the glass.
5 Do not stick labels over the element on the inside of the glass.
6 A voltmeter or ohmmeter can be used to assist with location of defects. Look for an open-circuit or a sudden change in voltage, reading along the resistance wires.
7 If the element grids do become damaged, a special conductive paint is available from most motor factors to repair it.
8 To repair, degrease the affected area and wipe dry.
9 Apply tape at each side of the conductor to mask the adjacent area.
10 Shake the repair paint thoroughly, and apply a thick coat with a fine paint brush. Allow to dry between coats, and do not apply more than three.
11 Allow to dry for at least one hour before removing the tape.
12 Rough edges may be trimmed with a razor blade if necessary, after a drying time of some hours.
13 Do not leave the heated rear window switch on unnecessarily, as it draws a high current from the electrical system.

Fan and heated rear window switch (from 1991 model year) – removal and refitting
14 The switch can be removed from the facia as follows. Move the switch to the 'ON' position, and release the retaining clips on either side of the switch using a screwdriver or a similar tool. Refitting is a reversal of removal.

Electrically-operated exterior mirrors
15 The method of removal and refitting for this type of mirror is similar to the operation described for the manual type in Chapter 12, Section 32. Be sure to disconnect the battery, and before removing the mirror, disconnect its wiring plug.

Dim-dip headlamps
16 In line with recent legislation, this system is fitted to all 1987-on models.
17 Its purpose is to supplement the side lamps with dimmed headlamps when driving the vehicle in built-up areas.
18 If the sidelamps are switched on when the ignition is switched on, the dimmed headlamps come on automatically. Normal headlamp illumination is provided once the lighting switch is moved off the sidelamp position.
19 The headlamp dim-dip relay is located behind the right-hand footwell side trim panel. There have been several instances of the relay self-activating due to water or condensation contaminating the multi-plug terminals. Where this has occurred, the windscreen should be checked for water leaks, and if none are found, the relay should be re-positioned as shown in Fig. 13.48 with the multi-plug pointing forwards. This is the new position for the relay as from early 1988.

Headlamps-on warning buzzer
20 This is fitted to later models, to warn the driver that the headlamps have not been switched off if the door is opened prior to leaving the vehicle.

Headlamp switch (from 1991 model year) – removal and refitting
21 Manufacture two tools (ideally from steel plate) as shown in Fig. 13.49, to the following dimensions; thickness 2 mm (0.08 in); width 5 mm (0.20 in) and make a 90° bend approximately 4 mm (0.16 in) from the end.
22 Ensure that the battery earth lead has been disconnected, then turn the rotary switch knob to the dipped-beam 'ON' position. Release the knob by inserting a small screwdriver or a suitable rod through the hole in the bottom of the knob, and pull the knob from the switch.

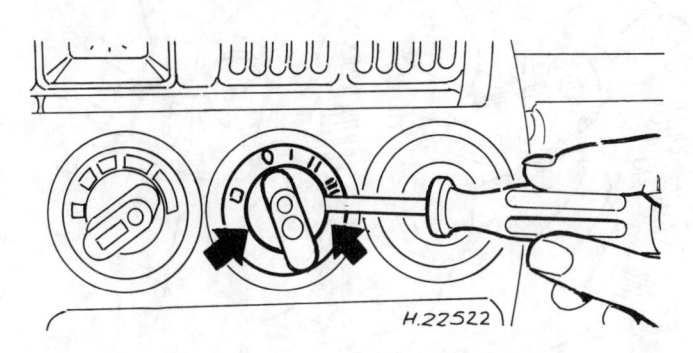

Fig. 13.47 Releasing the combined fan and heated rear window switch (Sec 16)

Fig. 13.48 Revised headlamp dim-dip relay position – from early 1988 (Sec 16)

16.3 Multi-tone horn relay and fuse

16.28A Headlamp securing bolts (arrowed)

16.28B Headlamp base-locating clip (arrowed)

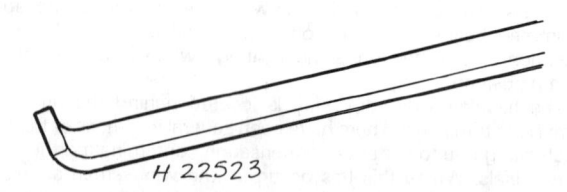

Fig. 13.49 Headlamp switch removal tool (Sec 16)
For dimensions see text

23   Release the spring retainers at each side of the switch, using the manufactured tools, then withdraw the switch and disconnect the wiring.
24   Refitting is a reversal of removal.

Headlamp (from 1991 model year) – removal and refitting
25   Disconnect the battery earth lead.
26   Working at the back of the headlamp unit (under the bonnet), disconnect the wiring from the main and dipped beam bulbs, the sidelamp bulb and the headlamp range control motor. Also disconnect the wiring from the adjacent direction indicator lamp.
27   Remove the direction indicator lamp as described later in this Section.
28   Undo the bolts securing the top of the headlamp to the front panel, then detach the headlamp from its base-locating clip and remove it (photos). For improved access, it may be necessary to remove the radiator grille.
29   Refitting is a reversal of the removal procedure.
30   Although provision is made for basic alignment adjustment, accurate headlamp alignment can only be achieved using optical beam setting equipment. Entrust this to your dealer or other competent specialist.

Headlamp range-control motor – removal and refitting
31   With the relevant headlamp removed, turn the range-control motor through 90° (refer to Fig. 13.52) then remove it from the ball-end (photo).
32   Refitting is a reversal of removal.
33   The headlamp range-control switch should be set to position 0 to give the basic adjustment setting.

Headlamp range-control motor switch – removal and refitting
34   Disconnect the battery earth lead.
35   Remove the rear foglamp switch as described later in this Section.
36   Carefully lever the headlamp range-control switch out from the facia, using two screwdrivers, then disconnect the wiring.
37   Refitting is a reversal of the removal procedure.

Headlamp washer system
38   A headlamp washer system may be fitted. This is automatically activated when the windscreen washer switch is activated with the headlamps switched on.

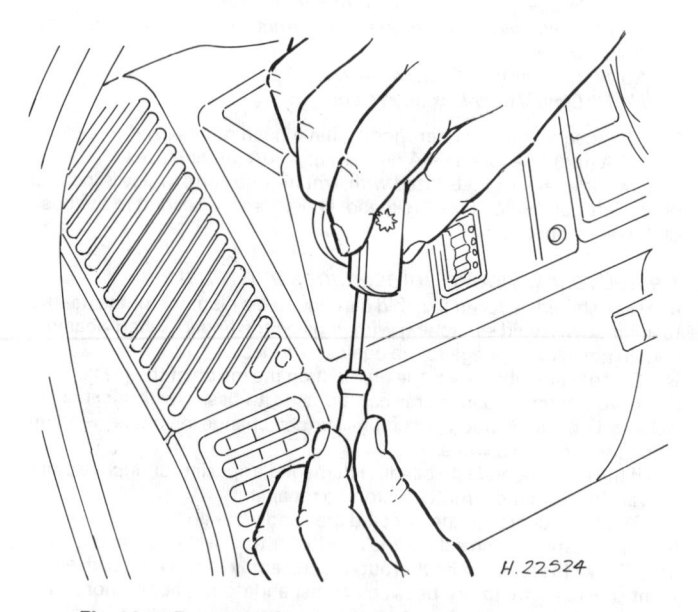

Fig. 13.50 Removing the headlamp switch knob (Sec 16)

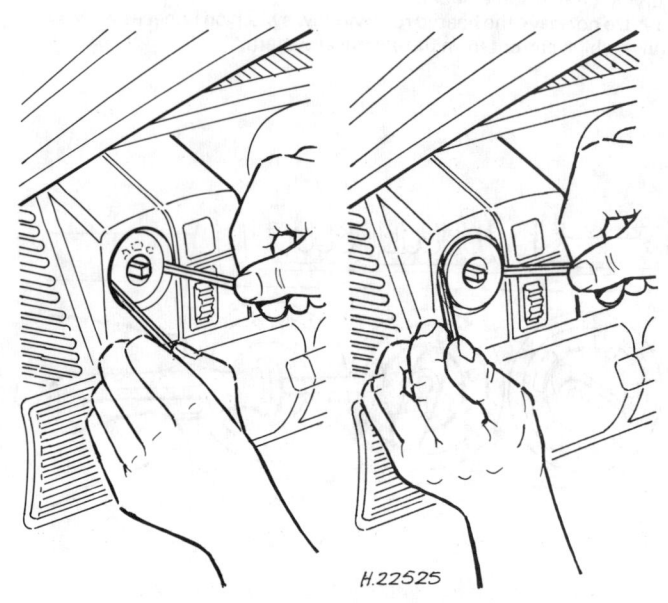

Fig. 13.51 Using the manufactured tools to release then withdraw the headlamp switch (Sec 16)

Chapter 13 Supplement: Revisions and information on later models

16.31 Withdrawing a headlamp range-control motor

16.41 Removing a front direction indicator lamp

16.43 Withdrawing a hazard warning lamp switch

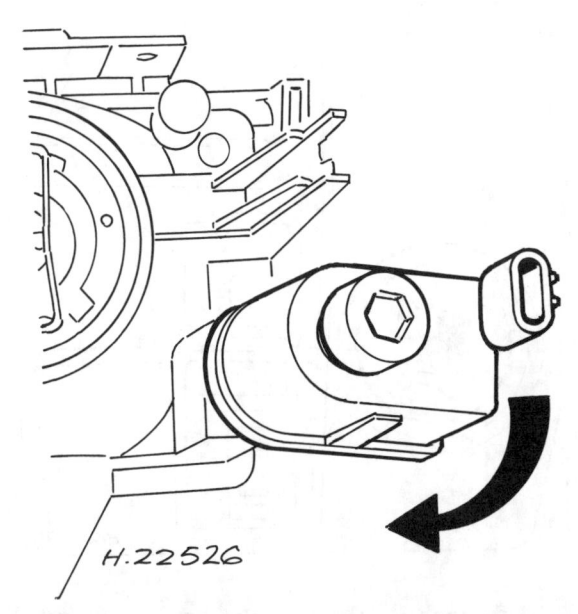

Fig. 13.52 Turn the headlamp range-control motor through 90° before releasing it from the ball-end (Sec 16)

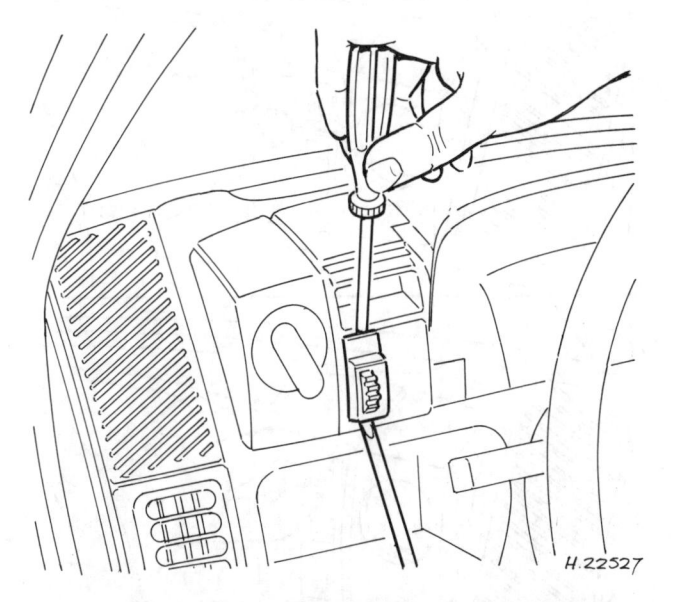

Fig. 13.53 Removing the headlamp range-control motor switch (Sec 16)

39 The system comprises a reservoir, an electrically-operated pump (located on the side of the reservoir) and nozzles. The reservoir is located at the front of the engine compartment, on the right-hand side.

Front direction indicator lamp (from 1991 model year) – removal and refitting

40 Working inside the engine compartment, disconnect the wiring from the relevant front direction indicator lamp.
41 Release the lamp retaining spring, then withdraw the lamp unit from the outside of the vehicle (photo).
42 Refitting is a reversal of the removal procedure.

Hazard warning lamp switch (from 1991 model year) – removal and refitting

43 Disconnect the battery earth lead, and move the switch to the 'ON' position. The switch may now be pulled carefully from its socket (photo). Note that if attempts are made to withdraw the switch in the 'OFF' position, damage will result.
44 Refitting is a reversal of the removal procedure.

Rear foglamp switch (from 1991 model year) – removal and refitting

45 Remove the switch cover, then carefully withdraw the switch from its socket.
46 Refitting is a reversal of the removal procedure.

Cigarette lighter (from 1991 model year) – removal and refitting

47 Remove the ashtray lower cover, undo the four cigarette lighter/ashtray panel securing screws, then disconnect the wiring as the panel assembly is withdrawn (photo).
48 Refitting is a reversal of the removal procedure.

Clock (from 1991 model year) – removal and refitting

49 Remove the hazard warning lamp switch (as described previously in this Section), then carefully pull the clock from its location and disconnect the wiring (photo).
50 Refitting is a reversal of the removal procedure.

Instrument panel and instruments (from 1991 model year) – removal and refitting

51 Undo the four screws securing the instrument panel cover (photo).
52 Remove the headlamp switch as described previously in this Section.
53 Undo the securing screw, and remove the switch housing (refer to Fig. 13.55). Disconnect the remaining switch wiring as the switch is withdrawn.
54 Undo the three screws securing the instrument panel, then partially withdraw the panel to allow the disconnection of the speedometer cable and the multiplugs (photo). The speedometer cable is released by simultaneously depressing the spring retainer and pulling on the cable end fixing.

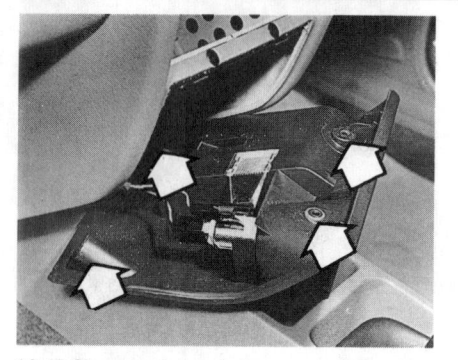

16.47 Cigarette lighter/ashtray panel securing screw locations (arrowed)

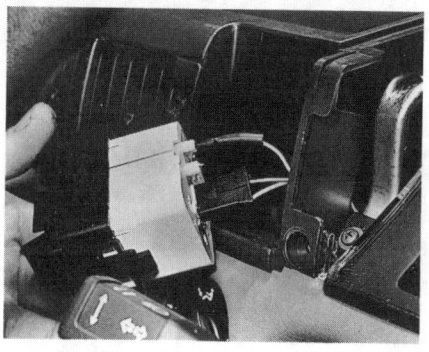

16.49 Withdrawing the clock from its location

16.51 Two of the instrument panel cover securing screws (arrowed)

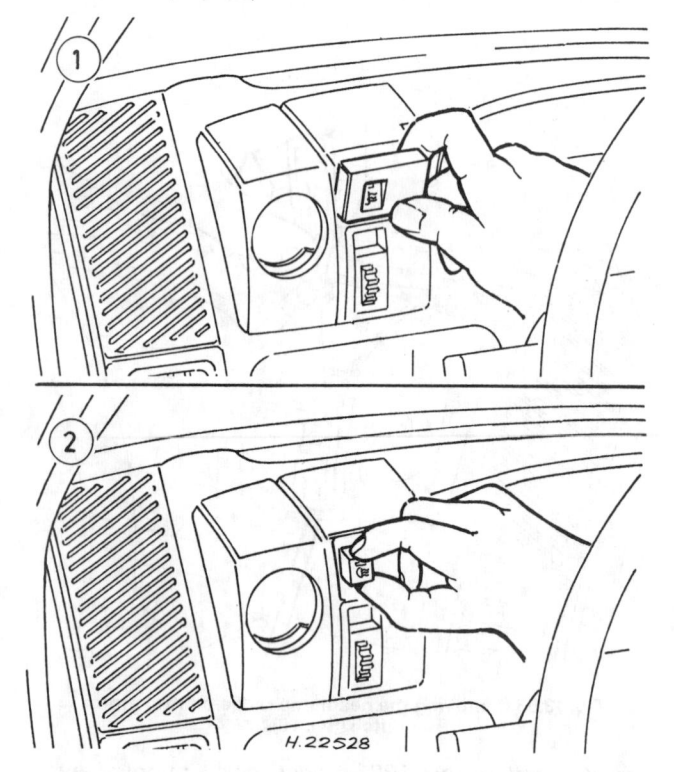

Fig. 13.54 Remove the cover (1), then withdraw the rear foglamp switch (2) (Sec 16)

Fig. 13.55 Headlamp switch housing securing screw (arrowed) (Sec 16)

55 The instruments are secured to the rear of the instrument panel by screws, and the instruments can be removed after removing the trip meter reset button and unclipping the cover.
56 The bulbs are of bayonet-fitting type (photo).
57 Refitting is a reversal of the removal procedure.

Heater controls (from 1991 model year) – removal and refitting

58 Remove the centre left-hand heater vent from the facia (photo).
59 Remove the hazard warning lamp switch and the clock as described previously in this Section.
60 Undo the securing screws then, disconnecting the wiring as it becomes accessible, remove the centre vent cluster trim (photos).
61 Remove the fan and heated rear window switch as described previously in this Section.
62 Remove the knobs from the temperature and air distribution switches, then undo the screws securing the switch cover.
63 The control unit and the air distribution control cable can be withdrawn after removing the parcel shelves from the footwell, and disconnecting the other cables.
64 Refitting is a reversal of the removal procedure. If necessary, for adjustment, the positions of the cables in their lever clips (on the heater housing) can be altered.

Underbonnet lamp

65 This utilises a 12V, 10W bulb. The lamp comes on automatically whenever the vehicle exterior lamps are switched on.

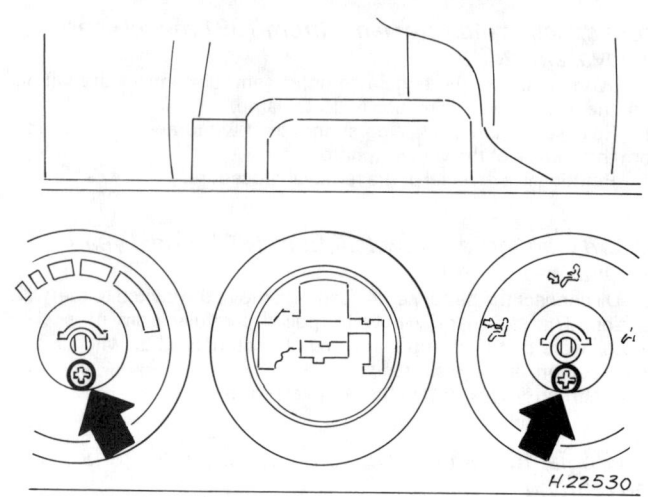

Fig. 13.56 Temperature and air distribution switch cover securing screws (arrowed) (Sec 16)

Chapter 13 Supplement: Revisions and information on later models

16.54 Two of the three instrument panel securing screws (arrowed)

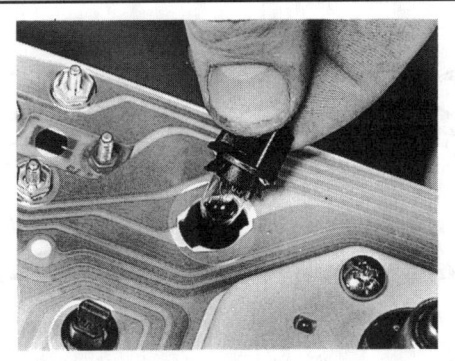

16.56 Removing an instrument panel illumination bulb

16.58 Removing the centre left-hand heater vent

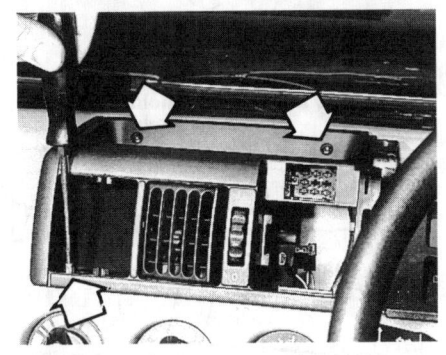

16.60A Undo the centre vent cluster trim securing screws (arrowed) ...

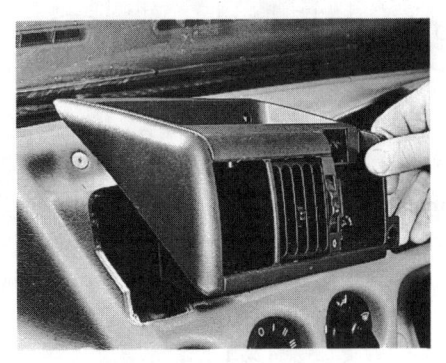

16.60B ... then withdraw the trim

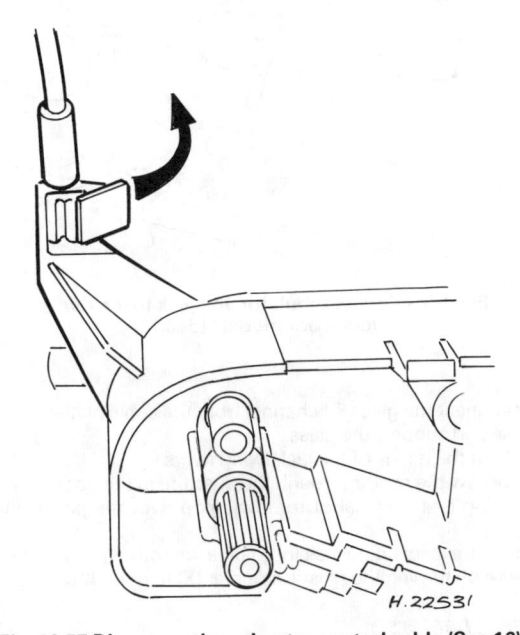

Fig. 13.57 Disconnecting a heater control cable (Sec 16)

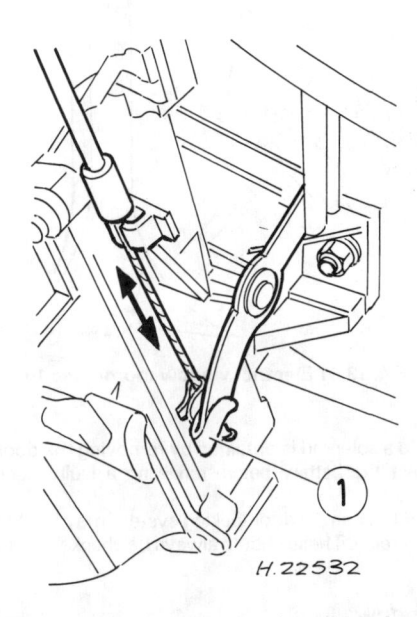

Fig. 13.58 The heater control cables can be adjusted at their lever clips (1) on the heater housing (Sec 16)

Luggage compartment lamp

Saloon
66 The lamp is mounted in the side panel of the luggage boot. Access to the bulb is obtained by prising out the lamp using a screwdriver inserted under the edge of the lens.
67 The switch is very similar to the type used at the hinged edge of the door to actuate the courtesy lamps, and completes the circuit as the boot lid is opened.

Hatchback
68 The lamp is mounted in the luggage area side panel, access to the bulb being obtained as on the Saloon.

69 The switch is held in position by the lower mounting of the tailgate support strut.
70 As soon as the tailgate is opened, the switch plunger is released and the circuit is completed.

Central door locking system
71 This system may be fitted to high-specification models.
72 The locking system does not include the tailgate.
73 The doors are locked by the action of a solenoid, controlled by the locking plunger on the driver's door, or by turning the key in the door lock.

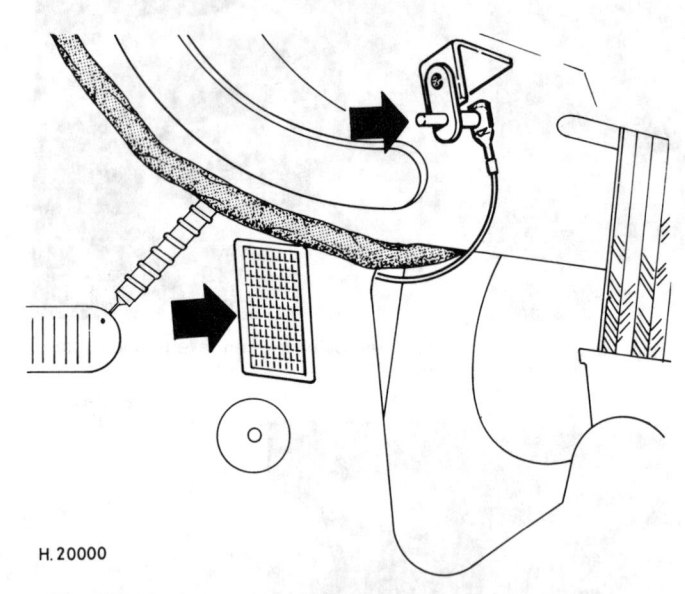

Fig. 13.59 Luggage compartment lamp and switch (arrowed) – Saloon (Sec 16)

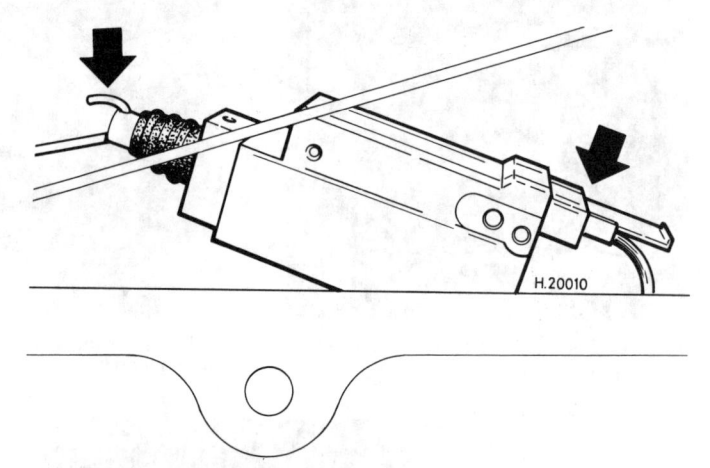

Fig. 13.60 Door lock solenoid. Wiring plug and actuating rod (arrowed) (Sec 16)

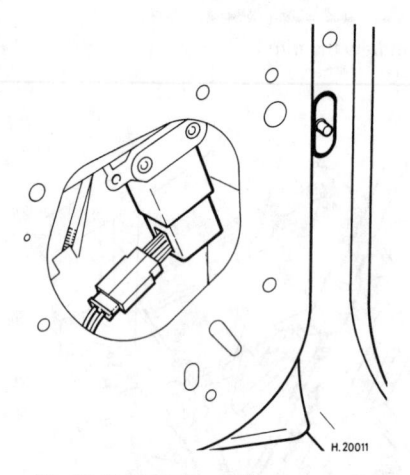

Fig. 13.61 Electric window motor (Sec 16)

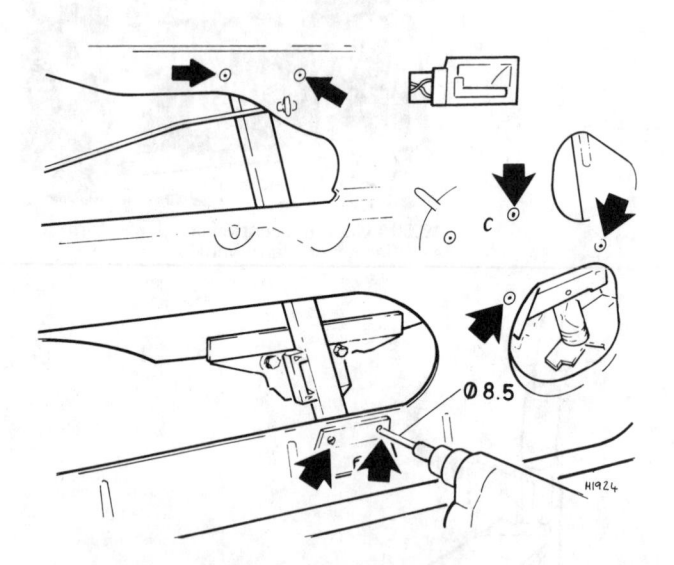

Fig. 13.62 Electric window regulator and jack fixing rivets (arrowed) – four-door models (Sec 16)

74 Access to a solenoid is obtained by removing the door trim panel.
75 Disconnect the battery before renewing a faulty door lock solenoid.
76 A fault in the central door locking system may be due to a blown fuse or a faulty relay. These should always be checked first when a fault occurs.

Electric windows
77 These may be fitted to high-specification models.
78 Raising or lowering of the front windows is controlled by switches mounted on the centre console. These switches are removed by prising them up, then disconnecting the wiring.
79 Disconnect the battery before removing a window lift motor, which should be renewed as a unit if faulty. However, note that a fault in the electric windows may be due to a blown fuse or faulty relay. These should always be checked first when a fault occurs.
80 Access to a window lift motor is obtained by removing the door trim panel. On two-door models, the regulator and motor can then be removed as described in Chapter 12, Section 18, in addition disconnecting the wiring plug. The motor can be unbolted from the regulator.
81 On four-door models, the regulator and motor cannot be separated. With the door trim removed, disconnect the wiring plug.
82 Disconnect the glass lift channel from the cable plate.
83 Raise and support the glass.
84 Drill out the heads of the jack fixing rivets.
85 Withdraw the motor through the aperture in the door.
86 Refitting is a reversal of removal, pop-rivet the jack using steel rivets.
87 Adjust the cable if necessary to give smooth and correct opening and closure by turning the plastic adjuster (X) (Fig. 13.63).

Front seat heaters
88 Heating pads are fitted to the front seats of cars destined for use in certain territories. Before attempting to remove a seat so equipped, disconnect the battery and the leads from the heating pad.

Radio equipment (factory-fitted) – removal
89 On some later models, the radio/cassette unit is retained by spring clips. This will be confirmed if two small holes are visible on each side of the trim panel.
90 A pair of U-shaped removal tools will be required. These are available from motor accessory stores, or can be made from lengths of stiff wire.
91 Insert the tools in the holes to fully depress the clips, and then use the tools to withdraw the unit from the facia panel. At the same time, disconnect the leads from the rear of the unit.

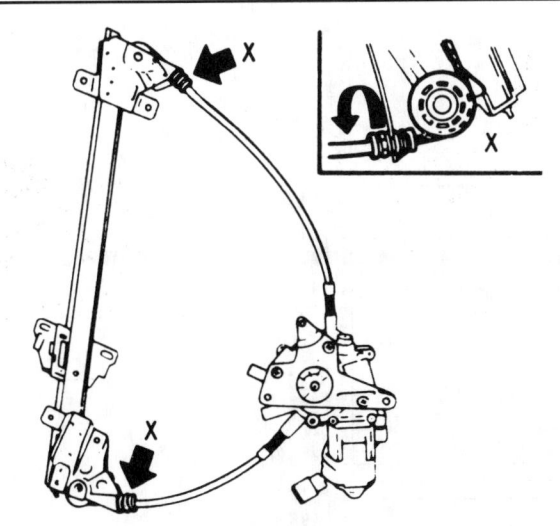

Chapter 13 Supplement: Revisions and information on later models

Fig. 13.63 Electric window cable plastic adjusters (X) – four-door models (Sec 16)

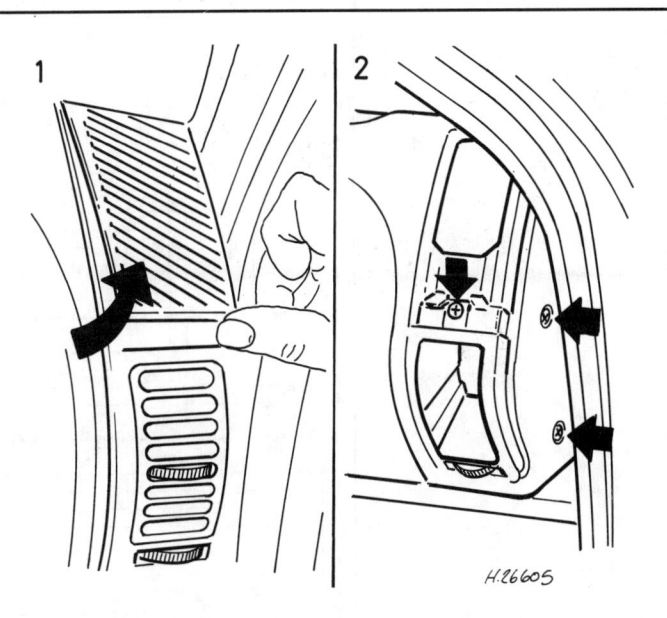

Fig. 13.64 Removing the speaker grille panel (1), and side window vent housing and facia side securing screws (2) (Sec 16)

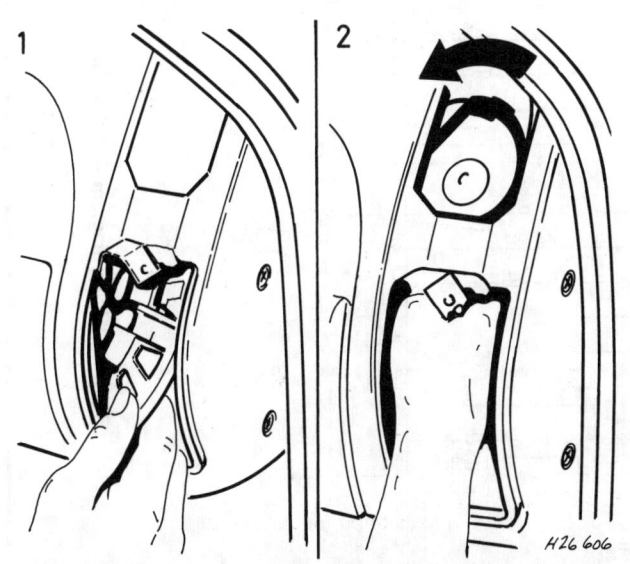

Fig. 13.65 Inserting a speaker through the side window vent housing aperture (1), into its mounted position (2) (Sec 16)

92 When refitting the radio, make sure that the lower rear mounting claw engages with the slide as the unit is pushed into position.
93 The standard location for the aerial is on the right-hand front wing.
94 Front speakers may be mounted in the front footwell side panels or, on vehicles from 1991 model year, in the facia behind removable grille panels. Rear speakers are fitted to the rear shelf or side panels depending on model.
95 To remove a front speaker on vehicles from 1991 model year, proceed as follows.
96 Unclip the relevant grille panel above the side window vent. When working on the left-hand speaker, the glovebox and switch strip must be removed. Undo the screws securing the side window vent housing, and remove the housing. Undo the screws securing the side of the facia and pull it carefully outwards at the top. Access to the speaker can now be gained through the side window vent housing aperture.
97 Refitting is a reversal of the removal procedure.

Direction indicator side repeater lamp – bulb renewal

98 To renew the bulb, twist the lamp lens anti-clockwise and withdraw it.
99 The bulb is a push fit in the bulbholder.
100 Refitting is the reverse of the removal procedure, but ensure that the rubber sealing ring is correctly seated between the lens and the wing.

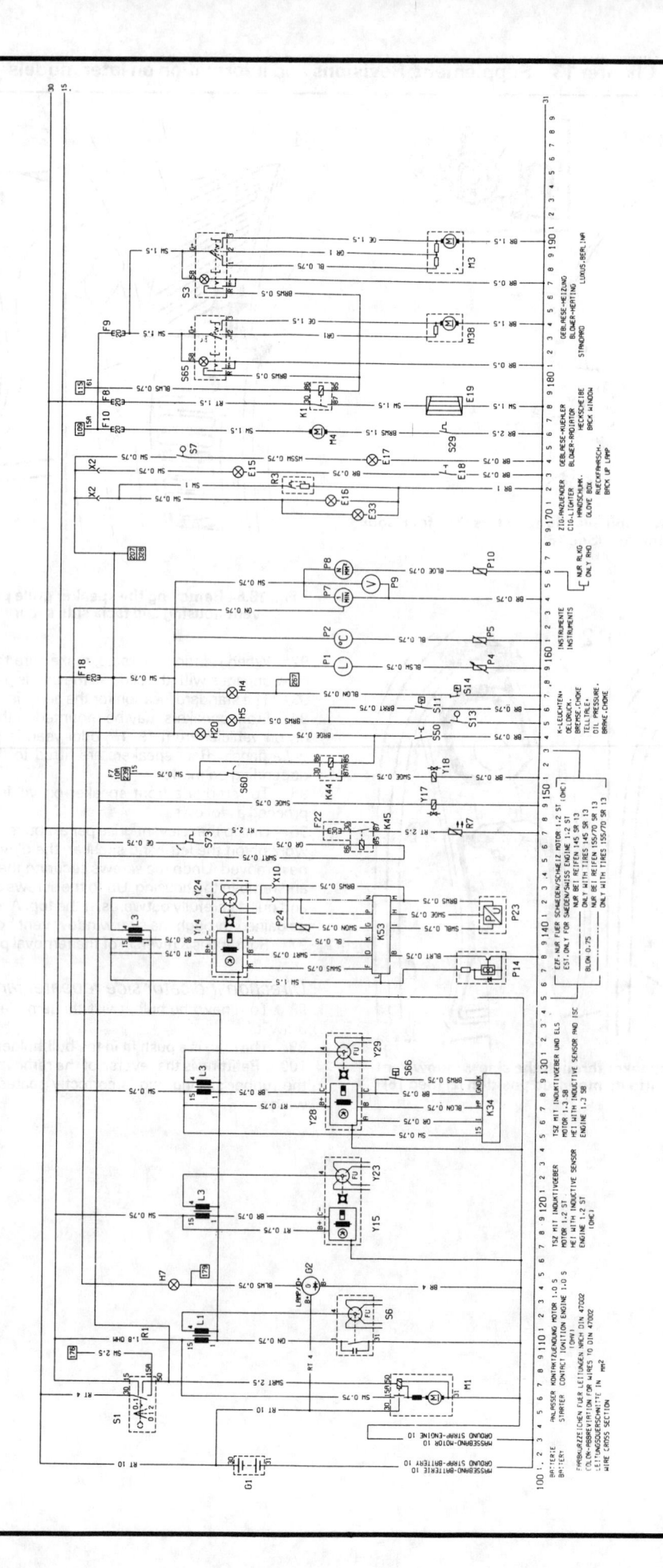

Fig. 13.66 Wiring diagram – 1985 models

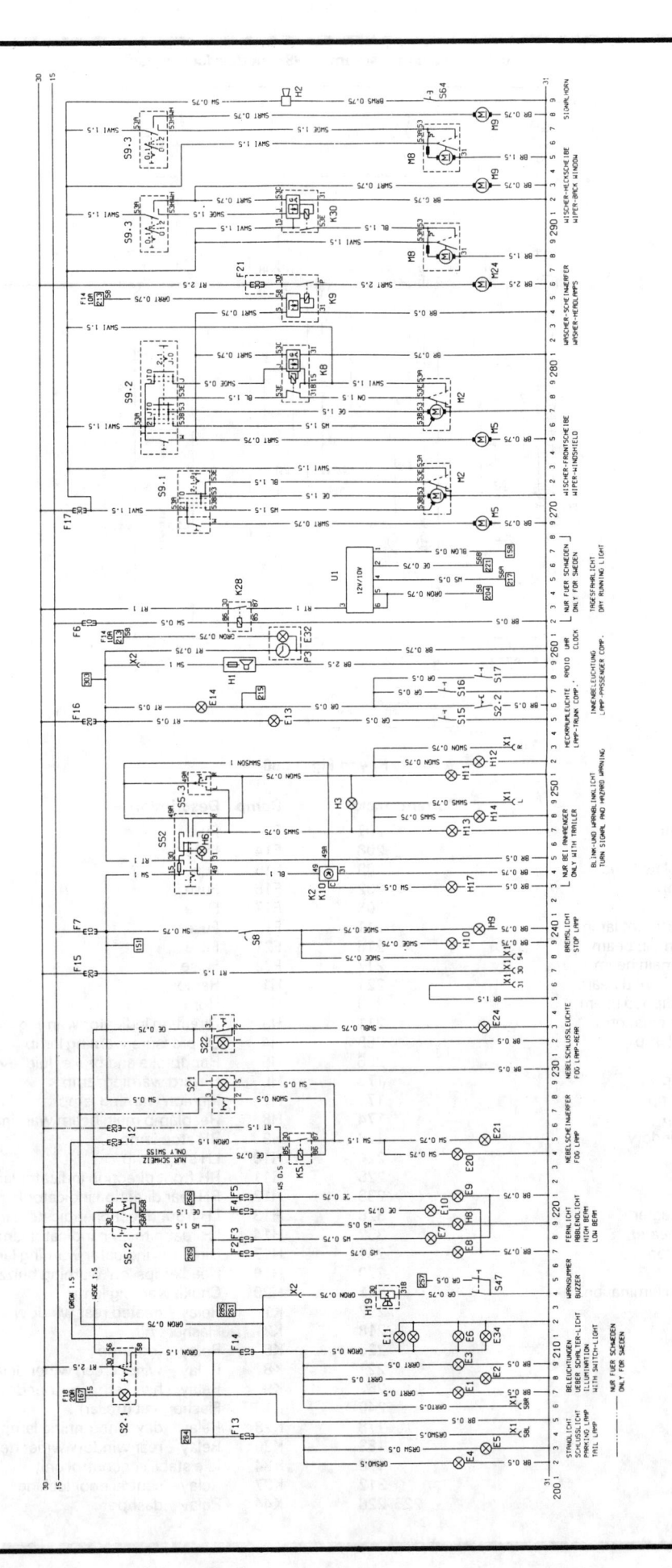

Fig. 13.66 Wiring diagram – 1985 models (continued)

Fig. 13.66 Wiring diagram – 1985 models (continued)

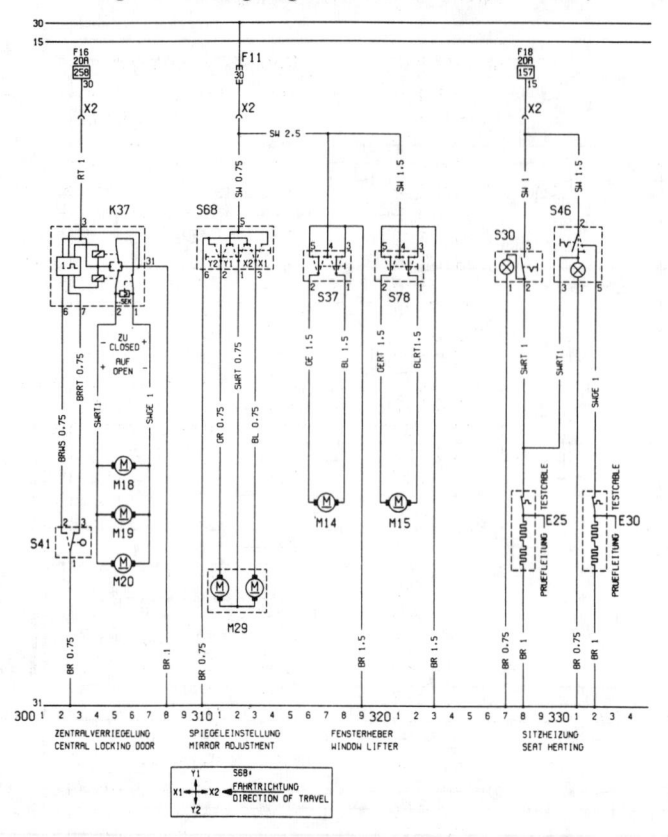

Key to Fig. 13.66

Comp	Description	Current track	Comp	Description	Current track
E1	RH parking lamp	207	F13	Fuse	204
E2	RH tail lamp	208	F14	Fuse	210
E3	Rear number plate lamp	209	F15	Fuse	237
E4	LH parking lamp	202	F16	Fuse	255
E5	LH tail lamp	203	F17	Fuse	270
E6	Engine compartment lamp	211	F18	Fuse	158
E7	RH headlamp main beam	218	F21	Fuse	286
E8	LH headlamp main beam	217	F22	Fuse	147
E9	RH headlamp dipped beam	221	H1	Radio	259
E10	LH headlamp dipped beam	220	H2	Horn	299
E11	Instrument illumination	211	H3	Direction indicator warning lamp	249
E13	Luggage boot lamp	255	H4	Oil pressure warning lamp	157
E14	Interior lamp	256	H5	Handbrake and brake fluid level warning lamp	155
E15	Glove box lamp	173	H6	Hazard warning lamp	246
E16	Cigar lighter lamp	171	H7	Ignition warning lamp	114
E17	Reversing lamp	174	H8	Headlamp main beam warning lamp	219
E19	Heated rear window	178	H9	RH stop lamp	240
E20	LH fog lamp	224	H10	LH stop lamp	239
E21	RH fog lamp	225	H11	RH front direction indicator lamp	251
E24	Rear fog lamp	233	H12	RH rear direction indicator lamp	252
E25	LH front seat heater	328	H13	LH front direction indicator lamp	247
E30	RH front seat heater	332	H14	LH rear direction indicator lamp	248
E32	Clock illumination	261	H17	Direction indicator warning lamp (trailer)	243
E33	Ashtray lamp	170	H19	Headlamps 'on' warning buizzer	214
E34	Heater control illumination	211	H20	Choke warning lamp	154
F2	Fuse	217	K1	Relay – heated rear window	178, 179
F3	Fuse	218	K2	Flasher unit	244
F4	Fuse	220	K5	Relay – fog lamp	224, 225
F5	Fuse	221	K8	Relay – windscreen wiper delay	278 to 281
F6	Fuse	262	K9	Relay – headlamp washer delay	284 to 286
F7	Fuse	240	K10	Flasher unit (trailer)	243, 244
F8	Fuse	178	K28	Relay – daylight running lamps	262, 263
F9	Fuse	183	K30	Relay – rear window wiper delay	290 to 292
F10	Fuse	176	K34	Idle stabiliser control unit	125 to 129
F11	Fuse	312	K37	Relay – central door locking	302 to 308
F12	Fuse	225, 226	K44	Relay – dashpot	151, 152

Key to Fig. 13.66 (continued)

Comp	Description	Current track
K45	Relay – mixture preheating	146, 147
K53	Timing control	138 to 142
L1	Ignition coil	110, 111
L3	Ignition coil (breakerless system)	119, 120, 128, 129, 139, 140
M1	Starter	105 to 107
M2	Windscreen wiper motor	270 to 273, 276 to 279
M3	Heater blower	188 to 190
M4	Radiator cooling fan	176
M5	Windscreen washer pump	269, 275
M8	Rear window wiper motor	288 to 290, 295 to 297
M9	Rear window washer pump	293, 298
M14	LH front window lift motor	316, 318
M15	RH front window lift motor	320, 322
M18	RH front door lock solenoid	304, 307
M19	LH rear door lock solenoid	304, 307
M20	RH rear door lock solenoid	304, 307
M24	Headlamp washer pump	286
M29	Exterior mirror adjuster motor	311 to 313
M38	Heater blower	183, 184
O1	Battery	101
O2	Alternator	114
P1	Fuel gauge	159
P2	Coolant temperature gauge	161
P3	Clock	260
P4	Fuel gauge sender unit	159
P5	Coolant temperature gauge sender unit	161
P7	Tachometer	164
P8	Oil pressure gauge	166
P9	Voltmeter (battery condition)	165
P10	Oil pressure gauge sender unit	166
P14	Distance sensor	137, 138
P23	Vacuum sensor	140 to 142
P24	Oil temperature gauge	140
R1	Resistor cable	110
R3	Cigar lighter	172
R7	Mixture preheating	147
S1	Ignition switch	106, 107
S2.1	Light switch	209, 210
S2.2	Interior lamp switch	256
S3	Heater blower/heated rear window switch	186 to 190
S5.2	Dip switch	219, 220
S5.3	Direction indicator switch	250, 251
S6	Distributor (breaker type)	110, 112
S7	Reversing lamp switch	174
S8	Stop lamp switch	240
S9.1	Windscreen wiper switch	269 to 272
S9.2	Windscreen wiper interval switch	275 to 279
S9.3	Rear window wiper interval switch	291, 292, 297, 298
S11	Brake fluid level switch	156
S13	Handbrake 'on' warning switch	155
S14	Oil pressure switch	157
S15	Luggage boot lamp switch	256
S16	RH front door contact switch	257
S17	LH front door contact switch	258
S18	Glove box lamp switch	173
S21	Fog lamp switch	228 to 230
S22	Rear fog lamp switch	232, 233
S29	Radiator fan switch	176
S30	LH front seat heater switch	327, 329
S37	LH front window lift switch	316 to 319
S41	Central door locking switch	302 to 303
S46	Front seat heaters switch	330 to 332
S47	Door closure with headlamps 'on' warning switch	214, 215
S50	Choke warning switch	154
S52	Hazard warning switch	244 to 248
S60	Clutch pedal dashpot switch	151
S64	Horn	299
S65	Heater blower/heated rear window switch	180 to 184
S66	Vacuum switch	129
S68	Exterior mirror switch	310 to 313
S73	Mixture preheating switch	146
S78	RH front window lift switch	320 to 323
U1	Daylight running lights transformer	263 to 267
X1	Trailer socket	204, 206, 236 to 238, 249, 253
X2	Connectors (auxiliaries)	171, 173, 214, 259, 303, 312, 328
X10	Connector (ignition adjustment)	142, 143
Y14	Inductive sensor	138, 139
Y15	Inductive sensor	118, 119
Y17	Idle fuel cut-off valve	149
Y18	Dashpot solenoid valve	151
Y23	Ignition distributor (breakerless)	122
Y24	Ignition distributor (breakerless)	142
Y28	Inductive sensor	127, 128
Y29	Ignition distributor (breakerelss)	131

Note: *Refer to Chapter 10, Section 42 for explanation of use. As diagram covers all models, alternative wiring is shown for some components where differences occur within the range.*

Colour codes

BL Blue	GE Yellow	RT Red	LI Lilac
HBL Light blue	GR Grey	WS White	VI Violet
BR Brown	GN Green	SW Black	

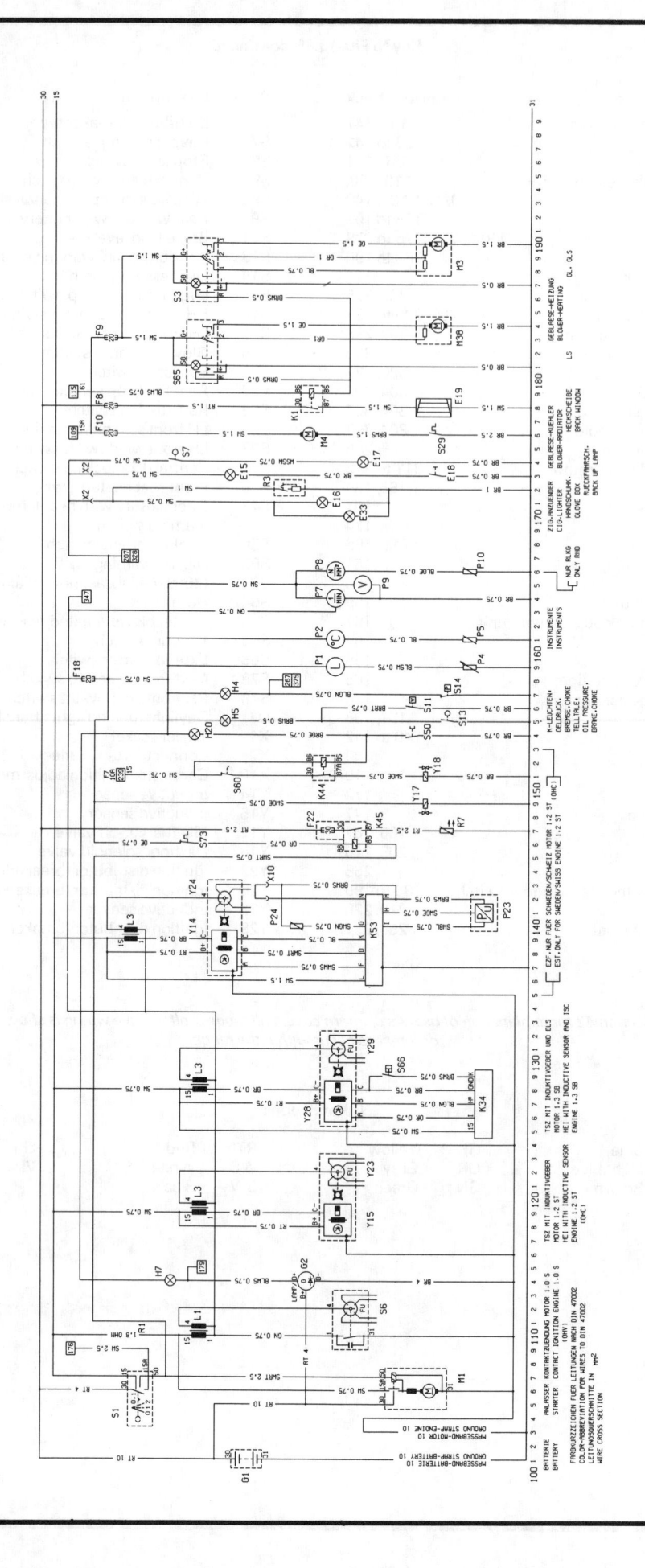

Fig. 13.67 **Wiring diagram – 1986 models**

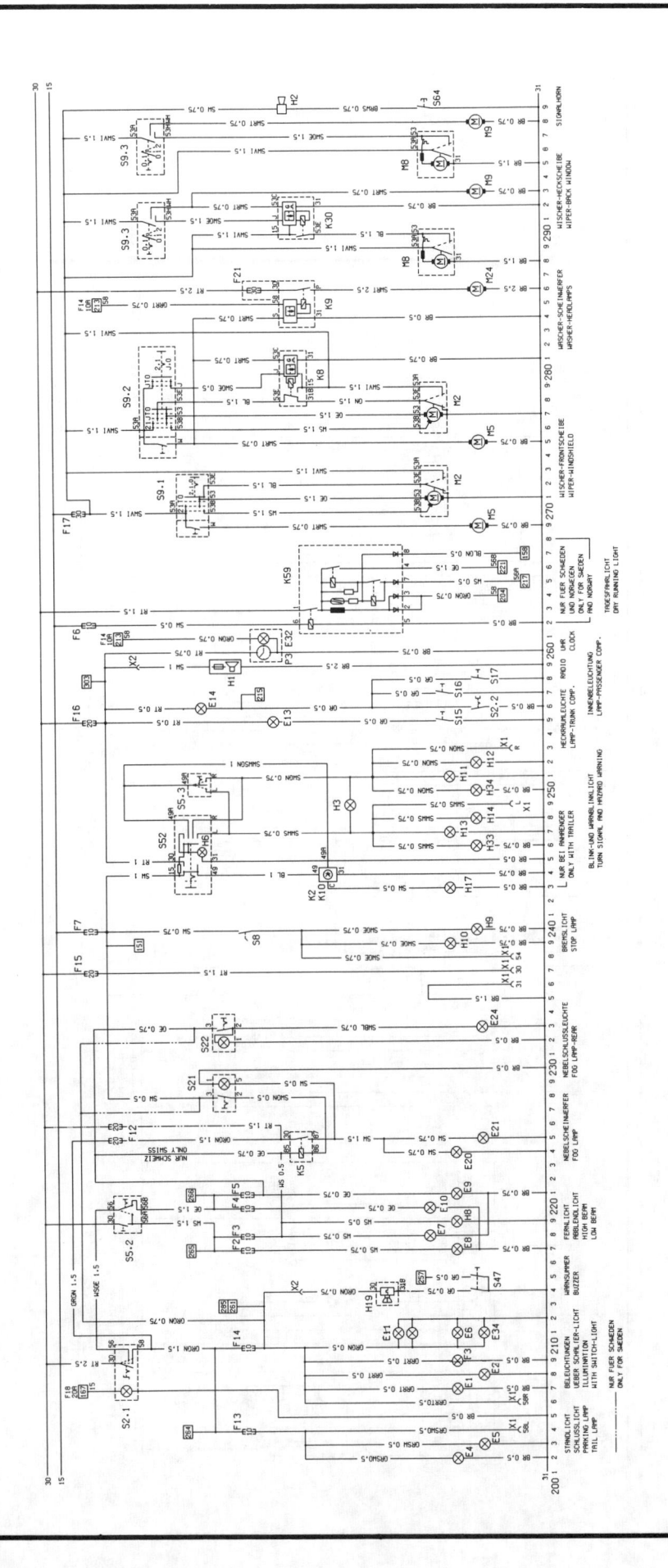

Fig. 13.67 Wiring diagram – 1986 models (continued)

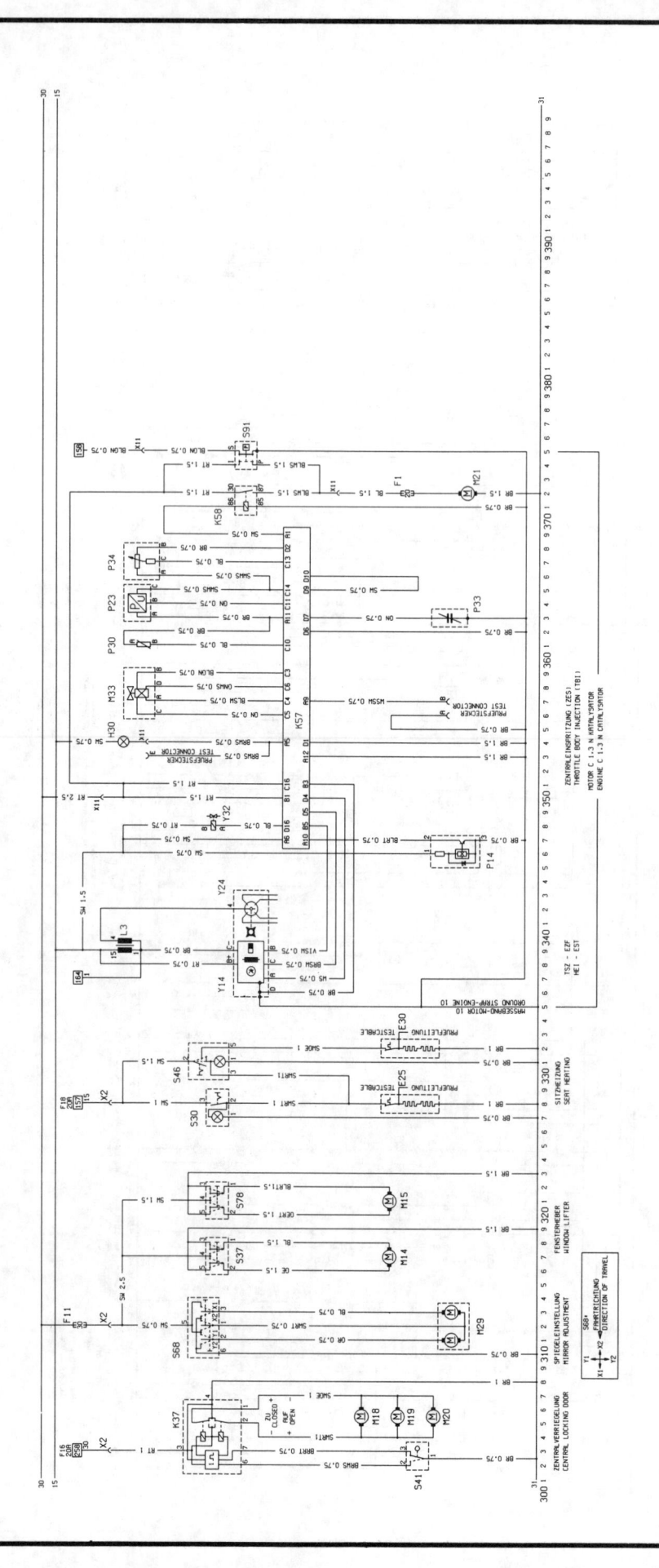

Fig. 13.67 Wiring diagram – 1986 models (continued)

Key to Fig. 13.67

Comp	Description	Current track
E1	RH parking lamp	207
E2	RH tail lamp	208
E3	Rear number plate lamp	209
E4	LH parking lamp	202
E5	LH tail lamp	203
E6	Engine compartment lamp	211
E7	RH headlamp main beam	216
E8	LH headlamp main beam	217
E9	RH headlamp dipped beam	221
E10	LH headlamp dipped beam	220
E11	Instrument panel lamps	211
E13	Luggage boot lamp	255
E14	Interior lamp	253
E15	Glovebox lamp	173
E16	Cigar lighter lamp	171
E17	Reversing lamp	174
E19	Heated rear screen	178
E20	LH foglamp	224
E21	RH foglamp	225
E24	Rear fog warning lamp	233
E25	LH front seat heater	328
E30	RH front seat heater	332
E32	Clock illumination	261
E33	Ashtray lamp	170
E34	Heater illumination control	211
F1	Fuse	372
F2	Fuse	217
F3	Fuse	218
F4	Fuse	220
F5	Fuse	221
F6	Fuse	262
F7	Fuse	240
F8	Fuse	178, 183
F9	Fuse	176
F11	Fuse	312
F12	Fuse	225, 226
F13	Fuse	204
F14	Fuse	210
F15	Fuse	237
F16	Fuse	255
F17	Fuse	270
F18	Fuse	158
F21	Headlamp washer fuse	286
F22	Fuse, mixture preheating	147
G1	Battery	101
G2	Alternator	114
H1	Receiver	259
H2	Horn	299
H3	Direction indicator warning lamp	249
H4	Oil pressure warning lamp	157
H5	Handbrake/brake fluid level warning lamp	155
H6	Hazard warning system warning lamp	246
H7	Charge indicator light	114
H8	Headlamp main beam warning lamp	219
H9	RH stop-lamp	240
H10	LH stop-lamp	239
H11	Front RH direction indicator lamp	251
H12	Rear RH direction indicator lamp	252
H13	Front LH direction indicator lamp	247
H14	Rear LH direction indicator lamp	248
H17	Direction indicator warning lamp (trailer)	243
H19	Headlamps left on warning buzzer	214
H20	Choke warning lamp	154
H30	Ignition warning lamp (electronic ignition)	354
K1	Heated rear screen relay	178, 179
K2	Flasher unit	244
K5	Fog lamp relay	224, 225
K8	Wiper (intermittent) relay	278, 281
K9	Washer (delay) relay for headlamps	284, 286
K10	Trailer flasher unit	243, 244
K28	Day running lights relay	262, 263
K30	Rear screen wiper (delay) relay	290, 292
K34	Control unit, idle stabilizer	125, 129
K37	Central door lock relay	302, 308
K44	Dashpot relay	151, 152
K45	Mixture preheating relay	146, 147
K53	Timing control	138, 142
K57	Control unit (electronic ignition)	347, 389
K58	Fuel pump relay	371, 372
L1	Ignition coil	110, 111
L3	Ignition coil (transistorized) inductive sensor	128, 129, 140, 339
M1	Starter	105, 107
M2	Windscreen wiper motor	270, 273, 276, 279
M3	Heater blower motor	188, 190
M4	Radiator cooling fan	176

Key to Fig. 13.67 (continued)

Comp	Description	Current track
M5	Windscreen washer pump	269, 275
M8	Rear screen wiper motor	295, 297
M9	Rear screen washer pump	293, 298
M14	LH front window motor	316, 318
M15	RH front window motor	320, 322
M18	RH front central door lock solenoid	304, 307
M19	LH rear central door lock solenoid	304, 307
M20	RH rear central door lock solenoid	304, 307
M21	Fuel pump	372
M24	Headlamp washer pump	286
M29	RH exterior mirror	311, 313
M33	Idling control	356, 359
M38	Heated blower motor	183, 184
P1	Fuel contents gauge	159
P2	Coolant temperature gauge	161
P3	Clock	260
P4	Fuel contents sender unit	159
P5	Coolant temperature sender unit	161
P7	Tachometer	164
P8	Oil pressure gauge	166
P9	Battery condition gauge (voltmeter)	165
P10	Oil pressure sensor	166
P14	Distance sensor	346, 247
P23	Vacuum sensor	140, 142, 363, 365
P24	Oil temperature sensor	140
P30	Coolant temperature sensor	361, 362
P33	Lambda sensor	363
P34	Throttle valve position sensor	366, 368
R1	Resistor cable	110
R3	Cigar lighter	172
R7	Mixture preheating	147
S1	Ignition switch	106, 107
S2	Switch	
S2.1	Light switch	209, 210
S2.2	Interior lamp switch	256
S3	Heater blower and heated rear screen switch	186, 190
S5	Direction indicator switch	250, 251
S5.2	Headlamp dipswitch	219, 220
S6	Ignition distributor	110, 112
S7	Reversing lamp switch	174
S8	Stop-lamp switch	240
S9	Wiper switch	
S9.1	Windscreen wiper switch	269, 272
S9.2	Windscreen wiper (intermittent) switch	275, 279
S9.3	Rear screen wiper (intermittent) switch	291, 292, 297, 298
S11	Brake fluid warning switch	156
S13	Handbrake warning switch	155
S14	Oil pressure switch	157
S15	Luggage boot lamp switch	255
S16	RH door closure switch	
S17	LH door closure switch	258
S18	Glovebox lamp switch	173
S21	Foglamp switch	228, 230
S22	Rear foglamp switch	232, 233
S29	Radiator fan thermal switch	176
S30	LH front seat heater switch	327, 328
S37	LH front window motor switch	316, 319
S41	Central door locking switch	302, 303
S46	Front heated seat switch	330, 332
S47	Courtesy lamp switch and headlamps 'on' warning	214, 215
S50	Choke warning switch	154
S52	Door closure switch warning lamp	244, 248, 257
S60	Clutch pedal dashpot switch	151
S64	Horn switch	299
S65	Heater blower and heated rear screen switch	180, 184
S66	Vacuum switch	129
S68	Exterior mirror switch	310, 313
S73	Mixture preheating temperature switch	146
S78	RH front window motor switch	320, 323
S91	Oil pressure switch	374, 375
X1	Trailer socket	Various
X2	Connector – auxiliaries	Various
X10	Connector, ignition adjustment	142, 143
X11	Wiring harness plug	
Y14	Inductive sensor	138, 139, 335, 339
Y15	Inductive sensor	122
Y17	Solenoid valve, idle cut-off	149
Y18	Dashpot solenoid valve	151
Y24	Distributor (transistorized)	142
Y28	Inductive sensor	127, 128
Y29	Distributor (transistorized)	131
Y32	Injection valve	348
Y42	Ignition distributor (EST)	342

Note: *Refer to Chapter 10, Section 42 for explanation of use. Wiring code as for 1985 models*

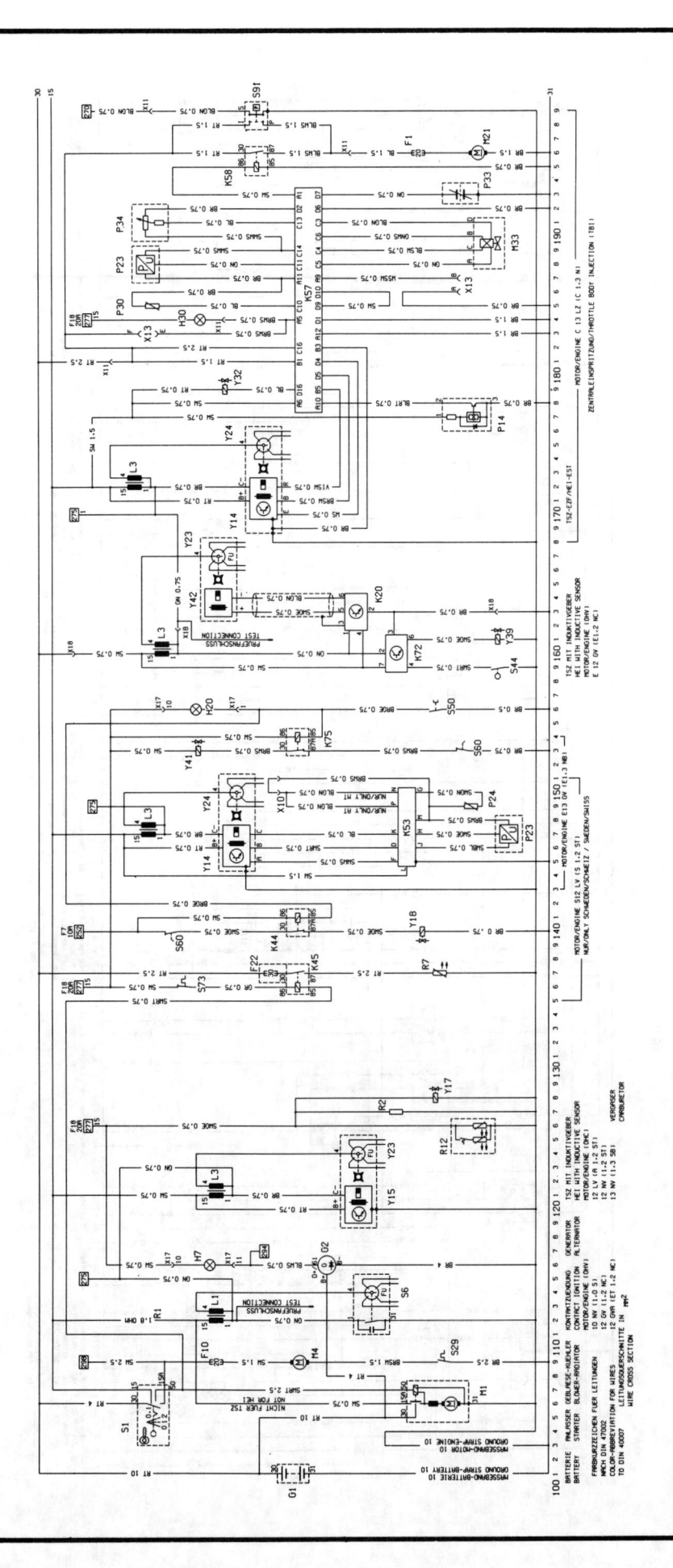

Fig. 13.68 Wiring diagram – 1987 models

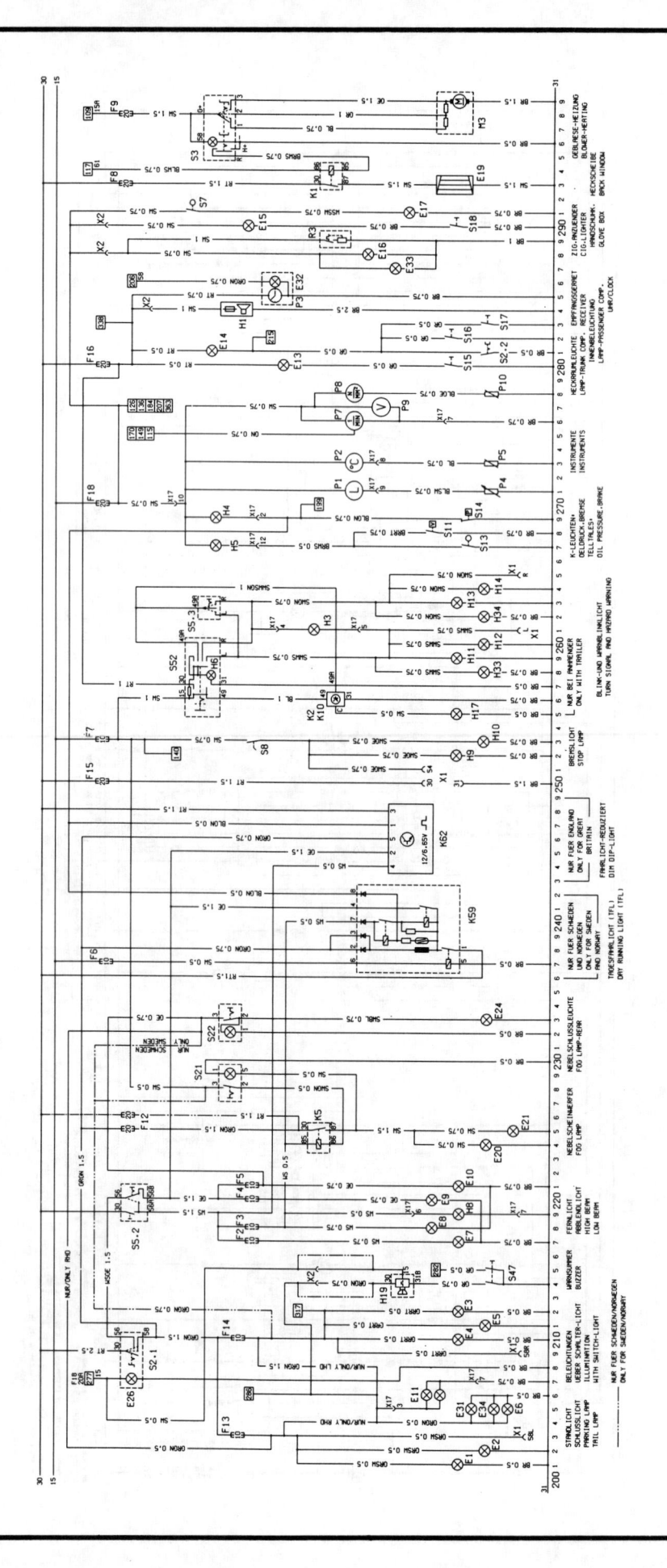

Fig. 13.68 Wiring diagram – 1987 models (continued)

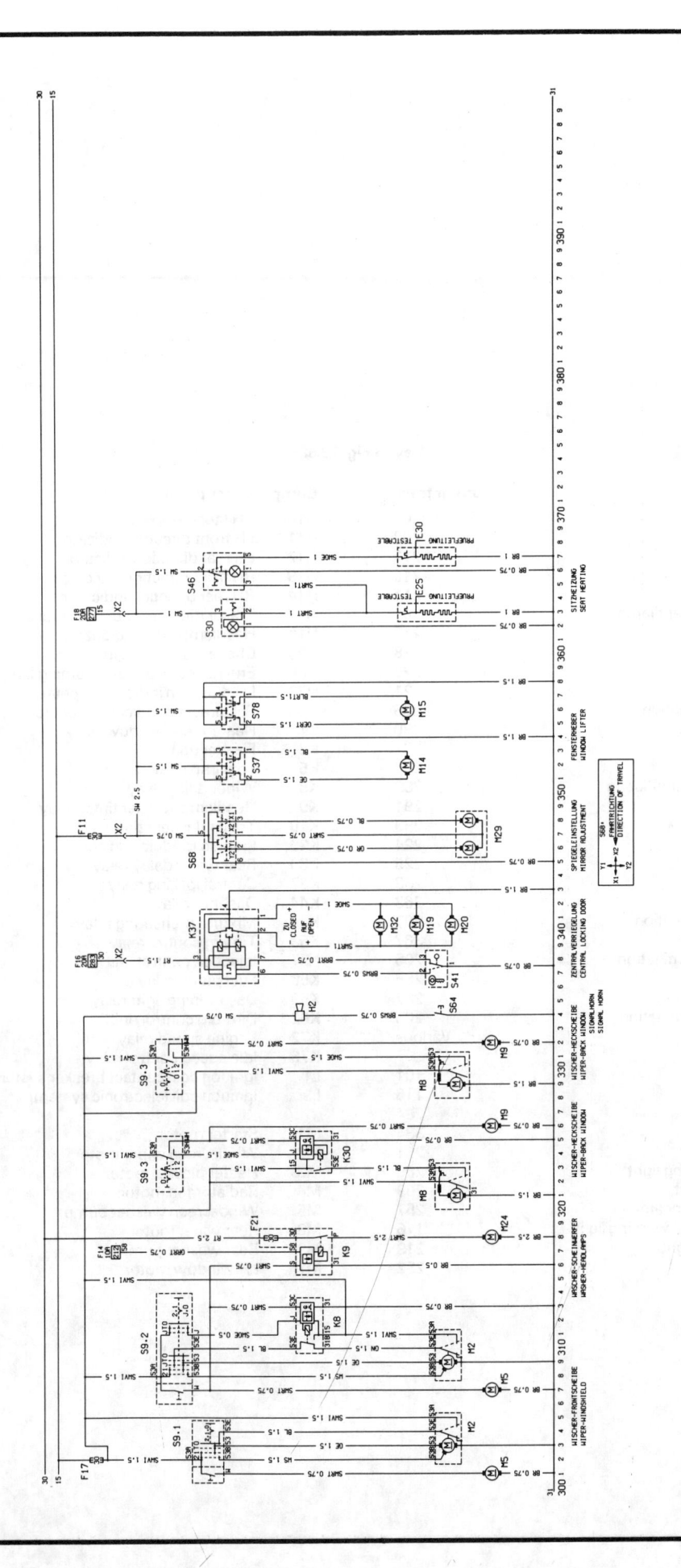

Fig. 13.68 Wiring diagram – 1987 models (continued)

Key to Fig. 13.68

Comp	Description	Current track
E1	LH parking lamp	201
E2	LH tail lamp	202
E3	Number plate lamp	212
E4	RH parking lamp	210
E5	RH tail lamp	211
E6	Engine compartment lamp	205
E7	LH main beam	217
E8	RH main beam	218
E9	LH dipped beam	220
E10	RH dipped beam	221
E11	Instrument illumination	206
E13	Luggage area light	280
E14	Courtesy light	281
E15	Glovebox light	290
E16	Cigarette lighter illumination	288
E17	LH reversing lamp	291
E19	Heated rear window	293
E20	LH front foglamp	224
E21	RH front foglamp	225
E24	Rear foglamp	233
E25	LH seat heater	363
E26	Light switch illumination	207
E30	RH seat heater	367
E31	Control switch illumination	205
E32	Clock illumination	286
E33	Ashtray illumination	287
E34	Heater control illumination	205
F1 to F22	Fuses	Various
G1	Battery	101
G2	Alternator	116
H1	Radio	284
H2	Horn	334
H3	Direction indicator	261
H4	Oil pressure warning light	269
H5	Brake warning light	269
H6	Hazard warning repeater	257
H7	No charge (ignition) warning light	116
H8	Main beam pilot light	219
H9	LH stop-lamp	252
H10	RH stop-lamp	253
H11	LH front direction indicator	259
H12	LH rear direction indicator	260
H13	RH front direction indicator	263
H14	RH rear direction indicator	264
H17	Trailer direction indicator repeater	255
H19	Headlamp warning buzzer	214, 215
H20	Choke warning lamp	156
H30	Engine management warning lamp	184
H33	LH direction indicator repeater	258
H34	RH direction indicator repeater	262
K1	Heated rear window relay	293, 294
K2	Flasher unit	256
K5	Foglamp relay	224, 225
K8	Wiper delay relay	310 to 313
K9	Headlamp washer timer relay	316 to 318
K10	Trailer flasher relay	255, 256
K20	Ignition module and coil	162 to 164
K30	Rear wiper delay relay	323 to 325
K37	Central locking relay	337 to 343
K44	Dashpot relay	140, 141
K45	Mixture preheating relay	136, 137
K53	Timing control relay	144 to 151
K57	Injection control unit	178 to 193
K58	Fuel pump relay	195, 196
K59	Day running light relay	236 to 242
K62	Dim-dip control unit	244 to 248
K72	Engine speed relay	159 to 161
K75	Idle speed control relay	153, 154
L1	Ignition coil (contact breaker system)	112, 113
L3	Ignition coil (electronic system)	121, 122, 147, 148, 160, 161, 172, 173
M1	Starter motor	105 to 107
M2	Windscreen wiper motor	302 to 305, 308 to 311
M3	Heater blower motor	297 to 299
M4	Radiator fan motor	109
M5	Windscreen washer pump	301, 307
M8	Rear wiper motor	321 to 323, 329 to 331
M9	Rear washer pump	326, 332
M14	LH window motor	351, 353

Key to Fig. 13.68 (continued)

Comp	Description	Current track
M15	RH window motor	355, 357
M19	LH rear door lock motor	339, 342
M20	RH rear door lock motor	339, 342
M21	Fuel pump	196
M24	Headlamp washer pump	318
M29	Mirror adjustment motor	346 to 348
M32	Front passenger door lock motor	339, 342
M33	Idle speed control unit	188 to 191
P1	Fuel gauge	271
P2	Temperature gauge	273
P3	Clock	285
P4	Fuel gauge sender	271
P5	Temperature gauge sender	273
P7	Tachometer	276
P8	Oil pressure gauge	278
P9	Voltmeter	277
P10	Oil pressure sensor	278
P14	Distance sensor	177, 178
P23	Vacuum sensor	146, 148, 187 to 189
P24	Oil temperature sensor	149
P30	Coolant temperature sensor	185
P33	Lambda sensor	193
P34	Throttle position sensor	190 to 192
R1	Resistor cable	112
R2	Carburettor heater	127
R3	Cigarette lighter	289
R7	Mixture preheater	137
R12	Automatic choke	126
S1	Ignition/starter switch	106, 107
S2.1	Lighting switch	209, 210
S2.2	Interior light switch	281
S3	Heater blower/heated rear window switch	295 to 299
S5.2	Dipswitch	219, 220
S5.3	Direction indicator switch	262, 263
S6	Distributor (contact breaker)	112 to 114
S7	Reversing light switch	291
S8	Stop-lamp switch	253
S9.1	Windscreen wiper switch (continuous)	301 to 304
S9.2	Windscreen wiper switch (intermittent)	307 to 311
S9.3	Rear wiper switch	324, 325, 331, 332

Comp	Description	Current track
S11	Brake fluid level switch	268
S14	Oil pressure switch	269
S15	Luggage area light switch	280
S16	Driver's door switch	282
S17	Front passenger door switch	283
S18	Glovebox light switch	290
S21	Front foglamp switch	228 to 230
S22	Rear foglamp switch	232, 233
S29	Coolant temperature switch	109
S30	Seat heater switch (driver only)	362, 363
S37	LH window switch	351 to 354
S41	Driver's door lock switch	337, 338
S44	Throttle switch	159
S46	Seat heater switch (both front seats)	365 to 367
S47	Headlamp buzzer switch	214, 215
S50	Choke switch	156
S52	Hazard warning switch	256 to 260
S60	Clutch pedal switch	140, 153
S64	Horn switch	334
S68	Mirror adjustment switch	345 to 348
S73	Temperature switch (mixture preheating)	136
S78	RH window switch	355 to 358
S91	Oil pressure switch	198, 199
Y14	Ignition sensor	145 to 147, 169 to 172
Y15	Ignition sensor with module	120, 121
Y17	Idle cut-off solenoid	128
Y23	Distributor (electronic)	124, 167
Y24	Distributor	179
Y39	Fuel overrun cut-off valve	161
Y41	Idle control valve	153
Y42	Ignition sensor	163, 164
X1	Trailer socket	203, 209, 250, 251, 261, 265
X2	Auxiliary connector	214, 215, 284, 288, 290, 238, 247, 363
X10	Ignition adjustment connector	150, 151
X11	Fuel injection connector	181, 184, 196, 199
X13	Test connector	183, 186, 187
X17	Instrument connector	Various
X18	Ignition connector	160, 162, 163

Not all items fitted to all models

Note: *Refer to Chapter 10, Section 42 for explanation of use. Wiring code as for 1985 models*

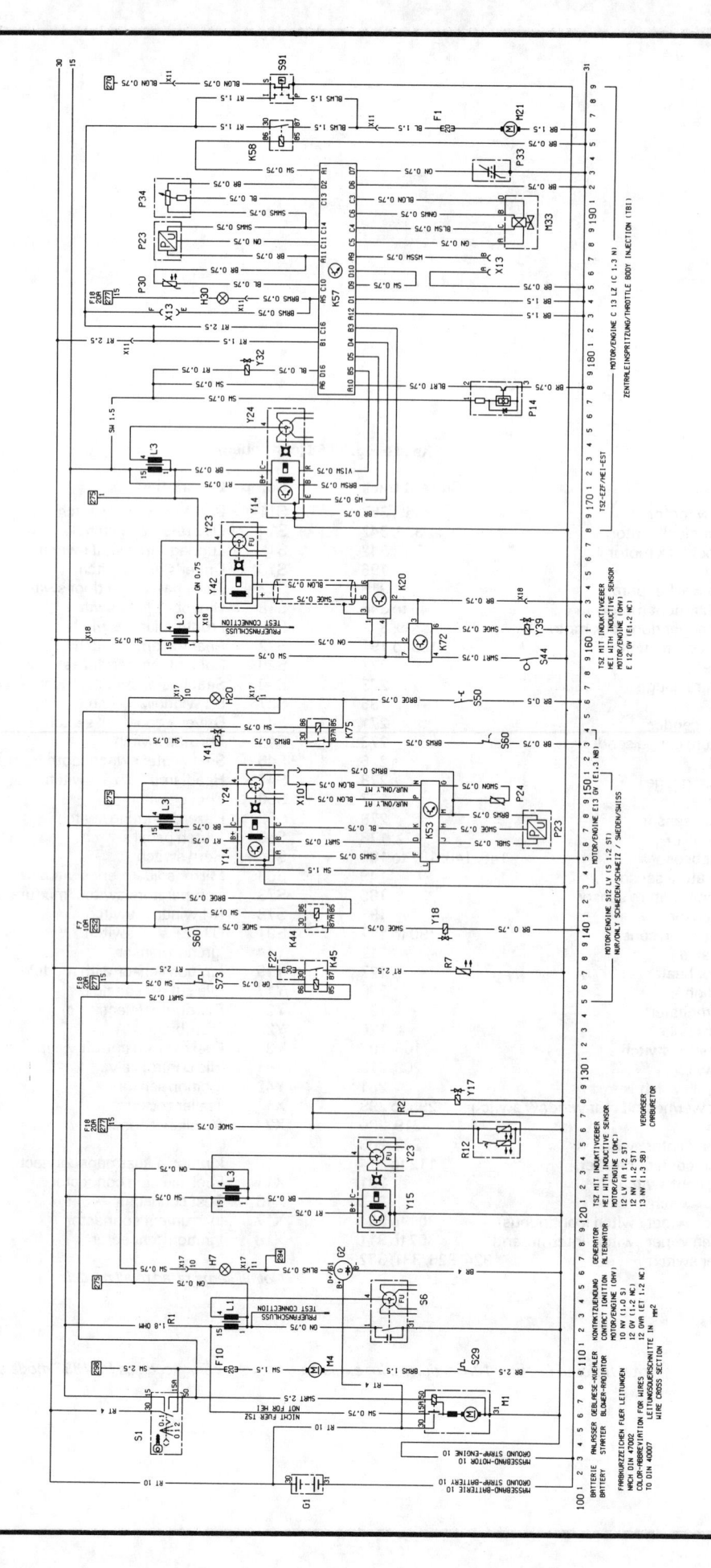

Fig. 13.69 Wiring diagram – 1988 to 1989 models

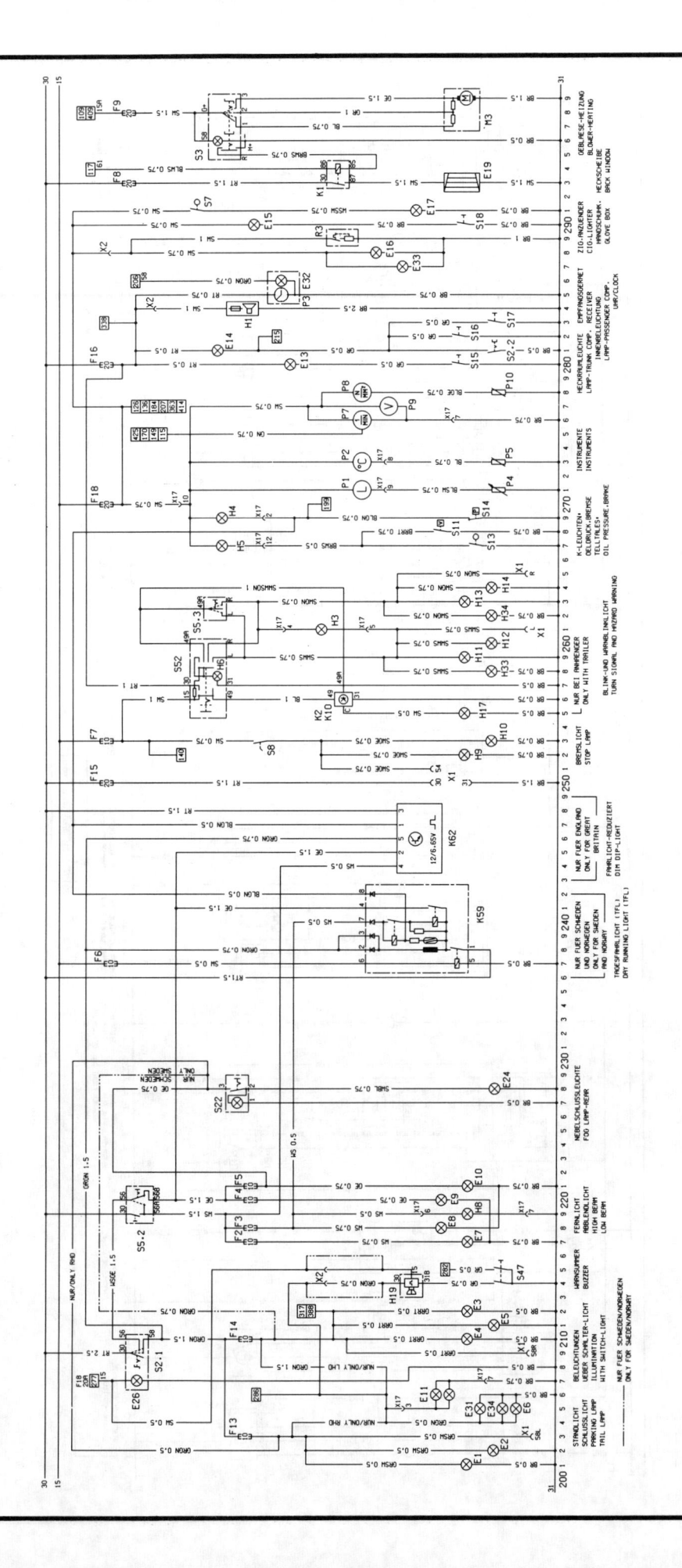

Fig. 13.69 Wiring diagram – 1988 to 1989 models (continued)

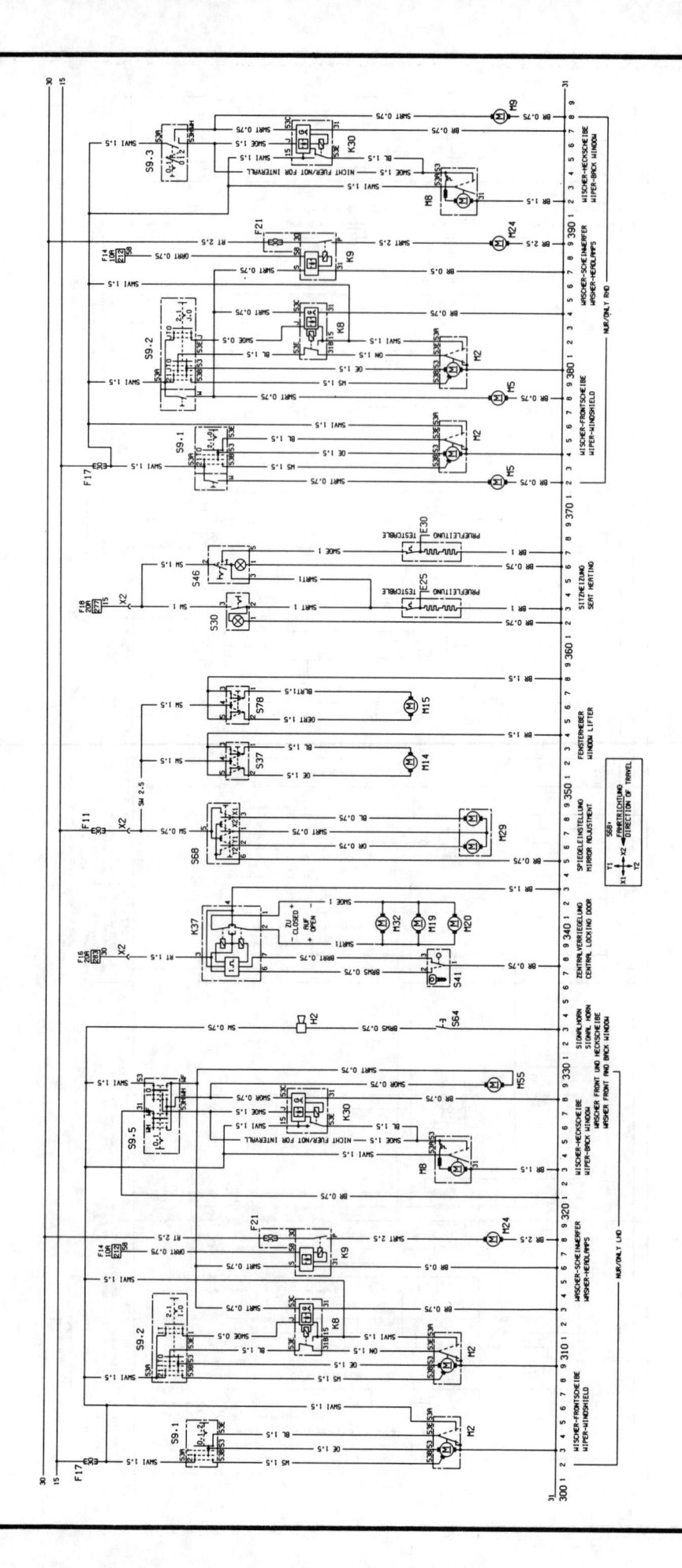

Fig. 13.69 Wiring diagram – 1988 to 1989 models (continued)

297

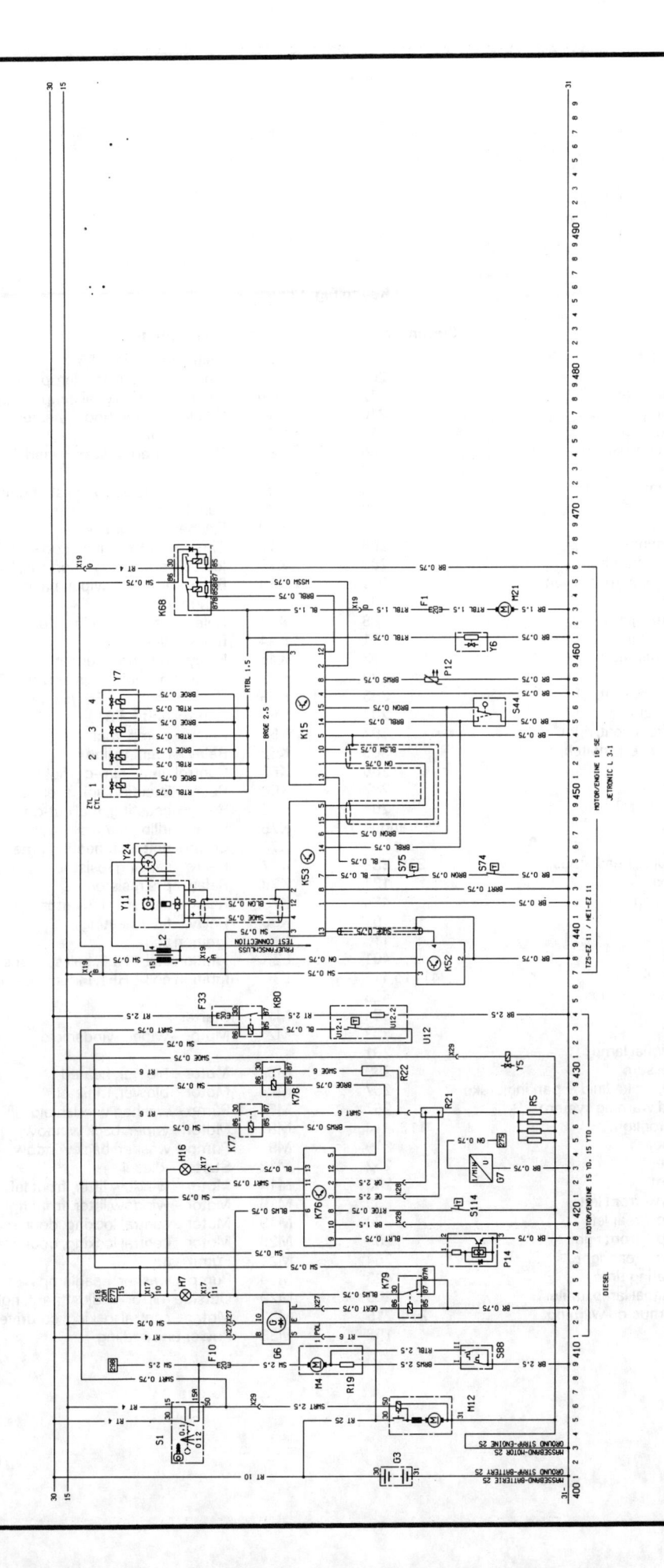

Fig. 13.69 Wiring diagram – 1988 to 1989 models (continued)

Key to Fig. 13.69

Comp	Description	Current track
E1	Parking lamp – left	201
E2	Tail lamp – left	202
E3	Lamp – number plate	212
E4	Parking lamp – right	210
E5	Tail lamp – right	211
E6	Lamp – engine compartment	205
E7	High beam – left	217
E8	High beam – right	218
E9	Low beam – left	220
E10	Low beam – right	221
E11	Lights – instrument	206
E13	Lamp – trunk compartment	280
E14	Lamp – passenger compartment	281
E15	Lamp – glovebox	290
E16	Lamp – cigarette lighter	288
E17	Reversing lamp – left	291
E19	Back window – heated	293
E24	Fog lamp – rear	229
E25	Heating mat – front seat, left	363
E26	Lamp – switch, light	207
E30	Heating mat – front seat, right	367
E31	Lamp – symbol insert, switch	205
E32	Lamp – clock	286
E33	Lamp – ashtray	287
E34	Lamp – heater control	205
F1 to F18	Fuse in fusebox	Various
F21	Fuse – washer unit, headlamps	318, 389
F22	Fuse (in fusebox)	137
F36	Fuse – filter heating	434
G1	Battery	101
G2	Alternator	116
G3	Battery – diesel	401
G6	Generator – diesel	411 to 413
G7	Sensor – engine revolution	423
H1	Receiver	284
H2	Signal horn	333
H3	Talltale – turn signal lamp	261
H4	Telltale – oil pressure	269
H5	Telltale – clutch, brake fluid & parking brake	267
H6	Telltale – hazard warning system	257
H7	Charging indicator light	116, 414
H8	Telltale – high beam	219
H9	Stop-lamp – left	252
H10	Stop-lamp – right	253
H11	Turn signal lamp – front left	259
H12	Turn signal lamp – rear left	260
H13	Turn signal lamp – front right	263
H14	Turn signal lamp – rear right	264
H16	Telltale – preheating time	423
H17	Telltale – turn signal lamp, trailer	255
H19	Buzzer – headlamps-on warning	214, 215
H20	Telltale – choke	156

Comp	Description	Current track
H30	Telltale – engine, TBI	184
H33	Auxiliary turn signal lamp – left	258
H34	Auxiliary turn signal lamp – right	262
K1	Relay – back window, heated	293, 294
K2	Flasher unit	256
K8	Relay – interval, wiper windshield	310 to 313, 381 to 384
K9	Relay – time delay, washer unit, headlamps	316 to 318, 387 to 389
K10	Flasher unit – trailer	255, 256
K15	Timing control – injection unit, fuel	451 to 460
K20	Ignition model – ignition coil, HE1	162 to 164
K30	Relay – interval, wiper, back window	326 to 328, 395 to 297
K37	Relay – central locking, door	337 to 343
K44	Relay – dashpot	140, 141
K45	Relay – mixture preheating	136, 137
K52	Ignition module – ignition coil, EZ 11	437 to 440
K53	Timing control – EST/EZ 11	144 to 151, 440 to 449
K57	Control unit – TB1	178 to 193
K58	Relay – pump, fuel	195, 196
K59	Relay – running light	236 to 242
K62	Control unit – dim-dip light	244 to 248
K68	Relay – injection unit, fuel	463 to 466
K72	Relay – coasting, fuel cut-off	159 to 161
K75	Relay – idling control	153, 154
K76	Control unit – preheating time	418 to 424
K77	Relay – sensing resistor	426, 427
K78	Relay – pre resistor	429, 430
K79	Relay – charging indicator	413 to 418
K80	Relay – filter heating	433, 434
L1	Ignition coil	112, 113
L2	Ignition coil – HE1, hall sensor system, EZ 11	438, 439
L3	Ignition coil – HE1, inductive sensor	121, 122, 147, 148, 160, 161, 172, 173
M1	Starter	105 to 107
M2	Motor – wiper, windshield	302 to 305, 308 to 311, 373 to 376, 379 to 382
M3	Motor – blower, heated	297 to 299
M4	Motor – blower, radiator	109, 409
M5	Pump – washer, windshield	372, 378
M8	Motor – wiper, back window	323 to 325, 392 to 394
M9	Pump – washer, back window	398
M12	Starter – diesel	405 to 406
M14	Motor – window lifter, front left	351, 353
M15	Motor – window lifter, front right	355, 357
M19	Motor – central locking, door, rear left	339, 342
M20	Motor – central locking, door, rear right	339, 342
M21	Pump – fuel	196, 463
M24	Pump – washer, headlamps	318, 389
M29	Outside mirror – adjustment, right	346 to 348
M32	Motor – central locking, co-driver door	339, 342
M33	Power unit – idling	188 to 191

Key to Fig. 13.69 (continued)

Comp	Description	Current track
M55	Pump – washer, windshield and back window	329
P1	Fuel indicator	271
P2	Temperature indicator – cooling agent	273
P3	Clock	285
P4	Sensor – fuel	271
P5	Sensor – temperature, cooling agent	273
P7	Tachometer	276
P8	Gauge – oil pressure	278
P9	Voltmeter	277
P10	Sensor – oil pressure	278
P12	Sensor – temperature, cooling agent	458
P14	Sensor – distance	177, 178, 417, 418
P23	Vacuum sensor – intake manifold	146 to 148, 187 to 189
P24	Sensor – oil temperature, EST	149
P30	Temperature sensor – cooling agent	185
P33	Lambda sensor	193
P34	Sensor – throttle valve position	90 to 192
R1	Resistor cable	112
R2	Carburettor preheating	127
R3	Cigarette lighter	289
R5	Glow plugs	425 to 427
R7	Mixture – preheating	137
R12	Automatic choke	126
R19	Preresistor – blower, radiator	409
R21	Sensing resistor	427
R22	Pre-resistor – glow plugs	430
S1	Switch – starter	106, 107, 405, 406
S2	Switch assembly – light	
S2.1	Switch – light	209, 210
S2.2	Switch – light, passenger compartment	281
S3	Switch – blower & back window, heated	295 to 299
S5	Turn signal switch assembly	
S5.2	Switch – low beam	219, 220
S5.3	Switch – turn signal	262, 263
S6	Ignition distributor	112 to 114
S7	Switch – reversing lamp	291
S8	Switch – stop-lamp	253
S9	Switch assembly – wiper unit	
S9.1	Switch – wiper, windshield	302 to 306, 372 to 375
S9.2	Switch – wiper, windshield, interval	308 to 311, 378 to 382
S9.3	Switch – wiper, back window, interval	396, 397
S9.5	Switch – wiper, back window and washer unit	327 to 329
S11	Control switch – brake fluid	268
S13	Switch – parking brake	267
S14	Switch – oil pressure	269
S15	Switch – lamp, trunk compartment	280
S16	Contact switch – door driver	282
S17	Contact switch – door, co-driver	283
S18	Switch – lamp, glove-box	290
S22	Switch – foglamp, rear	228, 229
S29	Switch – temperature, cooling agent	109
S30	Switch – heating mat, front seat left	362, 363
S37	Switch – window lifter, front left	351 to 354
S41	Switch – central locking, driver's door	337, 338
S44	Switch – throttle valve	159, 455, 456
S46	Switch – heating mats, front seats	365 to 367
S47	Contact switch – door and headlamps-on warning	214, 215
S50	Switch – bowden cable, choke	156
S52	Switch – hazard warning	256 to 260
S60	Switch – clutch pedal	140, 153
S64	Switch – signal horn	333
S68	Switch – outside mirror adjustment	345 to 348
S73	Switch – temperature, mixture preheating	136
S74	Switch – engine (housing)	444
S75	Switch – temperature, engine oil	444
S78	Switch – window lifter, front right	355 to 358
S88	Switch – temperature, cooling agent, radiator	409, 410
S91	Switch – oil pressure, TBI	198, 199
S114	Switch – temperature, cooling agent, engine	420
U12	Filter heating assembly	433, 434
U12.1	Switch – temperature	433
U12.2	Heating resistor	434
X1	Socket – trailer	Various
X2	Connector – auxiliary users	Various
X10	Connector – ignition adjustment	150, 151
X11	Connector – wiring harness, TBI, 5 pins	181, 184, 196, 199
X13	Test connector	183, 186, 187
X17	Connector – instrument	Various
X18	Connector – wiring harness, HEI, 3 pins	160, 162, 163
X19	Connector – wiring harness, Jetronic, 7 pins	437, 438, 463, 466
X27	Plug connection – alternator, 3 pins	411 to 413
X28	Plug connection, 3 pins	419 to 422
X29	Plug connection, 2 pins	406, 431
Y5	Solenoid valve – fuel	431
Y6	Slide valve – auxiliary air	461
Y7	Solenoid valves – fuel injection	450, 457
Y11	Hall sensor	441, 443
Y14	Inductive sensor – EST	145 to 147, 169 to 172
Y15	Inductive sensor with ignition module	120, 121
Y17	Solenoid valve – idle cut off	128
Y18	Solenoid valve – dashpot	140
Y23	Ignition distributor – HEI, inductive sensor system	124, 167
Y24	Ignition distributor – EST	150, 175, 445
Y32	Injection valve – TBI	179
Y39	Solenoid valve – coasting, fuel cut off	161
Y41	Solenoid valve – idling control	153
Y42	Inductive sensor – HEI	163, 164

Not all items fitted to all models

Note: *No specific diagram for 1.6 GTE/GSi models was available at time of writing*

Refer to Chapter 10, Section 42 for explanation of use – Wiring code as for 1985 models

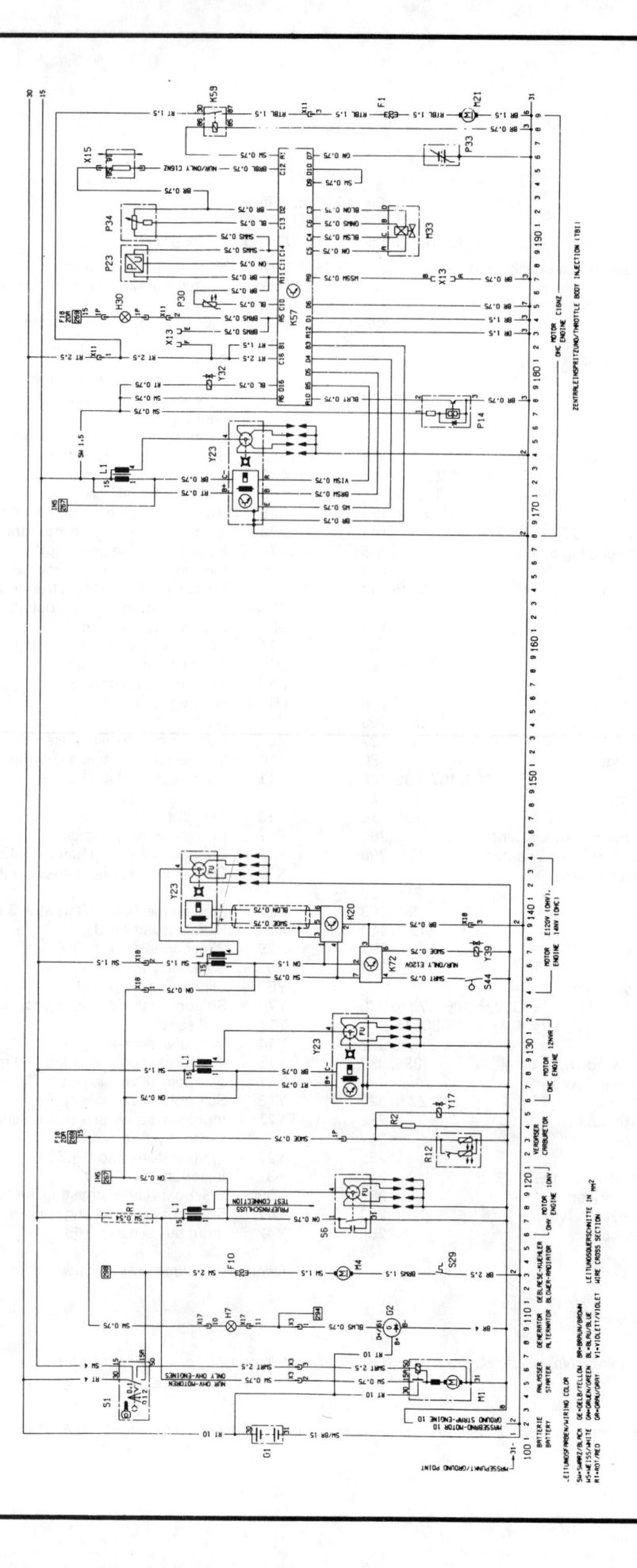

Fig. 13.70 Wiring diagram – 1990 models

301

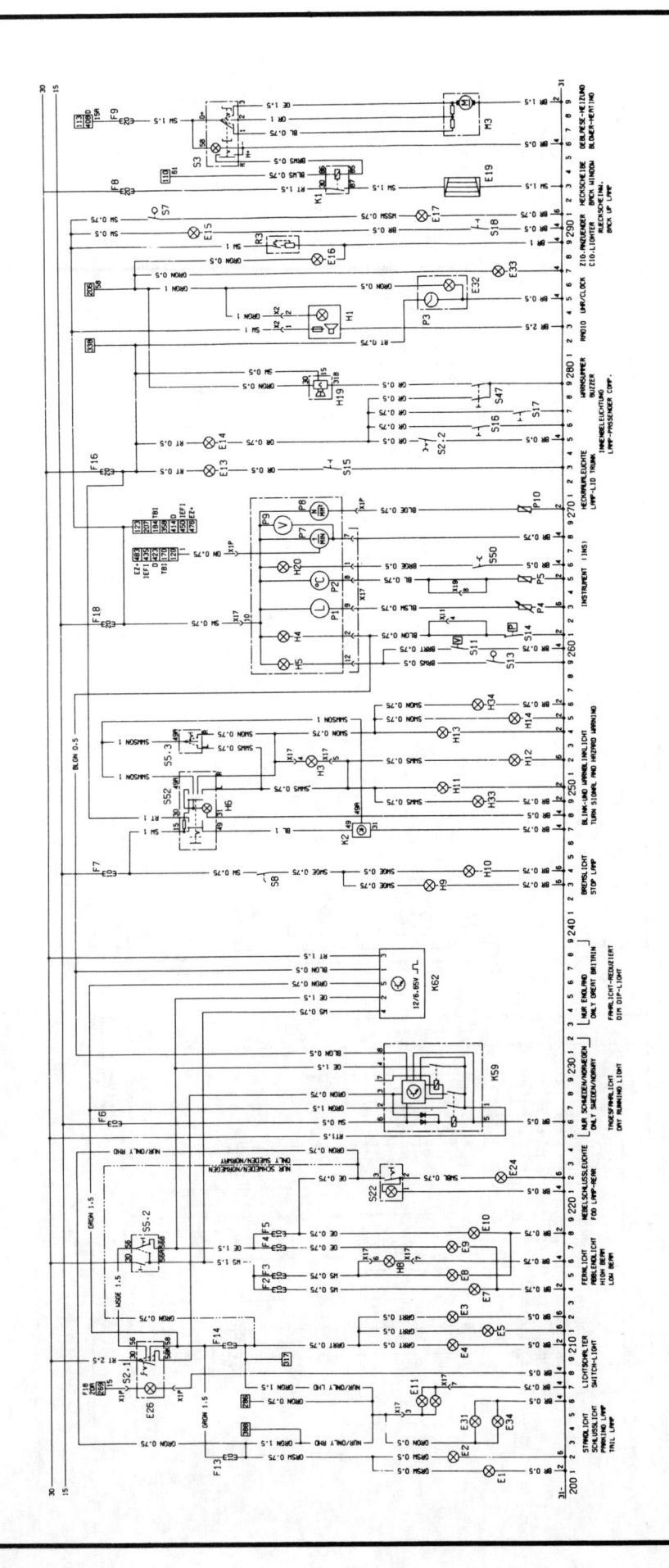

Fig. 13.70 Wiring diagram – 1990 models (continued)

302

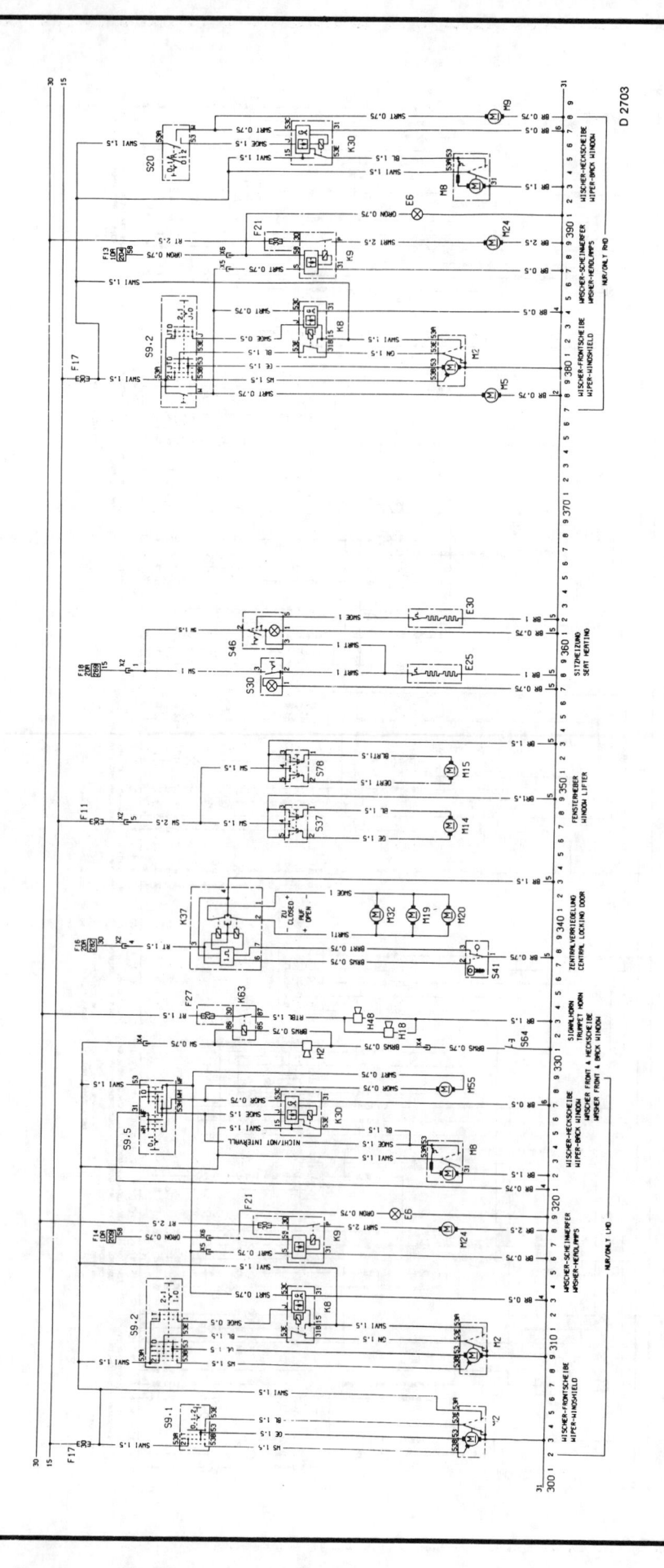

Fig. 13.70 Wiring diagram – 1990 models (continued)

Key to Fig. 13.70

Comp	Description	Current track
E1	LH parking lamp	201
E2	LH tail lamp	202
E3	Number plate lamp	212
E4	RH parking lamp	210
E5	RH tail lamp	211
E6	Engine compartment lamp	319, 391
E7	LH main beam	214
E8	RH main beam	215
E9	LH dipped beam	217
E10	RH dipped beam	218
E11	Instrument illumination	206
E13	Luggage area light	273
E14	Courtesy light	275
E15	Glovebox light	290
E16	Cigarette lighter illumination	288
E17	LH reversing lamp	291
E19	Heated rear window	293
E24	Rear foglamp	222
E25	LH seat heater	358
E26	Light switch illumination	207
E30	RH seat heater	362
E31	Control switch illumination	204
E32	Clock illumination	286
E33	Ashtray illumination	287
E34	Heater control illumination	204
F1 to F11, F13, F14, F16 to F18	Fuses (in fuse box)	Various
F21	Fuse – headlamp washers	318, 389
F27	Fuse – horn	333
F36	Fuse – filter heating (Diesel)	432
G1	Battery	101
G2	Alternator	109
G3	Battery – Diesel	401
G6	Generator – Diesel	411 to 413
G7	Sensor – engine speed/crankcase position	423
H1	Radio	283, 284
H2	Horn	331
H3	Direction indicator warning lamp	252
H4	Oil pressure warning light	261
H5	Brake warning light	259
H6	Hazard warning flasher repeater light	248
H7	No charge (ignition) warning light	109, 414
H8	Main beam warning light	216
H9	LH stop-lamp	243
H10	RH stop-lamp	244
H11	LH front direction indicator	250
H12	LH rear direction indicator	252
H13	RH front direction indicator	254
H14	RH rear direction indicator	255
H16	Glow plug preheating warning light (Diesel)	419
H18	Horn	332
H19	Headlamp warning buzzer	279
H20	Choke warning light	266
H30	Engine management warning light	184, 450, 476
H33	LH direction indicator repeater	249
H34	RH direction indicator repeater	256
H48	Horn	333
K1	Heated rear window relay	293, 294
K2	Flasher unit	247
K8	Wiper delay relay	310 to 313, 381 to 384
K9	Headlamp washer timer relay	316 to 318, 387 to 389
K15	Timing control unit	484 to 495
K20	Ignition module and coil	138 to 140, 435 to 437, 464, 465
K30	Rear wiper delay relay	325 to 327, 395 to 397
K37	Central locking relay	337 to 343
K57	Injection control unit	178 to 196
K58	Fuel pump relay	198, 199, 461, 462
K59	Day running light relay	225 to 231
K62	Dim-dip control unit	234 to 238
K63	Horn relays	332, 333
K68	Injection control unit relay	496 to 499
K72	Engine speed relay	135 to 137
K76	Glow plug control unit (Diesel)	418 to 424
K77	Sensor resistor relay	424, 425
K78	Pre-resistor relay	427, 428
K79	Charging indicator relay	413 to 416
K80	Filter heating relay	431, 432
K84	Ignition control unit	469 to 482
K100	Ignition control unit	440 to 459
L1	Ignition coil	117, 128, 136, 172, 437, 465
M1	Starter	104 to 106
M2	Windscreen wiper motor	302 to 305, 308 to 311, 379 to 382
M3	Heater blower motor	297 to 299
M4	Radiator fan motor	113, 408
M5	Windscreen washer pump	378
M8	Rear wiper motor	322 to 324, 393 to 395
M9	Rear washer pump	398
M12	Starter – Diesel	404, 405
M14	LH front window motor	346, 348
M15	RH front window motor	350, 352

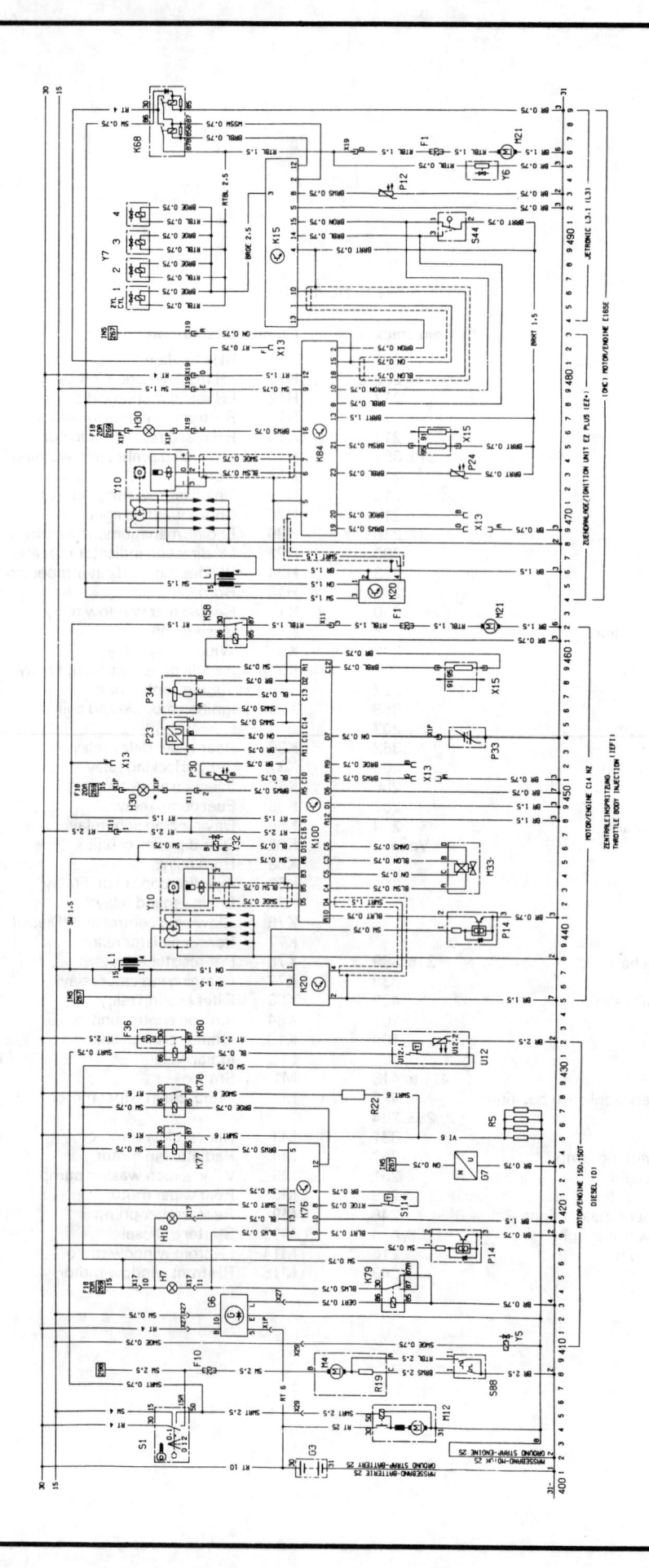

Fig. 13.70 Wiring diagram – 1990 models (continued)

Key to Fig. 13.70 (continued)

Comp	Description	Current track
M19	LH rear door locking motor	339, 342
M20	RH rear door locking motor	339, 342
M21	Fuel pump	199, 462, 496
M24	Headlamp washer pump	318, 389
M32	Front passenger door lock motor	339, 342
M33	Idle speed control unit	189 to 192, 443 to 446
M55	Windscreen and tailgate washer pump	328
P1	Fuel gauge	263
P2	Temperature gauge	265
P3	Clock	285
P4	Fuel gauge sender	263
P5	Temperature gauge sender	265
P7	Tachometer	268
P8	Oil pressure gauge	270
P9	Voltmeter	269
P10	Oil pressure sensor	270
P12	Coolant temperature sender	493
P14	Distance sensor	177, 178, 417, 418, 440, 441
P23	Vacuum sensor	187 to 189, 453 to 455
P24	Oil temperature sensor	473
P30	Coolant temperature sensor	185, 451
P33	Exhaust oxygen sensor	196, 454
P34	Throttle position sensor	190 to 192, 456 to 458
R1	Resistor cable	117
R2	Carburettor heater	124
R3	Cigarette lighter	289
R5	Glow plugs	425
R12	Automatic choke	123
R19	Cooling fan resistor	408
R22	Glow plug resistor (Diesel)	428
S1	Ignition/starter switch	105, 106, 404, 405
S2	Light switch	
S2.1	Lighting switch	209, 210
S2.2	Interior light switch	275
S3	Heater blower/heated rear window switch	295 to 299
S5	Multi-switch	
S5.2	Dipswitch	216, 217
S5.3	Direction indicator switch	253, 254
S6	Distributor (contact breaker)	117
S7	Reversing light switch	291
S8	Stop-lamp switch	244
S9	Wash/wipe switch	
S9.1	Windscreen wiper switch	302 to 304
S9.2	Windscreen wash/wipe switch	308 to 311, 378 to 382
S9.5	Tailgate wash/wipe switch	326 to 328
S11	Brake fluid level switch	260
S13	Handbrake switch	259
S14	Oil pressure switch	261
S15	Luggage area light switch	273
S16	Driver's door switch	276
S17	Front passenger door switch	277
S18	Glovebox light switch	290
S20	Tailgate wash/wipe switch	396, 397
S22	Rear foglamp switch	221, 222
S29	Coolant temperature switch	113
S30	Seat heater switch (driver only)	357, 358
S37	LH front window switch	346 to 348
S41	Driver's door lock switch	337, 338
S44	Throttle valve switch	135, 490, 491
S46	Seat heater switch (both front seats)	360 to 362
S47	Headlamp buzzer switch	278, 279
S50	Choke switch	266
S52	Hazard warning switch	247 to 251
S64	Horn switch	331
S78	RH window switch	350 to 352
S88	Coolant temperature sensor	408, 409
S114	Coolant temperature sensor	420
U12	Filter heating (Diesel)	
U12.1	Temperature switch (Diesel)	431
U12.2	Heating resistor (Diesel)	432
Y5	Fuel solenoid valve	410
Y6	Auxiliary air valve	495
Y7	Fuel injection solenoids	485 to 492
Y10	Distributor (Hall sensor)	438 to 444, 468 to 474
Y17	Idle cut-off solenoid	125
Y23	Distributor (electronic)	127, 128, 139, 140, 168 to 174
Y32	Fuel injector	179, 446
Y39	Coasting fuel cut-off solenoid	137
X2	Auxiliary user (in fuse box)	283, 284, 338, 347, 356, 358
X3	Main wiring harness and starter connector	105, 106, 109
X4	Main wiring harness and horn connector	331
X5, X6	Main wiring harness and headlamp washer connector	316, 318, 387, 388
X8	LH electric window connector	346 to 349
X9	RH electric window connector	350 to 353
X11	Main wiring harness and ignition system connector	181, 184, 199, 262, 447, 450, 462
X13	Test connector	182, 183, 187, 451, 452, 469, 470, 482
X15	Octane number connector	195, 458, 459, 475
X17	Instruments connector	109, 205, 207, 216, 252, 259 to 268, 414, 419
X18	Main wiring harness and ignition system connector	134, 136, 139
X19	Main wiring harness and fuel injection system connector	264, 476 to 483, 496
X27	Main wiring harness and alternator connector	411 to 413
X29	Main wiring harness and starter connector	405, 410

Not all items fitted to all models

Note: *Refer to Chapter 10, Section 42 for explanation of use. Wiring code as for 1985 models*

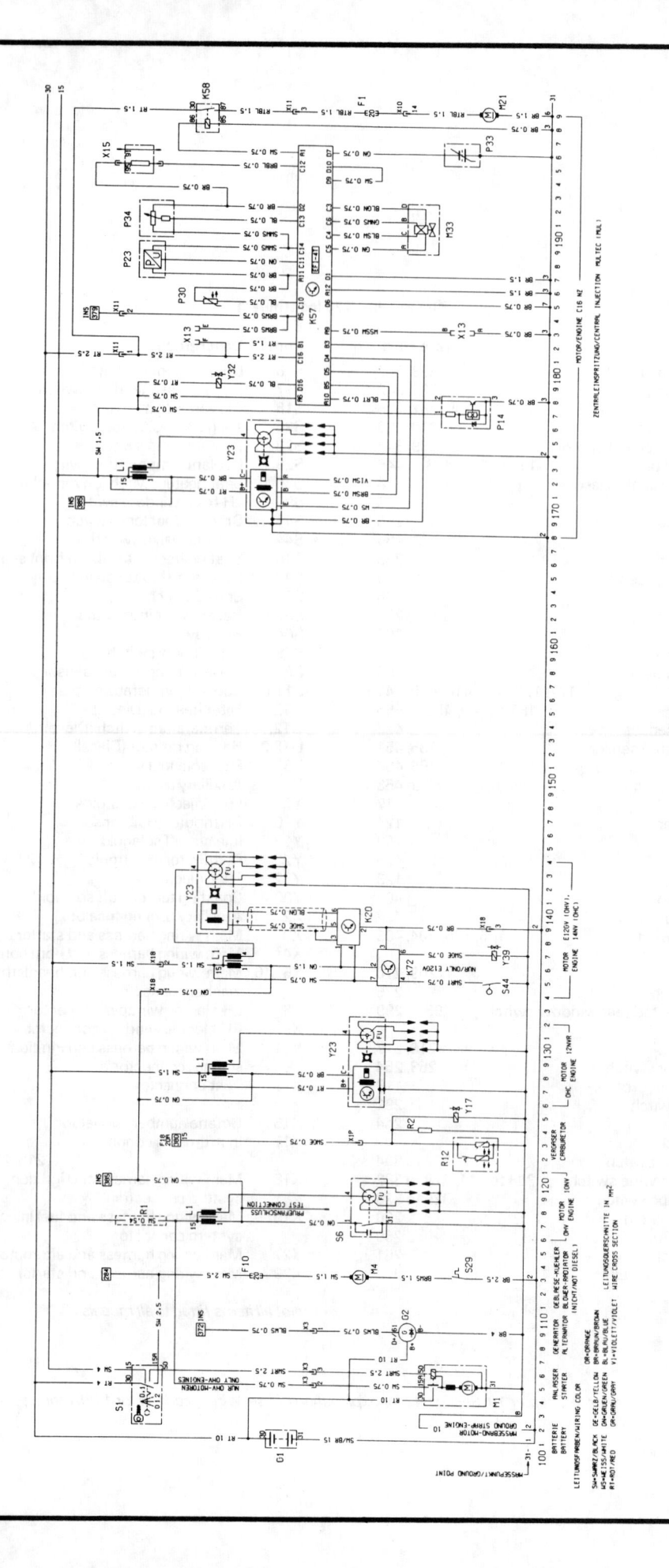

Fig. 13.71 Wiring diagram 1991 models

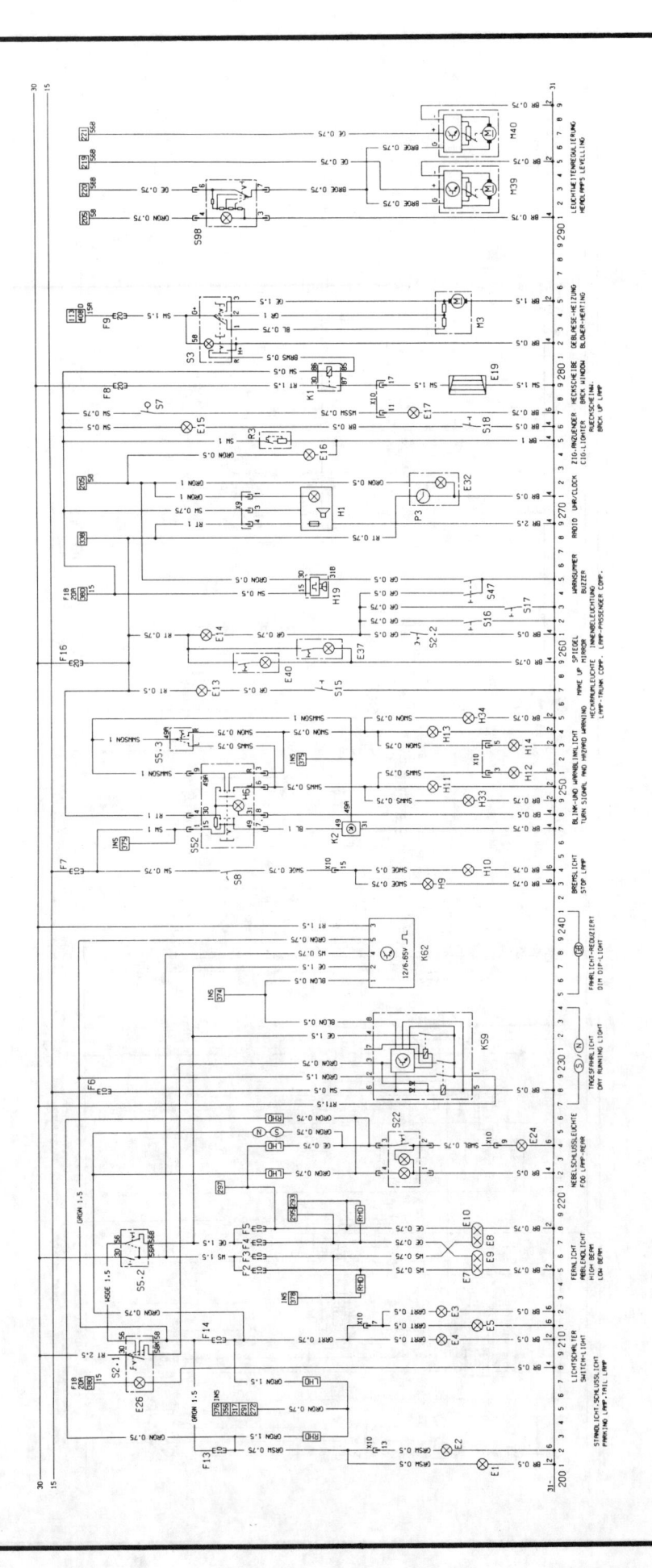

Fig. 13.71 Wiring diagram – 1991 models (continued)

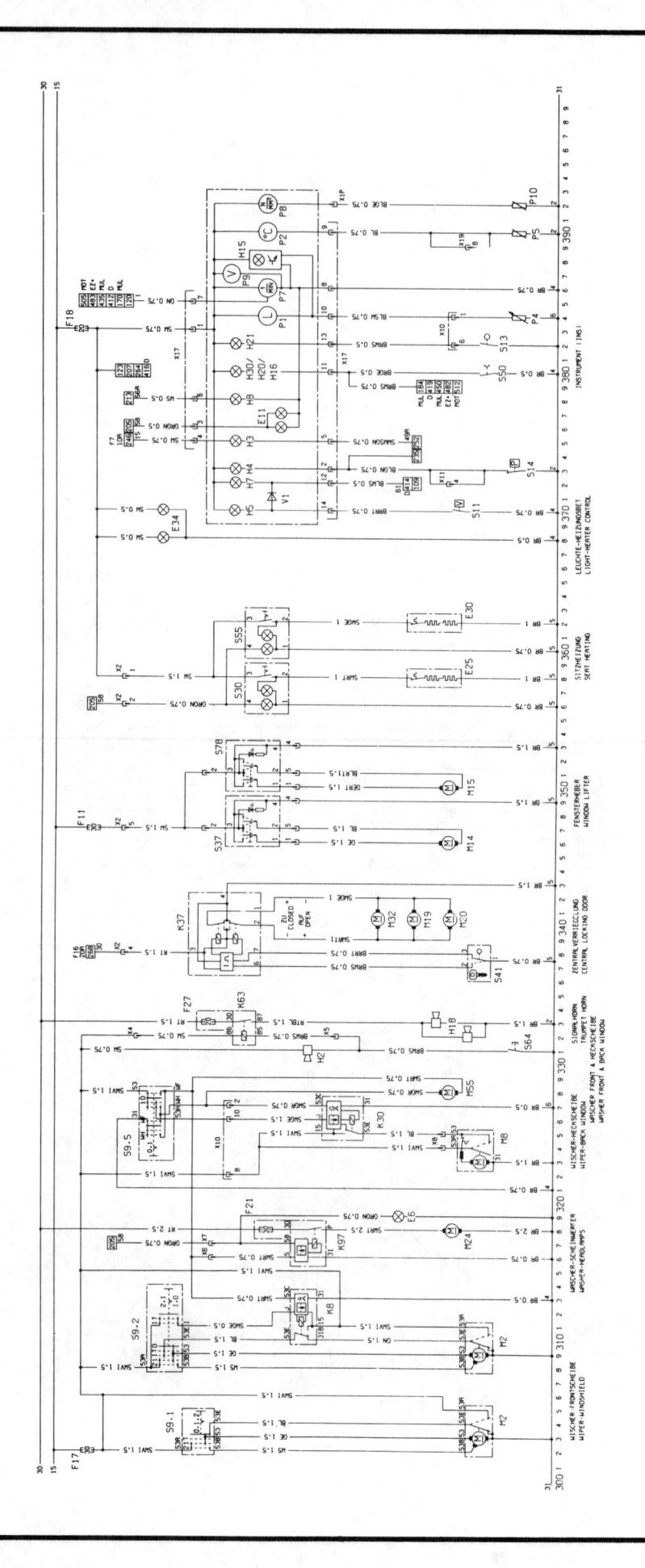

Fig. 13.71 Wiring diagram – 1991 models (continued)

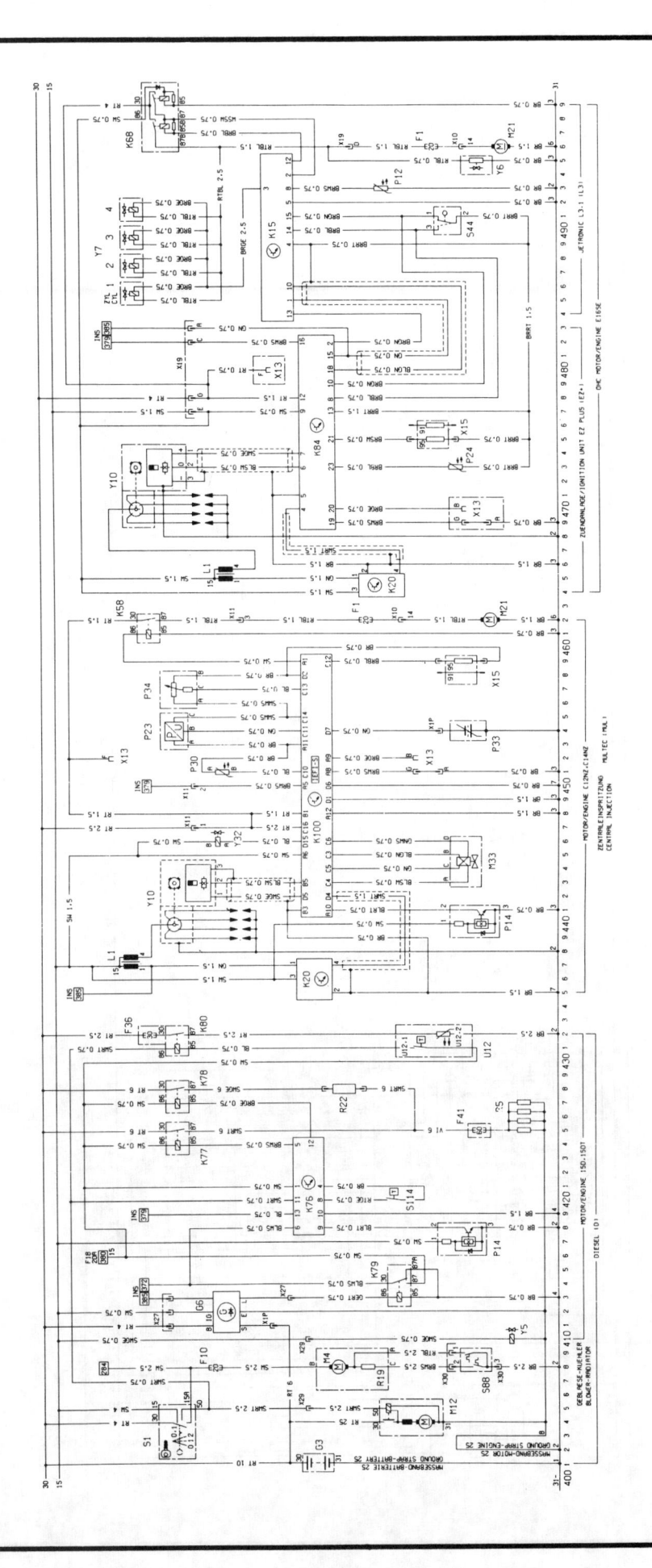

Fig. 13.71 Wiring diagram – 1991 models (continued)

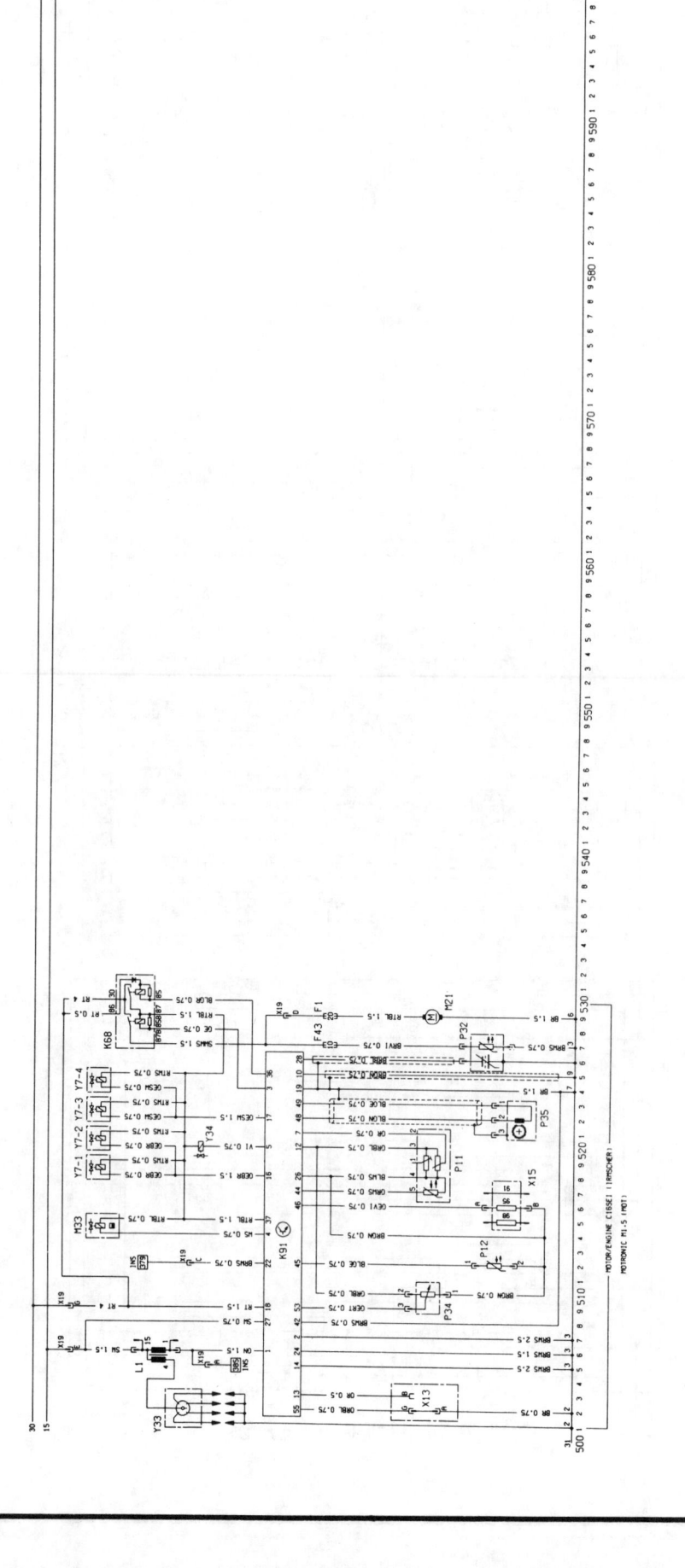

Fig. 13.71 Wiring diagram – 1991 models (continued)

Key to Fig. 13.71

Comp	Description	Current track
E1	LH parking lamp	201
E2	LH tail lamp	202
E3	Number plate lamp	212
E4	RH parking lamp	210
E5	RH tail lamp	211
E6	Engine compartment lamp	319
E7	LH main beam	215
E8	RH main beam	217
E9	LH dipped beam	216
E10	RH dipped beam	218
E11	Instrument illumination	376, 377
E13	Luggage area light	257
E14	Courtesy light	261
E15	Glovebox light	276
E16	Cigarette lighter illumination	274
E17	LH reversing lamp	277
E19	Heated rear window	279
E24	Rear foglamp	224
E25	LH seat heater	358
E26	Light switch illumination	207
E30	RH seat heater	362
E32	Clock illumination	272
E34	Ashtray illumination	368, 370
E37	Driver's vanity mirror illumination	260
E40	Passenger's vanity mirror illumination	259
F1 to F11, F13, F14, F16 to F18	Fuses (in fuse box)	Various
F21	Fuse – headlamp washers	318
F27	Fuse – horn	333
F36	Fuse – filter heating (Diesel)	432
F41	Fuse – glow plugs (Diesel)	425
F43	Fuse – oxygen sensor	527
G1	Battery	101
G2	Alternator	109
G3	Battery – Diesel	401
G6	Alternator – Diesel	411 to 413
H1	Radio	269 to 271
H2	Horn	331
H3	Direction indicator warning light	375
H4	Oil pressure warning light	373
H5	Brake warning light	370
H6	Hazard warning flasher repeater light	249
H7	No charge (ignition) warning light	372
H8	Main beam warning light	378
H9	LH stop-lamp	243
H10	RH stop-lamp	244
H11	LH front direction indicator	250
H12	LH rear direction indicator	251
H13	RH front direction indicator	254
H14	RH rear direction indicator	253
H15	Fuel level warning light	388
H16	Glow plug preheating warning light	380
H18	Horn	332, 333
H19	Headlight warning buzzer	264, 265
H20	Choke warning light	380
H21	Handbrake warning light	382
H30	Engine management warning lamp	380
H33	LH direction indicator repeater	249
H34	RH direction indicator repeater	255
K1	Heated rear window relay	279, 280
K2	Flasher unit	247
K8	Wiper delay relay	310 to 313
K15	Timing control unit	484 to 495
K20	Ignition module and coil	138 to 140, 435 to 437, 464, 465
K30	Rear wiper delay relay	325 to 327
K37	Central locking relay	337 to 343
K57	Injection control unit	178 to 196
K58	Fuel pump relay	198, 199, 461, 462
K59	Day running lamp relay	227 to 233
K62	Dim-dip control unit	236 to 240
K63	Horn relays	332, 333
K68	Injection control unit relay	496 to 499, 527 to 530
K72	Engine speed relay	135 to 137
K76	Glow plug control unit (Diesel)	418 to 424
K77	Sensor resistor relay	424, 425
K78	Pre-resistor relay	427, 428
K79	Charging indicator relay	413 to 416
K80	Filter heating relay	431, 432
K84	Ignition control unit	469 to 482
K91	Engine management control unit	502 to 526
K97	Headlight washer delay relay	316 to 318
K100	Engine management control unit	441 to 459

Key to Fig. 13.71 (continued)

Comp	Description	Current track
L1	Ignition coil	117, 128, 136, 172, 437, 465
M1	Starter	104 to 106, 206
M2	Windscreen wiper motor	302 to 305, 308 to 311
M3	Heater blower motor	283 to 285
M4	Radiator fan motor	113, 408
M8	Rear wiper motor	323 to 325
M12	Starter – Diesel	404, 405
M14	LH front window motor	346, 347
M15	RH front window motor	350, 351
M19	LH rear door locking motor	339, 342
M20	RH rear door locking motor	339, 342
M21	Fuel pump	199, 462, 496, 529
M24	Headlight washer pump	318
M32	Front passenger door lock motor	339, 342
M33	Idle speed control unit	189 to 192, 443 to 446, 514, 515
M39	LH headlamp levelling motor	292 to 294
M40	RH headlamp levelling motor	296 to 298
M55	Windscreen and tailgate washer pump	328
P1	Fuel gauge	384
P2	Temperature gauge	390
P3	Clock	271
P4	Fuel gauge sender	384
P5	Temperature gauge sender	390
P7	Tachometer	386
P8	Oil pressure	392
P9	Voltmeter	387
P10	Oil pressure sensor	392
P11	Air flow meter	517 to 521
P12	Coolant temperature sender	493, 512
P14	Distance sensor	177, 178, 417, 418, 440, 441
P23	Vacuum sensor	187 to 189, 453 to 455
P24	Oil temperature sensor	473
P30	Coolant temperature sensor	185, 451
P32	Exhaust oxygen sensor	526, 527
P33	Exhaust oxygen sensor	196, 454
P34	Throttle position sensor	190 to 192, 456 to 458, 509, 510
P35	Crankshaft position sensor	521 to 523
R1	Resistor cable	117
R2	Carburettor heater	124
R3	Cigarette lighter	275
R5	Glow plugs	425
R12	Automatic choke	123
R19	Cooling fan resistor	408
R22	Glow plug resistor	428
S1	Ignition/starter switch	105, 106, 404, 405
S2	Light switch	
S2.1	Lighting switch	207, 210
S2.2	Interior light switch	261
S3	Heater blower/heated rear window switch	281 to 285
S5	Multi-switch	
S5.2	Dipswitch	216, 217
S5.3	Direction indicator switch	253, 254
S6	Distributor (contact breaker)	117
S7	Reversing light switch	277
S8	Stop-light switch	244
S9	Wash/wipe switch	
S9.1	Windscreen wiper switch	302 to 304
S9.2	Windscreen wash/wipe switch	308, 311
S9.5	Windscreen wiper delay switch	308, 311
S11	Brake fluid level switch	370

Comp	Description	Current track
S13	Handbrake switch	382
S14	Oil pressure switch	373
S15	Luggage area light switch	257
S16	Driver's door switch	262
S17	Front passenger door switch	263
S18	Glovebox light switch	276
S22	Rear foglamp switch	222, 224
S29	Coolant temperature switch	113
S30	Seat heater switch (driver only)	357, 358
S37	LH front window switch	346 to 349
S41	Driver's door lock switch	337, 338
S44	Throttle valve switch	135, 490, 491
S47	Headlamp buzzer switch	264, 265
S50	Choke switch	380
S52	Hazard warning switch	247 to 251
S55	Front passenger seat heater switch	360 to 362
S64	Horn switch	331
S78	RH window switch	350 to 353
S88	Cooling fan switch	408, 409
S98	Headlamp levelling switch	291, 293
S114	Coolant temperature sensor	420
U12	Filter heating (Diesel)	
U12.1	Temperature switch (Diesel)	431
U12.2	Heating resistor (Diesel)	432
V1	Diode	371
Y5	Fuel solenoid valve	410
Y6	Auxiliary air valve	495
Y7	Fuel injection solenoids	485 to 492, 518 to 525
Y10	Distributor (Hall sensor)	438 to 444, 468 to 474
Y17	Idle cut-off solenoid	125
Y23	Distributor (electronic)	127 to 131, 139 to 143, 168 to 175
Y32	Fuel injector	179, 446
Y33	Distributor	501
Y34	Fuel tank vent valve	520
Y39	Coasting fuel cut-off solenoid	137
X2	Auxiliary user (in fuse box)	338, 347, 356, 358
X3	Main wiring harness and starter connector	105, 106, 109
X4, X5	Main wiring harness and horn connector	332
X6, X7	Main wiring harness and headlamp washer connector	316, 317
X8	Main wiring harness and tailgate wash/wipe connector	324, 325
X9	Main wiring harness and radio connector	269 to 271
X10	Main wiring harness plug connector	199, 202, 211, 224, 244, 251, 253, 277, 279, 322, 326, 327, 382, 384, 462, 496
X11	Main wiring harness engine management system connector	181, 184, 199, 372, 447, 450, 462
X13	Test connector	182, 183, 451, 452, 469, 470, 480, 502, 503
X15	Octane number connector	195, 458, 459, 475, 516
X17	Instruments connector	370 to 390
X18	Main wiring harness and ignition system connector	134, 136, 139
X19	Main wiring harness and fuel injection system connector	389, 477 to 483, 496, 505, 506, 509, 512, 529
X27	Main wiring harness and alternator connector	411 to 413
X29	Main wiring harness and starter connector	405, 410
X30	Coolant temperature switch	408, 409

Not all items fitted to all models

Note: Refer to Chapter 10, Section 42 for explanation of use. Wiring code as for 1985 models

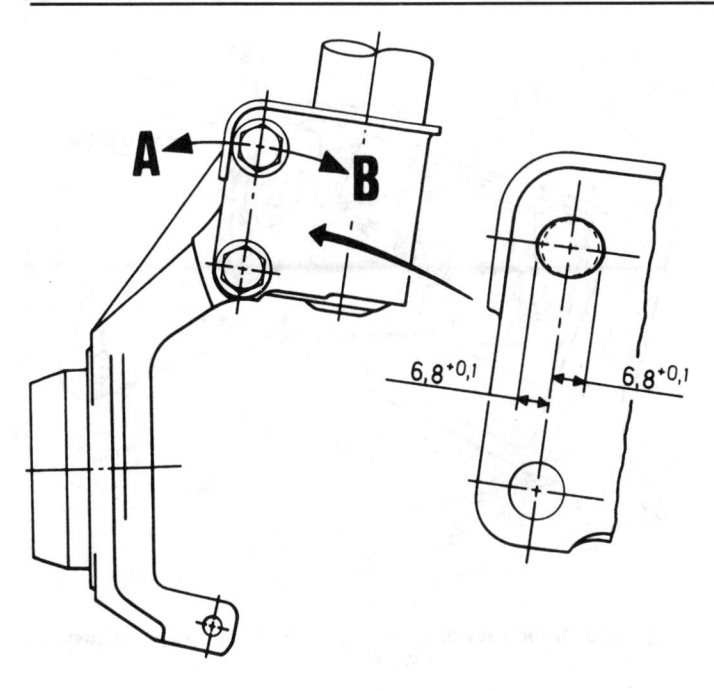

Chapter 13 Supplement: Revisions and information on later models 313

Fig. 13.72 Front suspension strut lower clamp bolt holes (Sec 17)

Dimensions in mm   A to B camber change

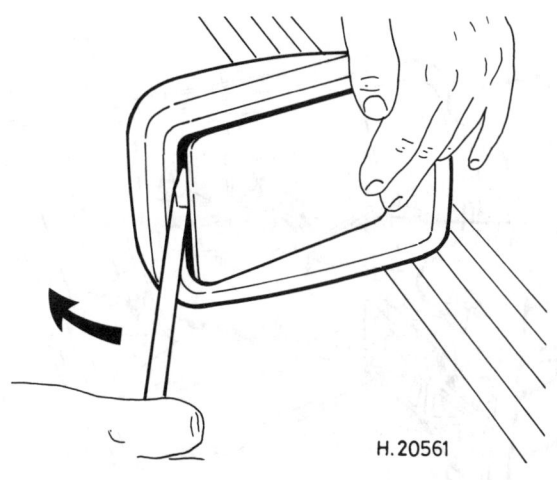

Fig. 13.73 Prising out exterior mirror glass (Sec 18)

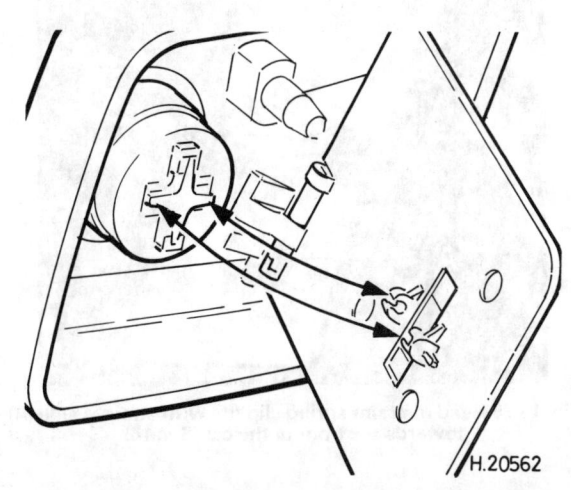

Fig. 13.74 Mirror glass-to-casing connection – early type (Sec 18)

by careful filing as shown in Fig. 13.72. Protect the filed edge of the hole by applying paint.
7 When the hub carrier is reconnected to the clamp at the base of the strut the extra movement now provided for the fixing bolts will result in a change of camber of $\pm$ 0° 40'.
8 An essential part of this operation is that after the camber has been adjusted, the strut-to-hub carrier clamp bolts must be tightened to 100 Nm (74 lbf ft).
9 Repeat the operations on the opposite front wheel if required.

18 Bodywork

Exterior mirror glass – renewal
1 The glass can be renewed without having to change the complete mirror.
2 Insert a wide plastic blade (such as a windscreen or refrigerator ice scraper) between the outer edge of the glass and the mirror casing. Prise the glass, complete with socket, from the casing.
3 Pull the adjustment segment from the adjustment rod.
4 When refitting the new mirror glass, connect the segment and rod, and press the glass into position to engage the knuckle joint.

Front seat head restraints – removal and refitting
Pre-1985 models
5 Prise out the spring clip from the outboard sleeve of the head restraint.
6 Pull the head restraint upwards from the seat back.
7 When refitting, make sure that the straight leg of the spring clip is towards the front of the car.
1985 models onwards
8 On these models, the head restraints can be removed simply by pushing back the retaining clip plunger (photo) to release the restraint.

Passenger grab handles – removal and refitting
9 These are fitted to later models. The handles can be removed from their locations by driving out the centre expansion pins from the heads of the retaining plugs.
10 Use a thin punch to do this. The pins will be lost in the body cavity, so obtain new ones in advance to carrying out the job (photos).
11 Pull the grab handle sharply to release the fixing plugs from their holes in the body.
12 When refitting, hold the grab handle tightly against its mounting surface, and drive the new pins in flush.

Rear door – dismantling and reassembly
13 On four-door Saloon or five-door Hatchback models, the rear door trim panel can be removed and the door dismantled in the following way.

17 Suspension

Modified components
1 The front struts and rear springs and shock absorbers have been modified on 1985 models to improve roadholding and comfort. The operations described in Chapter 11 are not affected.

Front wheel camber
2 As mentioned in Chapters 8 and 11 the camber angle is set in production and is normally not adjustable. However, it is possible for the camber angle to be found outside the specified tolerance, as a result of fitting replacement or pattern type front struts.
3 Where this condition is recognised, as the result of a front end check, it is possible to alter the camber angle in the following way.
4 Raise the front of the car and remove the roadwheel.
5 Remove the suspension strut as described in Chapter 11.
6 The two bolt holes at the base of the strut should now be elongated

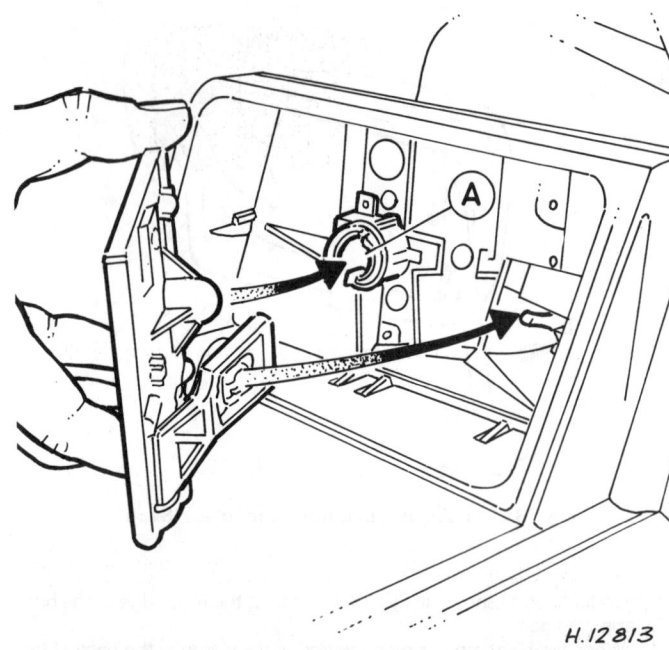

314　Chapter 13　Supplement: Revisions and information on later models

Fig. 13.75 Mirror glass-to-casing connection – later type (Sec 18)

A　Knuckle joint

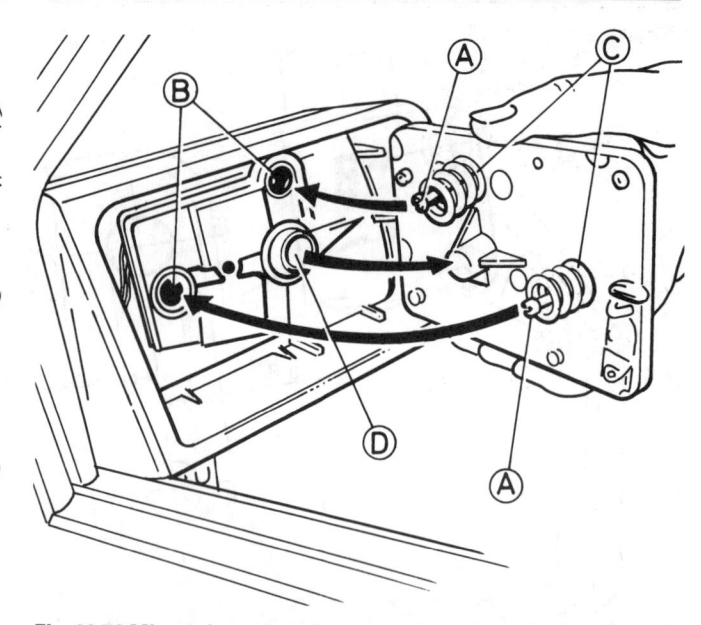

Fig. 13.76 Mirror glass-to-casing connection – electrically-adjusted type (Sec 18)

A　Rods　　C　Dust sleeves
B　Sockets　D　Knuckle joint

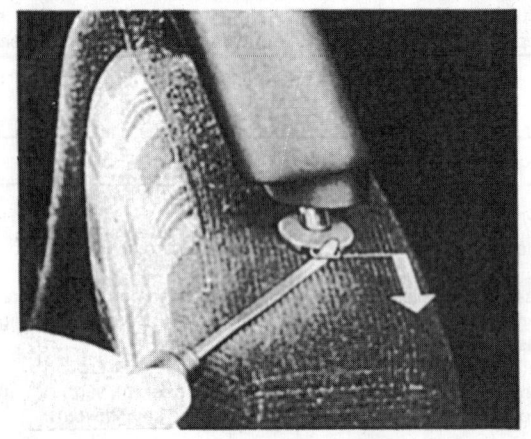

Fig. 13.77 Prising out a head restraint spring clip (pre-1985 models) (Sec 18)

Fig. 13.78 Head restraint spring clip fits with straight side (A) towards the front of the car (Sec 18)

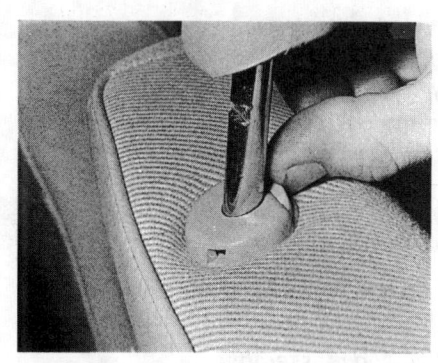

18.8 Removing a later-type head restraint

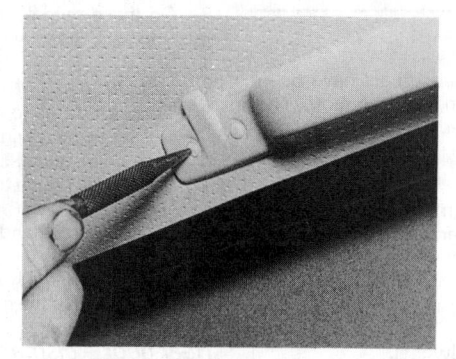

18.10A Driving out grab handle retaining plug centre pin

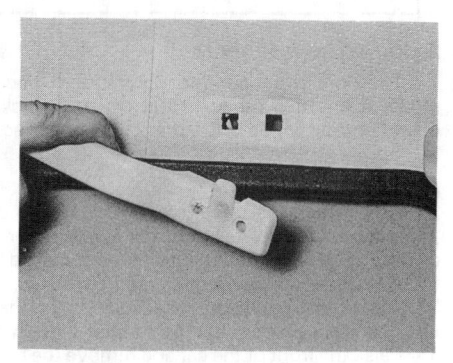

18.10B Grab handle released

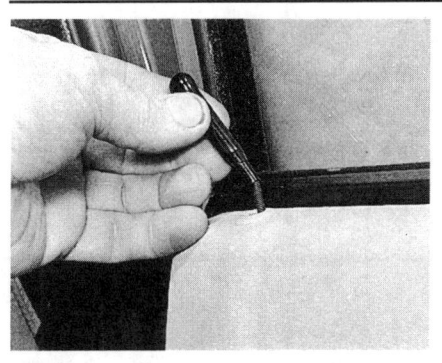

Chapter 13 Supplement: Revisions and information on later models 315

18.14 Removing a door lock plunger knob

18.16 Extracting a screw from the ashtray recess

18.17 Removing the remote control handle escutcheon

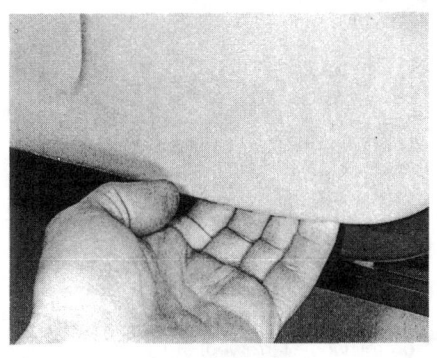

18.19A Releasing door trim panel

18.19B Trim panel retaining clip

18.20 Removing door panel waterproof sheet

14 Open the door fully, unscrew and remove the door lock plunger knob (photo).
15 Remove the ashtray from its recess.
16 Unscrew and remove the three Allen screws from within the ashtray recess (photo).
17 Carefully prise off the escutcheon which surrounds the lock remote control handle (photo).
18 Extract the retaining clip from the window winder handle. To do this, either insert a piece of wire with a hooked end between the handle and the escutcheon plate, or pull a strip of rag back and forth behind and under the handle boss.

19 Insert the fingers between the trim panel and the door frame, and jerk the panel clips from their holes. Move the fingers all round the edge of the panel until the panel can be removed (photos).
20 Peel away the waterproof sheet (photo).
21 The cable-operated window regulator is pop-riveted to the door, and its removal will require drilling out the securing rivets (photo). From 1989 model year, the cable-operated regulator was replaced by a single arm type regulator similar to that described in Chapter 12, Section 18.
22 Release the cable clamp from the glass channel, and withdraw the regulator through the lower aperture (photo).
23 If the door glass is to be removed, first disconnect the winding cable from the glass channel, and fully lower the glass.
24 Prise out the weatherstrips from the glass slot of the door waist.
25 Extract the two screws from the top edge of the door. These screws retain the glass divider channel (photo).
26 Remove the bolt from the lower end of the divider channel, push the upper end of the channel towards the front of the car and remove the quarter glass complete with rubber weatherseal (photo).
27 Raise the main glass, swivel it and pull it out of its slot in the door.
28 If the lock is to be removed from the door, the glass can remain in position, but it must be wound fully up.
29 Release the lock remote control handle by prising up its end nearest the hinge of the door. Unlock the handle from the control rod (photos).
30 The latch may be removed from the door edge after extracting the three retaining screws (photo).
31 Reassembly is a reversal of dismantling, but observe the following points.
32 When fitting the main glass, adjust it for smooth up-and-down operation by moving the divider glass channel within the limits of its lower elongated bolt hole.
33 When refitting the window winder handle, locate the spring clip fully in the handle, and then simply tap the handle onto its splined shaft with the hand (photo). Check that with the glass fully up, the handle knob is at the bottom and at 20° from the vertical towards the front of the car.

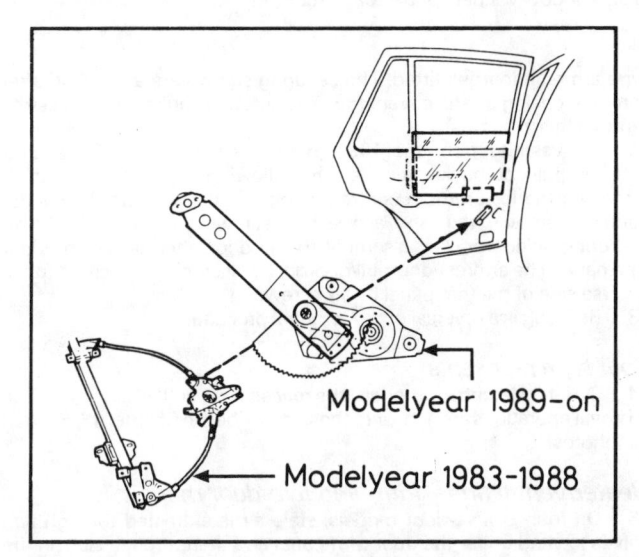

Fig. 13.79 Rear door window regulator modifications (Sec 18)

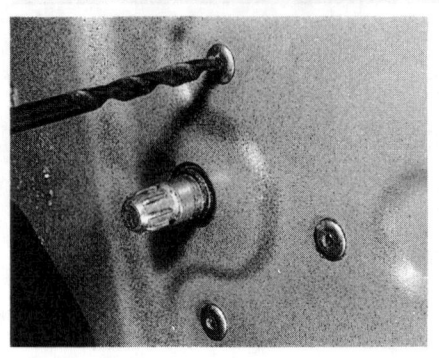

Chapter 13 Supplement: Revisions and information on later models

18.21 Drilling out a window regulator rivet

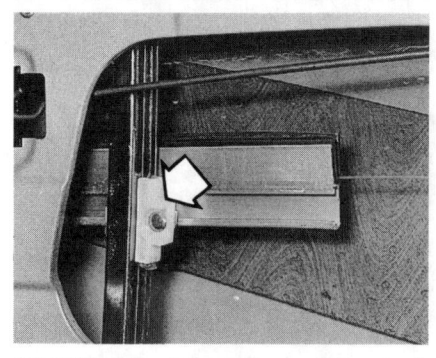

18.22 Window winder cable clamp (arrowed)

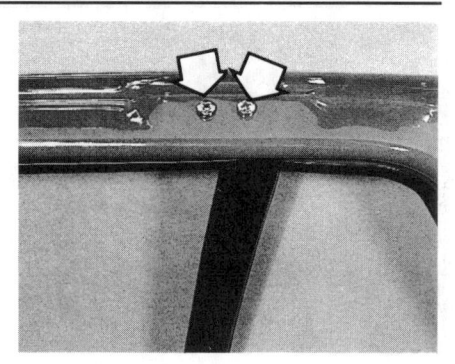

18.25 Glass divider channel upper retaining screws (arrowed)

18.26 Glass divider channel lower fixing bolt (arrowed)

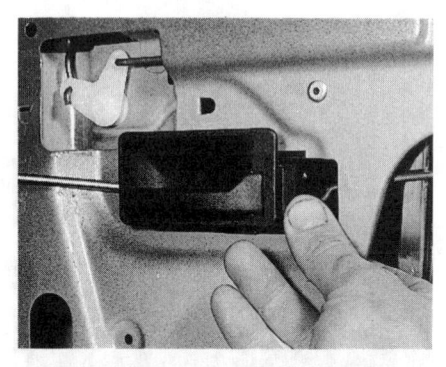

18.29A Prising out remote control handle

18.29B Control rod attachment to remote control handle (arrowed)

18.30 Door latch secured by three screws

18.33 Window winder handle ready for fitting

Rear quarter trim panel and window – removal and refitting

Saloon models

34 Remove the rear seat as described in Chapter 12, Section 30.

Hatchback models

35 Remove the baggage area cover.
36 Remove the rear seat as described in Chapter 12, Section 30.

All models

37 Remove the plastic clips and self-tapping screws which retain the upper quarter trim panel (photos).
38 If rear seat belts are fitted, the belt need not be removed, but the upper anchor bracket must be unbolted so that the trim panel can be withdrawn (photos).
39 The quarter glass and frame can now be removed if the plastic fixing nuts are unscrewed (photo).
40 Refitting is a reversal of removal.

Front door armrest/doorpull/doorpocket – removal and refitting

41 Later models up to the 1991 model year had a revised 'swan-neck' type armrest/doorpull fitted. The securing screws are accessible after carefully prising up the cover strip. The screws require an Allen key to remove them (photos).
42 On vehicles from the 1991 model year, removal of the armrest/doorpull/doorpocket requires the following action. In addition to removing the lock remote control handle escutcheon and the winder handle, remove the six accessible screws securing the armrest/doorpull/doorpocket assembly to the door, then unclip the door trim panel. The armrest/doorpull/doorpocket assembly is secured to the reverse side of the trim panel by six screws.
43 Refitting is a reversal of the removal procedure.

Split type rear seats

44 On later Hatchback models, the rear seat is of 60/40 split type. The removal operations are similar to those described in Chapter 12, Section 30 (photos).

Heater/ventilation – four- and five-door models

45 On four- or five-door models, stale air is extracted through flap valves located under the door trim panel and then through slots in the door edge (photos).

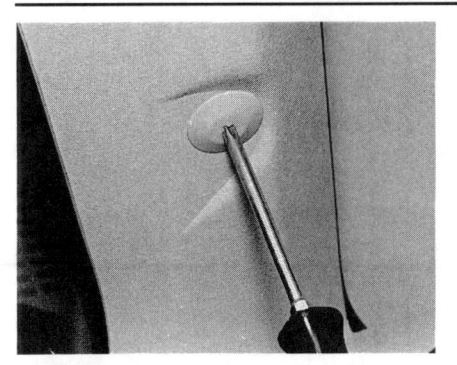

Chapter 13 Supplement: Revisions and information on later models

18.37A Removing a quarter trim panel retaining clip

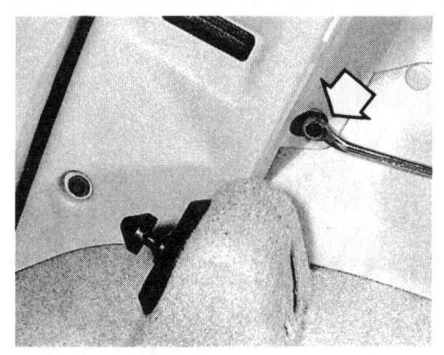

18.37B Removing a quarter trim panel retaining screw (arrowed)

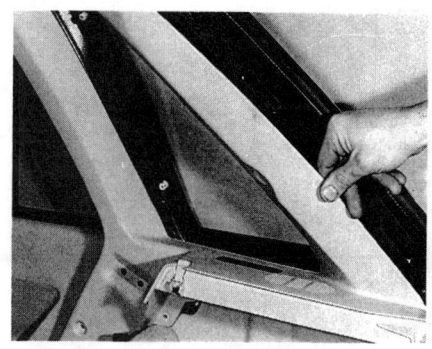

18.38A Removing a quarter trim panel upper section

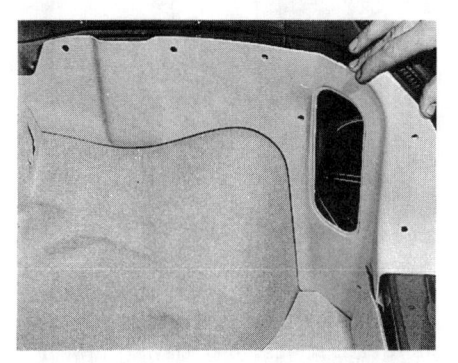

18.38B Removing a quarter trim panel lower section

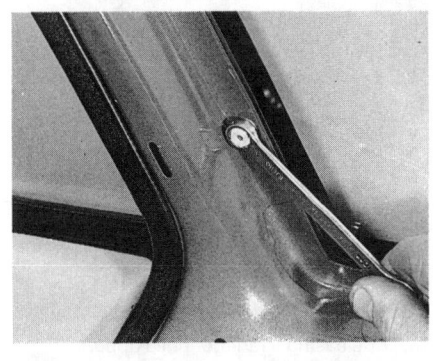

18.39 Unscrewing a quarter window glass fixing nut

8.41A Removing a front door armrest/doorpull cover strip

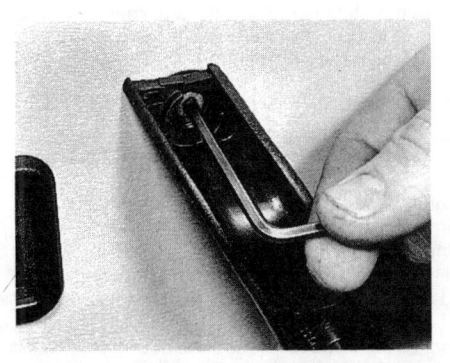

8.41B Extracting an armrest/doorpull screw

18.44A Removing a split-type rear seat cushion securing screw

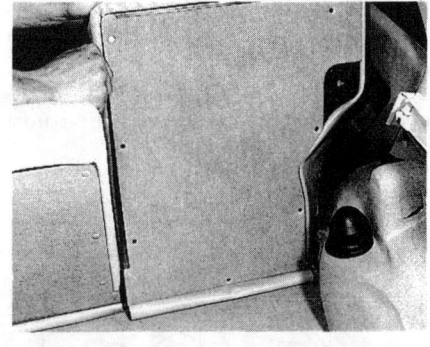

18.44B Unclipping the rear seat back trim

18.44C Removing a rear seat back pivot cover screw

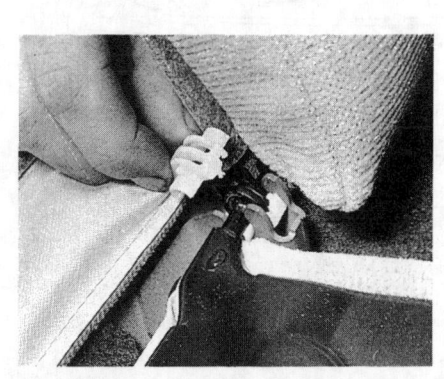

18.44D Rear seat back pivot spacer

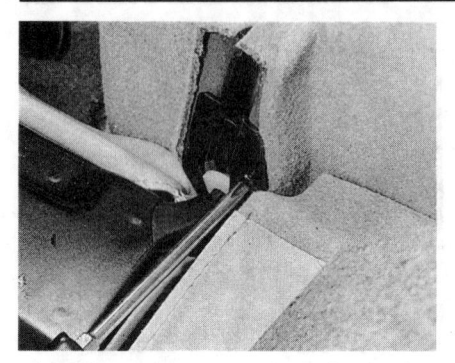

318　Chapter 13　Supplement: Revisions and information on later models

18.44E Removing rear seat back pivot bush retainer plate screw

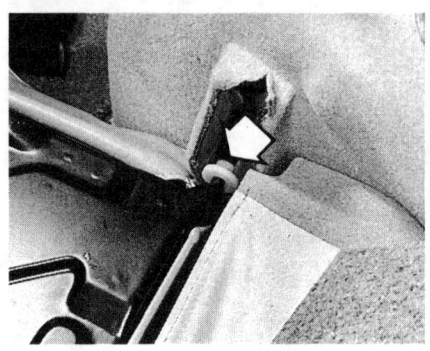

18.44F Rear seat back pivot bush (arrowed)

18.45A Rear door flap valve (door trim panel removed)

18.45B Rear door edge air-extraction slots

18.47 Releasing a radiator grille upper securing clip

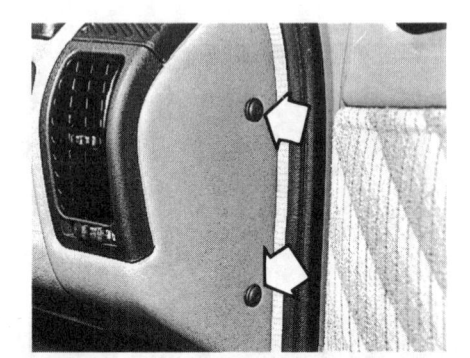

18.78 Facia side securing screws (arrowed)

46　Cutaways are incorporated at the base of the trim panel to provide a passage to the flap valves.

Radiator grille (from 1991 model year) – removal and refitting

47　Making reference to Fig. 13.80, unclip the radiator grille from the front panel and the lower edges of the front wings, then remove it (photo).
48　New radiator grilles are supplied in two sections, these being the surround and the insert. After the surround has been painted in the required colour, the insert is fitted, and its plastic pins melted to hold it in place. Special tool KM-205 must be heated and used to melt the pins, although it may be possible to improvise a suitable alternative.
49　Refitting is a reversal of the removal procedure.

Front bumper (from 1991 model year) – removal and refitting

50　The procedure is similar to that described in Chapter 12, Section 9, but note the following.
51　The radiator grille must be removed, along with the active carbon canister and the headlamp washer fluid reservoir (where applicable), before commencing bumper removal.
52　Two bolts are used to secure the bumper to each front wing. Note also that the body-bound rivets on the underside of the bumper must be removed (Fig. 13.81).
53　Refitting is a reversal of the removal procedure, noting the direction of the arrow on each of the bumper side-retaining blocks (where applicable) (Fig. 13.82).

Rear bumper (from 1991 model year) – removal and refitting

54　Refer to the procedure given in Chapter 12, Section 10, noting that on models from 1991 model year, the sides of the rear bumper are secured by bolts which are accessible from inside the rear wheelarches.

Front wing – removal and refitting

55　Refer to Chapter 12, Section 11, noting that the direction indicator side repeater lamp must (where fitted) be disconnected from the wing before the wing can be withdrawn.
56　On 1991-on models, the radiator grille must first be unclipped (as described above) and the front direction indicator lamp must be removed (Section 16 of this Chapter).
57　On all models, release the front bumper on the appropriate side (see above and/or Chapter 12).
58　On all models, work as described in Section 11 of Chapter 12; on 1991-on models, the wing is also secured (on the outside of the vehicle) by a bolt under the appropriate headlamp (Fig. 13.83).
59　Refitting is the reverse of the removal procedure.

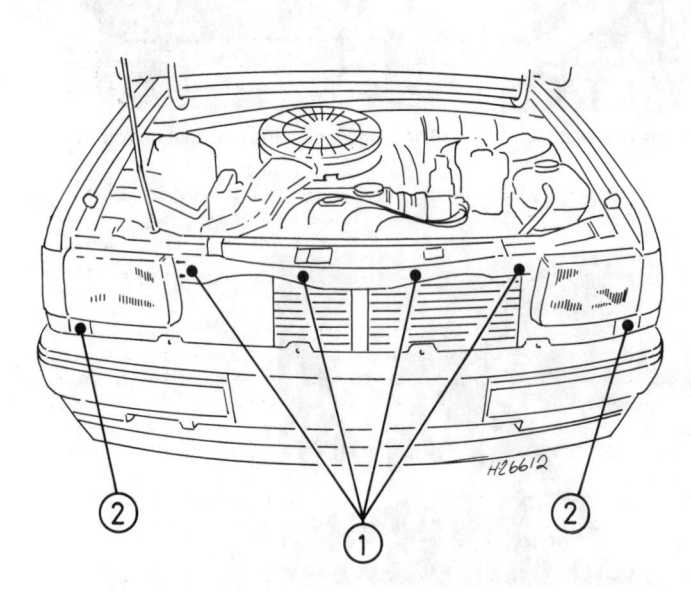

Fig. 13.80 Radiator grille fixing points on the front panel (1) and wings (2) – from 1991 model year (Sec 18)

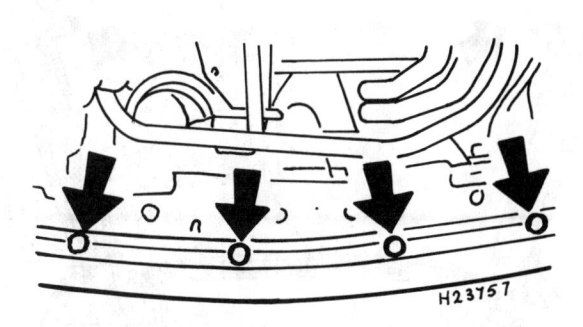

Fig. 13.81 Body-bound rivets (arrowed) on the underside of the front bumper (Sec 18)

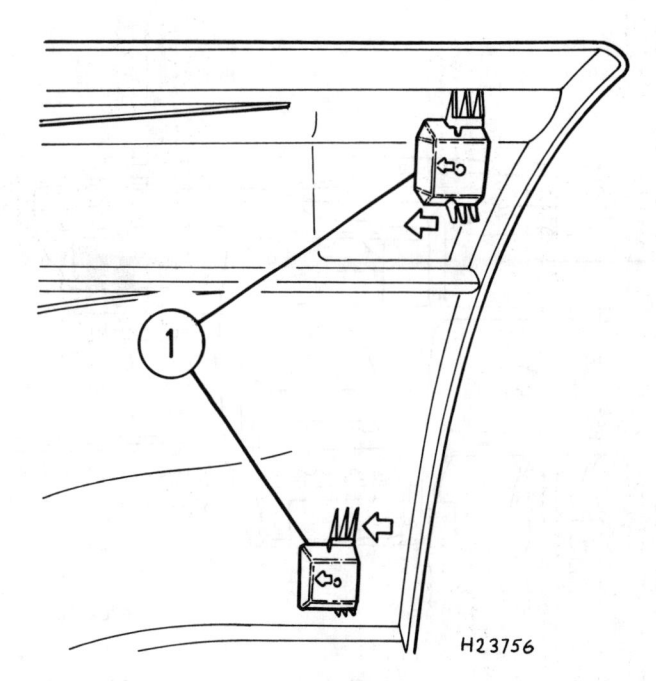

Fig. 13.82 Front bumper side-retaining block (1) arrows facing the correct way (Sec 18)

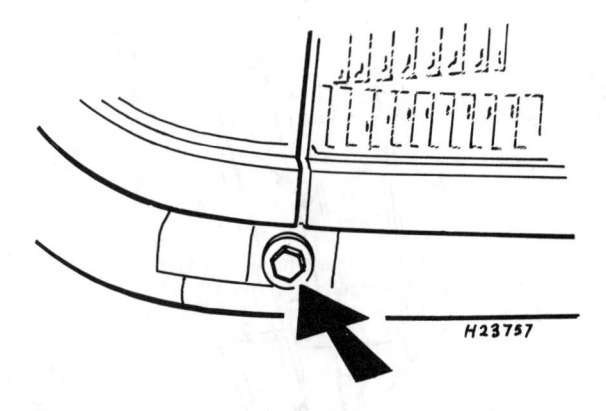

Fig. 13.83 Front wing securing bolt (arrowed) by the base of the adjacent headlamp – from 1991 model year (Sec 18)

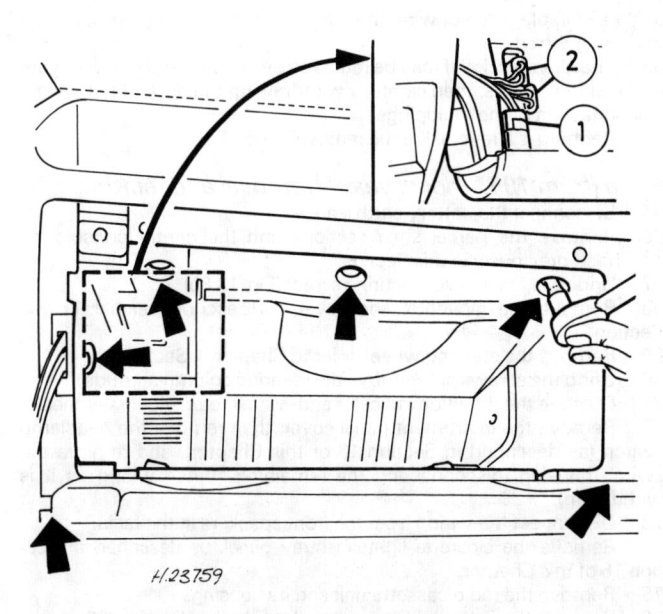

Fig. 13.84 Glovebox securing screws (arrowed). Inset shows the wiring multiplugs (1) and (2) (Sec 18)

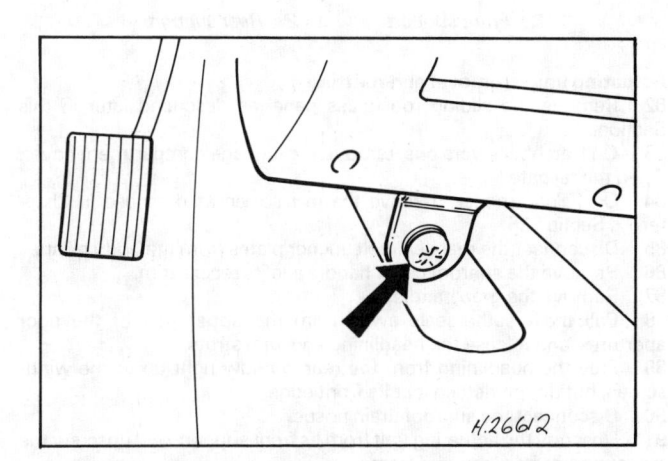

Fig. 13.85 Facia lower securing screw (arrowed) – from 1991 model year (Sec 18)

Centre console (from 1991 model year) – removal and refitting
60 Refer to Chapter 12, Section 33, but note that where a switch panel is fitted, this must first be released by undoing its securing screws and disconnecting the switch multiplugs. The centre console can be removed without disturbing the gear lever rubber boot.

Parcel shelf (from 1991 model year) – removal and refitting
61 On models from 1991 model year, a split-type parcel shelf is fitted. Each section engages in guides and is secured by two screws.

Glovebox and cover (from 1991 model year) – removal and refitting
62 To remove the glovebox cover, press out the hinge retaining pins

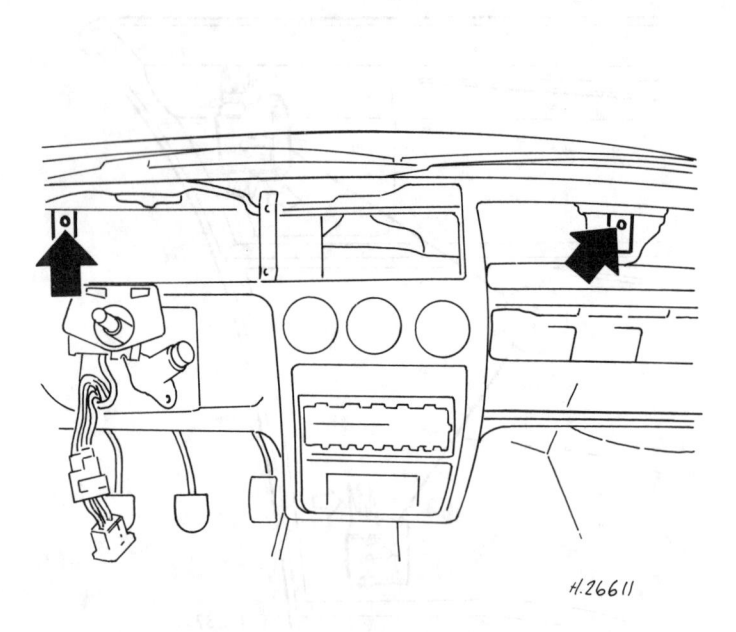

320 Chapter 13 Supplement: Revisions and information on later models

Fig. 13.86 Facia upper securing bolts (arrowed) – from 1991 model year (Sec 18)

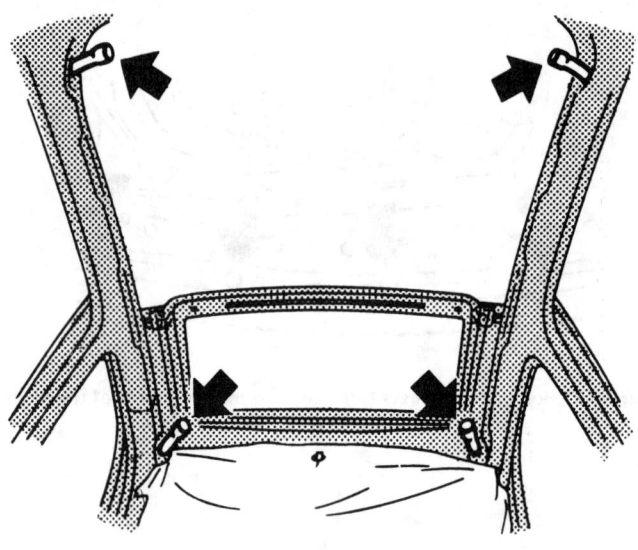

Fig. 13.87 Sliding sunroof drain hoses (arrowed) – models up to August 1989 (Sec 18)

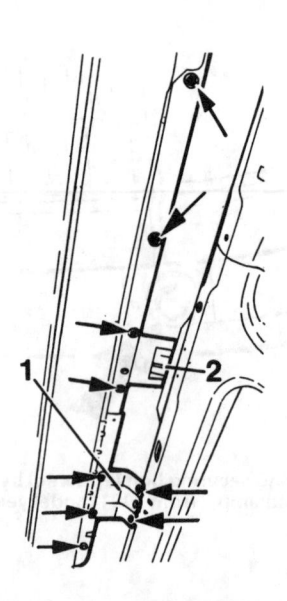

Fig. 13.88 Sliding sunroof actuating unit fixings (arrowed) – models up to August 1989 (Sec 18)

1 Front support   2 Rear support

using a suitable piece of wire, then depress the retaining arm pins and withdraw the cover.
63 The glovebox itself may be removed by undoing and removing the six securing screws, then carefully withdrawing the assembly to allow disconnection of the multiplugs.
64 Refitting is a reversal of the removal procedure.

Facia (from 1991 model year) – removal and refitting
65 Disconnect the battery earth lead.
66 Remove the parcel shelf sections and the centre console, as described previously in this Section.
67 Undo the facia lower securing screw (Fig. 13.85).
68 Remove the glovebox and cover, as described earlier in this Section.
69 Remove the steering wheel (refer to Chapter 8, Section 3).
70 Undo the screws and remove the steering column shrouds.
71 Remove the direction indicator and windscreen wiper switches.
72 Remove the instrument panel cover, then remove the headlamp switch (as described in Section 16 of this Chapter), and remove the switch housing (disconnecting the remaining switch wiring as it is withdrawn).
73 Disconnect the wiring from the front speakers in the facia.
74 Remove the cigarette lighter/ashtray panel, as described in Section 16 of this Chapter.
75 Remove the radio/cassette unit and its housing.
76 Remove the instrument panel as described in Section 16 of this Chapter.
77 Proceed as described in paragraphs 58 to 62 (inclusive) of Section 16 in this Chapter.
78 Undo the two securing screws on either side of the facia (photo).
79 Remove the facia upper securing bolts (Fig. 13.86), then carefully pull the facia squarely from its location and remove it. The speakers, centre connecting housing and vent nozzles may be removed if required.
80 Refitting is a reversal of the removal procedure, ensuring that all cables and wiring are correctly routed and that the heater controls and ventilation ducting are correctly located. Use a new locking plate when installing the steering wheel.

Sliding sunroof (up to August 1989) – removal and refitting
Operation
81 The sunroof on later models is operated by means of a crank handle instead of the handwheel described in Chapter 12.

Actuating unit – removal and refitting
82 Remove the sliding roof glass panel as described later in this Section.
83 On Hatchback versions, remove the luggage compartment cover. Open the tailgate.
84 On Saloon models, remove the rear screen as described in Chapter 12, Section 26.
85 Disconnect the rear seat belt anchor plates from the body pillars.
86 Remove the sunroof crank handle and its recess trim.
87 Remove the grab handles.
88 Pull the weatherseals away from the upper part of the door apertures, and release the headlining and wire struts.
89 Free the headlining from the rear window right up to the windscreen, but do not detach it at its front edge.
90 Disconnect the sunroof drain hoses.
91 Unscrew the actuating unit from its front support, and unscrew the rear supports from the roof frame.
92 Remove the actuating unit through the rear screen aperture (Saloon) or the tailgate aperture (Hatchback).

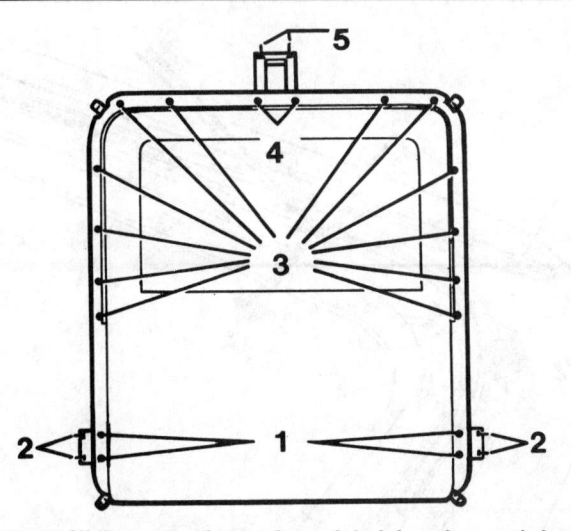

Chapter 13 Supplement: Revisions and information on later models

Fig. 13.89 Sliding sunroof actuating unit bolt locations and sizes – models up to August 1989 (Sec 18)

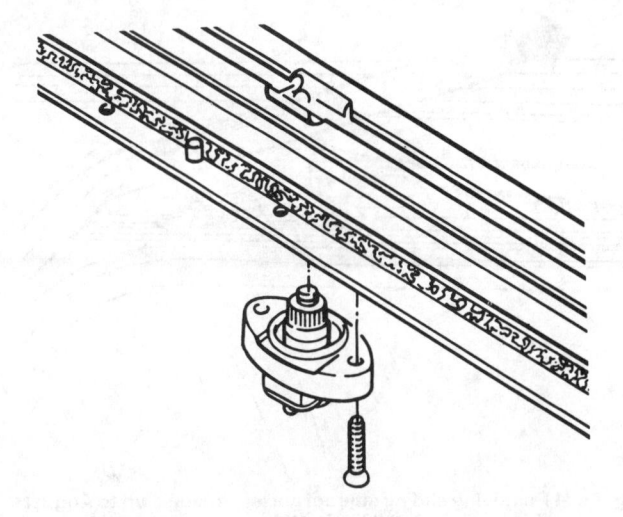

1 M5 x 15   4 M5 x 10
2 M5 x 15   5 M5 x 10
3 M5 x 10

Fig. 13.90 Removing the sliding sunroof crank drive (Sec 18)

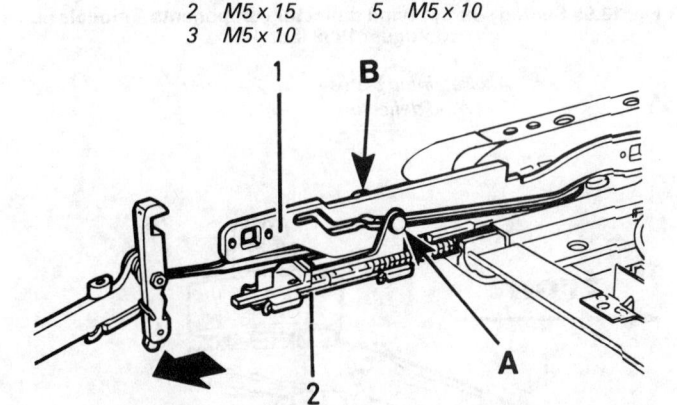

Fig. 13.91 Sliding sunroof operating cable components – models up to August 1989 (Sec 18)

A Bolt   1 Slide block guides
B Retaining ring   2 Cable with guide

93 When refitting the actuating unit, refer to Fig. 13.89, and note carefully the location of the different length bolts.
94 The drain hoses are a push-fit on the pipe stubs – do not use adhesive.

Operating cables – renewal
95 The operating cables should always be renewed as a pair, even if only one is frayed or broken.
96 Remove the actuating unit as previously described.
97 Remove the gutter from the actuating unit, and also the crank drive.
98 Pull the plugs from the ends of the roof guides.
99 Withdraw the cables with part of the guide assembly from the roof guide. Prise off retaining ring (B) (Fig. 13.91) from the bolt, and separate the cables with the guides from the slide block guides.
100 Refitting is a reversal of removal, but adjust the crank drive as described later in this Section.

Slide blocks and guides
101 These can be removed and refitted after withdrawing the actuating arm as described earlier.

Glass panel – removal and refitting
102 Close the glass panel and push the sun screen to the rear.
103 Extract the self-locking bolts and disconnect the panel from both slide block guides. Remove the glass panel.
104 Before fitting the panel, check the distance between the screw flanges (Fig. 13.93). This should be 772.0 mm (30.4 in). Adjust by hand if necessary.
105 Locate the panel, noting that the safety glass sign is to the rear, and the chequered edge at the front.

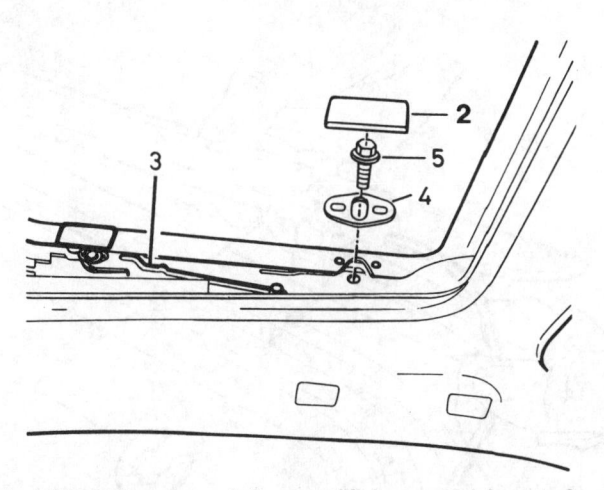

Fig. 13.92 Sliding sunroof glass panel fixings – models up to August 1989 (Sec 18)

2 Clip   4 Lock washer
3 Slide block guide   5 Self-locking bolt

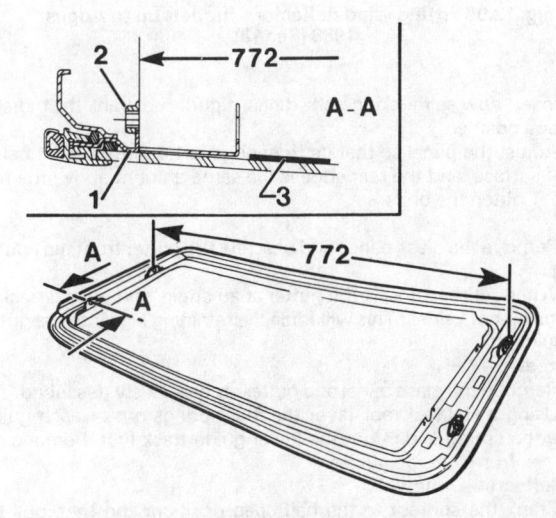

Fig. 13.93 Sliding sunroof panel alignment details – models up to August 1989 (dimensions in mm) (Sec 18)

1 Screw flange   3 Temporary protective sheeting for paintwork
2 Welded (captive) nut

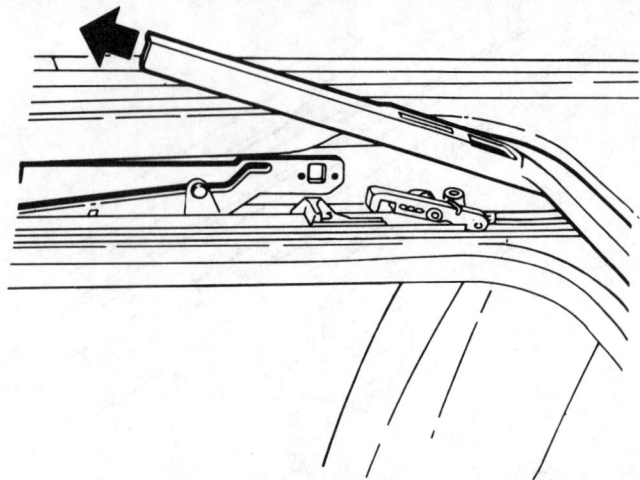

Fig. 13.94 Removing sliding sunroof gutter – models up to August 1989 (Sec 18)

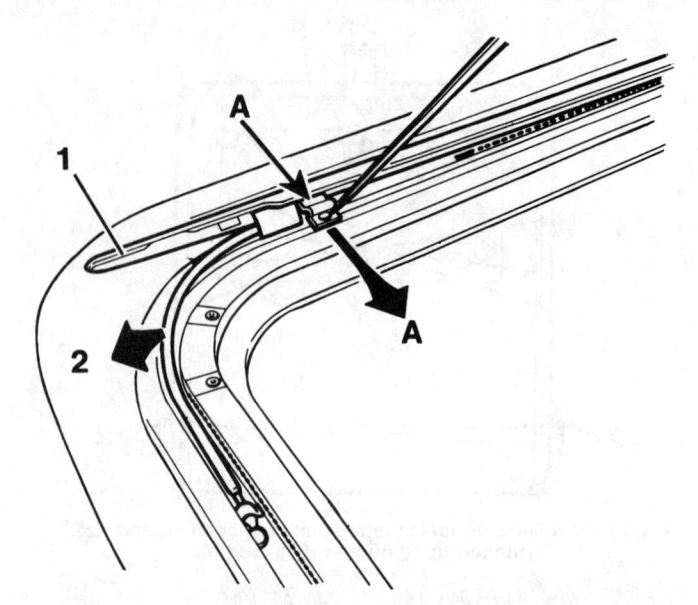

Fig. 13.95 Sliding sunroof wind deflector components – models up to August 1989 (Sec 18)

A   Retaining plates
1   Wind deflector
2   Lifter

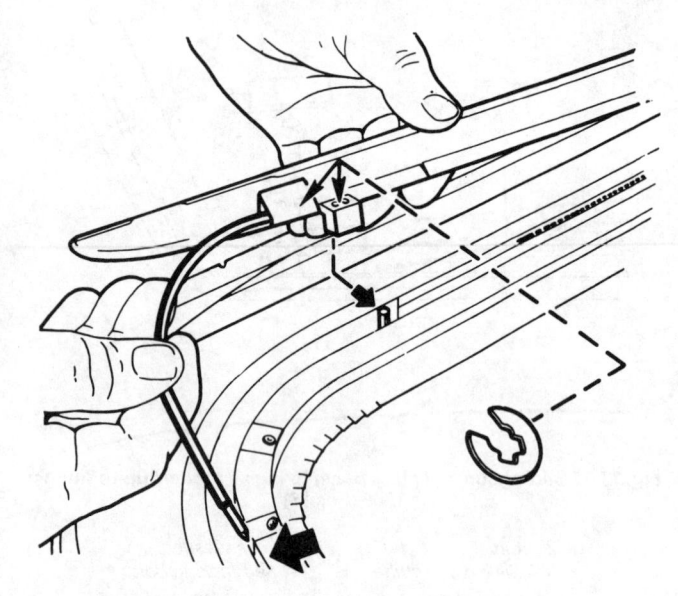

Fig. 13.96 Fitting wind deflector – models up to August 1989 (Sec 18)

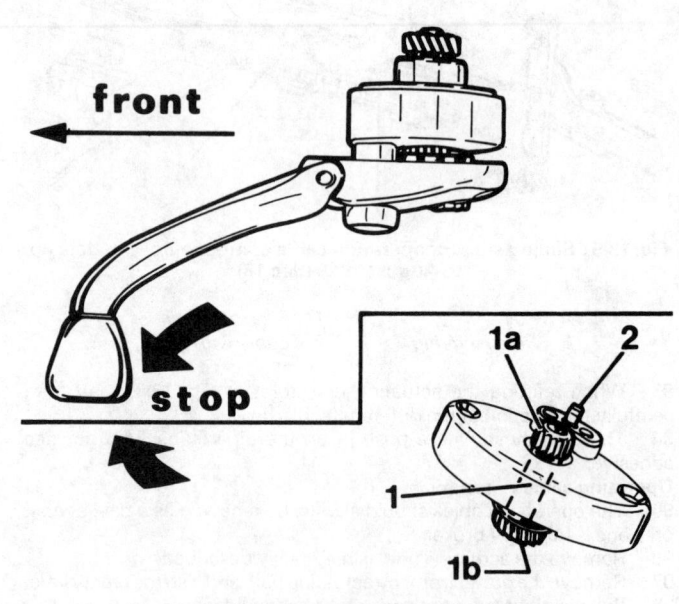

Fig. 13.97 Sliding sunroof crank drive (Sec 18)

1   Pinion
1a Straight splines (for crank handle)
1b Angled splines (for cables)
2   Pin (towards front of vehicle)

106 Insert new self-locking bolts (finger-tight), and crank the panel to the closed position.
107 Adjust the panel so that its front edge is 1.0 mm (0.04 in) below the roof surface, and the rear edge is the same amount above the roof surface. Tighten the bolts.

Gutter
108 Remove the glass panel, and unscrew the gutter from the actuating unit.
109 When refitting, insert the gutter at an angle, and push it back to the stops on both sides. This will force the retaining lugs into the gutter outer guide.

Sun screen
110 Remove the glass panel and gutter, as previously described.
111 Using a suitable tool, lever the four springs out of each guide. Release the spring ends from the lower guide track first. Remove the sun screen from the side guides.

Wind deflector – renewal
112 Crank the sunroof to the half-open position, and then pull the wind deflector from the actuating unit.
113 The new wind deflector unit with pre-mounted lifter should be inserted in the guide, and the retaining plates fixed in position. Use a retaining ring to prevent the bearing block in the wind deflector from springing out.

Crank drive – adjustment
114 Whenever the crank drive has been disturbed, it must be adjusted in the following way.
115 Hold the crank drive unit so that the pin (2) (Fig. 13.97) points towards the front of the car.
116 Turn the crank handle clockwise until it stops. Now turn it anti-clockwise three complete turns to its stop.
117 Fit the crank drive unit with handle (roof closed), checking that the crank handle folds centrally into its dished recess.
118 To tilt the roof, press the unlocking button (roof closed) and turn

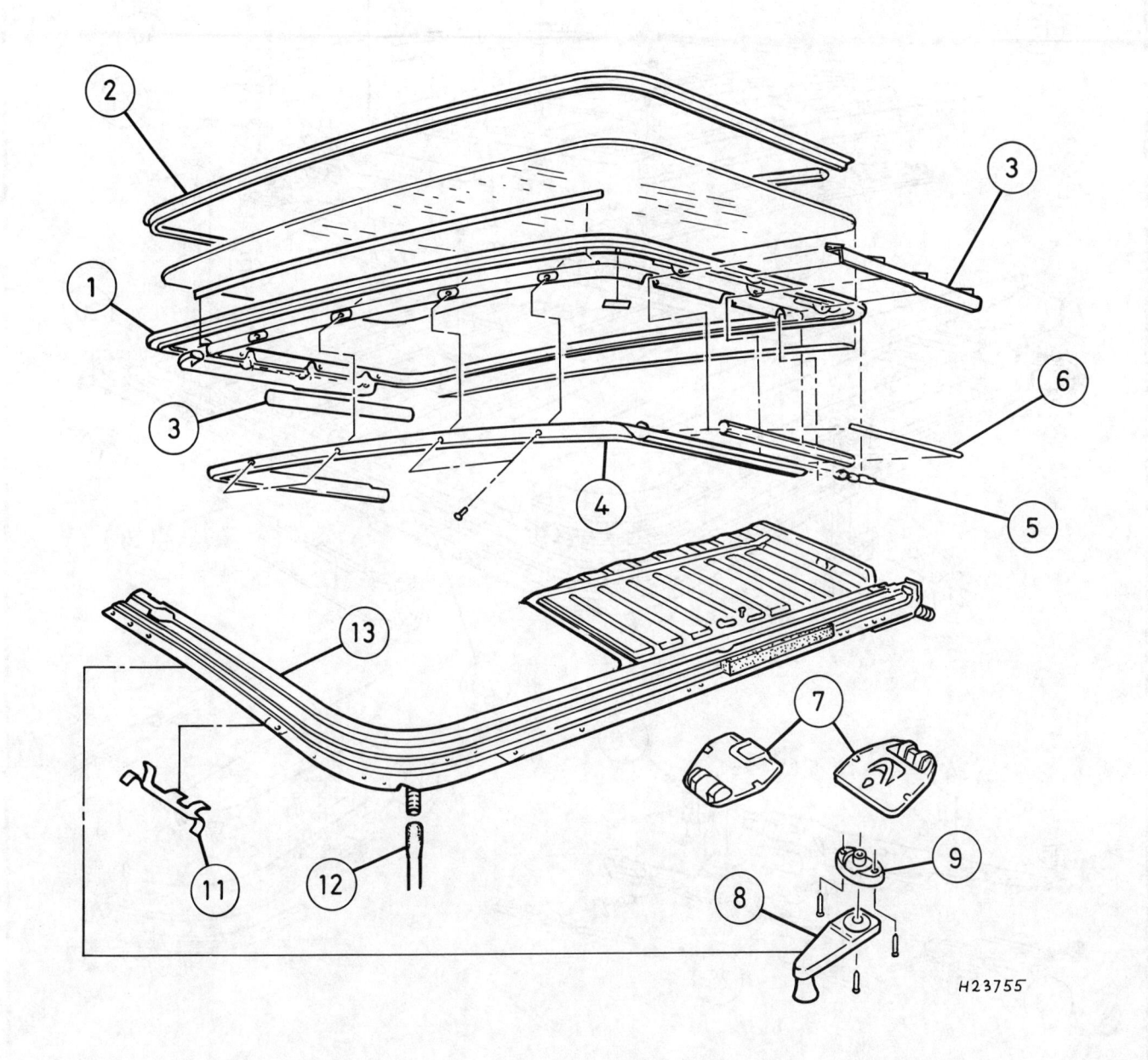

Chapter 13 Supplement: Revisions and information on later models 323

Fig. 13.98 An exploded view of the sliding sunroof components – models from August 1989 (Sec 18)

1 Shell (glass panel)
2 Controlled-gap seal
3 Cover (shell side)
4 Cover (screwed to shell)
5 Spring clip
6 Brush
7 Cover in headlining (handle recess/interior light)
8 Crank
9 Crank drive
11 Front support
12 Water drain hose
13 Actuating unit

the crank handle clockwise. To slide the roof, turn the crank handle in an anti-clockwise direction.

119 The sun screen can be opened or closed with the sunroof closed or tilted. Should the sun screen be accidentally pushed right back, it will automatically re-appear if the sunroof is opened fully and then closed again.

Sliding sunroof (from August 1989) – removal and refitting

120 Reference should be made to the operations described previously in this Section for earlier sliding sunroofs as a guide to all procedures. Figs. 13.98 and 13.99 may be used to determine specific differences.

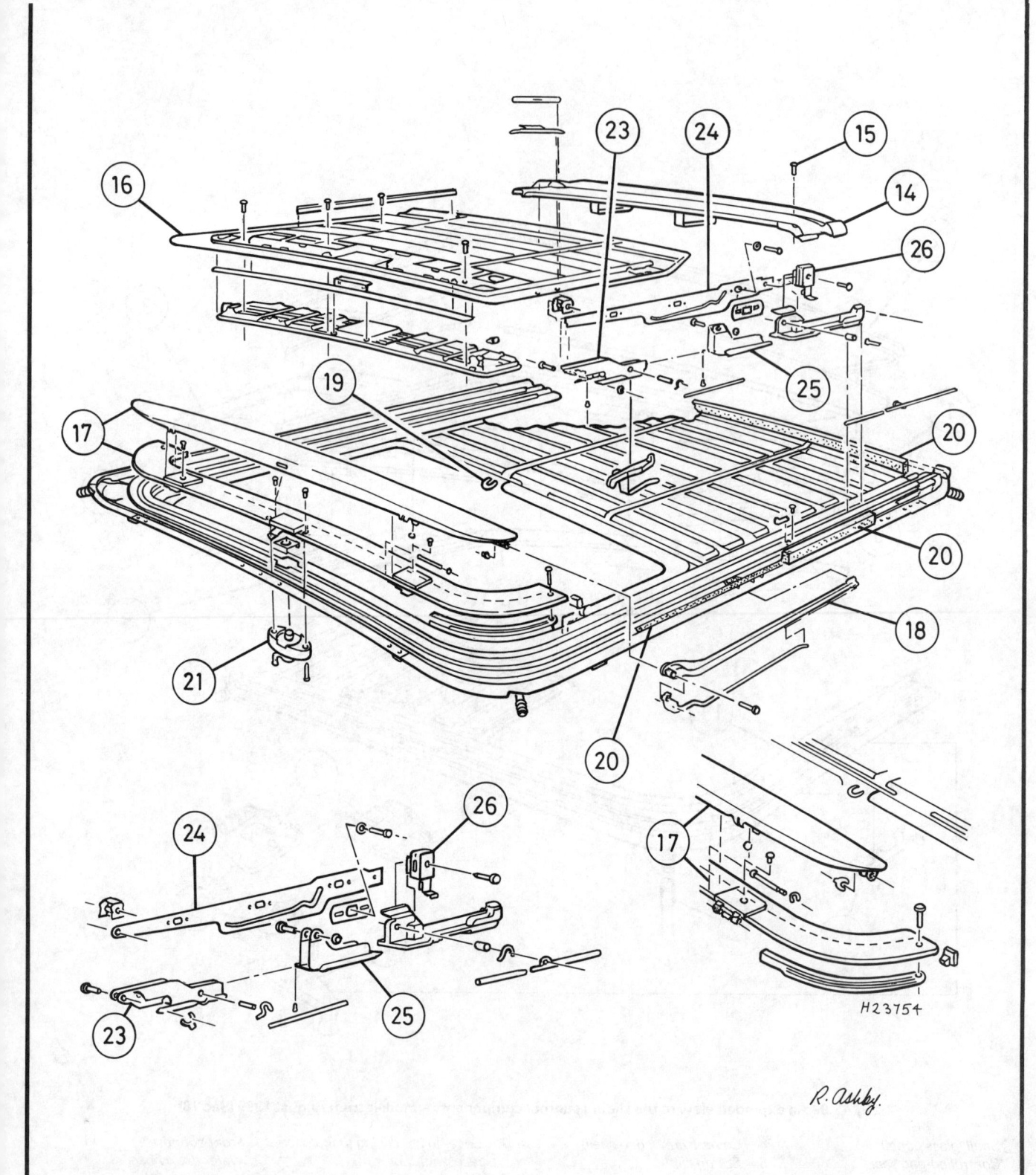

Fig. 13.99 An exploded view of the sliding sunroof actuating unit components – models from August 1989 (Sec 18)

14 Water deflector panel (gutter)
15 Allen bolt
16 Sun screen
17 Wind deflector (with cable channel cover plate)
18 Lifter arm (with spring pin and lock washers)
19 Weatherstrip (shell cut-out)
20 Frame weatherstrips
21 Crank drive
23 Front guide (retaining lug)
24 Slide block guide (with pin, lock washer and anti-rattle spring)
25 Rear guide (operating cable)
26 Hold-down clamp (for slide block)

325

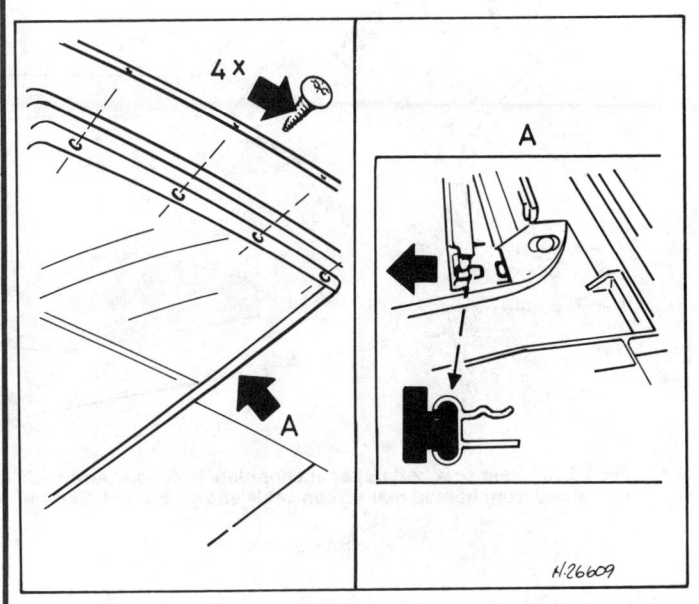

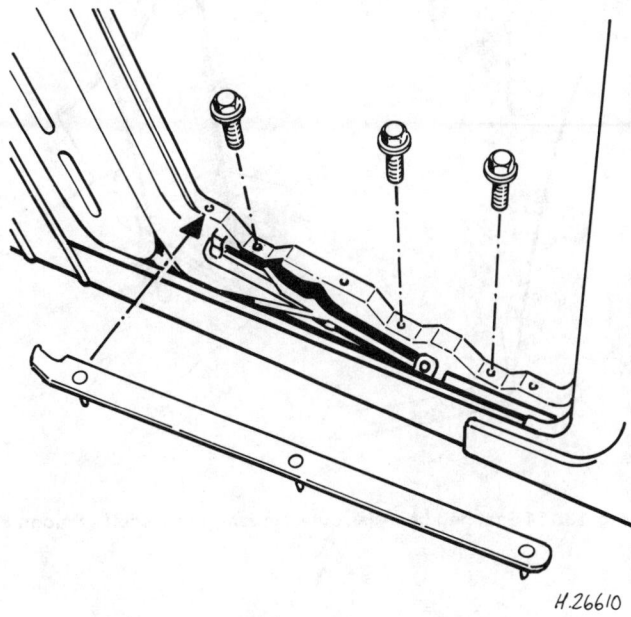

Fig. 13.100 Remove the four shell cover securing screws, and release the cover springs sideways (A) before unclipping the cover – models from August 1989 (Sec 18)

Fig. 13.101 Undo the three shell-to-slide block guide securing screws on each side – models from August 1989 (Sec 18)

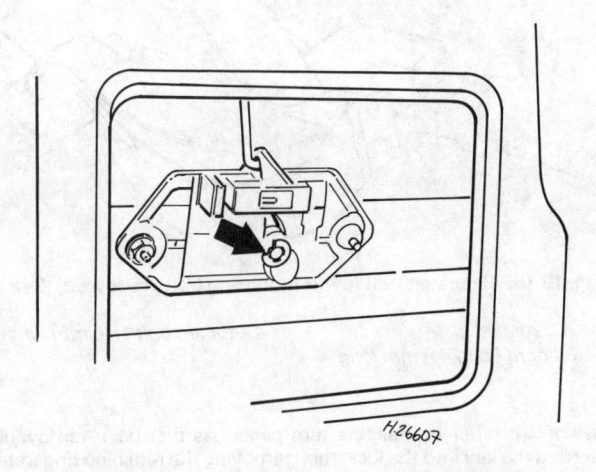

Fig. 13.102 Tailgate lock cylinder retaining ring (arrowed) – models from August 1988 (Sec 18)

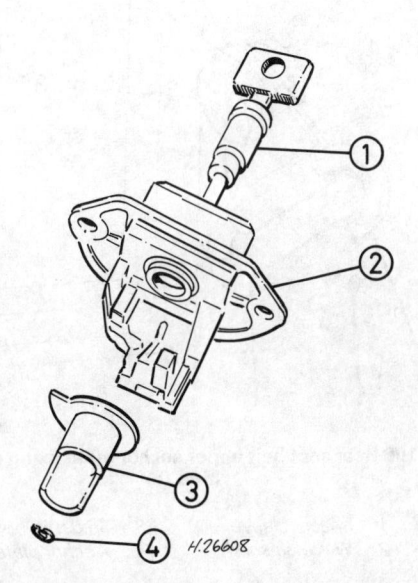

Fig. 13.103 An exploded view of the tailgate lock assembly components – models from August 1988 (Sec 18)

1 Lock cylinder
2 Housing
3 Lever
4 Retaining ring

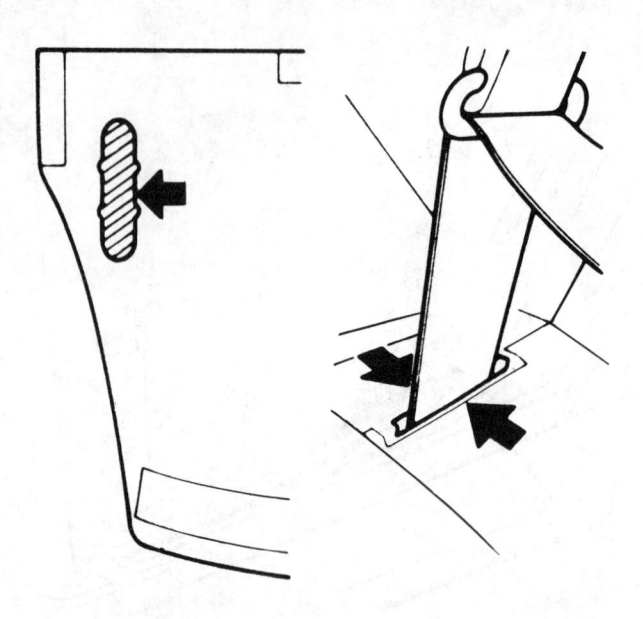

Fig. 13.104 Seat belt holes (arrowed) in rear parcel shelf – Saloon (Sec 18)

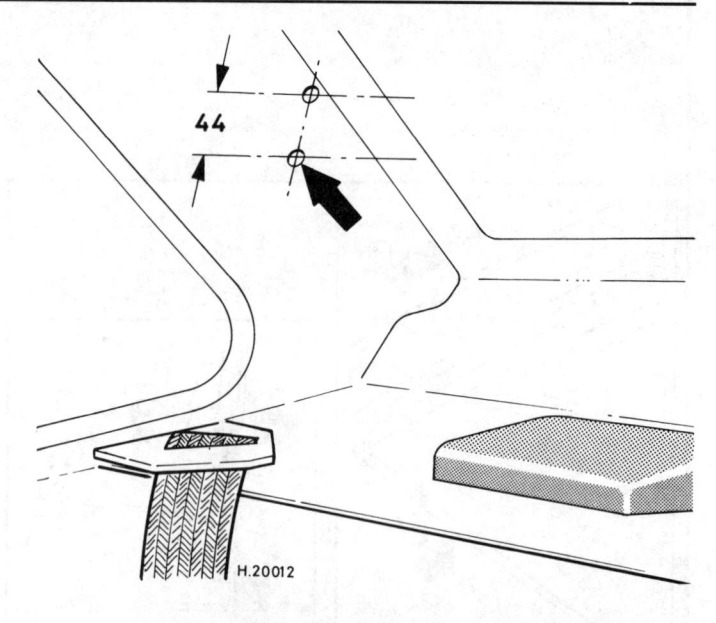

Fig. 13.105 Rear seat belt upper anchor plate bolt hole (arrowed) measured from heated rear screen cable entry – Saloon (Sec 18)

Dimension in mm

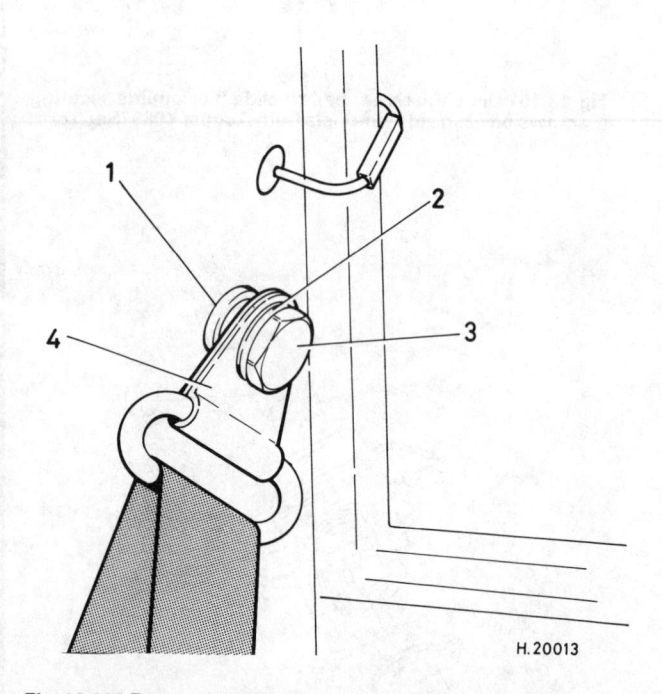

Fig. 13.106 Rear seat belt upper anchor plate fixing components (Sec 18)

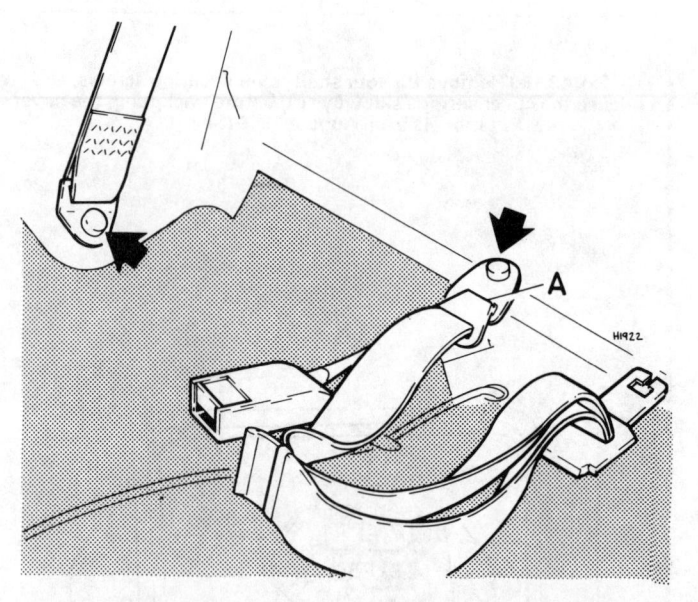

1 Spacer sleeve
2 Wave washer
3 35.0 mm bolt
4 Anchor plate

Fig. 13.107 Rear seat belt lower mounting bolts (arrowed) (Sec 18)

A Angled plate
Inboard bolt 24.0 mm long
Outboard bolt 19.0 mm long

121 When removing the glass panel, note that the panel should be in the half-open position, and that additional work is required to separate the components before extracting the self-locking bolts and disconnecting the panel from the slide block guides (see Fig. 13.100).
122 Crank drive adjustment is carried out as described for earlier sliding sunroofs, but note that the sliding roof moves fully rearwards after ten turns of the crank. After actuating the unlocking button, two additional turns are possible.

Tailgate lock (from August 1988)

123 The tailgate lock cylinder can be removed from the revised lock mechanism (after the tailgate trim panel has first been removed) by inserting the key into the lock, then removing the retaining ring (refer to Fig. 13.102). The lock can then be withdrawn from the outside of the vehicle using the key.
124 Refitting is a reversal of the removal procedure, having first greased the components.

Seat belts – maintenance, removal and refitting

125 Regularly check the seat belts for fraying or other damage.
126 The belts may be cleaned using warm water and detergent only.
127 If the belts have been subjected to strain in a collision, then they should be renewed.
128 A belt is removed simply by unbolting the reel, anchor brackets and floor stalk.

Chapter 13 Supplement: Revisions and information on later models

129 When refitting a seat belt, it is important that the original fitted sequence of anchor plate, spacer and washers is retained.

Rear seat belts – after-market fitting

130 Rear seat belts are fitted to 1987-on models as standard equipment. The belt reels are located vertically on the rear quarter panel within the luggage boot on Saloon models, and behind the luggage area side trim panels on Hatchback versions.

131 Rear seat belts may be fitted to older vehicles, but observe the following points.

Saloon

132 Remove the rear seat (Chapter 12) and clear the holes for the belts in the rear parcels shelf. On some models the holes are pre-cut.

133 Mount the belt reel vertically on the quarter panel so that the alignment pin engages in its notch. Bolt on the reinforcement plate supplied with the belt.

134 Remove the plugs from the belt anchor plate bolt tapped holes, and attach the belts to the seat pan and rear side panel using the reinforcements supplied. Make sure that the upper anchor plate is secured using the parts supplied, fitted in their correct sequence – spacer against body, then the wave washer and anchor plate, all secured by the 35.0 mm bolt. Note the angled anchor plates for the lower mountings.

Hatchback

135 When fitting the belt reels behind the luggage area quarter trim panels, set them vertically with the alignment pin in its notch.

136 In order to bolt the belt anchor plate into position, on some models, the trim panels and luggage area side covers may have to be cut. On other models, the apertures are pre-cut.

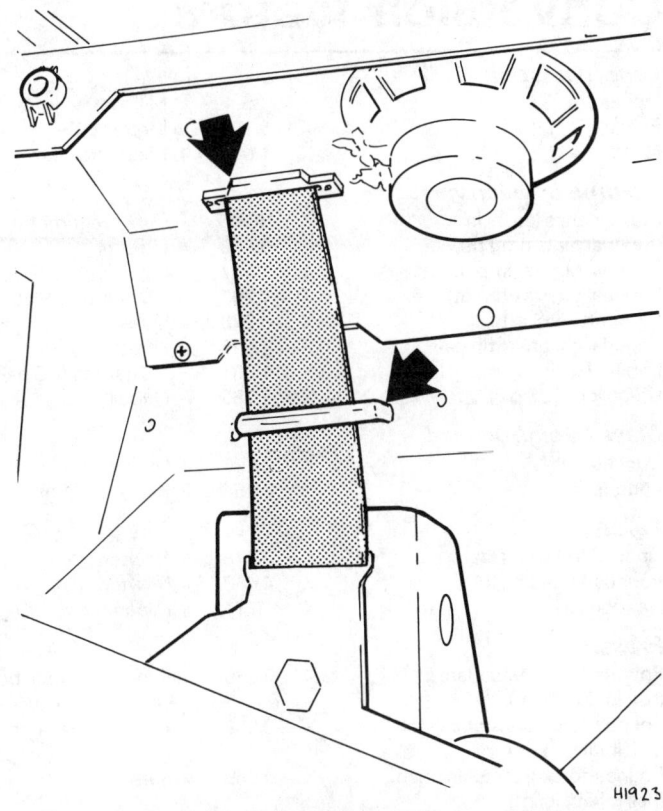

Fig. 13.108 Rear seat belt feed guides (arrowed) – Hatchback (Sec 18)

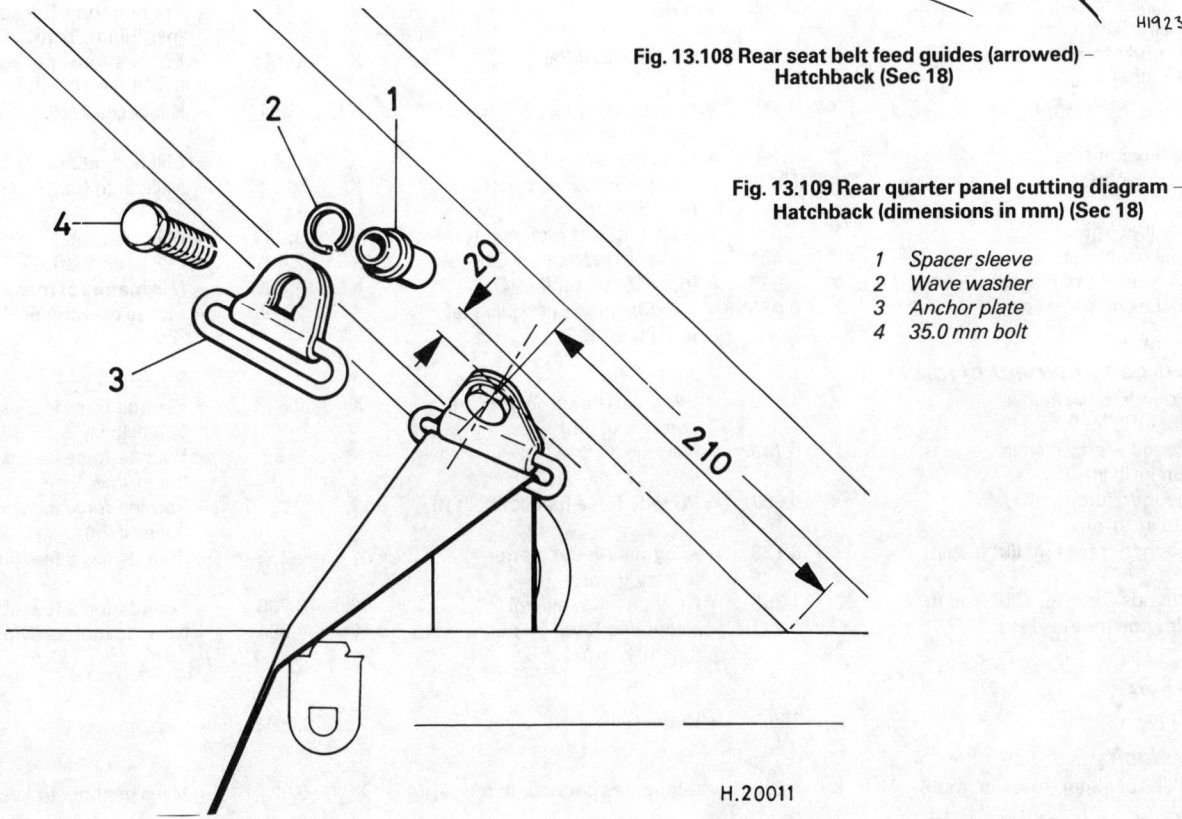

Fig. 13.109 Rear quarter panel cutting diagram – Hatchback (dimensions in mm) (Sec 18)

1 Spacer sleeve
2 Wave washer
3 Anchor plate
4 35.0 mm bolt

Conversion factors

Length (distance)
Inches (in)	X	25.4	= Millimetres (mm)	X 0.0394	= Inches (in)
Feet (ft)	X	0.305	= Metres (m)	X 3.281	= Feet (ft)
Miles	X	1.609	= Kilometres (km)	X 0.621	= Miles

Volume (capacity)
Cubic inches (cu in; in^3)	X	16.387	= Cubic centimetres (cc; cm^3)	X 0.061	= Cubic inches (cu in; in^3)
Imperial pints (Imp pt)	X	0.568	= Litres (l)	X 1.76	= Imperial pints (Imp pt)
Imperial quarts (Imp qt)	X	1.137	= Litres (l)	X 0.88	= Imperial quarts (Imp qt)
Imperial quarts (Imp qt)	X	1.201	= US quarts (US qt)	X 0.833	= Imperial quarts (Imp qt)
US quarts (US qt)	X	0.946	= Litres (l)	X 1.057	= US quarts (US qt)
Imperial gallons (Imp gal)	X	4.546	= Litres (l)	X 0.22	= Imperial gallons (Imp gal)
Imperial gallons (Imp gal)	X	1.201	= US gallons (US gal)	X 0.833	= Imperial gallons (Imp gal)
US gallons (US gal)	X	3.785	= Litres (l)	X 0.264	= US gallons (US gal)

Mass (weight)
Ounces (oz)	X	28.35	= Grams (g)	X 0.035	= Ounces (oz)
Pounds (lb)	X	0.454	= Kilograms (kg)	X 2.205	= Pounds (lb)

Force
Ounces-force (ozf; oz)	X	0.278	= Newtons (N)	X 3.6	= Ounces-force (ozf; oz)
Pounds-force (lbf; lb)	X	4.448	= Newtons (N)	X 0.225	= Pounds-force (lbf; lb)
Newtons (N)	X	0.1	= Kilograms-force (kgf; kg)	X 9.81	= Newtons (N)

Pressure
Pounds-force per square inch (psi; lbf/in^2; lb/in^2)	X	0.070	= Kilograms-force per square centimetre (kgf/cm^2; kg/cm^2)	X 14.223	= Pounds-force per square inch (psi; lbf/in^2; lb/in^2)
Pounds-force per square inch (psi; lbf/in^2; lb/in^2)	X	0.068	= Atmospheres (atm)	X 14.696	= Pounds-force per square inch (psi; lbf/in^2; lb/in^2)
Pounds-force per square inch (psi; lbf/in^2; lb/in^2)	X	0.069	= Bars	X 14.5	= Pounds-force per square inch (psi; lbf/in^2; lb/in^2)
Pounds-force per square inch (psi; lbf/in^2; lb/in^2)	X	6.895	= Kilopascals (kPa)	X 0.145	= Pounds-force per square inch (psi; lbf/in^2; lb/in^2)
Kilopascals (kPa)	X	0.01	= Kilograms-force per square centimetre (kgf/cm^2; kg/cm^2)	X 98.1	= Kilopascals (kPa)
Millibar (mbar)	X	100	= Pascals (Pa)	X 0.01	= Millibar (mbar)
Millibar (mbar)	X	0.0145	= Pounds-force per square inch (psi; lbf/in^2; lb/in^2)	X 68.947	= Millibar (mbar)
Millibar (mbar)	X	0.75	= Millimetres of mercury (mmHg)	X 1.333	= Millibar (mbar)
Millibar (mbar)	X	0.401	= Inches of water (inH$_2$O)	X 2.491	= Millibar (mbar)
Millimetres of mercury (mmHg)	X	0.535	= Inches of water (inH$_2$O)	X 1.868	= Millimetres of mercury (mmHg)
Inches of water (inH$_2$O)	X	0.036	= Pounds-force per square inch (psi; lbf/in^2; lb/in^2)	X 27.68	= Inches of water (inH$_2$O)

Torque (moment of force)
Pounds-force inches (lbf in; lb in)	X	1.152	= Kilograms-force centimetre (kgf cm; kg cm)	X 0.868	= Pounds-force inches (lbf in; lb in)
Pounds-force inches (lbf in; lb in)	X	0.113	= Newton metres (Nm)	X 8.85	= Pounds-force inches (lbf in; lb in)
Pounds-force inches (lbf in; lb in)	X	0.083	= Pounds-force feet (lbf ft; lb ft)	X 12	= Pounds-force inches (lbf in; lb in)
Pounds-force feet (lbf ft; lb ft)	X	0.138	= Kilograms-force metres (kgf m; kg m)	X 7.233	= Pounds-force feet (lbf ft; lb ft)
Pounds-force feet (lbf ft; lb ft)	X	1.356	= Newton metres (Nm)	X 0.738	= Pounds-force feet (lbf ft; lb ft)
Newton metres (Nm)	X	0.102	= Kilograms-force metres (kgf m; kg m)	X 9.804	= Newton metres (Nm)

Power
Horsepower (hp)	X	745.7	= Watts (W)	X 0.0013	= Horsepower (hp)

Velocity (speed)
Miles per hour (miles/hr; mph)	X	1.609	= Kilometres per hour (km/hr; kph)	X 0.621	= Miles per hour (miles/hr; mph)

Fuel consumption
Miles per gallon, Imperial (mpg)	X	0.354	= Kilometres per litre (km/l)	X 2.825	= Miles per gallon, Imperial (mpg)
Miles per gallon, US (mpg)	X	0.425	= Kilometres per litre (km/l)	X 2.352	= Miles per gallon, US (mpg)

Temperature
Degrees Fahrenheit = (°C x 1.8) + 32
Degrees Celsius (Degrees Centigrade; °C) = (°F − 32) x 0.56

* It is common practice to convert from miles per gallon (mpg) to litres/100 kilometres (l/100km), where mpg (Imperial) x l/100 km = 282 and mpg (US) x l/100 km = 235

Index

A

About this manual – 4
Accelerator cable – 69
Accelerator pedal – 69
Acknowledgements – 4
Active carbon canister (Multec central fuel injection) – 254
Air cleaner – 67, 249, 256
Air intake temperature switch (Bosch L3.1-Jetronic) – 259
Airflow sensor unit (Bosch L3.1-Jetronic) – 256, 258
Alternator – 62, 162, 163
Anti-roll bar – 202, 205
Antifreeze mixture – 62
Armrest – 316
Axle (rear) – 207, 208

B

Battery – 162
Bearings
    big-end – 40
    clutch release – 103
    hub – 200, 203
    main – 39, 40, 51, 242
Big-end bearings – 40
Bleeding (brakes) – 144
Bodywork – 209 *et seq*, 313
Bodywork repair – *see colour pages between pages 32 and 33*
Bonnet – 212
Boot lid – 222
Brake pedal – 156, 268, 271
Braking system – 142 *et seq*, 268
Bulb renewal
    courtesy light – 183
    facia panel illumination – 183
    headlamp – 183
    indicators – 183, 275
    instrument panel – 183
    interior lamp – 183
    luggage compartment – 277
    number plate lamp – 183
    rear lamp – 183
    sidelamp – 183
    underbonnet – 276
Bumpers – 212, 318

C

Cables
    accelerator – 69
    bonnet release – 212
    choke – 69
    clutch – 97
    handbrake – 157, 271
    speedometer – 180
Caliper – 146, 272
Camber – 313
Camshaft
    1.0 models – 38, 41
    1.2, 1.3 and 1.6 models – 43, 46, 53, 239
Camshaft toothed belt
    1.4 models – 240, 241
    1.6 models – 242
Capacities – 6
Carburettors
    Pierburg 1B1 – 70, 78, 80, 243
    Pierburg 2E3 – 243, 244, 245
    Weber 32 TL – 70, 74, 75, 243
Catalytic converter – 254
Central door locking system – 277
Centre console – 223, 319
Choke cable – 69
Cigarette lighter – 178, 275
Clock – 177, 275
Clutch – 96 *et seq*, 266
Coasting air valve (Bosch L3.1-Jetronic) – 260
Coil (ignition) – 94
Coil spring – 205
Condenser – 87, 262
Connecting rods
    1.0 models – 37, 40
    1.2, 1.3 and 1.6 models – 40, 49
    1.4 models – 242
Console – 223, 319
Contact breaker points – 86, 87, 262
Conversion factors – 328
Coolant temperature sensor
    Bosch L3.1-Jetronic – 259
    Multec central fuel injection – 253
Cooling fan – 62, 242
Cooling system – 57 *et seq*, 242
Courtesy light – 177, 183
Crankcase ventilation system – 54
Crankshaft
    1.0 models – 38 to 40
    1.2, 1.3 and 1.6 models 40, 51
    1.4 models – 242
Crankshaft pulley
    1.2, 1.3 and 1.6 models – 240
    1.4 models – 242, 263
    1.6 models – 242
Cylinder head
    1.0 models – 29, 32
    1.2, 1.3 and 1.6 models – 45, 46, 48
    1.4 models – 241

D

Decarbonising engine
    1.0 models – 32
    1.2, 1.3 and 1.6 models – 48
Dim-dip headlamps – 273
Dimensions – 6, 235
Direction indicators – 173, 176, 183, 275, 279
Discs – 149, 271, 272
Distributor – 263
    1.4 models – 263
    1.6 GTE/GSi models – 264
    Bosch (1.0 models) – 262
    Delco-Remy – 88
Doors – 214 to 219, 278, 313, 316
Draining (cooling system) – 58
Drivebelt (alternator/water pump) – 62, 242
Driveshafts – 127 *et seq*, 268

E

EGR valve (Bosch L3.1-Jetronic) – 261
Electric windows – 278
Electrical system – 160 *et seq*, 273
Electronic Control Unit (Multec central fuel injection) – 252
Engine – 23 *et seq*, 239, 240, 242
Exhaust gas oxygen sensor (Multec central fuel injection) – 252
Exhaust gas recirculation (EGR) system – 247, 261
Exhaust manifold – 82
Exhaust system – 82
    Bosch L3.1-Jetronic – 261
    Multec central fuel injection – 254

F

Facia – 183, 320
Fan switch – 273
Fault diagnosis – 20
    braking system – 158
    clutch – 102
    cooling system – 63
    driveshafts – 130
    electrical system – 163, 186
    engine – 21, 56
    fuel and exhaust systems – 83
    ignition system – 94, 95
    steering – 141
    suspension – 208
    transmission – 126
Filling (cooling system) – 59, 242
Flushing (cooling system) – 58
Flywheel
    1.0 models – 37, 41
    1.2, 1.3 and 1.6 models – 41, 51
Foglamp switch – 275
Fuel and exhaust systems – 64 *et seq*, 243, 247, 254
Fuel filler neck (Multec central fuel injection) – 248
Fuel filter
    Bosch L3.1-Jetronic – 256
    Multec central fuel injection – 249
Fuel gauge sender unit – 69, 256
Fuel injectors
    Bosch L3.1-Jetronic – 256
    Multec central fuel injection – 249
Fuel pressure check (Multec central fuel injection) – 249
Fuel pressure regulator
    Bosch L3.1-Jetronic – 259
    Multec central fuel injection – 249
Fuel pump – 66
    L3.1-Jetronic – 254, 256, 259
    Multec central fuel injection – 248
Fuel tank – 68
    Multec central fuel injection – 248
Fuses – 172, 273

G

Gaiter (steering rack) – 137
Gauges
    fuel – 69
    temperature – 62
Gear lever – 107
Gearbox *see* **Transmission**
Gearchange linkage – 106, 266
Glove compartment – 223, 319
Grab handles – 313
Grille (radiator) – 212

H

Handbrake – 156 to 158, 270, 271
Hazard flashers – 173, 275
Head restraints – 313

Headlamp range-control motor – 274
Headlamps – 183, 185, 273, 274
Heated rear window – 173, 273
Heated seats – 278
Heater – 225, 228, 276, 316
Horn – 177
HT leads – 94
Hubs – 200, 203
Hydraulic brake pipes and hoses – 154
Hydraulic timing chain tensioner (1.0 ohv) – 239

I

Idle air control stepper motor (Multec central fuel injection) – 250
Idle speed adjustment (Bosch L3.1-Jetronic) – 260
Ignition coil
    1.4 models – 263
    1.6 GTE/GSi – 266
Ignition/fuel injection control unit (1.6 GTE/GSi) – 264
Ignition system – 85 *et seq*, 136, 262
Ignition timing – 91, 266
    1.6 GTE/GSi – 264
    Multec central fuel injection – 264
Indicators – 173, 176, 183, 275
Inlet manifold – 82
Input shaft – 118
Instrument panel – 275
Instrumentation – 178, 183, 275
Introduction to the Vauxhall Nova – 4

J

Jacking – 11

L

Locks
    central locking – 277
    door – 215, 217
    steering column – 136
    tailgate – 220, 326
Lubricants and fluids – 12
Luggage compartment lamp – 277

M

Main bearings
    1.0 models – 39, 40
    1.2, 1.3 and 1.6 models – 40, 51
    1.4 models – 242
Mainshaft – 118
Manifold absolute pressure sensor
    Multec central fuel injection – 253
Manifolds – 82
Master cylinder – 134, 153, 271
Mirror – 223, 273, 313
Mixture adjustment (Bosch L3.1-Jetronic) – 260
Mountings (engine/transmission)
    1.0 models – 38
    1.2, 1.3 and 1.6 models – 38, 51

N

Number plate light – 183

Index

O

Octane number plug (Multec central fuel injection) – 252
Oil filter (1.2, 1.3 and 1.6 models) – 240
Oil pressure regulator valve (1.2, 1.3 and 1.6 models) – 43
Oil pump
    1.0 models – 34
    1.2, 1.3 and 1.6 models – 49
    1.4 models – 241
Oil seals
    crankshaft front (1.4 models) – 242
Oil temperature sensor (Bosch L3.1-Jetronic) – 259
Oil temperature switch (Bosch L3.1-Jetronic) – 259

P

Pads 145, 271
Parcel shelf – 223, 319
Pedals
    accelerator – 69
    brake 156, 268, 271
    clutch – 97
Pistons
    1.0 models – 32, 37, 40
    1.2, 1.3 and 1.6 models – 40, 48, 49
    1.4 models – 242
Pressure regulating valve (braking system) – 154
Pushrods (1.0 models) – 41

Q

Quarter lights – 219

R

Radiator 59, 62
    electric cooling fan (1.4 models) – 242
Radiator grille – 62, 318
Radio – 186, 278
Rear axle – 207, 208
Rear lamp cluster – 183
Relays – 173, 273
    fuel pump control
        Bosch L3.1-Jetronic – 259
        Multec central fuel injection – 249
Release bearing (clutch) – 103
Repair procedures – 14
Reversing lamp switch – 176
Road speed sensor (Multec central fuel injection) – 254
Rockers
    1.0 models – 41
    1.2, 1.3 and 1.6 models – 53
Routine maintenance – 14, 235
    bodywork – 209 to 211
    braking system – 143
    clutch – 97
    cooling system – 58
    driveshafts – 127
    electrical system – 161
    engine – 27
    fuel and exhaust systems – 65
    ignition system – 86
    suspension – 194
    steering – 131
    transmission – 105

S

Safety first – 13
Seat belts – 326, 327
Seats – 222, 278, 316
Servo unit – 155, 156
Shock absorbers – 196, 204
Shoes – 150, 268, 270
Sidelamps – 183
Signal amplifier/module (1.4 models) – 263
Spare parts – 7
Spark plugs – 94
Spark plug conditions – *see colour pages between pages 32 and 33*
Speedometer cable – 180
Starter motor – 163, 167, 168, 170, 171
Steering – 131 *et seq*
Steering column – 132, 136, 176
Steering knuckle – 200
Steering wheel – 132
Stop-lamp switch – 176
Stub axle – 205
Sump
    1.0 models – 33
    1.2, 1.3 and 1.6 models – 48
Sunroof – 320, 323
Supplement: Revisions/information on later models – 229 *et seq*
Supplementary air valve (Bosch L3.1-Jetronic) – 260
Suspension – 193 *et seq*, 313
Switches
    air intake temperature (Bosch L3.1-Jetronic) – 259
    cooling fan (1.4 models) – 242
    courtesy light – 177
    facia – 173
    fan – 273
    foglamp – 275
    handbrake – 177
    hazard warning – 275
    headlamp – 273
    headlamp range-control motor switch – 273
    heated rear window – 173, 273
    horn – 177
    ignition – 136
    indicator – 176
    oil temperature (Bosch L3.1-Jetronic) – 259
    steering column – 176
    stop-lamp – 176
Synchroniser units – 121

T

Tailgate – 173, 185, 220, 326
Tappets
    1.0 models – 38, 41
TDC sensor (1.2, 1.3 and 1.6 models) – 94
Temperature gauge transmitter – 62
Thermostat – 59, 60, 242
Throttle housing (Bosch L3.1-Jetronic) – 258
Throttle position sensor (Multec central fuel injection) – 250
Throttle valve housing (Multec central fuel injection) – 252
Throttle valve switch (Bosch L3.1-Jetronic) – 258
Tie-bar – 202
Timing (ignition) – 91, 264, 266
Timing gear components (1.0 models) – 34, 41
Tools – 9
Towing – 11
Transmission – 38, 51, 104 *et seq*, 266
Trim panels – 214, 316
Tyres – 208
Underbonnet lamp – 276
Unleaded fuel – 246, 252, 266

V

Vacuum servo unit – 154, 156

Index

Valve clearances (1.0 models) – 28
Valves
   1.0 models – 41
   1.2, 1.3 and 1.6 models – 53
Vehicle identification numbers – 7
Ventilation – 316
Vents – 228

W

Washer system – 186, 274

Water pump – 62, 242
Weights – 6, 235
Wheel alignment – 208
Wheel cylinder (braking system) – 152, 270
Wheels and tyres – 208
Windows – 218, 220, 278, 316
Windscreen – 219
Wing – 213, 318
Wiper blades and arms – 184
Wiper motor – 185
Wiring diagrams – 188 to 192, 280 to 312
Working facilities – 10